Integrated Circuits and Systems

Series editor

Anantha P. Chandrakasan, Massachusetts Institute of Technology
Cambridge, MA, USA

More information about this series at http://www.springer.com/series/7236

Sylvain Clerc • Thierry Di Gilio • Andreia Cathelin
Editors

The Fourth Terminal

Benefits of Body-Biasing Techniques
for FDSOI Circuits and Systems

🐎 Springer

Editors
Sylvain Clerc
STMicroelectronics
Crolles, France

Thierry Di Gilio
STMicroelectronics
Crolles, France

Andreia Cathelin
STMicroelectronics
Crolles, France

ISSN 1558-9412 ISSN 1558-9420 (electronic)
Integrated Circuits and Systems
ISBN 978-3-030-39495-0 ISBN 978-3-030-39496-7 (eBook)
https://doi.org/10.1007/978-3-030-39496-7

© Springer Nature Switzerland AG 2020
This work is subject to copyright. All rights are reserved by the Publisher, whether the whole or part of the material is concerned, specifically the rights of translation, reprinting, reuse of illustrations, recitation, broadcasting, reproduction on microfilms or in any other physical way, and transmission or information storage and retrieval, electronic adaptation, computer software, or by similar or dissimilar methodology now known or hereafter developed.
The use of general descriptive names, registered names, trademarks, service marks, etc. in this publication does not imply, even in the absence of a specific statement, that such names are exempt from the relevant protective laws and regulations and therefore free for general use.
The publisher, the authors, and the editors are safe to assume that the advice and information in this book are believed to be true and accurate at the date of publication. Neither the publisher nor the authors or the editors give a warranty, expressed or implied, with respect to the material contained herein or for any errors or omissions that may have been made. The publisher remains neutral with regard to jurisdictional claims in published maps and institutional affiliations.

This Springer imprint is published by the registered company Springer Nature Switzerland AG.
The registered company address is: Gewerbestrasse 11, 6330 Cham, Switzerland

To our families.

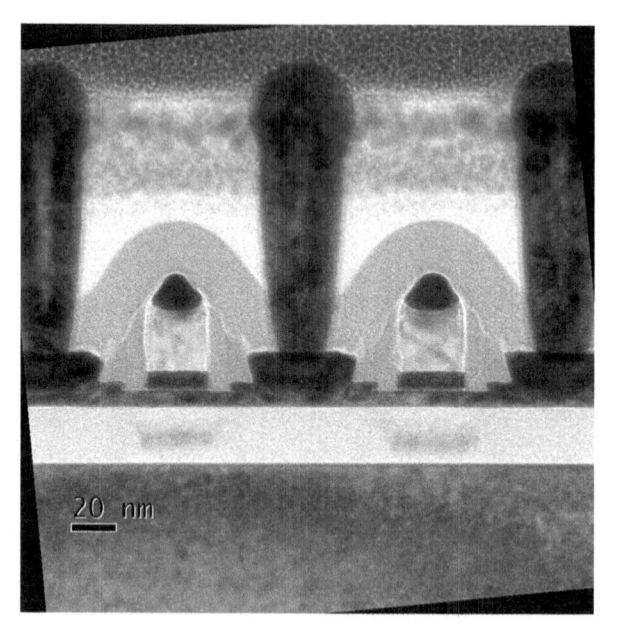

FD-SOI transistor TEM, the transistors' channel is the volume between the gates and the white colored buried oxide. Copyright STMicroelectronics. Used with permission

Foreword

This book deals with applying a voltage potential to the volume below the buried oxide in the TEM adjacent photography, and this modulates transistor's V_T; hence, the motto "FD-SOI Body-Bias enables software defined V_T."

Crolles, France Sylvain Clerc
Crolles, France Thierry Di Gilio
Crolles, France Andreia Cathelin
December 2019

Acknowledgements

The authors are indebted to the many people who supported, helped, contributed, or reviewed this book. We specially thank Robin Wilson, Martin Cochet on top of his own chapter writing, Yann Carminati, Frederic Paillardet, Frederic Hasbani, Dominique Bousquet, Phillipe Larré, Lise Doyen, Philippe Cathelin, Laurent Le-Pailleur, Sylvain Biard, Mourad Djouder, Stéphane Hanriat, Jean-Michel Mirabel, Philippe Quinio, Cyril Colin-Madan, Laurent Malier, and Patrick Aidoune. Last but not least, the authors would like to acknowledge the leading contribution of Joel Hartmann in FD-SOI technology deployment.

We also acknowledge the contribution of many STMicroelectronics Colleagues involved beside the course of their regular duties; we thank you all gentlewomen and gentlemen.

Contents

1	**Introduction**	1
	Andreia Cathelin and Sylvain Clerc	

Part I Device Level and General Studies for Analog and Digital

2	**FD-SOI Technology**	9
	Franck Arnaud	
3	**Body-Bias for Digital Designs**	59
	Sylvain Clerc and Ricardo Gomez Gomez	
4	**Body-Biasing in FD-SOI for Analog, RF, and Millimeter-Wave Designs**	85
	Andreia Cathelin	
5	**SRAM Bitcell Functionality Under Body-Bias**	93
	Lorenzo Ciampolini	

Part II Design Examples: From Analog RF and mmW to Digital. From Building Blocks and Circuits to SoCs

6	**Coarse/Fine Delay Element Design in 28 nm FD-SOI**	119
	Ilias Sourikopoulos, Andreia Cathelin, Andreas Kaiser, and Antoine Frappé	
7	**Millimeter-Wave Distributed Oscillators in 28 nm FD-SOI Technology**	135
	Raphaël Guillaume, Andreia Cathelin, and Yann Deval	
8	**Millimeter-Wave Power Amplifiers for 5G Applications in 28 nm FD-SOI Technology**	169
	Florent Torres, Andreia Cathelin, and Eric Kerhervé	

9	**An 802.15.4 IR-UWB Transmitter SoC with Adaptive-FBB-Based Channel Selection and Programmable Pulse Shape**	223
	David Bol and Guerric de Streel	
10	**Body-Bias Calibration Based Temperature Sensor**	243
	Martin Cochet, Ben Keller, Sylvain Clerc, Fady Abouzeid, Andreia Cathelin, Jean-Luc Autran, Philippe Roche, and Borivoje Nikolić	
11	**System Integration of RISC-V Processors with FD-SOI**	263
	Ben Keller, Borivoje Nikolić, Brian Zimmer, Martin Cochet, Yunsup Lee, Jaehwa Kwak, Alberto Puggelli, Milovan Blagojević, Ruzica Jevtić, Pi-Feng Chiu, Stevo Bailey, Palmer Dabbelt, Colin Schmidt, Hanh-Phuc Le, Po-Hung Chen, Nicholas Sutardja, Rimas Avizienis, Andrew Waterman, James Dunn, Brian Richards, Philippe Flatresse, Andrei Vladimirescu, Andreia Cathelin, Elad Alon, and Krste Asanović	

Part III   Body-Bias Deployment in Mixed-Signal and Digital SoCs

12	**Timing-Based Closed Loop Compensation**	305
	Ricardo Gomez Gomez and Sylvain Clerc	
13	**Open Loop Compensation**	327
	Sylvain Clerc and Ricardo Gomez Gomez	
14	**Compensation and Regulation Solutions' Synthesis**	333
	Ricardo Gomez Gomez	
15	**Body-Bias Voltage Generation**	351
	Thierry Di Gilio	
16	**Digital Design Implementation Flow and Verification Methodology**	385
	Sébastien Marchal, Damien Riquet, and Sylvain Clerc	
A	**FD-SOI Process Flow**	395
B	**Digital Implementation Flow and Terminology**	399
C	**IEEE 1801 UPF Example**	413
Index		419

Acronyms

ABB	Adaptive body-biasing
AC	Alternative current
ADC	Analog to Digital Converter
ALU	Arithmetical and logical unit
ATPG	Automatic Test Pattern Generator
AVS	Adaptive voltage scaling, can be static of dynamic
BBCO	Body-biased-controlled oscillator
BBG or BBGEN	Body-bias generator
BEOL	Back-end-of-line
BIST	Built-in Self Test
BJT	Bipolar junction transistor
BL	Bit line
BLE	Bluetooth Low Energy
BLM	Bit line margin
BOX	Buried oxide
BPPM	Burst-pulse position modulation
BPSK	Binary Phase-Shift Keying modulation
BTI	Bias Temperature Instability
BW	Bandwidth
CAD	Computer-Aided Design
CG	Common gate
CPR	Critical path replica
CPU	Central processing unit
CPW-G	Grounded coplanar line
CS	Common source
CSR	Control and status register
CTAT	Complementary to absolute temperature
DAC	Digital to Analog Converter
DC	Direct current
DC-DC	DC to DC (converter)
DCO	Digitally controlled oscillator

DFT	Design for Test
DIBL	Drain-induced barrier lowering
DLL	Delay locked loop
DRM	Design Rules Manual
DUT	Device under test
DVFS	Dynamic voltage and frequency scaling, this is nonstatic AVS
ECO	Engineering change order
EDA	Electronic Design Automation
EM	Electromagnetic
ESD	Electrostatic discharge
EWS	Electrical Wafer Sort test
FA	Full-adder
FBB	Forward body-bias
FD-MIMO	Full Duplex Multiple Input Multiple Output
FF	Flip-Flop or Fast-Fast process
FIFO	First in first out
FOM	Figure of merit
FOx	Fan-Out-of-x (x can be 2, 3, 4, ...)
FS	Fast N, Slow P process corner
FSM	Finite state machine
GSG	Ground signal ground
HA	Half-adder
HCI	Hot carrier injection
HF	High frequency
HVM	High volume manufacturing
HVT	High VT
IIP2	Second-order Input Intercept Point
IO	Input/output
IP	Intellectual property (circuit)
IR	Impulse Radio
ITRS	International Technology Roadmap Standard for Semiconductors
L	Transistor Length
LDO	Low DropOut (regulator)
LFN	Low frequency noise
LFSR	Linear feedback shift register
LO	Local oscillator
LPDDR	Low Power Dual Data Rate (Memory)
LSB	Least significant bit
LUT	Look-up table
LVT	Low VT
MC	Monte Carlo
MHC	Model to Hardware Correlation
MIM	Metal-insulator-metal (capacitor)
MOM	Metal-oxide-metal (capacitor)

Acronyms

MS	Microstrip (line)
MSB	Most significant bit
NBTI	Negative Bias Temperature Instability
OTA	Operational transconductance amplifier
P.V.T.a. or PVTA	Process voltage temperature and temperature (variations)
PA	Power amplifier
PAE	Power-added efficiency
PB	Poly Bias, MOS length optical deshrink, PB4 means effective length is drawn L + 4 nm
P-Cell	Parameterized cell
PG	Pulse generator
PHY	Physical layer
PLL	Phase locked loop
PCB	Printed Circuit Board
PMU	Power management unit
PRF	Pulse repetition frequency
Psat	Saturated (output) power
PSD	Power spectral density
PSRR	Power supply rejection ratio
PTAT	Proportional to absolute temperature
PVT	Process Voltage Temperature operating point
QoR	Quality of Result
RA	Read ability
RBB	Reverse Body-Bias
RF	Radio-frequency mmW millimeter-wave
RO	Ring oscillator
RS	Read stability
RVT	Regular VT
SC	Switched-capacitor
SF	Slow N, Fast P process corner
SiGe	Silicon Germanium
SNM	Static Noise Margin
SOA	Safe-operating area
SoC	System on a Chip
SOF	Sign-off
SS, SF, FS, FF	Slow-slow, slow-fast, fast-slow, fast-fast (process corners)
SS	Slow N, Slow P process corner
STA	Static Timing Analysis
TA	Application temperature
TCR	Temperature coefficient of resistivity
TD	Thermal diffusion
TDC	Time-to-digital converter
TDDB	Time Dependent Dioxide Breakdown
THz	Terahertz (frequency)
Tj	Junction temperature

TL	Transmission line
TRC	Tunable replica circuit
TTF	Time To Failure
TX	Transmitter
ULP	Ultra low power
USL	Upper specification level
UTBB	Ultra-thin body and box (FD-SOI)
UWB	UltraWide band
V_{Tinv}	Voltage of temperature inversion same as ZTC
Vbb	Body-Bias voltage
VCDL	Voltage-controlled delay line
VMU	Vector memory unit
VRU	Vector roundhead unit
VSWR	Voltage standing wave ratio
VT	Threshold voltage
VXU	Vector execution unit
W	Transistor width
WM	Write margin
WS	Write stability
ZTC	Zero-temperature coefficient

Chapter 1
Introduction

Andreia Cathelin and Sylvain Clerc

1.1 Foreword

The CMOS integration race has reached limitations for planar silicon process starting from the 40 nm node. The transistor channel was more and more difficult to control and specific process integration methods such as pocket implant, silicon strain, and lightly doped drain were introduced to enable devices' good carrier mobility and electrostatic control, are moreover this type of process integration could not be successfully continued after the 20 nm node. Starting from the 28 nm node a consensus solution emerged consisting in the use of fully depleted active devices either fully depleted silicon on insulator (FD-SOI) or Fin-FET. While the fundamental physics laws are similar for these two big families of devices, the process integration is much different and had to bring the process engineers from the well-known planar technologies (applies also for FD-SOI) to fully 3D structures (for Fin-FET).

As we will largely discuss all over the different chapters of this book, all FD-SOI technologies are *planar* and offer to designers a brand new operation playground by the fact that the transistors in this technology have now *4 effective terminals*: source, drain, gate, and body – the volume underneath the conduction channel. In FD-SOI, by applying a voltage bias on this latest, one can efficiently vary the transistor's threshold voltage. While this technique is not new [1] it is the first time it can be deployed with a large electrical impact into commercial CMOS technology.

All design techniques existing in classical planar CMOS technologies since more than 30 years can now be revisited and brought to a higher level, by adding a new and efficient tuning knob, the transistor's body tie, inside all the schematics.

A. Cathelin (✉) · S. Clerc
STMicroelectronics, Crolles, France
e-mail: andreia.cathelin@st.com; sylvain.clerc@st.com

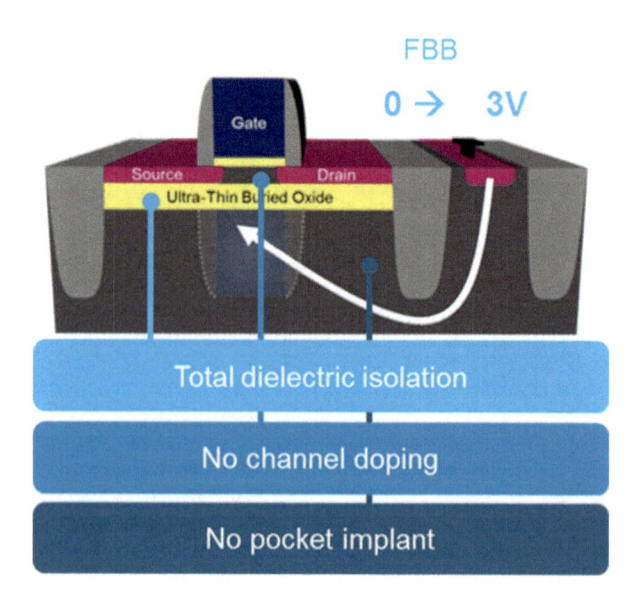

Fig. 1.1 A generic cross-section of an UTBB FD-SOI transistor. ©2017 IEEE. Reprinted with permission

The circuit designer can now control on the fly the transistor's threshold voltage variation by software commands. It brings in to all types of schematics, from analog to RF, mmW to mixed signal, digital and memories, a new degree of design freedom. Its integration into low power CMOS technologies offers fantastic integration framework for all kinds of IoT, 5G and any other type of energy efficient circuits and systems.

The aim of this book is to introduce to the design community the straightforward design solutions in any modern FD-SOI planar CMOS technologies, by taking full advantage of body-biasing techniques to efficiently modulate on the fly SoC solutions from high performance operation to energy efficiency mode. All design techniques are based on the classical pillar of regular planar CMOS devices. As the first fully industrial solution has been the 28 nm FD-SOI CMOS technology from STMicroelectronics, all the design examples given in this book have been demonstrated within this process integration frame.

Figure 1.1 gives a generic cross-section of a FD-SOI CMOS device [2]. This technology is called ultra-thin body and BOX (UTBB) FD-SOI CMOS, as in the 28 nm node the active device has an ultra-thin conduction film (7 nm) and lays atop a 25 nm insulation layer of buried oxide (BOX). This planar topology's direct implications are the following: thanks to the SOI BOX layer, the transistor gets total dielectric isolation. No channel doping is needed as thanks to the thin silicon film, the channel is fully depleted. Also enabled by this topology, no pocket implants are needed for the source and drain, which enhances naturally the analog/RF transistor's behavior. Another implication of the thin BOX layer is the fact that the front-side transistor's electrostatics can be controlled from underneath the BOX (area called transistor body). By applying a voltage on the transistor body, one can change or modulate the threshold voltage of the main (front-side) transistor. We can see this device as well as a planar dual-gate device: the front-gate is the regular one (like in bulk technology) and the second one comes from the body tie, with the BOX as the back-side gate oxide. As the thickness ratio of the front and back gate oxides is about 10, the front-side transistor's transconductance gm is 10 times bigger than the one of the back-side gate.

Body-bias can be applied in two directions: either to lower devices threshold voltage, with PMOS body terminal driven to a lower voltage than source supply (VDD) and NMOS body driven to a higher voltage than source supply voltage (usually ground), this is forward body-bias. This leads to a *high performance* type of operation. Reversely, body-bias can be applied to increase devices threshold voltage, with PMOS body terminals drive to a higher voltage than VDD and NMOS body terminal drive to a lower voltage than supply (i.e., negative voltage), this is reverse body-bias. This enables a *low leakage* energy efficiency type of operation.

When applied to bulk technology, the forward body-bias voltage range is limited by junction diodes on-current and latch-up sensitivity, reverse body-bias is accessible to bulk technology without latch-up risk but is limited by gate induced drain leakage (GIDL). As will be exposed in Chap. 2 detailing the device microstructure, the FD-SOI technology enables a wide range of body-bias voltages because its intrinsic channel isolation overcomes the bulk technology bias range limitations of both GIDL and latch-up.

1.2 Analog Design Aspects

In the analog/RF design application domain, the historical nanometer downscaling roadmaps have accustomed analog designers to have transistors' analog behavior degraded from node-to-node because of digital low-power and high-speed objectives needing the above-mentioned process integration techniques. While this can be overcome in deep-submicron CMOS by introducing analog specific devices (like having no pocket implants), in FD-SOI technology the thin film structure inherently eliminates many of the analog performance limitations and permits to have better DC gain behavior and higher current drive capability [3]. As well, the transistors implementation with thin film layers permits to limit the important variability effects present in such nm nodes. The presence of the SOI topology naturally limits the parasitic effects, hence we get higher achievable bandwidth or a lower power consumption for a given operation bandwidth. As it will be largely developed over the next chapters, in FD-SOI voltage biasing range on the transistor's body is very wide (almost |3 V|), whereas in a regular CMOS process it is only of few tenth of mV. This, in conjunction with a higher body effect, enables an unprecedented tuning range of the threshold voltage of more than 250 mV.

In terms of high frequency capability, the 28 nm CMOS node permits to obtain transition frequencies of around 300 GHz, which hence enable operation from RF domain up to mmW frequencies, such as the recently investigated communication bands at 60 GHz and above, for 5G high data rate applications.

Regarding ultra-low power solutions tailored for IoT applications, FD-SOI technologies are as well the sweet spot for ultra-low voltage flexible and energy efficiency implementations.

Furthermore, body-biasing enables either statically or dynamically, via embedded voltage generators, to compensate for environmental conditions and/or reconfigure/tune analog design features depending on application and usage. As well, this huge optimization domain offered by the wide body-biasing range permits to propose new design schemes with much reduced design margins, hence with smaller power consumption and reduced chip area. Simple CMOS inverter based transconductor schematics can be efficiently revisited for class A or AB amplifiers, and the classical differential pair can now be given new dimensional design heights by simply tuning its body ties. New kinds of closed-loop tuning and trimming schemes can be proposed to ensure an efficient in specifications operation of full analog/RF/mixed signal SoCs.

1.3 Digital Design Aspects

Looking back at the past 20 years, the digital designers have faced the following trends:

- increased interconnect delay versus gate delay which lowers custom digital added value
- increased local mismatch affecting SRAM write margin, non-CMOS Flip-Flops robustness, and constraints closure of synchronous digital parts
- transition from 1D device to 2D device with increased influence of lateral electric field on channel control, causing V_T roll-off and drain-induced leakage (DIBL)
- increased power density as device dimensions shrink faster than power

These factors have some interdependence, for example, the increased local mismatch weakness has prevented the digital bitcell to shrink at the same rate as the gate length [4], and increased the leakage because of lower V_T needed. To address increased leakage and dynamic power density, various design approaches have been put in place: multi V_T design, retention and power gating, multi-VDD, and adaptive voltage scaling (AVS). Among these techniques, early body-bias usage can be traced back in memory design [5]. It has emerged in the years 1990–2000 in digital when some hardware operator and microprocessor designs were reported using this technique [6–8]. However, the body-bias tuning range has lowered with technology shrink in bulk technologies. The relative merit of AVS compared to bias was studied in [9] but with limitation of bias dynamic and PMOS tuning only. It will be exposed how FD-SOI technology recovers back body-bias tuning range.

1 Introduction

1.4 Book Overview

This book is structured in three parts as follows:

The first part presents a technology overview (Chap. 2), general considerations on what body-biasing can bring to the digital and analog designers (Chaps. 3 and 4) and SRAM bitcell design under body-biasing conditions (Chap. 5).

The second part presents a selection of circuits which illustrate the body-biasing usage in various fields, from analog to RF/mmW and digital, and from building blocks to circuits and SoC implementations. The choice of topics is non-exhaustive, they present some design fields where fully aware body-bias design can change significantly the game and permit to get straightforward competitive advantages. All presented design solutions, validated through silicon implementation, show best-in-class performance, and outperform state of the art by taking full benefit of body-biasing deployment.

Chapter 6 presents an analog/high-speed circuit block for wide range and fine grain programmable delay elements, very useful for high data rate communications both wired and wireless.

Chapters 7 and 8 bring highlight in mmW design solutions, with implementations showing some very high frequency mmW oscillators implementations (up to 200 GHz) and 30 GHz 5G mmW power amplifiers. Both active devices behavior at high signal level at mmW frequencies and passive elements integration with sufficient quality factor, in a dense back-end-of-line (BEOL) VLSI technology, are tackled in these two chapters.

Chapter 9 gives a detailed insight on an RF wireless sensor node IoT transmitter SoC with analog/RF/mixed signal and digital circuitry.

Monitoring is key for body-bias deployment as will be exposed in the following Part III, to this end, an energy efficient thermal sensing is presented in Chap. 10.

Full SoC implementations of several generations of RISC-V processors are then detailed in Chap. 11, demonstrating energy efficiency giga operations per seconds scale computing (GOPS).

The third section of this book presents the body-biasing deployment in mixed-signal and digital SoCs.

This part includes body-bias control designs based on either closed-loop (mid- to long-term solutions, Chap. 12) and open-loop (seen as short-term industrial perspective, Chap. 13).

Energy efficient design has been a design priority for decades now and it will see its full deployment through all the panel of applications covered by IoT implementations, in this sense modulating the devices Ion is a must have. Body-bias is an alternative to the well-known adaptive voltage scaling (AVS) technique, both have their pros and cons and can be eventually mixed, this subject will be covered in Chap. 14 together with a review on open-loop and closed-loop body-bias control solutions.

We study also body-bias voltage generation units for digital usage (Chap. 15), and review on the practical side, digital design flow methodology (DDFM) specific aspects (Chap. 16).

Through this set of design examples from small circuits up to SoCs, we are illustrating the power-performance-area (PPA) benefits which can be gained by body-biasing techniques in FD-SOI technologies and help the reader make the appropriate architectural choices for their own designs. We wish you an enjoyable journey through the art of body-biasing!

References

1. H.C. Wann, C. Hu, K. Noda, D. Sinitsky, F. Assaderaghi, J. Bokor, Channel doping engineering of MOSFET with adaptable threshold voltage using body effect for low voltage and low power applications, in *1995 International Symposium on VLSI Technology, Systems, and Applications. Proceedings of Technical Papers* (IEEE, Piscataway, 1995), pp. 159–163
2. A. Cathelin, Fully depleted silicon on insulator devices CMOS: the 28-nm node is the perfect technology for analog, RF, mmW, and mixed-signal system-on-chip integration. IEEE Solid-State Circuits Mag. **9**(4), 18–26 (2017)
3. A. Cathelin, RF/analog and mixed-signal design techniques in FD-SOI technology, in *2017 IEEE Custom Integrated Circuits Conference (CICC)* (IEEE, Piscataway, 2017)
4. B. Nikolić, Simpler, more efficient design, in *ESSCIRC Conference 2015-41st European Solid-State Circuits Conference (ESSCIRC)* (IEEE, Piscataway, 2015), pp. 20–25. https://doi.org/10.1109/ESSCIRC.2015.7313819. https://doi.org/10.1109/VTSA.1995.524654
5. H. Kawamoto, T. Shinoda, Y. Yamaguchi, S. Shimizu, K. Ohishi, N. Tanimura, T. Yasui, A 288 K CMOS pseudostatic RAM. IEEE J. Solid State Circuits **19**(5), 619 (1984). https://doi.org/10.1109/JSSC.1984.1052198
6. T. Kuroda, T. Fujita, S. Mita, T. Nagamatu, S. Yoshioka, F. Sano, M. Norishima, M. Murota, M. Kako, M. Kinugawa, M. Kakumu, T. Sakurai, A 0.9 V 150 MHz 10 mW 4 mm/sup 2/2-D discrete cosine transform core processor with variable-threshold-voltage scheme, in *1996 IEEE International Solid-State Circuits Conference. Digest of Technical Papers, ISSCC* (1996), pp. 166–167. https://doi.org/10.1109/ISSCC.1996.488555
7. J.W. Tschanz, J.T. Kao, S.G. Narendra, R. Nair, D.A. Antoniadis, A.P. Chandrakasan, V. De, Adaptive body bias for reducing impacts of die-to-die and within-die parameter variations on microprocessor frequency and leakage. IEEE J. Solid-State Circuits **37**(11), 1396 (2002). https://doi.org/10.1109/JSSC.2002.803949
8. T. Miyake, T. Yamashita, N. Asari, H. Sekisaka, T. Sakai, K. Matsuura, A. Wakahara, H. Takahashi, T. Hiyama, K. Miyamoto, K. Mori, Design methodology of high performance microprocessor using ultra-low threshold voltage CMOS, in *Proceeding of IEEE 2001 Conference on Custom Integrated Circuits* (2001), pp. 275–278. https://doi.org/10.1109/CICC.2001.929773
9. T. Chen, S. Naffziger, Comparison of adaptive body bias (ABB) and adaptive supply voltage (ASV) for improving delay and leakage under the presence of process variation. IEEE Trans. Very Large Scale Integr. VLSI Syst. **11**(5), 888 (2003). https://doi.org/10.1109/TVLSI.2003.817120

Part I
Device Level and General Studies for Analog and Digital

Chapter 2
FD-SOI Technology

Franck Arnaud

2.1 Introduction

Whatever the type of applications, from internet-of-things to automotive microcontrollers, we anticipate a strong demand in terms of power reduction, frequency increasing, and analog mixed signal co-integration. Consequently, advanced CMOS technologies have to adapt the offering with respect to customer demand with a limited rework in terms of design (library and complex IPs). A simple way, already well known and used, is the adaptive voltage scaling (AVS) technique, based on power supply (V_{dd}) modulation, depending on the type of application or need. Thus, with a same set of foundation IPs and an identical technology, we can extend the range of applications served. Even if AVS is very powerful, especially in terms of digital speed, we do see three major drawbacks: active power rising, severe transistor and wiring reliability degradation, and no intrinsic improvement for analog blocs.

Moving along from bulk-planar transistor toward FD-SOI-planar one and the FBB technique, we have the opportunity to provide the same level of design flexibility as AVS but without the drawbacks depicted previously. The main technical advantage of FD-SOI is the threshold voltage modulation of the device. It means we have a knob, thanks to gate bias, to change the intrinsic positioning of the device. Threshold voltage (V_T) reduction improves the speed of the device, especially at low voltage operation, feeding digital performance request. If no extra frequency is required, then we have the opportunity to drop the power supply itself to mitigate the power consumption. And for analog circuits, V_T lowering will bring higher gm without transistor mismatch degradation. Finally, threshold voltage reduction does

---

F. Arnaud (✉)
STMicroelectronics, Crolles, France
e-mail: franck.arnaud@st.com

© Springer Nature Switzerland AG 2020
S. Clerc et al. (eds.), *The Fourth Terminal*, Integrated Circuits and Systems,
https://doi.org/10.1007/978-3-030-39496-7_2

not hurt device reliability due to the fact that both lateral and vertical electrical fields are maintained.

In this chapter, we will first describe the physical structure of the device and provide the basic equations of the V_T modulation of FD-SOI transistor. The role of the back gate will be properly described and explained. Device structure comparison will be presented versus reference bulk-planar architecture. Section 2.3 will be dedicated to the native impact of the FBB on a single transistor. Basic transistor parameters will be reviewed such as V_T, drive current (I_{on}), or quiescent one (I_{off}) completed with a first dynamic analysis based on ring oscillator (RO) design. Transistor variability tightening will be discussed in Sect. 2.4 evidencing the superior of FBB to compensate intrinsic process variability. Sections 2.5 and 2.6 will focus on digital and analog figures of merit, respectively. Digital part will be dedicated to the power reduction thanks to FBB solution. FD-SOI architecture and body-biasing capability for 6T-SRAM bit-cell will be reviewed and deeply discussed in Sect. 2.7. Finally, reliability results will be presented in the last section of this chapter focusing on the technology.

Notation

Throughout this chapter, the classical notation applies [1, 2]:

ϕ_m	metal work-function
ϕ_{fb}	Flat band voltage
ψ_s	Silicon gate-side surface potential
ψ_{sb}	Silicon body-side surface potential
E_C	Silicon bottom conduction band energy level
E_V	Silicon top valence band energy level
E_i	Intrinsic energy level of channel carriers
V_{OX}	Voltage across gate oxide
V_{GS}	Gate to source voltage
V_{BS}	Body to source voltage

2.2 FD-SOI Technology Description and Basic Equations

FD-SOI transistor is a planar device, as bulk's one, it means the gate control is a two-dimensional electrode. However, the structure is based on silicon thin film, in the range of a few nanometers (≤ 10 nm actually) lying on a thin buried oxide (≤ 50 nm). Macroscopic transistor scheme [3] is depicted in Fig. 2.1

Thin and fully depleted silicon film will decrease significantly the parasitic effect of the source and drain junctions on the channel area, enabling a stronger influence of the front gate. The benefit of the superior electrostatic control of the device in case of FD-SOI device is well explained by the concept of drain induced barrier lowering (the so-called DIBL). As shown in Fig. 2.2, the role of the source and drain region on the channel is directly related to the junction depth (labeled x_j). By shortening the physical gate related to Moore's law, we observed on increasing of the

2 FD-SOI Technology

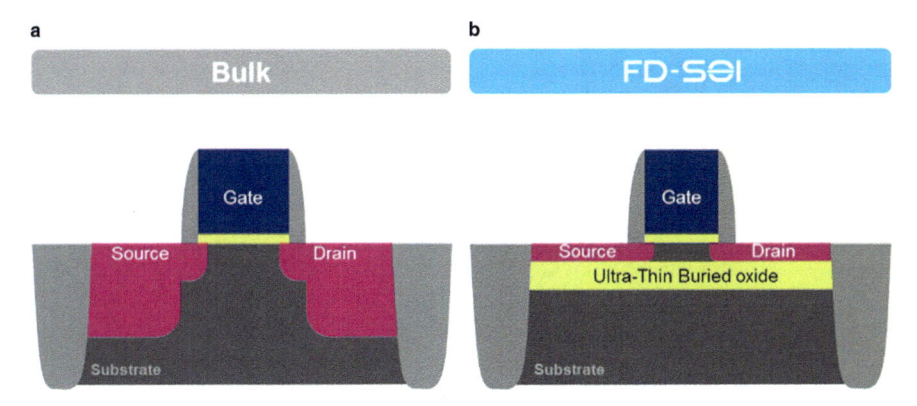

Fig. 2.1 General structure evolution of CMOS transistor from bulk scheme (**a**) toward FD-SOI transistor (**b**)

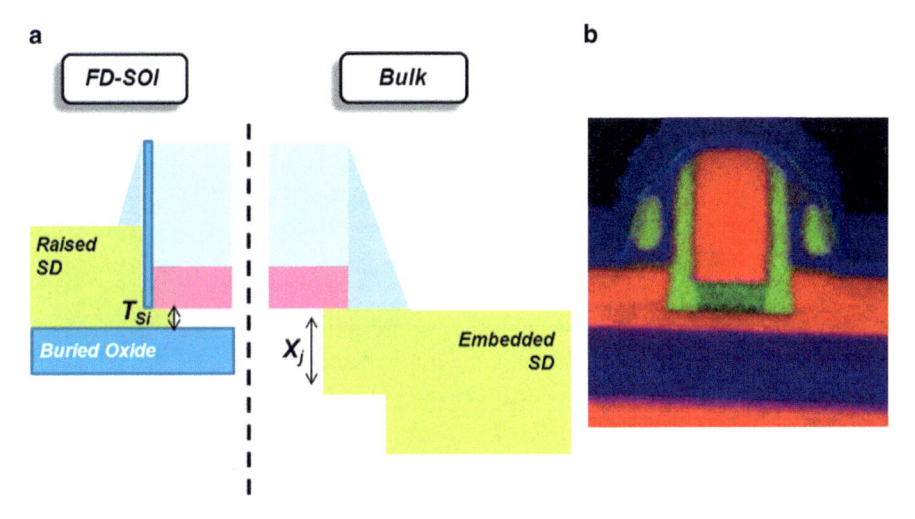

Fig. 2.2 Junction design differences between bulk transistor, left side of (**a**) and FD-SOI one (**b**) influencing the electrostatic control and DIBL behavior

parasitic region controlled by junctions. Process wise, the scalability of the x_j factor becomes more and more challenging because modulated by diffusion mechanism and consequently requiring very precise thermal budget at manufacturing side. To take into account such issue coming from the shrinkable, main parameter used to evaluate electrostatic performance of the transistor is x_j/L_{gate} ratio. This ratio is exhibited in the V_T Eq. (2.1) for short channel devices below [2].

$$\Delta V_T = \frac{qN_aW\left(\sqrt{1+\frac{2W}{x_j}}-1\right)}{C_{OX}} \cdot \frac{x_j}{L_{\text{gate}}} \quad (2.1)$$

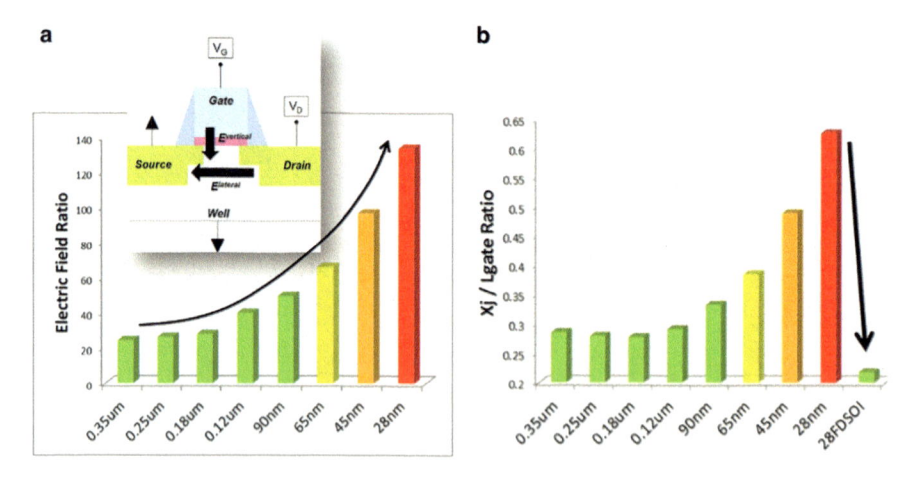

Fig. 2.3 Challenge of electrostatic control of the channel in advanced CMOS technology due to the shrink of lateral dimension overcome thanks to FD-SOI architecture (**a**). x_j/L_{gate} ratio significant improvement thanks to FD-SOI architecture (**b**)

In case of FD-SOI transistor, x_j factor is replaced by the silicon film thickness, with much superior capability to be scaled down, because not driven by the physics of the diffusion. Thus, moving from mature CMOS technology to most advanced node, the difficulty to control the channel of the transistor has been overcome thanks to FD-SOI scheme (Fig. 2.3).

Circuit performance at low voltage is directly linked to the DIBL value of the device. Lower is the DIBL, higher is the low voltage performance. As shown by Eqs. (2.3) and (2.3), low DIBL requires low x_j/L_{gate} ratio. In case of FD-SOI transistor, x_j/L_{gate} ratio is replaced by T_{Si}/L_{gate} [4].

$$DIBL_{\text{Bulk}} = 0.8 \cdot \frac{\epsilon_{Si}}{\epsilon_{OX}} \left(1 + \frac{X_j^2}{L_{el}^2}\right) \cdot \frac{T_{OX}}{L_{el}} \cdot \frac{T_{\text{dep}}}{L_{el}} \cdot V_{DS} \qquad (2.2)$$

$$DIBL_{\text{FD-SOI}} = 0.8 \cdot \frac{\epsilon_{Si}}{\epsilon_{OX}} \left(1 + \frac{T_{Si}^2}{L_{el}^2}\right) \cdot \frac{T_{OX}}{L_{el}} \cdot \frac{T_{Si}}{L_{el}} \cdot V_{DS} \qquad (2.3)$$

Aside the thin silicon film improving the control of the device, FD-SOI technology is based on thin buried oxide layer underneath the channel region. This thin oxide (<50 nm) allows a back side control of the channel. Well region play then the role of a back gate enabling a further biasing helping to create the inversion layer sooner. This pseudo-3D structure provides the opportunity for a back gate biasing as described in Fig. 2.4

Surface potential inside the channel represented by Ψs parameter, depends on the metal work-function (ϕm) and the front gate biasing as expected in bulk transistor. Surface potential represents the status of the channel in terms of inversion layer, i.e., the availability of the device to transport carrier from source to drain region.

2 FD-SOI Technology

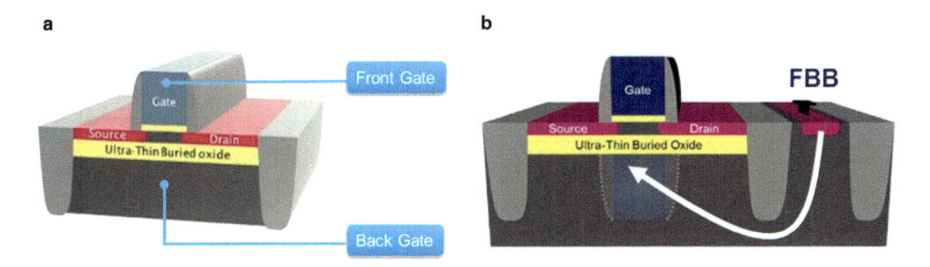

Fig. 2.4 FD-SOI transistor double gate structure enabling FBB technique. (**a**) 3D structure of FDSOI transistor. (**b**) Back gate control of the device

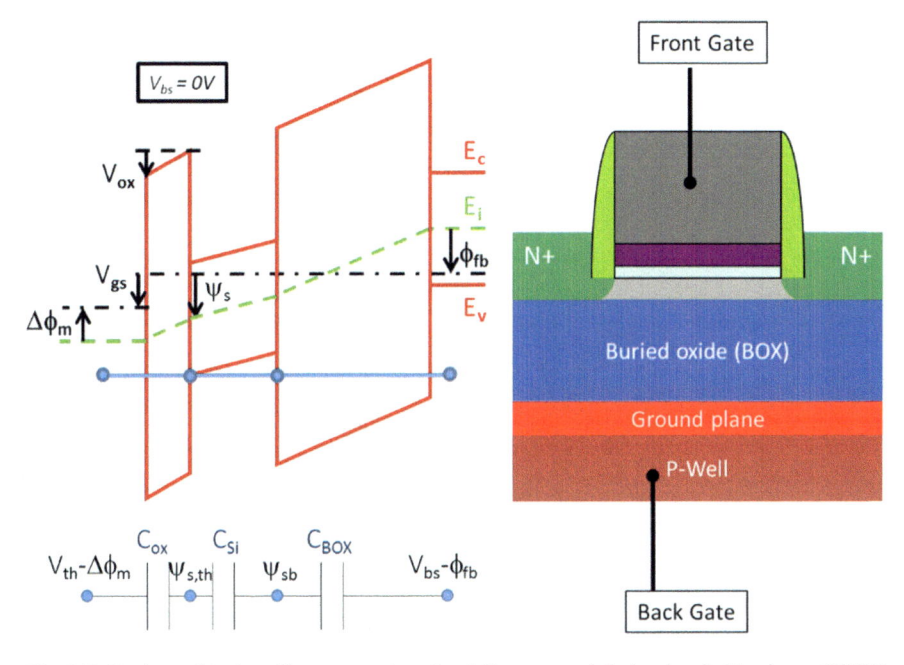

Fig. 2.5 Back gate biasing effect on transistor band diagram modulating threshold voltage. ©IEEE 2017 reprinted with permissions

By applying a bias at the front gate (V_{gs}), Ψs will increases and then free electrons will be generated in the channel. In the same way, a voltage applied at the rear of the buried oxide will modulate as well the surface potential, and the conduction channel. The setting and the adjustment of Ψs will then depend on both front and back gate biasing (V_{gs} and V_{bs}, respectively), as depicted in Figs. 2.5 and 2.6. Equivalent schematic is a coupling capacitance between C_{OX} (front gate capacitance) and C_{BOX} (back gate capacitance), the channel playing the role of an internal node. Morphological structure and band diagram illustrate this schematic [5].

This coupling allows a specific V_T modulation without carrier mobility or junction leakage degradation. The FD-SOI transistor' V_T equation is derived from bulk V_T inversion capacitance,

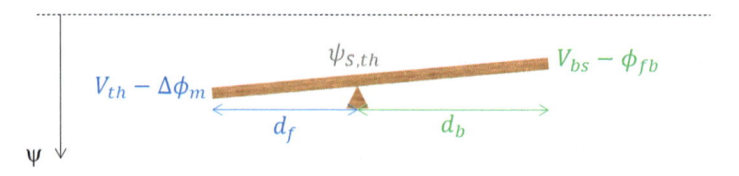

Fig. 2.6 Surface potential equilibrium between the front gate materialized by V_T and metal work-function and the back gate defined by body-bias and flat band voltage. ©2017 IEEE reprinted with permission

$$C_{inv} = C_{dep} + C_{OX} \quad (2.4)$$

Where in our case of full depletion,

$$C_{dep} = \frac{dQ_{dep}}{d\psi_S} = 0, \; C_{inv} = C_{OX} \quad (2.5)$$

From electrostatic potential:

$$C_{inv} = -\frac{dQ_{inv}}{d\psi_S} \quad (2.6)$$

$$C_{inv} = \frac{q \cdot n_i \cdot t_{Si}}{kT/q} \cdot e^{\psi_S/(kT/q)} \quad (2.7)$$

$$\psi_{S,th} = \frac{kT}{q} \cdot \ln\left(\frac{C_{OX}kT/Q}{qn_i t_{Si}}\right) \quad (2.8)$$

Considering the creation of the channel at the front interface, the threshold voltage expression is obtained from the capacitive divider displayed in Fig. 2.6:

$$\frac{(V_T - \Delta\phi_m) - \psi_{S,th}}{1/C_{OX}} = \frac{\psi_{S,th} - (V_{bs} - \phi_{fb})}{1/C_{Si} + 1/C_{BOX}} \quad (2.9)$$

Posing:

$$BF = \frac{\partial V_{gs}}{\partial V_{bs}} \bigg\|_{\text{fixed } \psi_S}$$

$$BF = \frac{1/C_{OX}}{1/C_{Si} + 1/C_{BOX}} \quad (2.10)$$

The capacitive coupling being denoted by BF factor depends on C_{BOX}/C_{OX} ratio [5] and represents the gate voltage overdrive/underdrive brought by body-bias. From the ratio formula of BF in Eq. (2.10), we can predict that body-biasing will be more efficient in case of ultrathin buried oxide and thick oxide devices, as for I/Os and analog blocks.

2 FD-SOI Technology

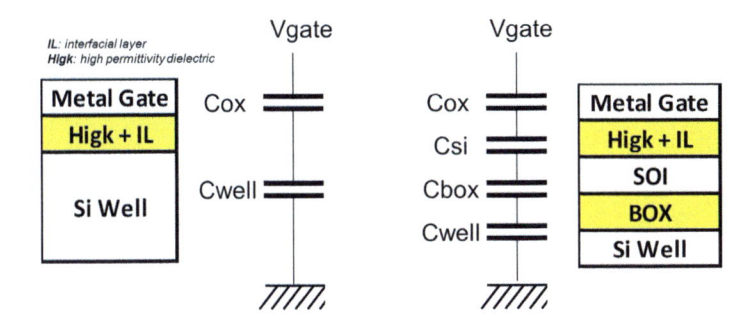

Fig. 2.7 Vertical structure of MOS device in case of bulk-planar device (left) and FD-SOI one (right)

Finally, the threshold voltage expression is

$$V_T = \Delta\phi_m + \psi_{S,th} + BF \cdot (\psi_{S,th} - (V_{BS} - \phi_{fb})) + \frac{h^2}{8q \cdot m_{\text{conf}} \cdot t_{SI}^2} \quad (2.11)$$

where m_{conf} represents the effective mass of confined carrier. The electrostatic behavior of the channel in case of FD-SOI device is different from what we use to observe in the case of the bulk-planar transistor. As described in Fig. 2.7, the vertical structure on FD-SOI transistor is completed by a pure coupling capacitance structure between the top oxide (C_{OX}) and the bottom one (C_{BOX}) and not by a depletion modulation.

The role of the coupling capacitance system inside FD-SOI scheme is represented in Fig. 2.8. We can model the evolution of the threshold voltage versus body-bias (here called V_{bb}). Since a reserve body-bias (RBB) is applied, the front channel is open, allowing a carrier transport at the front of Si channel. Channel conduction will shield the silicon film and then threshold voltage depends on C_{BOX}/C_{OX} ratio. In case of a forward body-bias (FBB), a second inversion layer appears at the rear of the silicon film at Si/Box interface. It means that threshold voltage reduction depends on (C_{BOX}, C_{OX}, C_{Si}) triplet.

Under usual front gate polarization ($V_g = V_{dd}$ and $V_{bb} = 0$), inversion layer is generated only at the front gate interface as shown in red on Fig. 2.9 representing the charge density as a function of the position inside the silicon channel. If a RBB biasing is applied at the rear of the transistor, inversion charge is decreased corresponding to an increasing of threshold voltage. On the contrary, if a FBB is used at the back gate, a second inversion layer is observed at the Si/Box interface. This back conduction corresponds to a decreasing of the threshold voltage value of the transistor [6].

The charge density described above can be measured thanks to C-V characteristic of the MOS transistor. By changing the back gate voltage, we can modulate the

Fig. 2.8 Threshold voltage evolution versus body-bias (V_{bb}) as a function of C_{BOX}, C_{OX}, and C_{Si} coupling capacitances

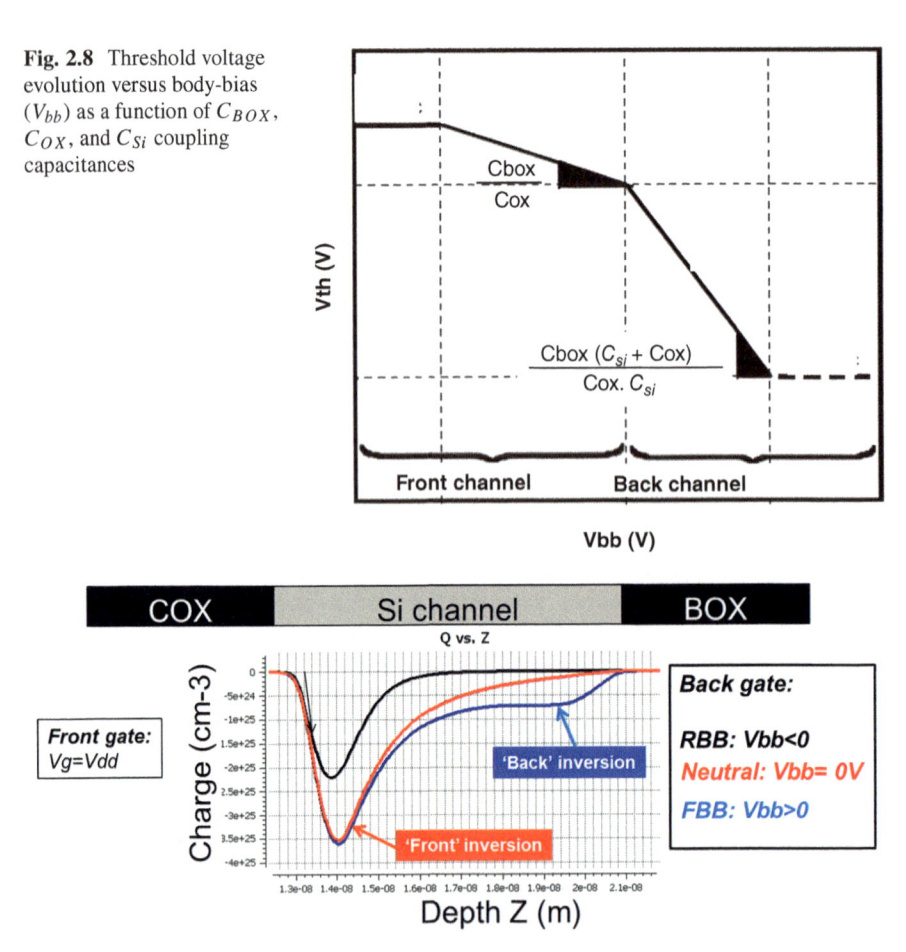

Fig. 2.9 Charge density inside silicon channel versus back gate biasing (RBB, NBB, and FBB)

generation of the second inversion layer and then the vertical capacitance of the structure, as depicted in Fig. 2.10. For nominal condition ($V_{bb} = 0$ V) and RBB, a maximum capacitance corresponding to the C_{OX} value is measured. It corresponds to a front conduction. In case of positive back biasing corresponding to FBB condition, and a back inversion layer appears dropping the capacitance value due to C_{OX} in parallel with C_{Si}. Thus the shape of the C-V is influenced by the back gate if FBB is applied.

The fact to change the body-bias value, a significant change occurred for the sub-threshold regime. Main effects are represented in Fig. 2.11. As expected, V_T goes up and down by applying negative or positive V_{bb}, respectively. As shown below, the sub-threshold slope is modulated by the body-bias as well. In the situation of RBB, we observed an improvement of the sub-threshold slope, as a sign of an improvement of the electrostatic control of the transistor. Actually, it can be explained by a better inversion layer localization under the front gate, as depicted

2 FD-SOI Technology

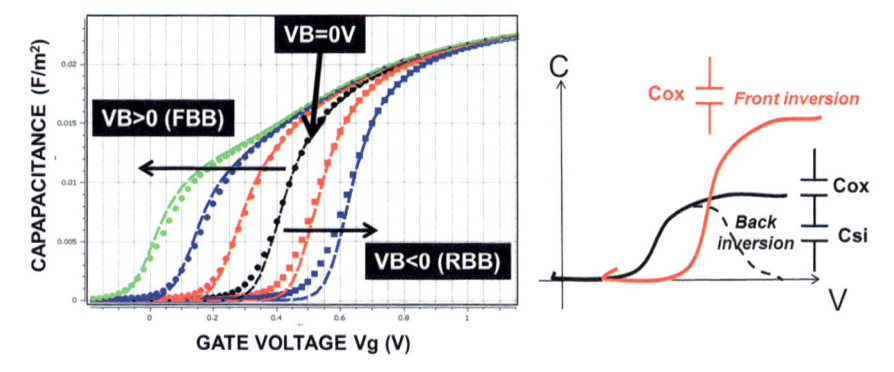

Fig. 2.10 Vertical MOS capacitance evolution versus front and back gate biasing condition evidencing the second conduction layer appearing at the rear of the channel under FBB

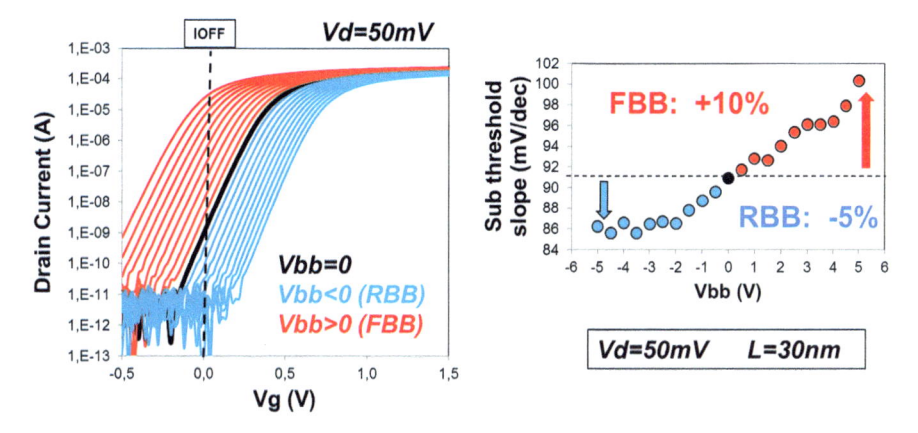

Fig. 2.11 Sub-threshold slope change versus body-bias applied to the back gate for short gate length and linear condition for the drain voltage

in Fig. 2.9. Thus, in case of RBB, leakage current of the device is reduced thanks to V_T increasing and sub-threshold slope reduction. On the opposite direction, with FBB voltage used on the back gate, the generation of a second channel at the Si/Box interface led to a weaker control of the carrier from the front gate. Electrostatic of the device is degraded to the remote channel control. Leakage of the transistor is increased due to both V_T lowering and sub-threshold slope rising.

We have just seen that back bias modulates the vertical behavior of the channel with the confinement of the inversion layer in case of RBB and the generation of a second inversion one at the rear interface with FBB. As a consequence, leakage-related parameters of the transistor have been modified, such as V_T and sub-threshold slope. In Fig. 2.12 we are presenting the impact of the body-bias on the lateral conduction between source and drain terminals. While FBB is applied, the dual conduction channel enhances significantly the drivability of the device. This current density improvement is materialized by an increasing of the carrier mobility

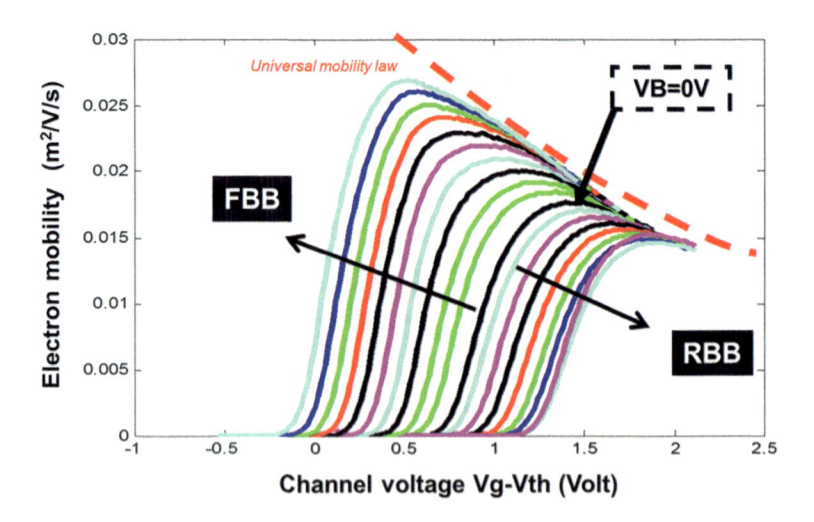

Fig. 2.12 Electron mobility enhancement thanks to FBB solution intensively used to increase circuit frequency especially at low voltage operation

as a function of the channel voltage $V_g - V_T$. More than +50% mobility boost has been measured for FBB at 3.5 V. On the contrary, due to inversion layer confinement at the front gate dielectric, electron transport degradation is detected due to coulomb scattering and oxide surface roughness impact. It is worth pointing out that body-bias is still fitting the universal mobility law modeled in the MOS transistor.

To summarize the body-bias impact on the basic element of the MOS transistor, we can say that FBB is dedicated to performance enhancement thanks to mobility increasing and V_T reduction whereas RBB is appropriate to mitigate the leakage current by increasing V_T and improving sub-threshold slope.

One of the main challenges facing by advanced CMOS technology is the threshold voltage adjustment and centering in order to serve a wide range of applications, from ultra-low leakage to ultimate speed. For technology beyond 40 nm, we clearly saw a major difficulty to properly tune the V_T only by fixing the channel doping level, through simple ion implantation technique. The migration to high-k dielectric complicated significantly V_T positioning, especially to deliver low V_T (LVT) and high V_T (HVT) with the same gate stack. The V_T fluctuation versus potential transistor architecture is summarized in Fig. 2.13.

As already published in 2000s years, the gate dielectric changes from SiO2 to high-k material to reduce the parasitic gate leakage by scaling down the thickness, induced a Fermi level pinning at the polysilicon interface [7]. Consequently, additional metal layer has been introduced between high-k and polysilicon. Classical metals like TiN, TaN, or W unfortunately have been measured as mid-gap materials with work-function close to 4.8 eV at the end of the process. To get rid of such issue, further elements such as La or Al have been implemented inside the stack reducing

2 FD-SOI Technology

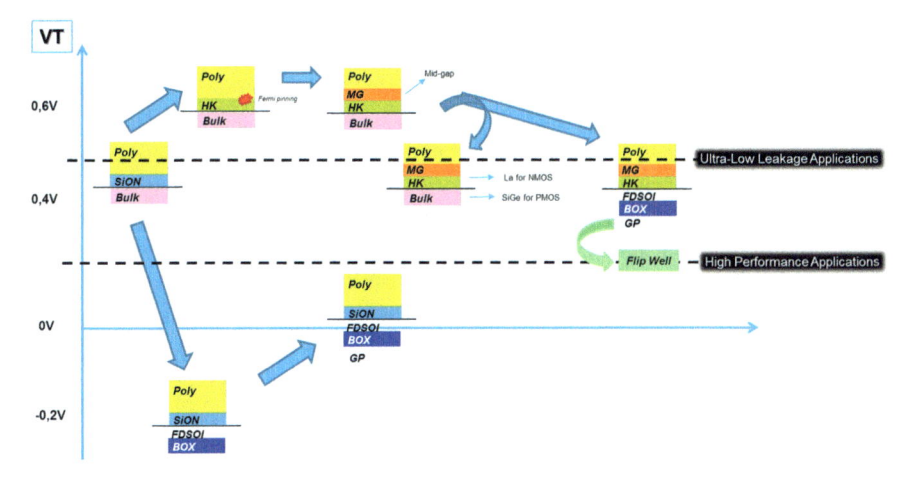

Fig. 2.13 Threshold voltage evolution versus transistor architecture and the challenge to fit ultra-low leakage and high performance application with only one scheme

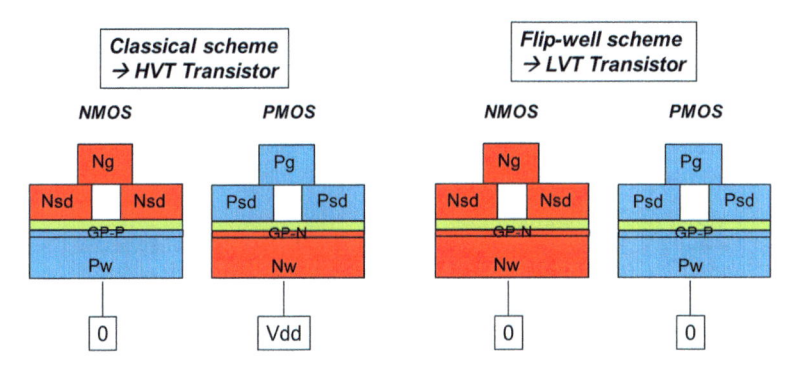

Fig. 2.14 Flip-well architecture allowing low-V_T transistor flavor suited for high performance circuit

the threshold of NMOS and PMOS, respectively. Those thin layers of metal can play the role of dipoles after a migration at the high-k/SiO2 interface [8].

In case of FD-SOI architecture, thanks to a fully depleted structure, HVT devices can be achieved without any dipoles introduction. Dipoles influencing negatively the carrier's mobility inside the channel, FD-SOI solution can deliver higher current. However, even if it was well adapted for ultra-low leakage applications, the transistor scheme had to be optimized to scale down the threshold voltage. The capability to lower the V_T has been provided by the so-called flip-well architecture. Flip-well device solution is depicted in Fig. 2.14.

Classical transistor architecture based on Nwell in a PMOS with V_{dd} polarization is used for high V_T devices, dedicated to blocs or circuit sensible to quiescent current. This architecture is aligned with former generations of MOS transistors.

On the other hand, as presented in Fig. 2.14, LVT option is based on the flip-well scheme, building PMOS with grounded Pwell. Actually, the threshold voltage reduction is explained by Eq. (2.11) and depends on the doping level at the rear interface of the buried oxide layer. By applying the same dopants in the back and front gate, the inversion layer inside the channel appeared sooner versus the gate biasing. It corresponds to higher drive current, and then threshold voltage lowering. This doping level is directly modulated by ground-plane implantation [PP3]; ground-plane P (GP-P) for PMOS and ground-plane N (GP-N) for NMOS device. In order to avoid a floating diode between the ground-plane and the well itself, the wells have been flipped between both transistors, giving the name of the LVT device family as a flip-well solution. And finally, to guaranty the perfect Nwell/Pwell diode isolation, wells biasing has been flattened to the ground.

This type of scheme will end up to different option in terms of body-bias. As drawn in Fig. 2.14, we can clearly see that both Nwell and Pwell voltage can be increased without any leakage risk, whereas in the opposite way, a limited biasing can be applied before the direct conduction of the inter-well diode. This is exactly the reverse in case of LVT option using flip-well design: only a negative bias can be applied at the rear of the Pwell to not turn on the well diode. The schematic of the body-bias capability is summarized in Fig. 2.15.

Usually, in FD-SOI design, we will talk about forward body-bias FBB in case of low-V_T transistors and reverse body-bias (RBB) for high-V_T options.

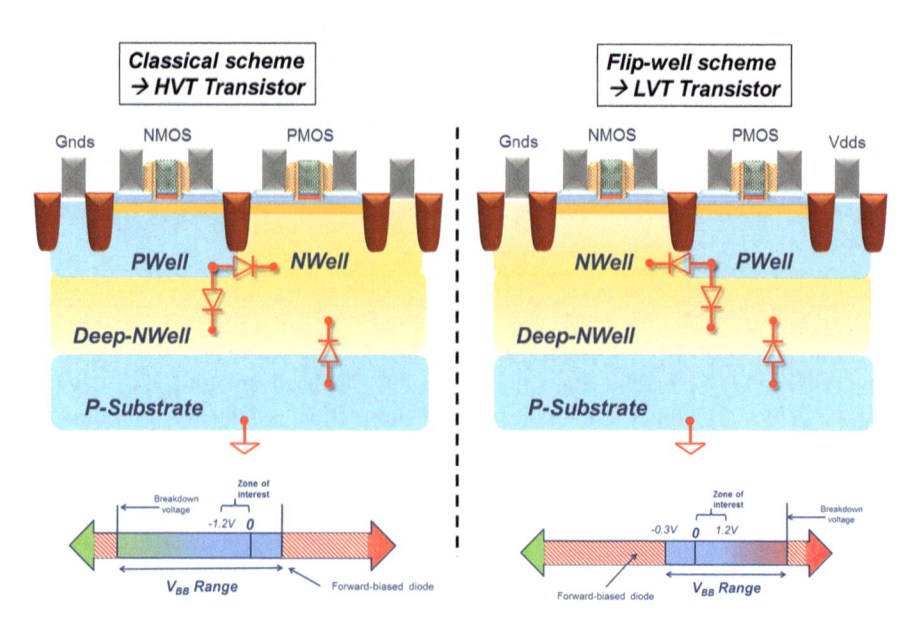

Fig. 2.15 Body-bias possibility in case of classical and flip-well transistor architecture

2.3 Transistor Parameters and Body-Bias

Following the intrinsic mechanisms explained in the previous section, we will describe here the influence of the body-bias opportunity on the macroscopic parameters of the transistor, such as the threshold voltage (V_T), the quiescent current (I_{off}), and the drive current (I_{on}). Finally, in most part of this section, we will present first effect on the AC performance (frequency and leakage) illustrated by ring oscillator (RO) design.

Threshold voltage behavior versus body-bias on isolated MOS transistors is described in Fig. 2.16. This factor is directly related to the buried oxide thickness. In this case, measurements have been performed on Tbox = 25 nm. About 80 mV/V of body-bias has been established for both NMOS and PMOS transistor. It means that 1 V body-bias can compensate one full V_T flavor, from standard V_T to low V_T, for example.

Following V_T variation across V_{bb}, in both direction, reverse and forward biasing, the drive current slope has been extracted for both NMOS and PMOS devices. Figure 2.17 illustrates those sensitivity. Basically, the behavior of the saturation current is symmetrical in RBB and FBB. It can be explained by the fact that mobility enhancement evidenced on MOS structure was completed in linear mode, i.e., at low V_{ds} voltage. Consequently, saturation is mainly influenced by V_T value, which is symmetrical on both biasing direction.

As body-bias modulates the threshold voltage in the range of 80 mV/V of V_{bb}, a specific care has to be taken for the standby current of the devices. As printed in Fig. 2.18, due to the exponential behavior of the sub-threshold current (see Eq. (2.12)), the leakage has been increased by 15× by applying 1 V at the back gate.

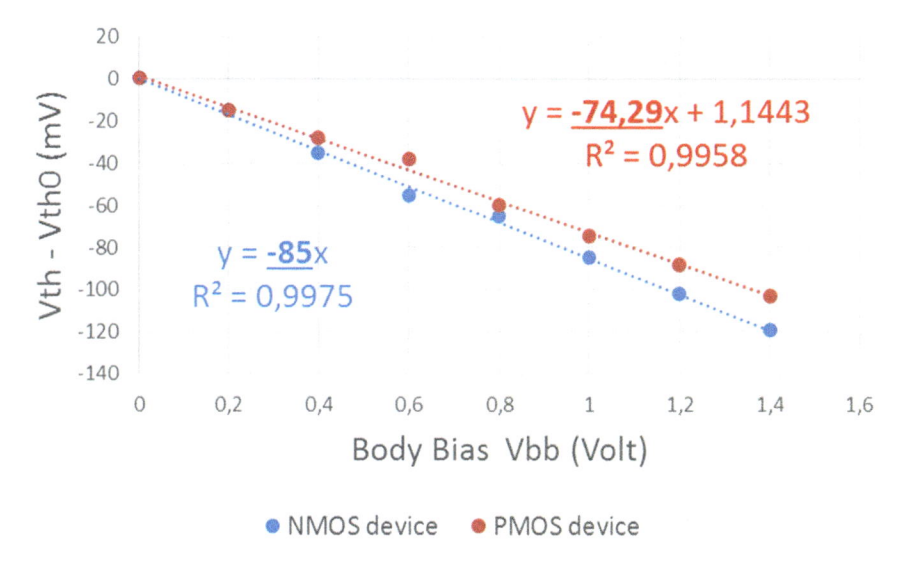

Fig. 2.16 Threshold voltage versus V_{bb} sensitivity for both NMOS and PMOS transistors

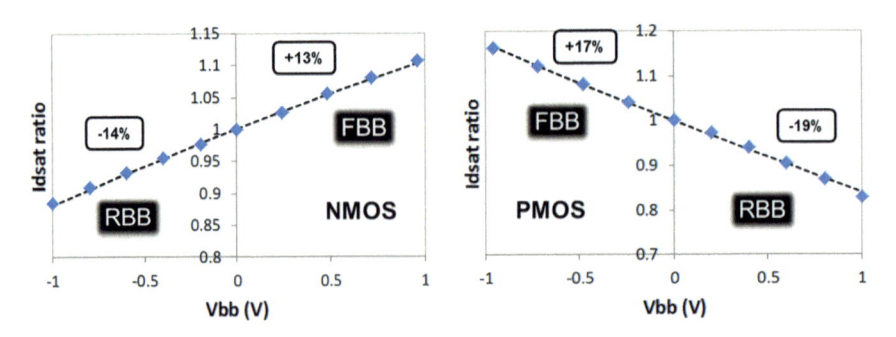

Fig. 2.17 Drive current modulation versus body-bias value for NMOS (left) and PMOS (right) transistors

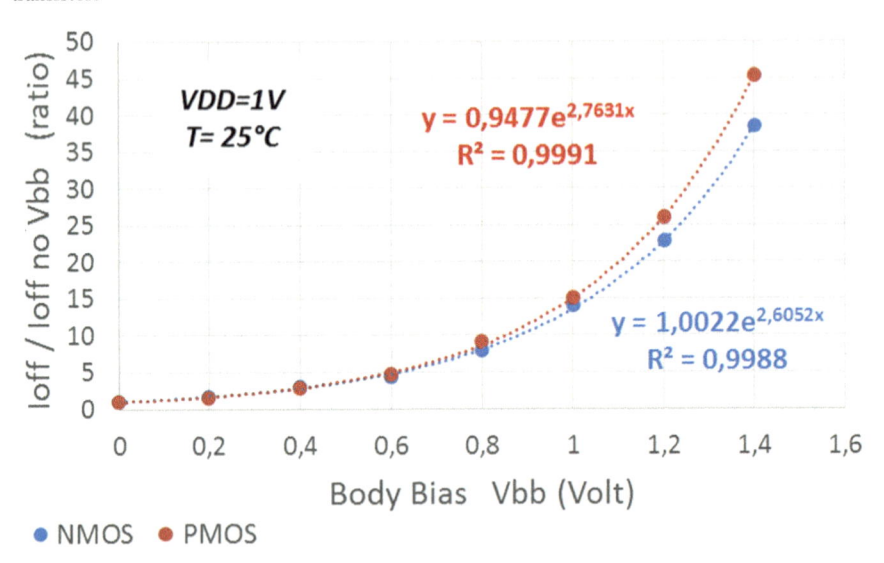

Fig. 2.18 Leakage current of transistor as a function of V_{bb} for NMOS and PMOS devices showing more than one decade of increasing for $V_{bb} = 1$ V

$$I_{\text{OFF}} = I_{th} \cdot \exp\left(\frac{-V_{th}}{\text{Slope}/\ln 10}\right) \quad (2.12)$$

with I_{th} as function of W, L, and carriers mobility.

Any further leakage degradation has been observed with FBB technique at high temperature. Actually, as described by Eq. (2.13), the variation of the threshold voltage versus temperature depends on physical features of the transistor [2], such as gate dielectric (C_{OX}, with thickness and permittivity), channel doping level (NA), or semi-conductor channel (ΨB, with Si or SiGe channel, for example).

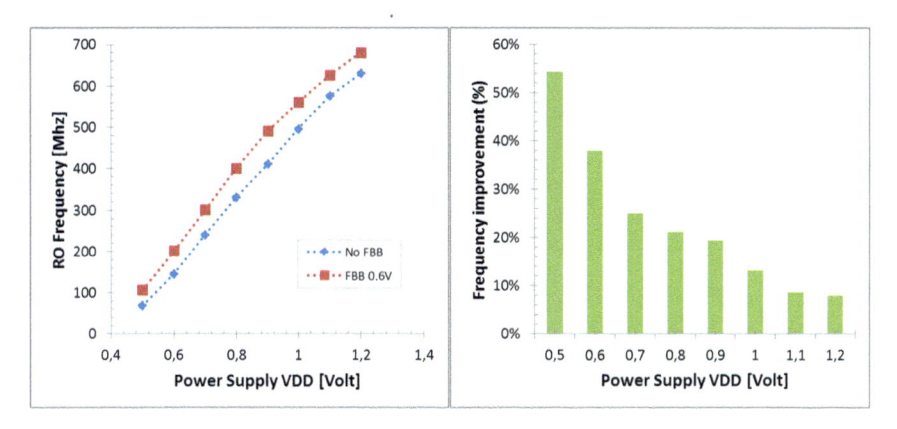

Fig. 2.19 Ring oscillator (RO) frequency enhancement with FBB = 0.6 V (left side). RO performance improvement versus the reference without FBB (right side) showing higher efficiency at low voltage

$$\frac{\partial V_T}{\partial T} \approx \frac{\partial \psi_B}{\partial T} \cdot \left(2 + \frac{1}{C_{OX}}\sqrt{\frac{\epsilon_{Si} q N_A}{\psi_B}}\right) \quad (2.13)$$

Consequently, the I_{off} ratio between 25 and 125 °C will be not affected by applying FBB option. This point is important for design community, especially during the sign-off phase where the high temperature is often used for leakage budget workout. Based on the signals collected from isolated NMOS and PMOS transistors for static characteristics, a first dynamic analysis has been performed using a ring oscillator (RO) design. Propagation time of the information is represented by the RO frequency as a function of both front gate basing (V_{dd}) and back gate biasing (V_{bb}). Results in terms of RO frequency are depicted in Fig. 2.19.

On the left plot, we can see a continuous improvement of the RO frequency versus V_{dd} and the extra boost put in evidence by applying specific FBB technique, here at $V_{bb} = 0.6$ V. Right plot presents the benefit of V_{bb} usage across V_{dd}. Major improvement has been measured at low voltage operation, confirming the great interest of FBB solution to enhance the performance at low voltage.

Power analysis with FBB usage will be intensively discussed in Sect. 2.5 of this chapter.

2.4 Transistor Variability

As shown in Eq. (2.3), FD-SOI architecture allows a significant reduction of the parasitic DIBL effect inherent to shortest gate length. We will see in the section, the advantage of this technology to improve the variability of the device while operating at low voltage. Actually, as soon as the power supply of the application is reduced,

Fig. 2.20 CMOS transistor I–V curve and current voltage dependency regimes

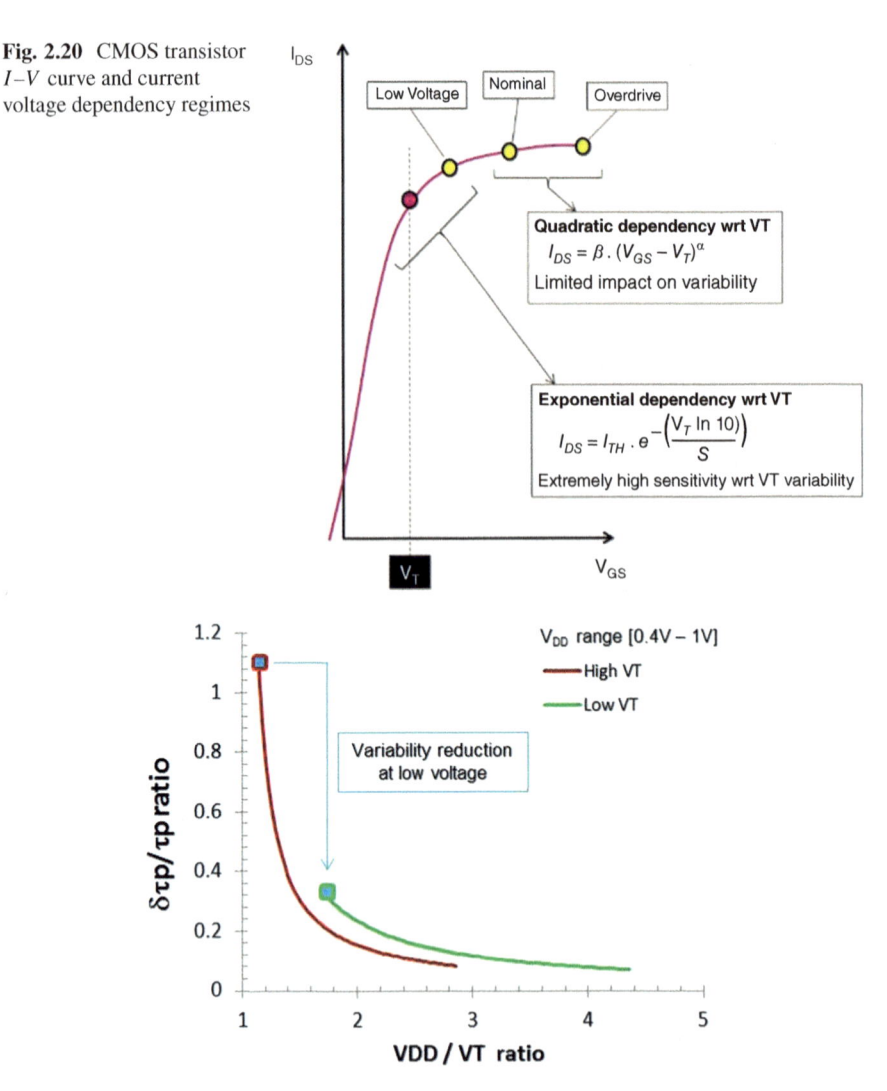

Quadratic dependency wrt VT
$$I_{DS} = \beta \cdot (V_{GS} - V_T)^\alpha$$
Limited impact on variability

Exponential dependency wrt VT
$$I_{DS} = I_{TH} \cdot e^{-\left(\frac{V_T \ln 10}{S}\right)}$$
Extremely high sensitivity wrt VT variability

Fig. 2.21 Ring oscillator propagation time dispersion versus V_{DD}/V_T ratio showing strong improvement while reducing threshold voltage of the device

getting closer to the threshold voltage, a significant increasing of the CMOS transistor variability appears. Exponential behavior of the spread is presented in Figs. 2.20 and 2.21.

This degradation is due to the change of the drain current (I_{DS}) dependency to the gate voltage (V_{GS}), moving from pseudo-linear behavior to exponential one. As a consequence, the spread of the propagation time of the signal thru CMOS stages will double by operating at low voltage, such as $V_{DD} = 0.6$ V.

2 FD-SOI Technology

Basic delay variability equations are reported below [9]. τ_P is the propagation delay metric and follows Eq. (2.14), its variation ratio is proportional to V_T variation ratio and follows Eq. (2.15)

$$I_{DS} = \beta.(V_{DD} - V_T)^\alpha$$

$$1 \leq \alpha \leq 2$$

$$\tau_P = \frac{C.V_{DD}}{I_{DS}}$$

$$\tau_P = \frac{C.V_{DD}}{\beta.(V_{DD} - V_T)^\alpha} \quad (2.14)$$

$$\frac{\partial \tau_P}{\partial V_T} = \frac{\alpha.C.V_{DD}}{\beta.(V_{DD} - V_T)^{\alpha+1}}$$

$$\frac{\delta \tau_P}{\tau_P} = \frac{\alpha.\delta V_T}{(V_{DD} - V_T)}$$

$$\frac{\delta \tau_P}{\tau_P} = \frac{\alpha}{\left(\frac{V_{DD}}{V_T} - 1\right)} \cdot \frac{\delta V_T}{V_T} \quad (2.15)$$

The main efficient way to mitigate this spread at low voltage is the decreasing of the threshold voltage, moving from high-V_T option to low-V_T one (illustrated in Fig. 2.21). The benefit is quite impressive, with an improvement of the slope of $\delta\tau_P/\tau_P$ by 2.5× at $V_{DD} = 0.5$ V by dropping the V_T by 100 mV, as depicted in Table 2.1.

In the case of low power applications requiring low voltage operations, the high transistor variability increasing can led to functionality risk or need for over-design jeopardizing the cost of the circuit. FD-SOI architecture can improve the situation thanks to electrostatic improvement (DIBL and sub-threshold slope) on one side, and FBB capability on the other hand. The key figure to contain the dispersion of the device at low power supply is the V_T reduction. FD-SOI technology can propose multiple process knobs allowing such threshold voltage decreasing with limited impact on leakage current. This V_T reduction is triggered by three steps, as detailed in Fig. 2.22. This plot represents the V_T increasing as V_{DD} is reduced due to parasitic DIBL effect. Reference with bulk technology is illustrated by blue line. Step one is materialized by a lower initial threshold voltage allowed thanks to a better sub-threshold slope. Step two is an improvement of the V_T dependency with

Table 2.1 Variation of $\delta\tau_P/\tau_P$ according to voltage and V_T

V_{DD} (V)	V_T (V)	$\delta\tau_P/\tau_P$ ratio
0.9	0.38	10.4%
0.6	0.38	24.5%
0.6	0.23	14.6%

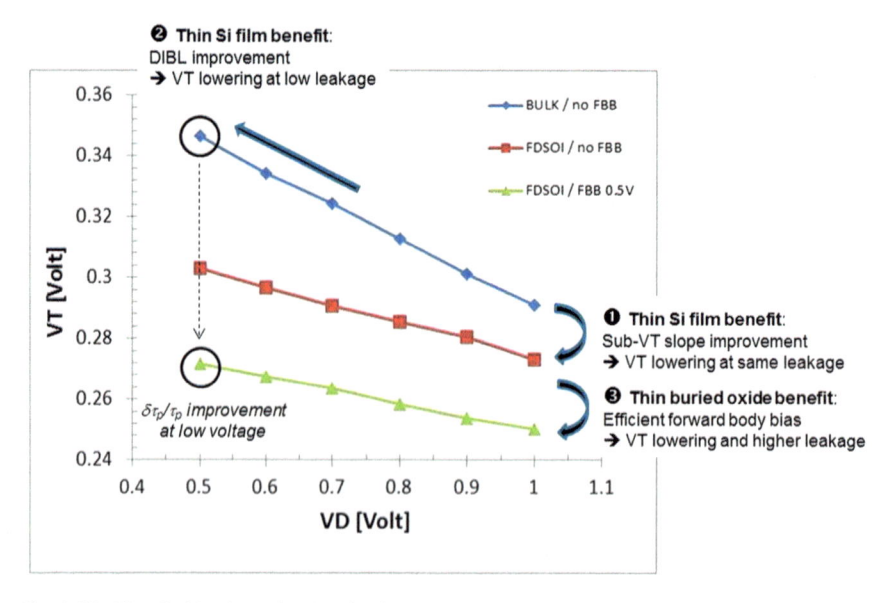

Fig. 2.22 Threshold voltage (V_T) reduction path enabled by FD-SOI technology (thin Si film and thin buried oxide interest)

respect to power supply thanks to DIBL effect lowering. Finally, third step of V_T reduction at low V_{DD} is provided by FBB solution. By combining the three steps together, a significant V_T reduction by -30% has been observed on CMOS devices between regular bulk architecture and FD-SOI scheme.

The total $\frac{\delta \tau_P}{\tau_P}$ benefit versus V_{DD}/V_T ratio is presented in Fig. 2.23 in case of bulk technology and FD-SOI one. Advantages brought by FBB are also proposed. In the inserted bares chart, we represented the breakdown of the gain measured in terms of variability versus bulk solution at low voltage (here for $V_{DD} = 0.5$ V). Basically 28% reduction of process spread is achieved thanks to the usage of FD-SOI architecture and up to 43% less dispersion have been demonstrated by applying FBB.

As previously discussed, key leverage to drop ring oscillator variability at low voltage is the threshold voltage reduction. If it is very efficient (as shown in Fig. 2.23), it comes along with a quiescent current increasing (the so-called I_{off}). In order to mitigate the leakage penalty due to V_T lowering, physical gate length has been increased thanks to multiple poly bias (PB) technique. Thus, by combining both PB and FBB on FD-SOI technology a spectacular transistor dispersion improvement can be reached at ultra-low voltage operation, as described in Fig. 2.24.

After a full review of the interest of FD-SOI architecture and FBB technique to mitigate the intrinsic transistor variability at low voltage operation, the second part of this Sect. 2.3 will cover the process dispersion tightening thanks to the body-bias technique [9]. For the sake of clarity, in this section we will use the term

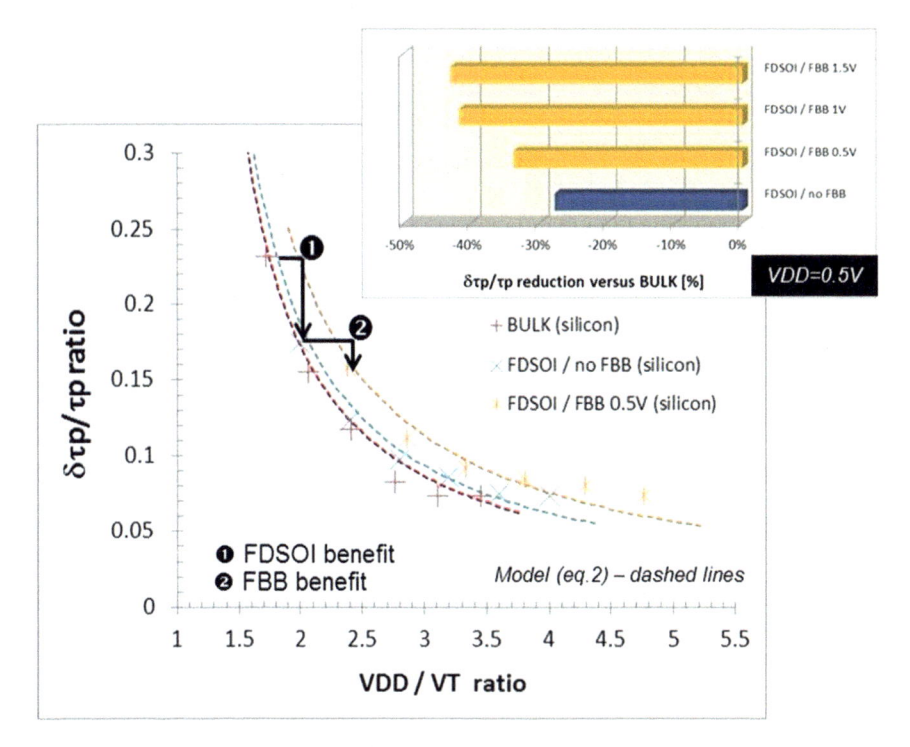

Fig. 2.23 Ring oscillator propagation time variability tightening at low voltage operation thanks to FD-SOI architecture and FBB technique deployment. Up to 43% less dispersion measured versus bulk reference

adaptive body-bias (ABB) covering both RBB and FBB side. Nominal drive current transistors spread measured in production, including typical, fast, and slow corners, is illustrated in Fig. 2.25. By applying a RBB and a FBB at the back gate of fast and slow devices, respectively, we can demonstrate a significant transistor dispersion reduction versus nominal conditions.

This spectacular saturation current dispersion reduction thanks to ABB technique has been observed as well for leakage current of both NMOS and PMOs devices. As shown in Sect. 2.2, standby current of the CMOS devices varying exponentially with respect to the threshold voltage, the V_T spread can be easily compensated through back biasing. Benefit of ABB in terms of leakage spread is illustrated in Fig. 2.26.

As described in Eq. (2.13), V_T and then leakage current of MOS devices are very sensitive to the temperature. Basically, parasitic sub-threshold current is double every 10 °C temperature rising. Consequently, it can be very attractive to compensate the temperature elevation by increasing V_T after RBB at the back of the device. Benefit to compensate the leakage increasing at high temperature is presented in Fig. 2.27.

As for the quiescent current, the performance of the device can be optimized as a function of the temperature and process dispersion. Saturation current spread of

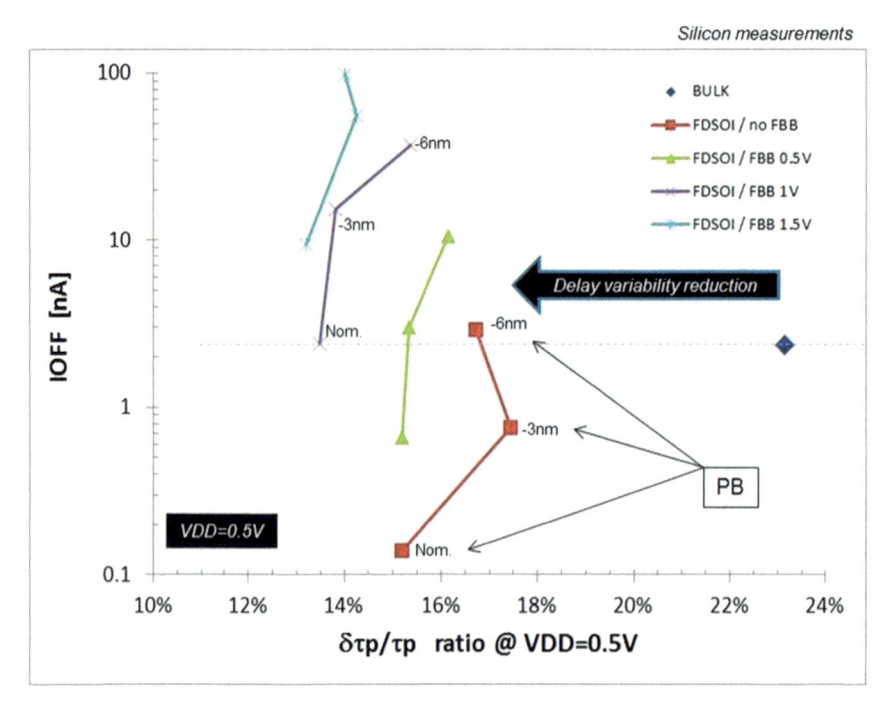

Fig. 2.24 Transistor leakage versus ring oscillator variability compromise. Physical gate length (the so-called PB) combined with FBB allows 2× dispersion reduction at $V_{DD} = 0.5$ V versus bulk reference

both NMOS and PMOS transistors is depicted in Fig. 2.28 with and without ABB solution.

Main advantages of ABB technique on the ring oscillator are summarized in Fig. 2.29. Both process and temperature compensation have been evidenced for RO frequency and leakage. Dramatic spread contraction has been verified thanks to ABB: frequency variability has been cut by 3× for both process and temperature effects, whereas half a decade reduction was extracted for IDDq.

As a conclusion, we can say that FD-SOI technology is perfectly suited for advanced and complex system-on-chip design by providing ultra-low variability capability across process, temperature, and voltage. After arguing about the intrinsic variability, both digital and analog performance are now going to be discussed in the following Sects. 2.5 and 2.6 of this chapter dedicated to the technology features.

2.5 Digital Performance Enhancement with FBB

In this section, we will present an analysis of the total power optimization as a function of main technology elements available in FD-SOI: power supply (V_{dd}),

2 FD-SOI Technology

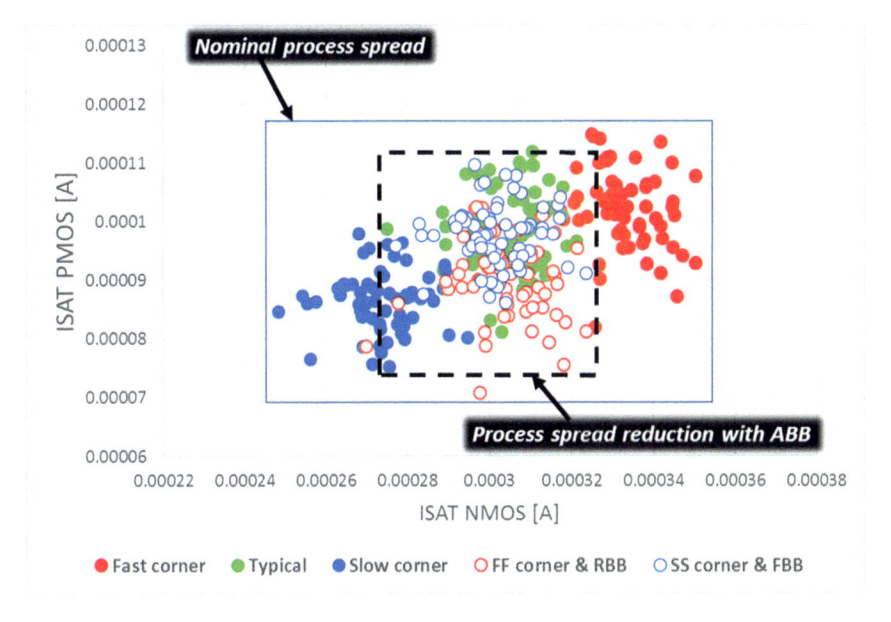

Fig. 2.25 NMOS and PMOS drive current transistor spread tightening thanks to adaptative body-bias technique. ©IEEE 2017 reprinted with permissions

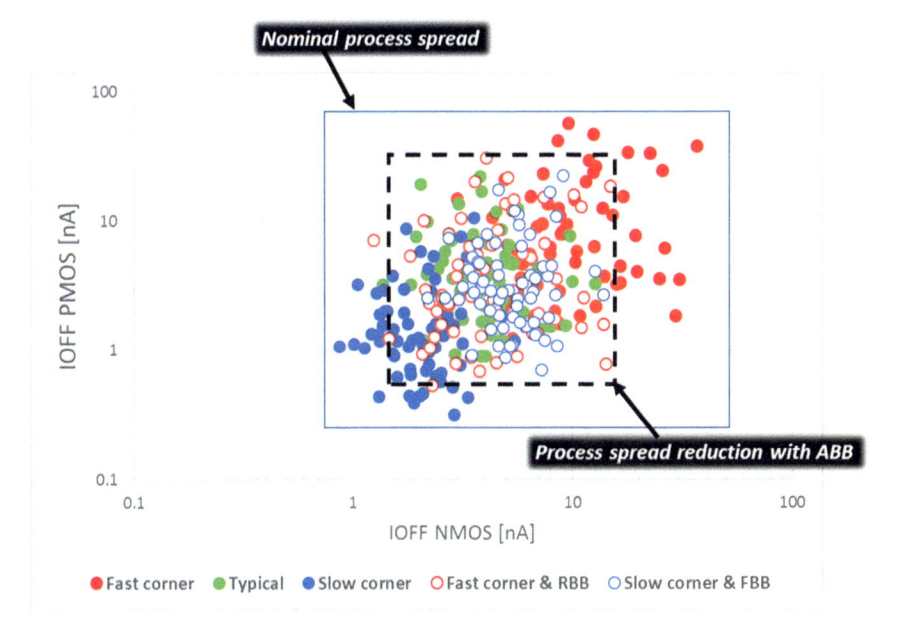

Fig. 2.26 NMOS and PMOS leakage current transistor spread tightening thanks to adaptative body-bias technique. ©IEEE 2017 reprinted with permissions

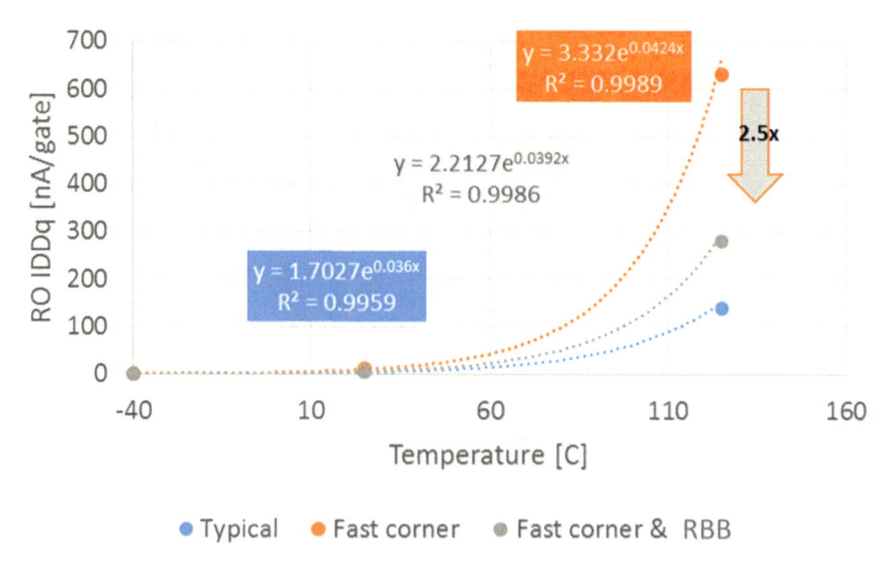

Fig. 2.27 Ring oscillator (RO) quiescent current (IDDq) across temperature. Significant reduction of the leakage budget at 125 °C by applying RBB on fast dice. ©IEEE 2017 reprinted with permissions

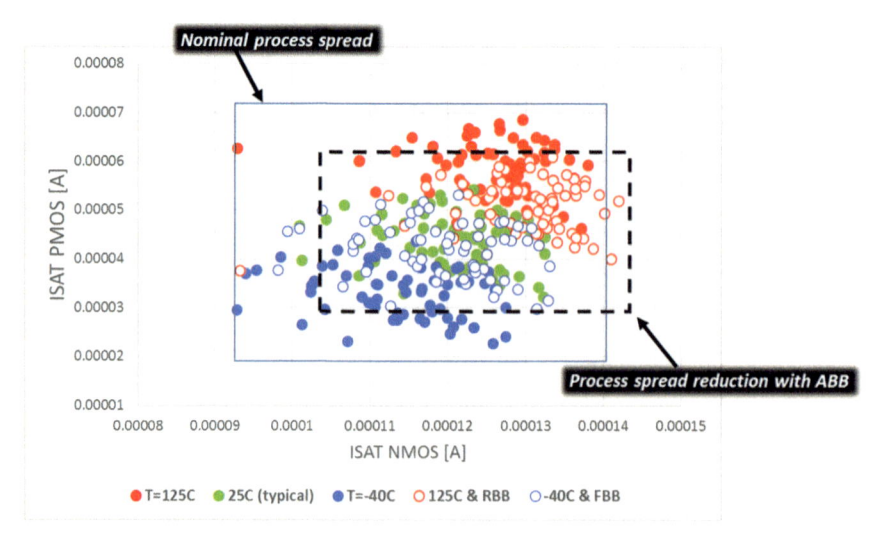

Fig. 2.28 NMOS and PMOS saturation current dispersion reduction by applying RBB at high temperature and FBB at −40 °C. Those data have been measured for $V_{dd} = 1$ V

back bias (V_{bb}), physical gate length (Lg), and threshold voltage. The power analysis will deal with both static power coming from both V_{dd} and quiescent current of the device and dynamic power related to V_{dd} and effective capacitance. Main power formulas have been summed up in Eq. (2.16) [10].

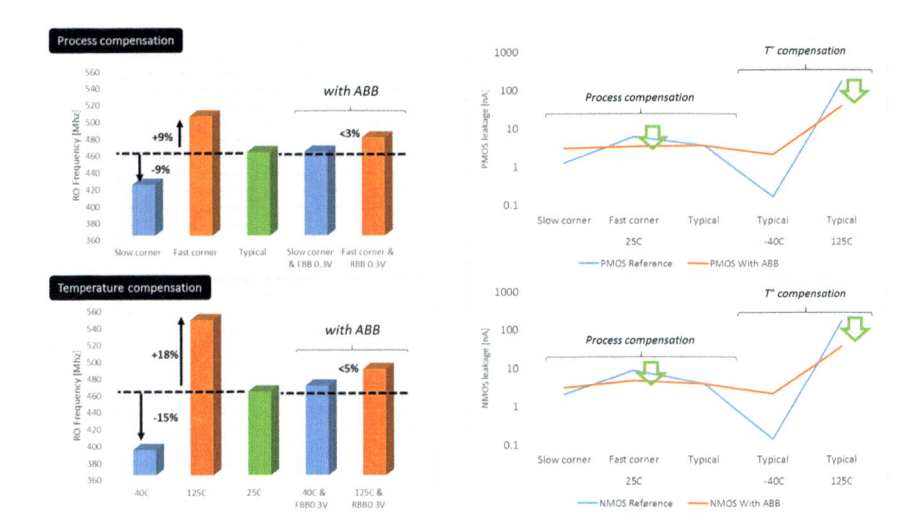

Fig. 2.29 Ring oscillator variability mitigation across temperature and process thanks to ABB technique. RO frequency (left side) and leakage current (right side) benefits are reported. ©IEEE 2017 reprinted with permissions

$$P_{\text{Total}} = V_{DD} \cdot I_{\text{STATIC}} \cdot \left(1 - \frac{\tau}{T}\right) \cdot (1-n) + V_{DD} \cdot I_{\text{DYNAMIC}} \cdot \frac{\tau}{T} \cdot n \quad (2.16)$$

with T representing the clock frequency, τ the switching slope rate, and n the duty cycle of the gate. Dynamic and static currents are summarized in equations below:

$$P_{\text{Static}} = \left(I_{\text{off}}^{\text{source}} + I_{\text{off}}^{\text{bulk}} + I_{\text{off}}^{\text{gate}}\right) \cdot V \quad (2.17)$$

$$P_{\text{Dynamic}} = 0.5 \cdot \alpha \cdot C_{\text{LOAD}} \cdot V^2 \cdot f \quad (2.18)$$

For the static power description, we can consider $I_{\text{off}}^{\text{Bulk}}$ and $I_{\text{off}}^{\text{gate}}$ as extrinsic parasitic leakages. In the rest of the study, only the so-called $I_{\text{off}}^{\text{source}}$ will be consider as modulated by threshold voltage, and then V_{bb}. Figure 2.30 gives the RO propagation time (the so-called delay) as a function of power supply V_{dd} and back bias conditions. It illustrates the strong effect of the back gate polarization to recover the performance as lowering V_{dd}. Thus, the delay at 0.7 V operation can be aligned with the one reached under $V_{dd} = 1$ V by applying 1.5 V FBB.

From design adjustment point of view, we can consider a full power improvement thanks to (V_{dd}; V_{bb}) pair. As depicted in Fig. 2.31, identical delay can be reached for different (V_{dd}; V_{bb}) couples. Then meaningful dynamic power reduction is demonstrated thanks to massive power supply decreasing, from 1 to 0.5 V.

Thus, we can establish a kind of abacus co-optimizing both dynamic power and ring oscillator delay in the same time after an optimization of the biasing pair [11]. As presented in Fig. 2.32, a unique V_{dd} lowering will hurt severely propagation time

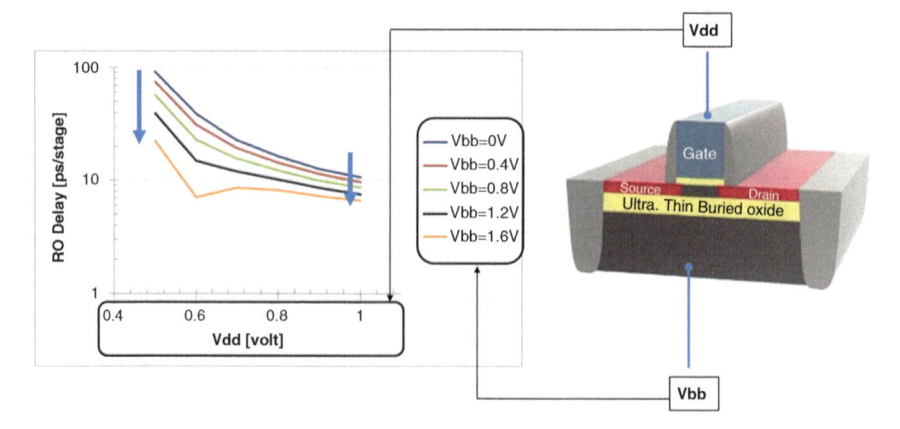

Fig. 2.30 Ring oscillator performance evolution versus gate back (V_{dd}) and back bias (V_{bb})

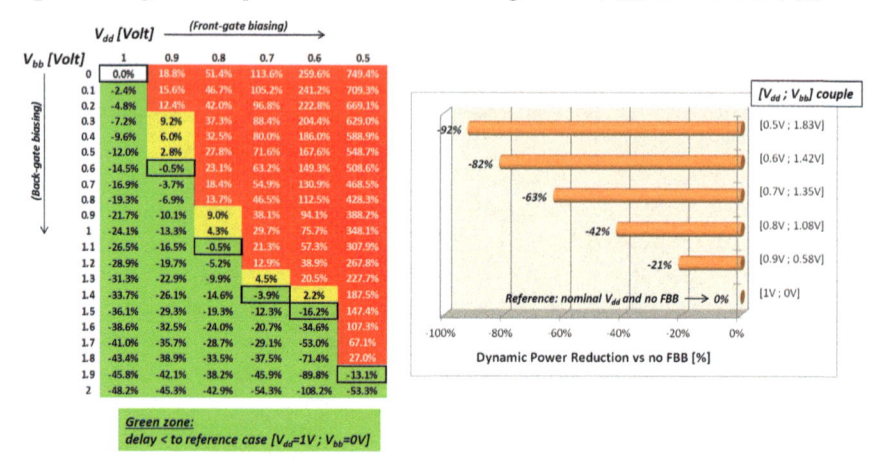

Fig. 2.31 Ring oscillator delay ratio versus original point ($V_{dd} = 1$ V; $V_{bb} = 0$ V) in the right table, with boxed black cells representing same delay. Dynamic power improvement measured after (V_{dd}; V_{bb}) pairs

(following the blue line). However, if an increasing of the back gate biasing V_{bb} is used, delay can be maintained and dynamic power strongly mitigated, as shown with the purple or orange lines.

As shown, there is no trade-of trying to optimize in the same time dynamic power and delay of the ring oscillator. Both of them will required to minimize the power supply V_{dd} and maximize the back bias V_{bb}. However, in a real circuit, designer has to take care of the leakage responsible of the static power consumption. It can impact the reliability of the circuit or induces thermal runaway effect, especially when operating at high temperature. FBB technique providing lower V_T will definitively degrade the off-state current of the transistor as explained by Eq. (2.12). Static power of the RO has been measured as function of both V_{dd} and V_{bb}, and reported in Fig. 2.33. Plot on the top, evidence a half decade RO leakage

2 FD-SOI Technology

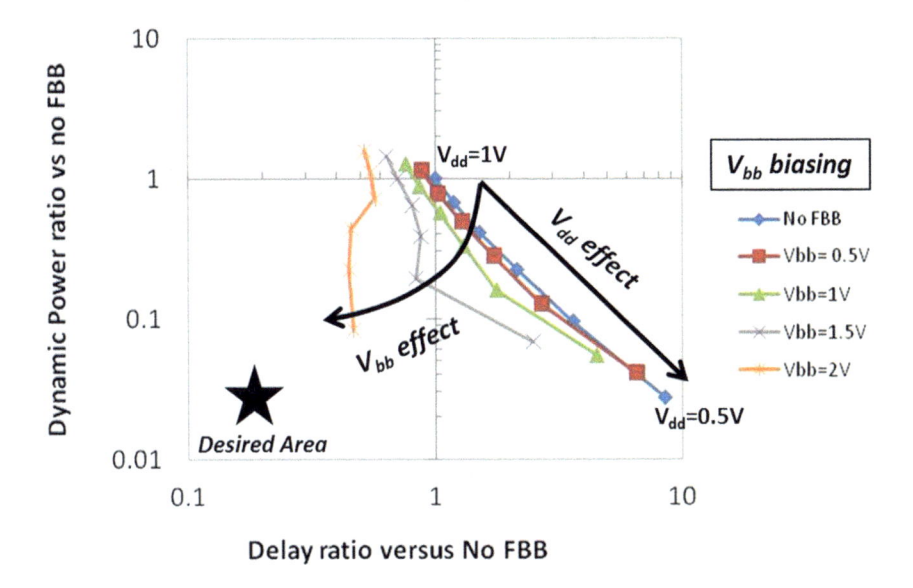

Fig. 2.32 Dynamic power and propagation time co-optimization thanks to (V_{dd}; V_{bb}) pair adjustment

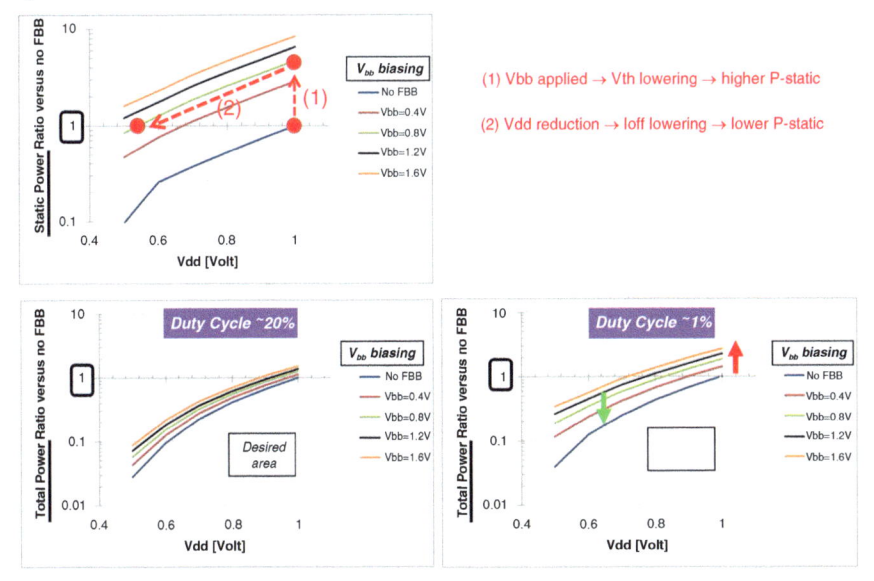

Fig. 2.33 Static power modulation (top plot) versus V_{bb} and V_{dd}. Total power optimization for high and low duty cycle after (V_{dd}; V_{bb}) pair adjustment

increasing while applying $V_{bb} = 0.8$ V. But static leakage from nominal point can be totally recovered by reducing the power supply V_{dd} down to 0.55 V. Bottom figures illustrate a more realistic case including both static and dynamic power. Those graphs put in evidence the larger efficiency of FBB to drop the total power in case of higher circuit activity (duty cycle above 10%) corresponding to more dynamic power. Thus, with 20% activity rate, one decade of power can be persevered with $V_{bb} = 1.5$ V and $V_{dd} = 0.5$ V versus nominal biasing conditions.

For the last part of this section dedicated to power reduction, we will review how to co-optimize both total power and the frequency. The key device metric combining those two critical parameters for the digital design is the Energy.Delay product. This product is described by Eq. (2.20) [11].

$$E \cdot \tau_P = CV^2 \cdot \frac{CV}{I_{EFF}}$$

$$E \cdot \tau_P = CV_{dd}^2 \cdot \frac{C(V_{dd} - V_T)}{\xi V_{GT}^\alpha}$$

$$E \cdot \tau_P = CV_{dd}^2 \cdot \frac{C(V_{dd} - V_T)}{\xi (V_{dd} - V_T)^\alpha} \tag{2.19}$$

$$E \cdot \tau_P = \frac{C^2 V_{dd}^2}{\xi} \cdot (V_{dd} - V_T)^{1-\alpha} \text{ with } 1 < \alpha << 2 \tag{2.20}$$

where ξ represents the transistor features (W, L, C_{OX}, and carrier mobility), supposing to be constant in this exercise. Main parameter modulated V_{dd} and V_{bb} polarization is the V_T of the device. As explain in Sect. 2.1 of this chapter dedicated to the technology, in one hand DIBL will induce a V_T rising as V_{dd} is lowered and in the other hand back bias will increase (RBB) or decrease (FBB) the V_T of the transistor through the back gate coupling capacitance. Description of the phenomena and basic equations are summarized in Fig. 2.34. Following those V_T laws, Energy. Delay product can be directly described by V_{dd} and V_{bb} biasing conditions, as shown in Eq. (2.21)

$$V_T(V_{dd}) = \frac{-\Delta}{V_0} \cdot V_{dd} + V_T^{Lin}$$

$$V_T(V_{bb}) = \frac{-n}{V_0} \cdot V_{bb} + \left[V_T(V_{dd})\right]$$

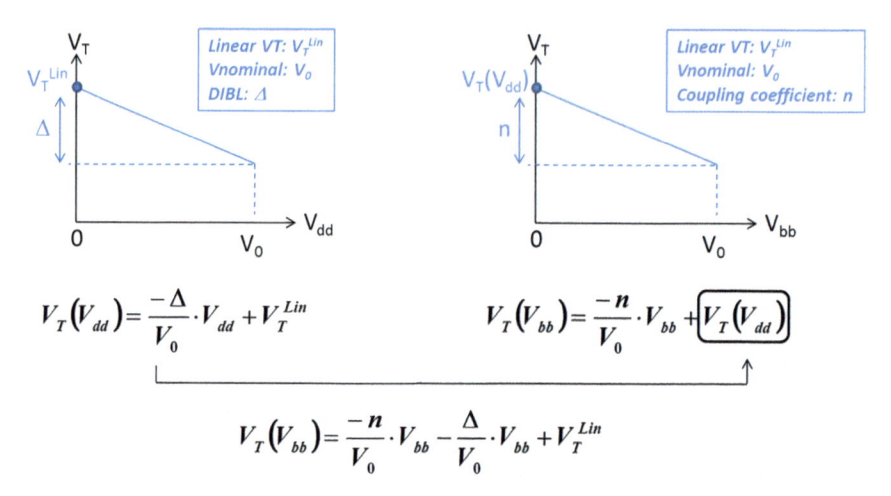

$$V_T(V_{bb}) = \frac{-n}{V_0} \cdot V_{bb} - \frac{\Delta}{V_0} \cdot V_{bb} + V_T^{Lin}$$

Fig. 2.34 MOS transistor threshold voltage modulation after V_{dd} and V_{bb} polarization, due to DIBL (left) and back gate capacitance (right)

$$E \cdot \tau_p = \frac{C^2 V_{dd}^2}{\xi} \cdot \left(V_{dd} \cdot \left(1 + \frac{\Delta}{V_0}\right) + V_{BS} \cdot \frac{n}{V_0} - V_T^{Lin} \right)^{1-\alpha} \quad (2.21)$$

where V_0 represents the nominal voltage of the technology, n the back gate coupling capacitance, and Δ the parasitic DIBL.

This Energy.Delay product corresponds to the switching energy efficiency coefficient. To define the best $(V_{dd}; V_{bb})$ pair in terms of energy efficiency, we will try minimize this E.τ product.

Minimum point of the switching energy efficiency plot can be obtained after annulling the first derivative of the function itself. This minimum point is given by Eq. (2.22) hereafter [11],

$$SEE = \frac{C^2}{\xi} \cdot \frac{4(\alpha - 1)^{1-\alpha}}{(\alpha - 3)^{3-\alpha}} \cdot \frac{\left(\frac{n}{V_0} \cdot V_{BS} - V_T^{Lin}\right)^{3-\alpha}}{\left(1 + \frac{\Delta}{V_0}\right)^2} \quad (2.22)$$

The V_{dd} value corresponding to this minimum can be worked out as shown in Eq. (2.23)

$$V_{\min} = \frac{2}{\alpha - 3} \cdot \frac{\left(\frac{n}{V_0} \cdot V_{BS} - V_T^{Lin}\right)}{\left(1 + \frac{\Delta}{V_0}\right)} \quad (2.23)$$

For the sake of clarity, the illustration the switching energy efficiency criteria is plotted in Fig. 2.35. Minimum and V_{dd} for the minimum are highlighted specifically as function of power supply V_{dd}

Both optimum point and V_{dd} value reaching the optimum point depend on back bias (V_{bb}) polarization. As depicted by Eqs. (2.22) and (2.23), body-bias increasing can shift the optimum V_{dd} biasing and significantly reduce the E.τ product. Behavior of the product is presented in Fig. 2.36.

The benefit for E.τ product by applying FBB technique is finally summarized in the bar graph of Fig. 2.37. In that case, we can clearly see the much superior efficiency in case of low voltage operation. It illustrates the stronger interest of FD-SOI technology for low power designs.

2.6 Analog Performance Enhancement with FBB

Leaving the digital part of the chapter, we will review in this section transistor parameters important for analog design. FD-SOI device architecture coupled with forward body-bias can provide substantial boost for the analog parameters of the

$$E \cdot \tau_p = \frac{C^2 V_{dd}^2}{\xi} \cdot \left(V_{dd} \cdot \left(1 + \frac{\Delta}{V_0}\right) + V_{bb} \cdot \frac{n}{V_0} - V_T^{Lin} \right)^{1-\alpha}$$

$$V_{min} = \frac{2}{\alpha - 3} \cdot \frac{\left(\frac{n}{V_0} V_{bb} - V_T^{Lin}\right)}{\left(1 + \frac{\Delta}{V_0}\right)}$$

$$SEE = \frac{C^2}{\xi} \cdot \frac{4(\alpha - 1)^{-\alpha}}{(\alpha - 3)^{-\alpha}} \cdot \frac{\left(\frac{n}{V_0} V_{bb} - V_T^{Lin}\right)^{3-\alpha}}{\left(1 + \frac{\Delta}{V_0}\right)^2}$$

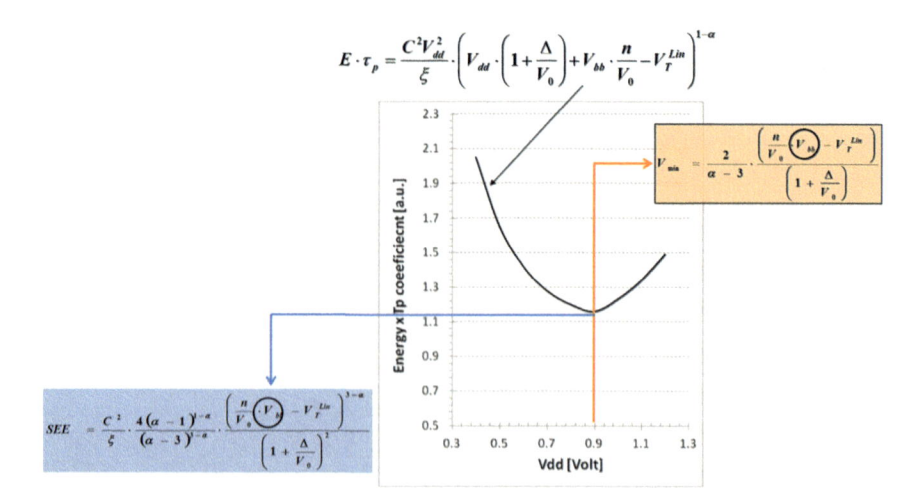

Fig. 2.35 Illustration of Eqs. (2.21), (2.22), and (2.23). E.τ product as a function of power supply (V_{dd}). Minimum point representing the most optimized (V_{dd}; V_{bb}) conditions in terms of switching energy efficiency

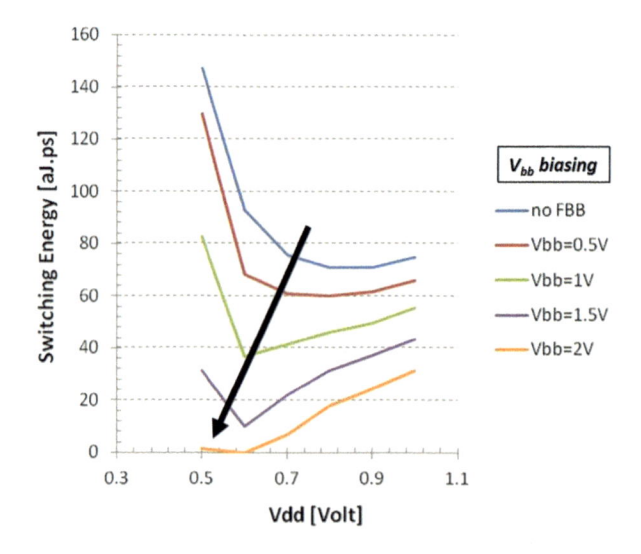

Fig. 2.36 Switching energy efficiency enhancement thanks to body-bias technique measured on ring oscillator

device. As identified in the section one of this technology description, carrier mobility is significantly enhanced versus classical bulk architecture thanks to fully depleted channel. Much less coulomb scattering appeared in case of FD-SOI device increasing the carrier transport between source and rain terminals. Mobility improvement will directly increase transistor transconductance (gm) as the first

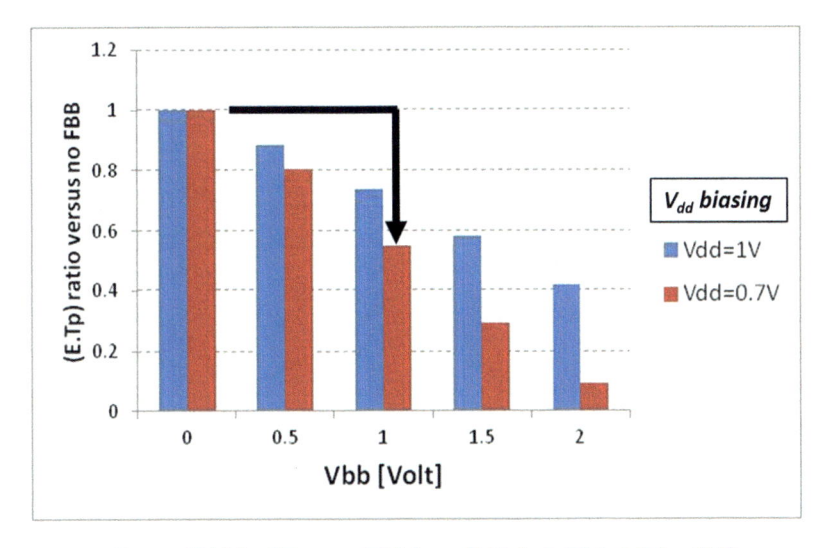

Fig. 2.37 Summary of overall RO performance enhancement leveraging body-bias technique

derivative of the drain current I_{ds} versus gate voltage V_{gs}. This benefit is presented in Figs. 2.38 and 2.39.

Aside the channel depletion, FBB technique increased the electrons mobility as well, thanks to the second inversion layer at the rear interface of the channel. Up to 3× improvement has been observed with 1.5 V V_{bb} used in the case of NMOS transistor, nominal gate length [12].

gm performance increasing comes along with a better cut-off frequency, the so-called f_T. As shown by Eq. (2.24), f_T corresponds to a current gain equal to one and is directly related by gm value (measurement setup as displayed in Fig. 2.40).

$$f_T \approx \frac{g_m}{2\pi(C_{gs} + C_{gd})} \tag{2.24}$$

As expected, the mobility enhancement enables higher peak for cut-off frequency f_T. Peak frequency reaching about 330 GHz has been demonstrated on NMOS transistor with 28 nm FD-SOI technology, representing roughly +25% versus original bulk-planar reference, as shown in Fig. 2.41. By improving f_T frequency, Fmax performance is increased as well versus previous node achieving 330 GHz. Even if the vertical gate resistance has been degraded by the implementation of metal gate layer, Fmax figure of merit has been boosted thanks to enhanced gm.

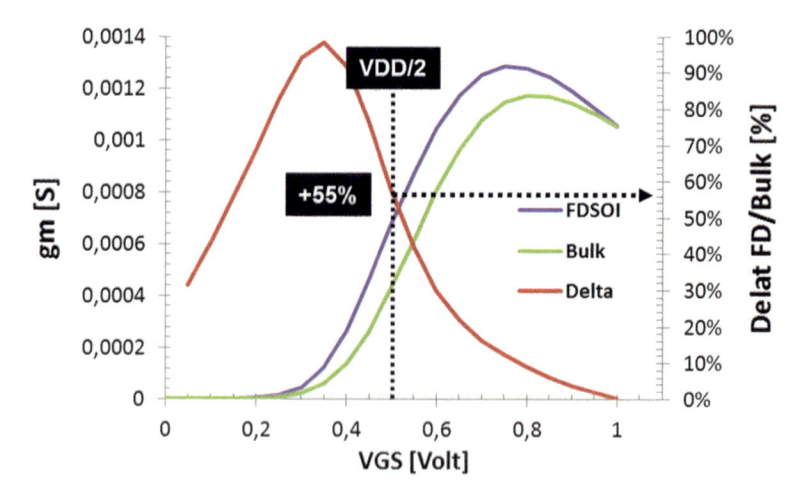

Fig. 2.38 Transconductance improvement thanks to FD-SOI architecture versus bulk reference. More than 50% measured at $V_{DD}/2$ conditions

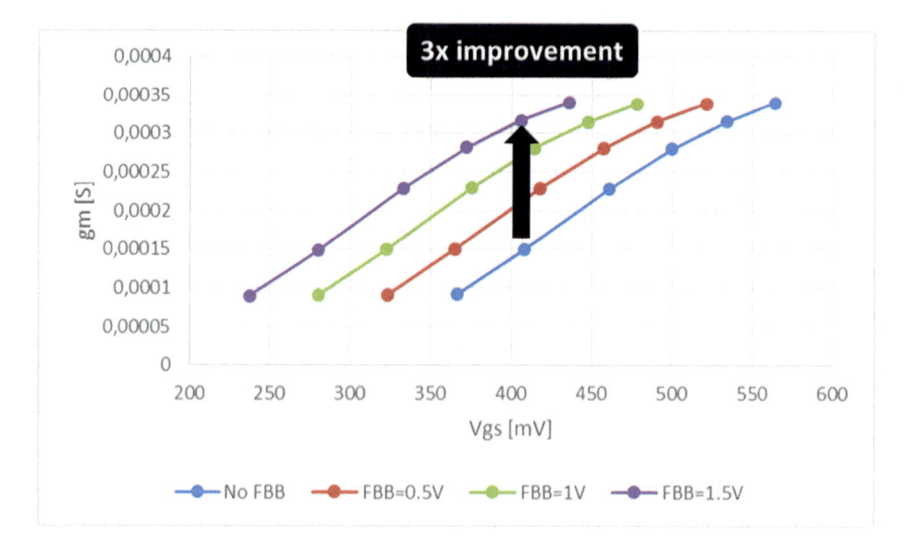

Fig. 2.39 Transconductance improvement by applying back gate biasing

Leveraging very short carrier transit time between source and drain terminals thanks to FD-SOI technology and body-bias, significant switch resistance reduction has been demonstrated versus bulk reference solution. RON value versus switch input voltage has been printed in Fig. 2.42, showing more 30× resistance decreasing at $|FBB| = 1.8$ V.

FD-SOI transistor is featuring a fully depleted channel. As explained during the Sect. 2.1, threshold voltage and electrostatic control are driven by silicon film thickness (the so-called Tsi). Thus, on the contrary of standard bulk transistor, FD-

2 FD-SOI Technology

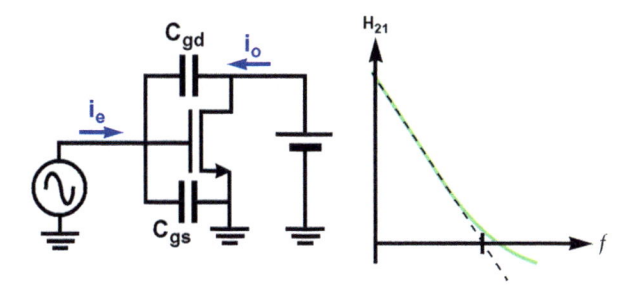

Fig. 2.40 Cut-off frequency extraction corresponding to current gain equal to one

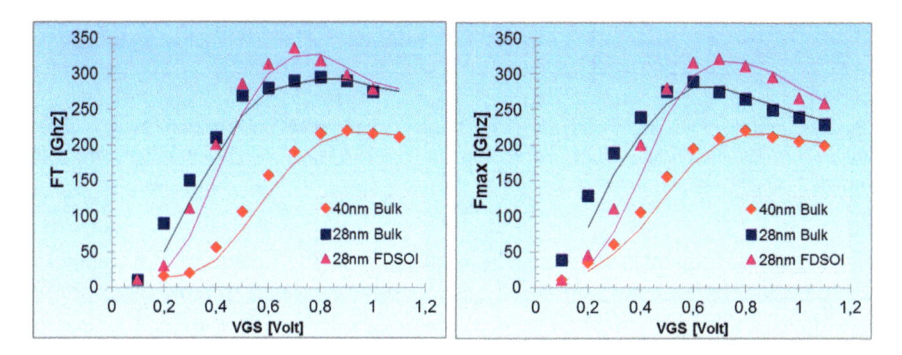

Fig. 2.41 NMOS transistor cut-off frequency for different technology. 330 GHz demonstrated with 28 nm FD-SOI node corresponding to +25% increasing versus former bulk solution

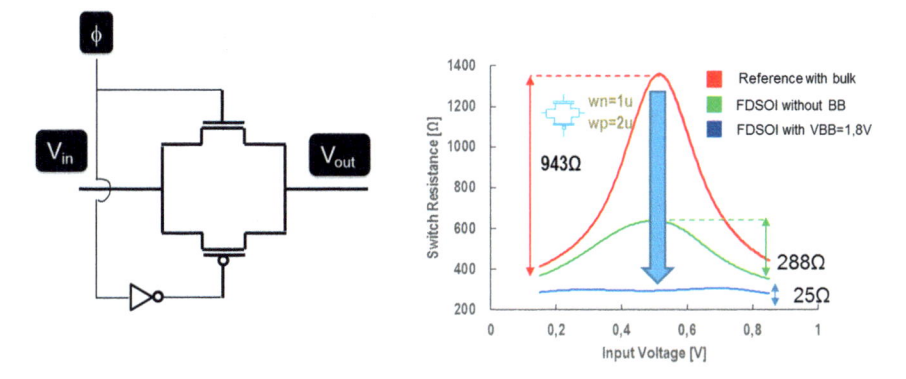

Fig. 2.42 RF switch schematic (left) and performance (right) versus technology and body-bias

SOI device does not use pocket or halos implantations. Those implantations are self-aligned with respect to gate electrode and will contribute to mitigate parasitic short-channel effect. Basic source–drain architecture proposed for MOS transistor is pictured in Fig. 2.43.

Those drawings illustrate the MOS junctions design for bulk and FD-SOI structure. In case of bulk transistor the natural channel doping profile shows a lack

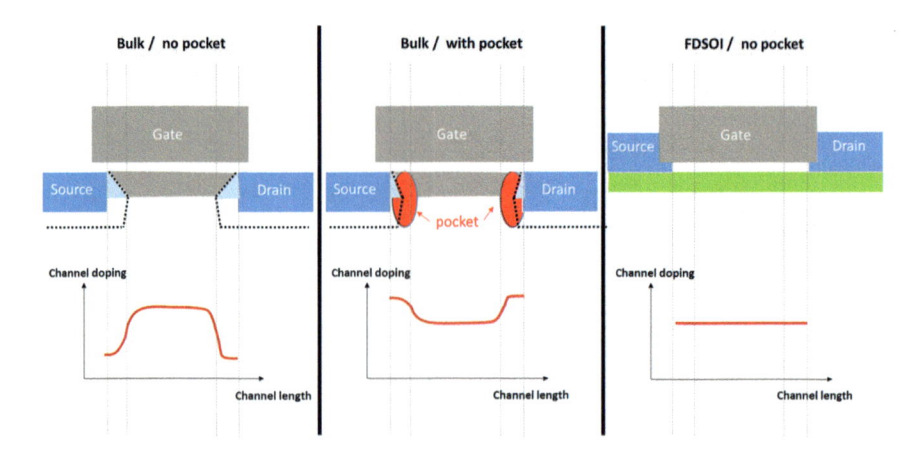

Fig. 2.43 Source–drain architectures for MOS transistor and pocket implantations usage compensating short-channel effect. Only FD-SOI structure can guaranty and uniform doping along the channel of the device

of effective dopants at source and drain sides. Near the junctions, the channel is not controlled by the gate itself but by the source/drain areas. It corresponds to the parasitic short-channel effect. In order to get rid of such junctions influence, specific pocket implantation has been defined to compensate the short-channel effect. With pocket implantation, doping profile along the channel is completely reversed with an excess of impurities at the junction sides. The non-uniformity of the doping profile along Lgate will generate a strong modulation of the channel pinch off point with respect to drain voltage (V_{ds}). This modulation will degrade the first derivative of the current as a function of V_{ds} and consequently degrade the output conductance, gd. Mainly driven by gd parameter, output voltage gain (V_{out}/V_{in}) is usually strongly reduced by pocket implantation, as shown by Eq. (2.25)

$$\left|\frac{V_{OUT}}{V_{IN}}\right| = \frac{\partial V_{DS}}{\partial V_{GS}} = \frac{\partial V_{DS}}{\partial I_{DS}} \cdot \frac{\partial I_{DS}}{\partial V_{GS}} = \frac{g_m}{g_d} \quad (2.25)$$

This further gain degradation is reported in [13], where it is shown the impact of pocket dosage. Because the short channel devices are mainly impacted by DIBL, this parasitic effect of the non-uniformity of the doping along the gate affects long channel transistors.

Usage of pocket implantations in case of bulk device degrading gd parameter induces a net penalty in terms of output voltage gain. As presented in Fig. 2.44, FD-SOI architecture allowing a steady dopant repartition inside the channel, a significant improvement of analog gain has been observed, especially for long channel transistor (up to 15 dB).

Aside gm/gd ratio degradation, usually pocket implants used for bulk devices increase the transistors pair mismatch as well. As explained in [14], the pocket implants will go through the grains boundary distributions located inside the gate

Output Voltage Gain (Gm/Gd)

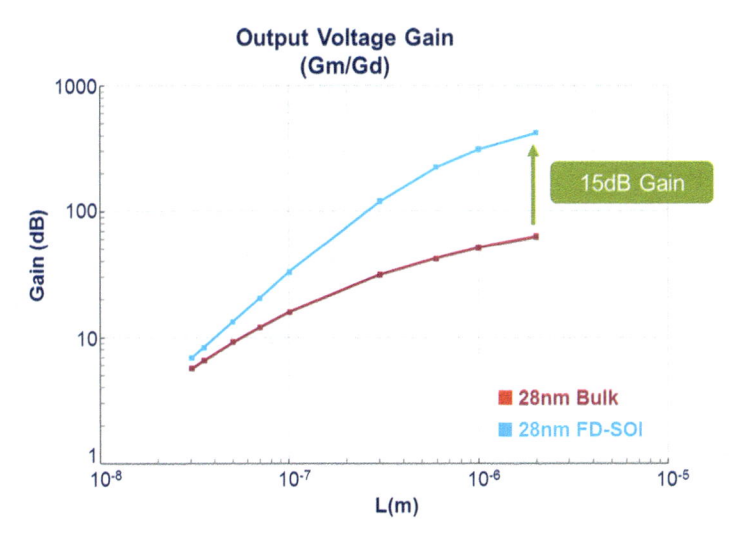

Fig. 2.44 Output voltage gain enhancement thanks to FD-SOI architecture versus bulk reference. Up to 2× gain ratio (i.e., ~15 dB) measured for long channel transistors

electrode of the device generating a dopant scattering in the channel. This local fluctuation of the dopants is responsible of the local V_T mismatch.

Usually, MOS transistor pair mismatch in advanced technology, based on high-k and metal gate electrode, depends on three origins: (1) the local channel dopant fluctuation, (2) the metal grains orientation, (3) line edge roughness of the gate stack. FD-SOI architecture being a pocket-free solution, an important source of the mismatch has been taken away versus bulk device. Consequently, threshold voltage mismatch (AVT parameter) is significantly improved with FD-SOI structure, as shown in Fig. 2.45.

The level of the transistor performance for RF applications can be summarized by (Gain.Bandwidth) product. Main parameters of this product are detailed in Eq. (2.26):

$$GBW \approx \frac{g_m}{2\pi(C_{gd} + C_{bd})} \quad (2.26)$$

The (Gain.Bandwidth) product description is provided in Fig. 2.46.

As for the cut-off frequency (f_T) parameter, this product is mainly driven by the transistor trans-conductance (g_m). As discussed previously in this section, g_m is strongly enhanced by applying FBB on the fourth terminal of the device. Up to 75% of improvement has been measured in case of 1 V body-bias, as illustrated in Fig. 2.47.

Finally to close the analog device parameters review in FD-SOI architecture, low frequency noise (LFN), the so-called Flicker noise, has been checked. Low

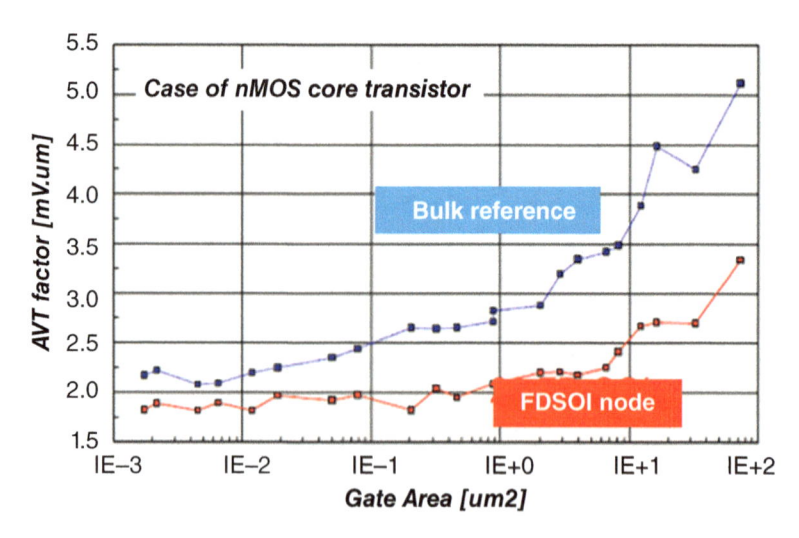

Fig. 2.45 Threshold voltage mismatch (AVT) of transistors pair reduced in case of FD-SOI

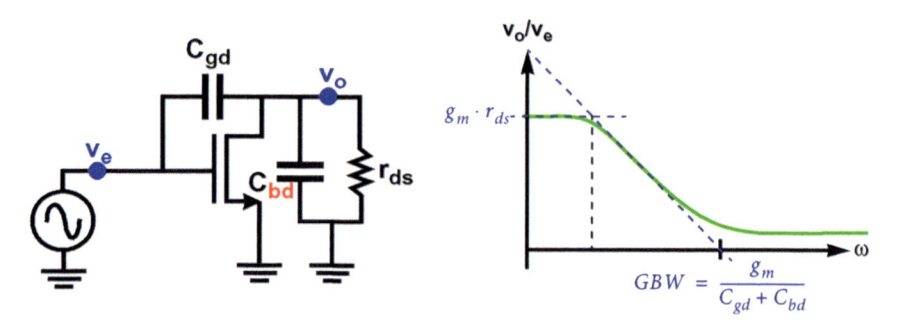

Fig. 2.46 Definition of (Gain.Bandwidth) product for MOS transistor

frequency noise of the drain current corresponds to the fluctuation of I_{ds} across time [15].

Extracted from $I(t)$ plot, the drain current spectral density as a function of the signal frequency can be established. The slope of the spectrum showing a shape of 1/f corresponds to the Flicker noise of the device. Equations (2.28) provide the basis of the drain current low frequency noise:

$$S_{I_d}(f) = \frac{1}{L^2} \cdot \int_0^L S_{\Delta I_d}(x, f) \Delta x\, dx \qquad (2.27)$$

$$S_{I_d}(f) = \frac{I_d^2 \alpha_H}{fWLN} \qquad (2.28)$$

$$\alpha_H = \frac{kT}{\nu} N_t(E_{fn}) \left(\frac{1}{N} + 2\alpha\mu + \alpha^2 \mu^2 N \right) \qquad (2.29)$$

2 FD-SOI Technology

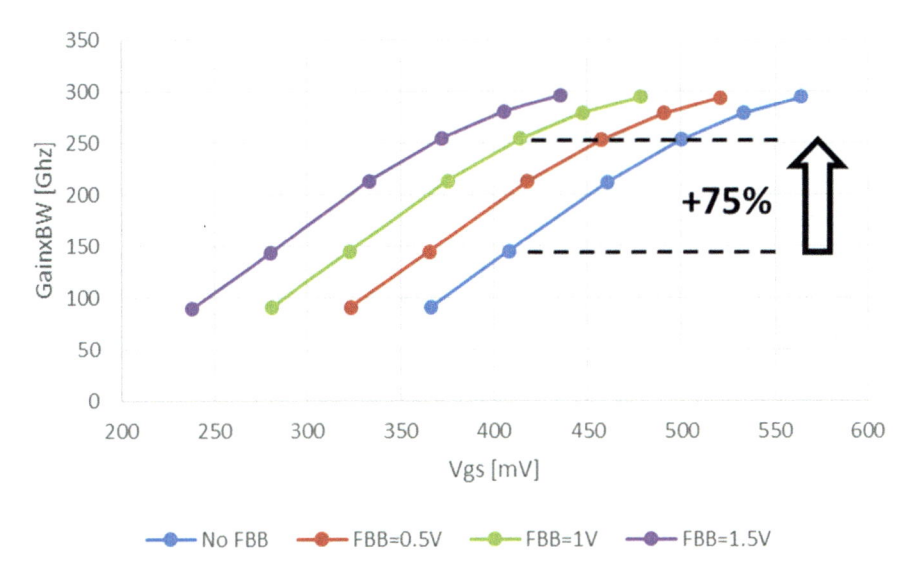

Fig. 2.47 Significant increasing of (Gain.Bandwidth) product thanks to FBB technique on MOS transistor. ©IEEE 2017 reprinted with permissions

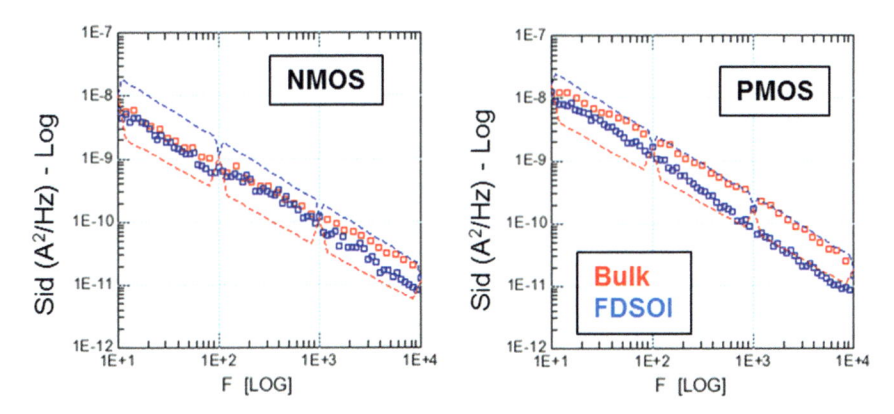

Fig. 2.48 Drain current spectral density ($1/f$ noise) for both bulk and FD-SOI architectures

where αH represents the Hooge constant. The aim of this constant is to model the physical mechanisms of the Flicker noise depicted in Fig. 2.48. Basically, the drain current fluctuation is related to carrier trapping/de-trapping phenomena. This mechanism is driven by doping density (N), traps density (N_t), carriers mobility (μ), and capture section of the carriers (σ).

As intensively published in the last decade, the carriers trapping/de-trapping efficiency is mainly modulated by gate stack structure. In advanced CMOS technology (28 nm and beyond) using high-k dielectric, it has been established that channel architecture is a second order effect on Flicker noise level [16, 17]. As a

Key parameters	FDSOI benefits	Body Bias interests
Transconductance (gm)	+ + (mobility)	+ + (further boost)
Output conductance (gd)	+ + (no pocket)	Neutral
Threshold voltage (Vth)	+ (better DIBL)	+ + (Vth lowering)
Mismatch (AVT)	+ + (no pocket)	- (back gate variability)
Low Frequency Noise (LFN)	neutral	neutral
Cut off frequency (Ft)	+ (mobility)	+ (further boost)
Gain.Bandwidth (GBW)	+ (gm & gd)	+ (gm)
Fmax	Neutral (same front gate)	neutral

+ : positive effect of FDSOI or BB   **Neutral** : no effect of FDSOI or BB   **-** : negative effect of FDSOI or BB

Fig. 2.49 Analog performance summary table, FD-SOI and bulk

consequence, as illustrated in Fig. 2.48, identical $1/f$ noise behavior between bulk and FD-SOI architecture has been observed.

The review of critical parameters for analog and RF applications in this section clearly highlighted the superior behavior of FD-SOI architecture versus bulk reference thanks to the thin silicon film and the FBB technique capability. The summary of this technology comparison is provided in the table of Fig. 2.49.

2.7 SRAM Bit-Cell

SRAM bit-cell being based on a top-to-tail CMOS inverter structure, the stability of the information stored in the internal node and the ability to write a data inside the structure will strongly depend on the strength of the six transistors used in the cell [18]. Both read and write margin effects are presented in Fig. 2.50, evolution of bit-cell is displayed in Fig. 2.51.

The main risk related to the read operation is to flip the internal node state while reading the data. The stability of the information will depend on the two NMOS transistors current, the so-called pull down (PD) and pass gate (PG), inside and at external to the CMOS inverter, respectively. Too strong PG will turn the internal node from 0 to 1, as depicted in Fig. 2.50. On the contrary, the internal node will be stuck at 0 anytime is the PD device is too strong. The cell will be too stable. For the read cycle, it is more the ratio between PD and PMOS Pull Up (PU) devices. Actually, a too strong PU or PD will keep the information to stick at 1 or 0, respectively. As illustrated in Fig. 2.50, cell margin will be modulated by NMOS and PMOS transistor centering. Moreover we know that PMOS transistor will suffer about NBTI reliability degradation, inducing a V_T rising, slowing down the PU drivability after aging (see Sect. 2.8 of this chapter). It means that a suitable NMOS/PMOS balance at time zero can lead a catastrophic situation after stress, especially if the circuit operates at low power supply. The key metric summarizing

2 FD-SOI Technology

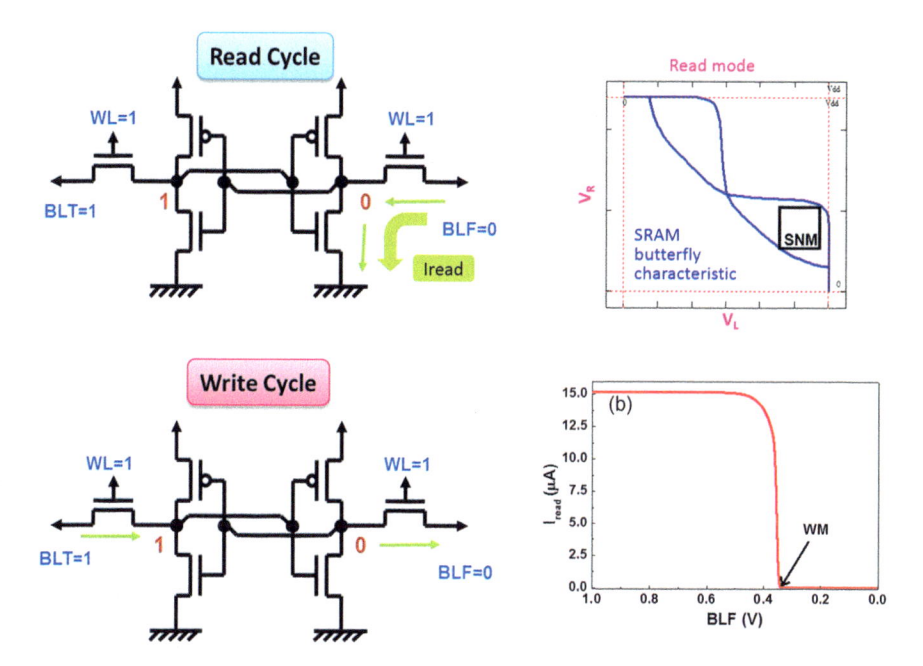

Fig. 2.50 Operations limitation inside SRAM bit-cell. Read margin (top) and write margin (bottom) representations

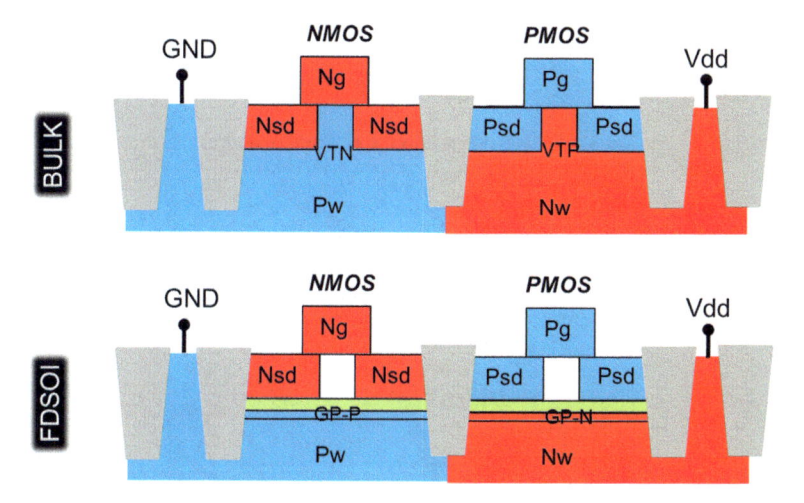

Fig. 2.51 Bit-cell architecture evolution from bulk to FD-SOI substrate

the bit-cell stability and capability in both direction, read and write, is the minimum voltage operation before a fail. This minimum voltage is called V_{min}. A lower V_{min} reduces the power consumption and better matches the digital library voltage capability.

As presented in the first section of this chapter, the benefit of FD-SOI transistor architecture in terms of carrier mobility and electrostatic control can be observed in 6T-SRAM bit-cell as well. Drivability enhancement is captured by read current of the cell (labeled I-read), whereas the off-state current is represented by standby leakage (IDDq). The usual figure of merit depicting the DC performance of the bit-cell is IDDq vs I-read, it was detailed in [19], where we put in evidence +25% benefit in terms of drive current versus the bulk reference at high voltage. This advantage can reach +80% by operating at low voltage thanks to a much lower DIBL (see Sect. 2.2) provided by fully depleted device.

The increasing of the read current of the bit-cell has been verified at the SRAM compiler level. The work in [19] show that an access time reduction of 27% without any degradation of IDDq has been characterized versus bulk architecture at 1 V. This number is coherent with the fact that I-read represents roughly one third of the access time.

Aside the pure digital performance of the cell, FD-SOI solution brought a significant advantage in terms of 6T-SRAM bit-cell stability and low voltage operation. For the same physical reason that the ones discussed in the analog section, the transistors pair mismatching is reduced by moving from bulk device to fully depleted one. The mismatch factor $A_{\Delta V_T}$ of the SRAM transistors has been improved by 40% [19]. As shown in Fig. 2.52, this strong pair matching enhancement has been observed at a 4Mb array with a V_{min} reduction of 100 mV with respect to the bulk reference, in 28 nm technology node.

The benefit of V_{min} described in Fig. 2.52 is mainly due to the physics of the fully depleted device. Let us see now how it can be further improved by adjusting

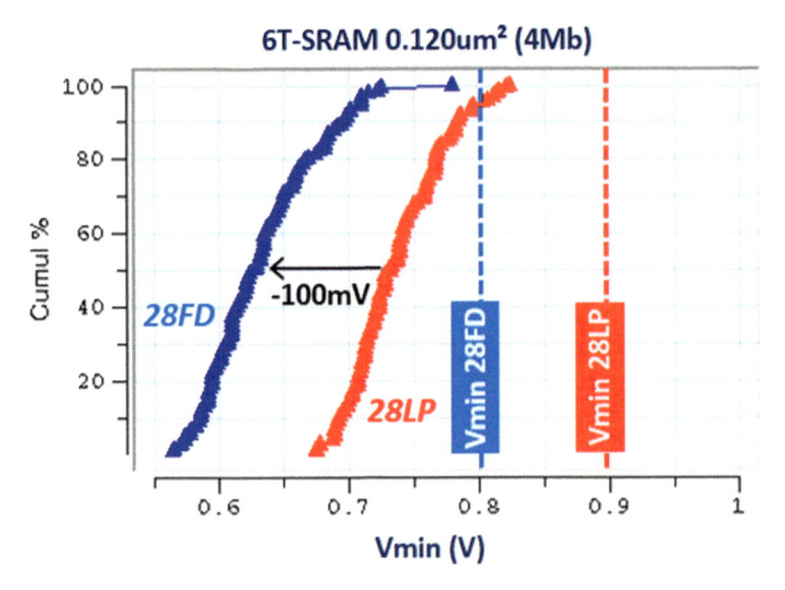

Fig. 2.52 SRAM transistor mismatch reduction and bit-cell V_{min} improvement in case of FD-SOI architecture

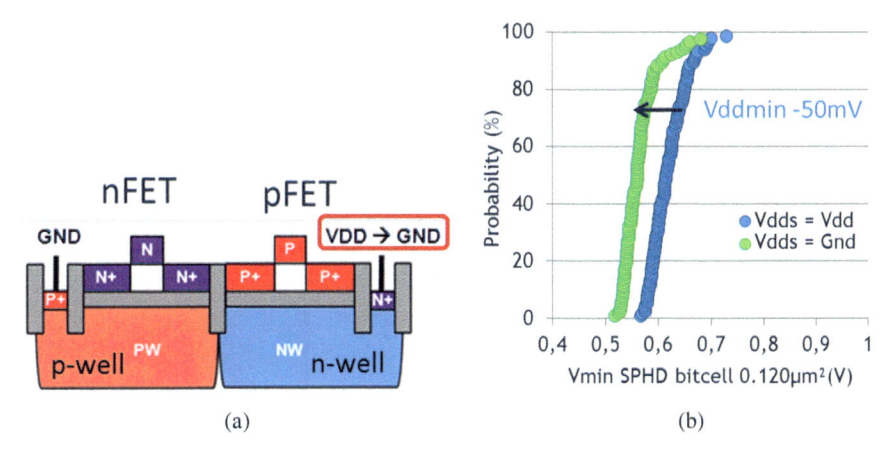

Fig. 2.53 PU transistor forward body-biasing improving the static noise margin (SNM) of the high density SRAM bit-cell thanks to a faster PMOS transistor. (**a**) PMOS/PU back biasing scheme. (**b**) HD SRAM V_{min} improvement in case of faster PU

voltages and layout. By turning the Nwell from V_{dd} to ground, the PMOS becomes naturally fast due to FBB. Faster PMOS increases the stability of the cell thanks to a higher static noise margin (SNM). As a consequence of the SNM improvement, the V_{min} of high density SRAM bit-cell has been reduced by 50 mV, as illustrated in Fig. 2.53.

FD-SOI SRAM bit-cell has a unique opportunity to further reduce the threshold voltage of PMOS transistor with the flip-well scheme. Actually, by implementing a so-called single-Pwell architecture, we can maintain NMOS transistors (PD and PG) with high V_T centering optimizing both leakage and SNM at high temperature, while shifting PU transistor toward lower V_T increasing cell stability and preventing bit-cell aging issue due to NBTI. Optimized single-Pwell solution is presented in Fig. 2.54. Leveraging the PU V_T reduction without any further degradation of the mismatch factor, V_{min} of the 6T-SRAM has been improved by 70 mV versus the reference cell based on dual well structure [20].

Aside the optimized well layout design and the native FBB proposed on PMOS device, additional cell biasing has been proposed to continue the V_{min} reduction, especially at low voltage (here at $-40\,°C$). As presented in Fig. 2.55, the rising of the Pwell biasing will contribute to further accelerate the PU transistor (FBB mode) and increase the NMOS V_T (RBB mode). As a consequence, bit-cell body-biasing will play a role of N/P transistors centering between the slow and the fast corner allowing a process compensation and then an optimization of the V_{min} versus process variability and temperature.

Figure 2.56 explains the static noise margin evolution of the SRAM bit-cell versus the body-biasing condition. In case of RBB applied on the NMOS, then the PMOS will be in FBB mode, by construction. The N/P ratio will change toward the N-slow/P-fast direction. It corresponds to the more stable bit-cell: SNM is strongly improved versus the nominal condition without body-bias. In the opposite direction,

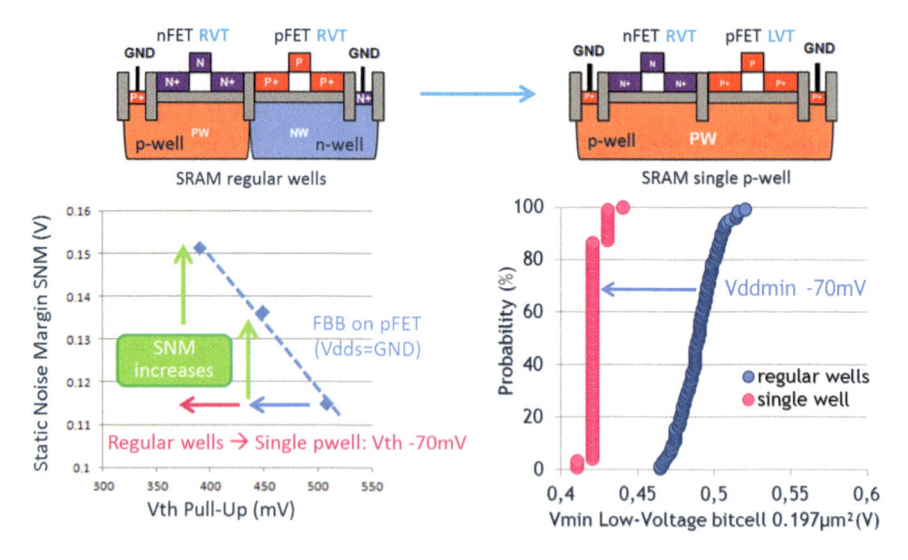

Fig. 2.54 Single Pwell SRAM bit-cell solution turning PMOS to LVT flavor and then enhancing SRAM V_{min}

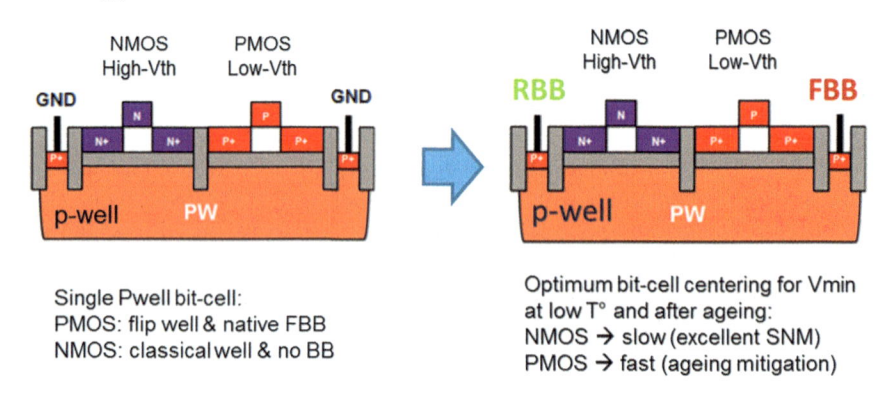

Fig. 2.55 Body-biasing structure in case of single Pwell bit-cell allowed with FD-SOI scheme

if FBB is applied on NMOS, then SNM is degraded due to a shift toward N-fast/P-slow process corner [21]. It worth pointing out that this behavior is similar whatever the power supply used, from 1 V down to 0.7 V.

Combination of single-Pwell cell and RBB/FBB applied on NMOS/PMOS, respectively, enables the more stable 6T-SRAM bit-cell with highest SNM at high temperature [22]. Actually the N-slow and P-fast device centering cumulates optimum cell advantages. First, at t0 (before stress) slow NMOS guarantees an excellent cell stability at high temperature. V_T lowering of the device observed at high temperature (see Eq. (2.13)) is absorbed by a higher V_T at room temperature. Moreover, aside the V_{min} improvement, it is worth to say that the standby leakage, being strongly linked to NMOS devices suite, is strongly reduced by using N-slow

2 FD-SOI Technology

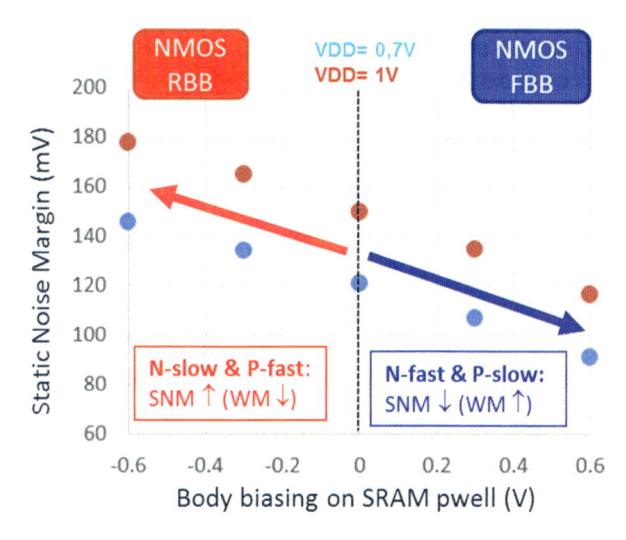

Fig. 2.56 6T-SRAM bit-cell SNM variation as a function of N/P transistors centering modulated by body-bias. Reference process is typical centering and room temperature, before aging

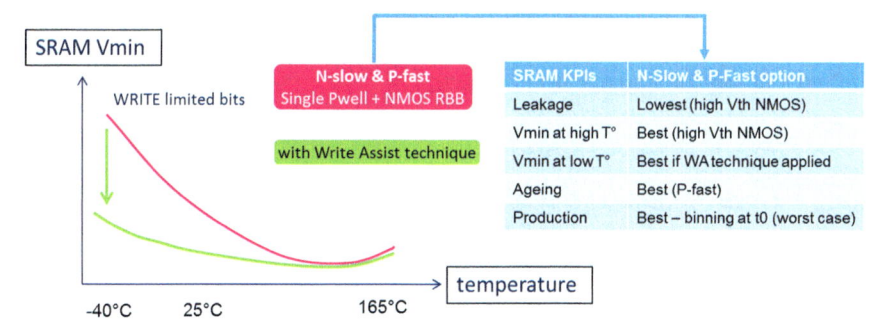

Fig. 2.57 V_{min} improvement thanks to FBB technique applied on single Pwell bit-cell

centering. Second, after aging and the NBTI effect, the PU's V_T rising is mitigated by a fast centering at t0. It means that the worst case situation in terms of V_{min} is not after stress but at fresh state. With a fast PU at t0, the SNM is actually improved after aging of the device. This type of centering can be attractive to set a production by putting in place a kind of dice binning directly at wafers sorting stage. Finally, the unique critical point of such solution is the write margin. At low temperature, because the threshold voltage of NMOS devices, VT will continue to rise leading to a too stable cell. Statistically, we may found some occurrences of fail bit during write operation. This situation is described in Fig. 2.57. This issue can be easily fixed by using a design technique called write assist. This type of design solutions have been already published and are intensively used in the advanced CMOS technology to offer a lower V_{min} capability, especially required for ultra-low power applications [23]. The effect of body-bias on bit-cell metrics is further detailed in Chap. 5.

2.8 Body-Bias Impact on Device Reliability

Well known and intensively CMOS transistor reliability weaknesses are reported in three categories: gate oxide breakdown (labeled time dependent dioxide breakdown, TDDB), negative bias temperature instability (NBTI), and hot carrier injection (HCI). The first two are strongly sensitive to the vertical electric field, driven by both gate oxide thickness and gate voltage applied. Thus, the worst case situation is the thinner oxide combined with higher gate biasing. The last one is modulated by the lateral electric field between source and drain regions. This degradation mechanism is amplified by physical gate length and drain biasing. Those intrinsic degradation modes occurring in advanced CMOS devices are summarized in Fig. 2.58.

TDDB failure mode can easily conduct to the destruction of the device due to the breakdown of the dielectric isolated the gate from the channel. In this sense, it is a hard fail inducing an abrupt stop of the functionality of the device and most likely the circuit. HCI and NBTI are related to charges trapping effect inside the gate oxide generating a shift of the threshold voltage. It corresponds to a soft failure with regular and irreversible aging of the transistor. Those failure modes being strongly activated by the electric field applied at the terminals of the transistors, they will be extremely depending on the design techniques used to enhance the speed of the circuit, such as AVS and FBB. In this section we will particularly evaluate the impact of those biasing solution of CMOS device.

2.8.1 Case of Gate Oxide Breakdown

As already reported, time to failure (TTF) obeys to a Weibull distribution activated by gate voltage V_{gs} as depicted in Eq. (2.30):

$$TTF = A_o \cdot V^P \cdot \left(\frac{1}{S}\right)^{\frac{1}{\beta}} \cdot e^{-\frac{E_a}{KT}} \qquad (2.30)$$

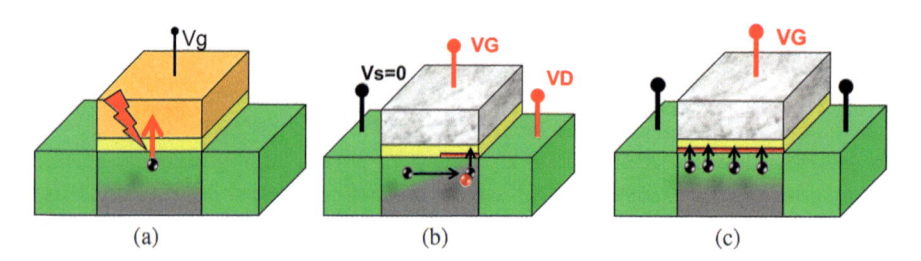

Fig. 2.58 Intrinsic failure modes occurring in advanced CMOS transistor inducing both gate dielectric integrity and threshold voltage of the device. (**a**) TDDB. (**b**) HCI. (**c**) BTI

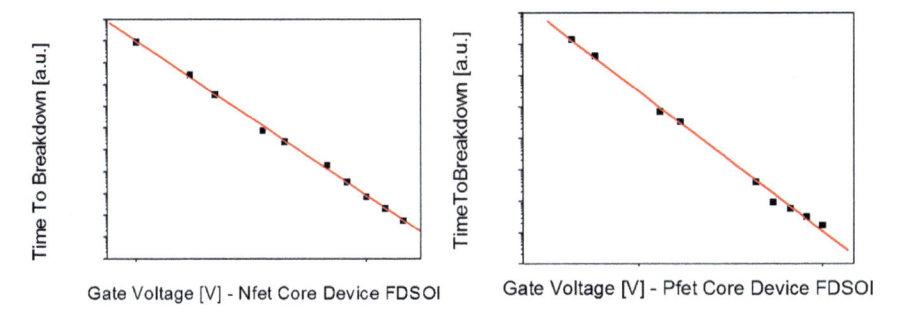

Fig. 2.59 Gate oxide time to break decreasing with V_{gs} applied (NMOS and PMOS transistors)

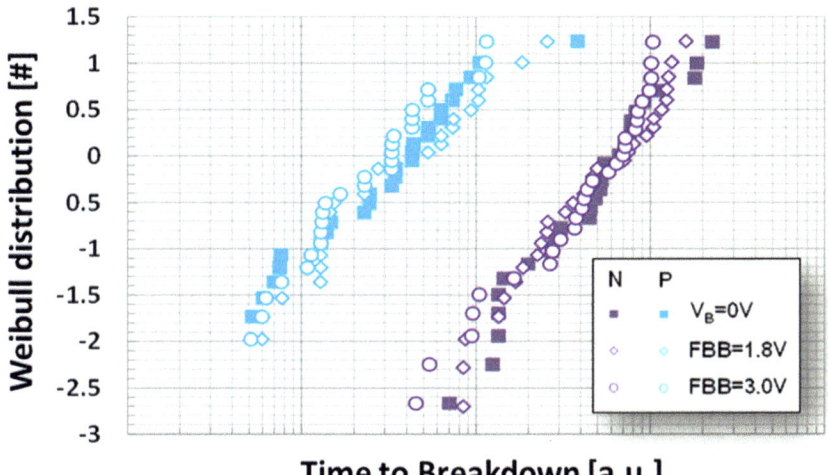

Fig. 2.60 Weibull distributions of gate oxide time to breakdown with (full symbols) and without (empty symbols) FBB in case of both NMOS and PMOS transistors. ©IEEE 2017 reprinted with permissions

β represents the slope of the Weibull distribution, S area of the gate, and V the gate voltage applied to the top electrode. E_a is the activation energy of the failure mechanism, and A_0 the initial value of the parameter before the stress sequence [24]. As presented in Fig. 2.59, gate oxide life time is reduced as a function of the gate voltage.

From those experimental data and theoretical model of the gate oxide breakdown, we can anticipate a dramatic gate oxide life time reduction by using overdrive option proposed in case of AVS technique. Thus AVS being very efficient to boost the performance at time0 will compromise the usage of the device in normal conditions for subsequent operation, due to a significant aging acceleration. On the contrary, back gate biasing increasing with FBB solution does not further damage the front gate oxide integrity. As shown in Fig. 2.60 printing the Weibull distributions of the time to breakdown of gate oxide with and without FBB, no difference has been measured for both NMOS and PMOS devices.

2.8.2 Case of NBTI

NBTI failure is related to vertical electric field and activated by temperature of the device. Under the electric field applied at the front gate, carrier trapping will occur in two different types of traps. First one is related to interface defect created next to the channel characterized by long emission time. This phenomenon is irreversible without any recovery possible. Second type of defects consists of trapping/de-trapping mechanism in gate oxide centers. This effect is characterized by short emission time with a fast recovery, as shown in Fig. 2.61.

NBTI failure mode will increase the threshold of the PMOS transistor across time and is accelerated by duty cycle and gate voltage. V_T shift is described by Eq. (2.31)

$$\delta Vt(t) = \sum_{i=1}^{i=N} g(\tau_c^i, \tau_e^i) U_i(t) \qquad (2.31)$$

where g factor represents the probability of both capture and emission time (respectively, τ_c and τ_e). As for TDDB issue, AVS technique based on power supply rising approach will severely accelerate the degradation of the threshold voltage. On the contrary, FBB solution will change only the back gate voltage using thick buried oxide, i.e., with very low vertical electric field. Consequently, as presented in Fig. 2.62, important V_T shift has been observed in case of AVS, whereas no V_T change was measured with FBB applied on back gate of the device.

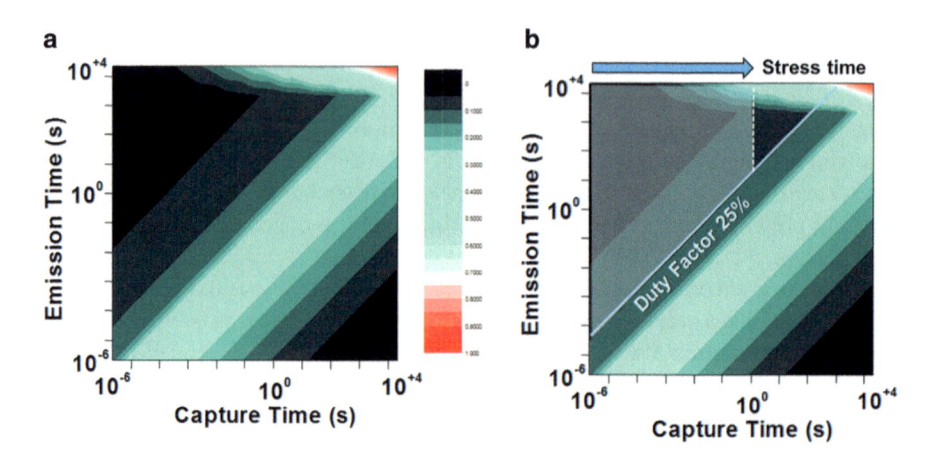

Fig. 2.61 NBTI failure modes (permanent and reversible) requiring AC (**a**) and DC measurements (**b**)

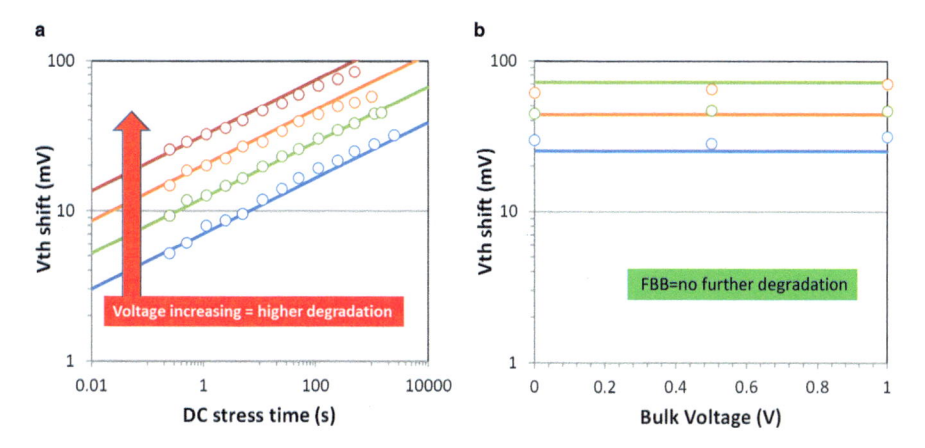

Fig. 2.62 Important V_T rising in case of AVS technique (**a**). No V_T degradation observed with FBB (**b**). ©IEEE 2017 reprinted with permissions

2.8.3 Case of HCI

HCI failure mechanism deals with lateral electric field related to drain voltage (V_{DS}). Under this stress conditions, carriers accelerate and acquire energy along the channel. Carriers hit and dissipate this energy at the drain side. Defect generation is directly related to both carrier energy and density interplay, as explained by Eq. (2.32)

$$R_{SE} = \int f(\epsilon) g(\epsilon) v(\epsilon) . S_{IT-SE} d\epsilon \tag{2.32}$$

Device parameters aging related to hot carrier is depicted by Eq. (2.33). P represents main transistor parameters under conductive channel influence, such as saturation and linear currents, threshold voltage, or transconductance. In this equation, A0 corresponds to the initial state of the parameter before stress, V the voltage applied at the drain of the device, Ea the activation energy of the phenomena, L the gate length of the device, and V_{bb} the body-bias.

$$\Delta P = A_o \cdot V^p \cdot \left(\frac{1}{L}\right)^m \cdot (t^n) \cdot e^{-\frac{E_a}{KT}} \cdot f(V_{BS}) \tag{2.33}$$

Physical mechanism is described in Fig. 2.63, showing the role of carrier density and the energy levels in the probability for a defect generation at drain region.

As for TDDB and NBTI, higher is the electric field, faster is the damage in the transistor. Consequently, AVS technique accelerates the degradation of the threshold voltage of the device due to the drain overdrive conditions used to enhance the speed of the circuit.

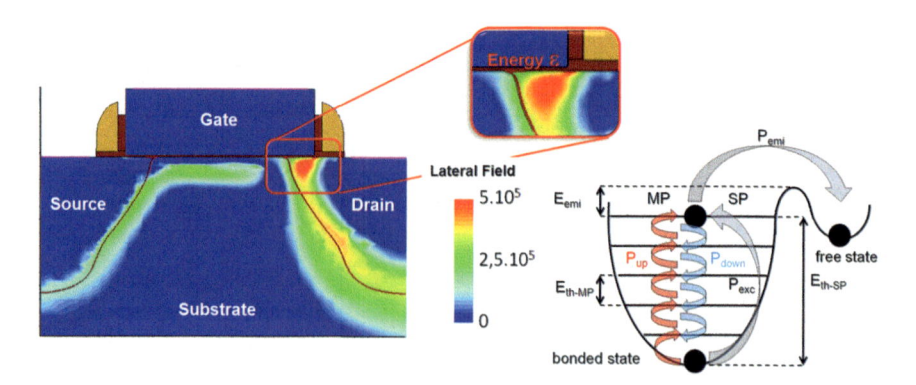

Fig. 2.63 HCI failure mechanism description

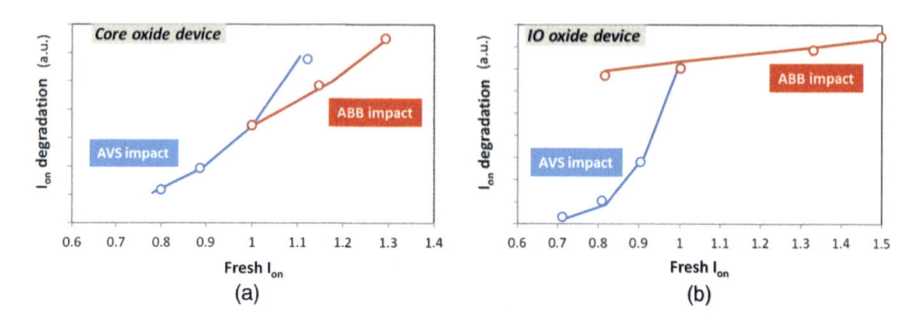

Fig. 2.64 HCI behavior versus voltage in case of AVS and FBB techniques for both core oxide and I/Os oxide transistors operating at 1 V and 1.8 V, respectively. (**a**) Core oxide device. (**b**) Thick oxide device. ©IEEE 2017 reprinted with permissions

By applying FBB biasing conditions, we do not amplify the root cause of the failure mode. Carrier's energy is maintained thanks to a constant lateral electric field. However, on the contrary to gate oxide related defects (TDDB and NBTI), we can anticipate a slight degradation of the device life time due to the V_T lowering itself, inducing more drive current, i.e., more carriers inside the channel of the transistor. The increasing of the number of carriers will moderately accelerate the aging of the device. Comparison between AVS and FBB techniques in terms of HCI behavior is depicted in Fig. 2.64

Figure 2.65 summarizes the comparison between the two design techniques used to boost the digital performance of the circuit, AVS and FBB in terms of transistor reliability impact. Clearly, FBB seems to be more suited for a smoother aging of the device [25]. It is worth pointing out that NBTI and HCI failure mode leads to an increase in the device threshold voltage, the penalty of the transistor aging will be much more sensitive for low voltage operation. Thus, AVS technique using intensively overdrive condition is not really appropriate for low power design requiring a high performance at low voltage, even after the aging of the device.

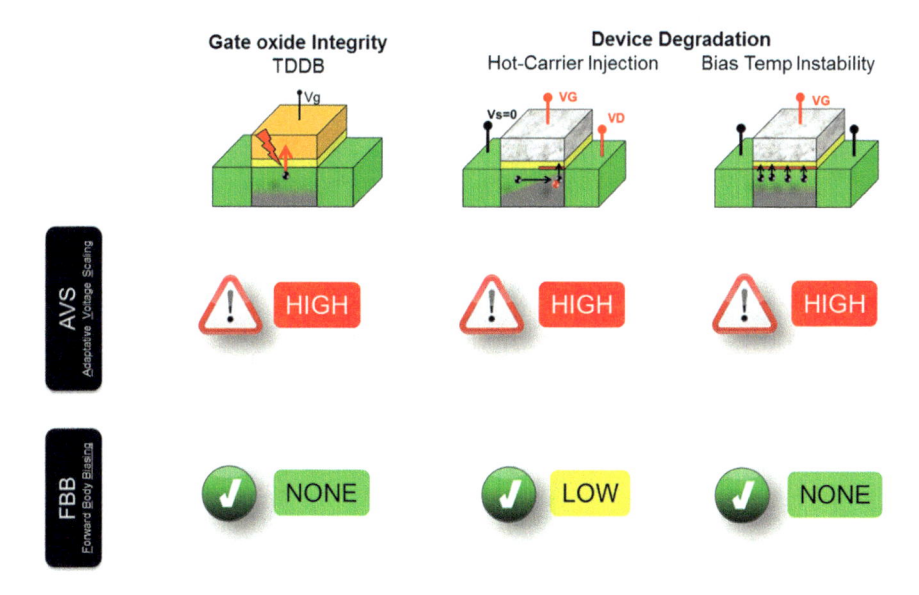

Fig. 2.65 CMOS transistor reliability comparison between AVS and FBB solutions

2.9 Conclusion

In this chapter, FD-SOI technology features have been presented and main transistor behaviors have been described highlighting the wide range of benefits for digital and analog applications. It is worth pointing out that all the enhancements demonstrated did not jeopardize reliability and did not accelerate the aging of the device. FD-SOI technology is very simple, based on planar solution, with limited number of masks and compatible for a low defect density technology road-map opening real opportunities for next CMOS nodes, even beyond 28 nm. The usage of all those advantages seen at the transistor level will be discussed and evidenced in the subsequent chapters at the design and application one.

References

1. Y. Taur, T.H. Ning, *Fundamentals of Modern VLSI Devices* (Cambridge University Press, New York, 1998)
2. S.M. Sze, *Physics of Semiconductor Devices* (Wiley, New York, 1981)
3. C. Gallon, C. Fenouillet-Beranger, A. Vandooren, F. Boeuf, S. Monfray, F. Payet, S. Orain, V. Fiori, F. Salvetti, N. Loubet, C. Charbuillet, A. Toffoli, F. Allain, K. Romanjek, I. Cayrefourcq, B. Ghyselen, C. Mazure, D. Delille, F. Judong, C. Perrot, M. Hopstaken, P. Scheblin, P. Rivallin, L. Brevard, O. Faynot, S. Cristoloveanu, T. Skotnicki, Ultra-thin fully depleted SOI devices with thin box, ground plane and strained liner booster, in *2006 IEEE International SOI Conference Proceedings*, Oct 2006, pp. 17–18

4. J. Lacord, J. Huguenin, T. Skotnicki, G. Ghibaudo, F. Boeuf, Simple and efficient MASTAR threshold voltage and subthreshold slope models for low-doped double-gate MOSFET. IEEE Trans. Electron Devices **59**(9), 2534–2538 (2012)
5. O. Rozeau, M. Jaud, T. Poiroux, M. Benosman, Surface potential based model of ultra-thin fully depleted SOI MOSFET for IC simulations, in *IEEE 2011 International SOI Conference*, Oct 2011, pp. 1–22
6. F. Monsieur, TCAD ST internal report on body bias for FDSOI structures, May 2012
7. C. Hobbs, L. Fonseca, V. Dhandapani, S. Samavedam, B. Taylor, J. Grant, L. Dip, D. Triyoso, R. Hegde, D. Gilmer, R. Garcia, D. Roan, L. Lovejoy, R. Rai, L. Hebert, H. Tseng, B. White, P. Tobin, Fermi level pinning at the polySi/metal oxide interface, in *2003 Symposium on VLSI Technology. Digest of Technical Papers (IEEE Cat. No.03CH37407)*, June 2003, pp. 9–10
8. S. Narasimha, P. Chang, C. Ortolland, D. Fried, E. Engbrecht, K. Nummy, P. Parries, T. Ando, M. Aquilino, N. Arnold, R. Bolam, J. Cai, M. Chudzik, B. Cipriany, G. Costrini, M. Dai, J. Dechene, C. DeWan, B. Engel, M. Gribelyuk, D. Guo, G. Han, N. Habib, J. Holt, D. Ioannou, B. Jagannathan, D. Jaeger, J. Johnson, W. Kong, J. Koshy, R. Krishnan, A. Kumar, M. Kumar, J. Lee, X. Li, C. Lin, B. Linder, S. Lucarini, N. Lustig, P. McLaughlin, K. Onishi, V. Ontalus, R. Robison, C. Sheraw, M. Stoker, A. Thomas, G. Wang, R. Wise, L. Zhuang, G. Freeman, J. Gill, E. Maciejewski, R. Malik, J. Norum, P. Agnello, 22nm high-performance SOI technology featuring dual-embedded stressors, epi-plate high-K deep-trench embedded DRAM and self-aligned via 15lm BEOL, in *2012 International Electron Devices Meeting*, Dec 2012, pp. 3.3.1–3.3.4
9. F. Arnaud, Enhanced low voltage digital & analog mixed-signal with 28 nm FDSOI technology, in *2015 IEEE SOI-3D-Subthreshold Microelectronics Technology Unified Conference (S3S)*, Oct 2015, pp. 1–4
10. S. Kang, Y. Leblebici, *CMOS Digital Integrated Circuits* (McGraw-Hill, New York, 2003)
11. F. Arnaud, N. Planes, O. Weber, V. Barral, S. Haendler, P. Flatresse, F. Nyer, Switching energy efficiency optimization for advanced CPU thanks to UTBB technology, in *2012 International Electron Devices Meeting*, Dec 2012, pp. 3.2.1–3.2.4
12. O. Weber, F. Andrieu, J. Mazurier, M. Casse, X. Garros, C. Leroux, F. Martin, P. Perreau, C. Fenouillet-Beranger, S. Barnola, R. Gassilloud, C. Arvet, O. Thomas, J.P. Noel, O. Rozeau, M.A. Jaud, T. Poiroux, D. Lafond, A. Toffoli, F. Allain, C. Tabone, L. Tosti, L. Brevard, P. Lehnen, U. Weber, P.K. Baumann, O. Boissiere, W. Schwarzenbach, K. Bourdelle, B.Y. Nguyen, F. Boeuf, T. Skotnicki, O. Faynot, Work-function engineering in gate first technology for multi-VT dual-gate FDSOI CMOS on UTBOX, in *2010 International Electron Devices Meeting*, Dec 2010, pp. 3.4.1–3.4.4
13. R.F.M. Roes, A.C.M.C. van Brandenburg, A.H. Montree, P.H. Woerlee, Implications of pocket optimisation on analog performance in deep sub-micron CMOS, in *29th European Solid-State Device Research Conference*, vol. 1, Sept 1999, pp. 176–179
14. J.M. Ginley, O. Noblanc, C. Julien, S. Parihar, K. Rochereau, R. Difrenza, P. Llinares, Impact of pocket implant on MOSFET mismatch for advanced CMOS technology, in *Proceedings of the 2004 International Conference on Microelectronic Test Structures (IEEE Cat. No.04CH37516)*, March 2004, pp. 123–126
15. K.K. Hung, P.K. Ko, C. Hu, Y.C. Cheng, A unified model for the flicker noise in metal-oxide-semiconductor field-effect transistors. IEEE Trans. Electron Devices **37**(3), 654–665 (1990)
16. T. Morshed, S.P. Devireddy, M.S. Rahman, Z. Celik-Butler, H. Tseng, A. Zlotnicka, A. Shanware, K. Green, J.J. Chambers, M.R. Visokay, M.A. Quevedo-Lopez, L. Colombo, A new model for 1/f noise in high-k MOSFETs, in *2007 IEEE International Electron Devices Meeting*, Dec 2007, pp. 561–564
17. G. Giusi, E. Simoen, G. Eneman, P. Verheyen, F. Crupi, K.D. Meyer, C. Claeys, C. Ciofi, Low-frequency (1/f) noise behavior of locally stressed HfO/sub 2//tin gate-stack PMOSFETs. IEEE Electron Device Lett. **27**(6), 508–510 (2006)
18. R.J. Baker, *CMOS Circuit Design Layout and Simulation*, ed. by S.K. Tewksbury J.E. Brewer (Wiley-Interscience, New York, 2008)

19. N. Planes, O. Weber, V. Barral, S. Haendler, D. Noblet, D. Croain, M. Bocat, P.O. Sassoulas, X. Federspiel, A. Cros, A. Bajolet, E. Richard, B. Dumont, P. Perreau, D. Petit, D. Golanski, C. Fenouillet-Beranger, N. Guillot, M. Rafik, V. Huard, S. Puget, X. Montagner, M.A. Jaud, O. Rozeau, O. Saxod, F. Wacquant, F. Monsieur, D. Barge, L. Pinzelli, M. Mellier, F. Boeuf, F. Arnaud, M. Haond, 28 nm FDSOI technology platform for high-speed low-voltage digital applications, in *2012 Symposium on VLSI Technology (VLSIT)*, June 2012, pp. 133–134
20. R. Ranica, N. Planes, O. Weber, O. Thomas, S. Haendler, D. Noblet, D. Croain, C. Gardin, F. Arnaud, FDSOI process/design full solutions for ultra low leakage, high speed and low voltage SRAMS, in *2013 Symposium on VLSI Circuits*, June 2013, pp. T210–T211
21. Y. Yamamoto, H. Makiyama, H. Shinohara, T. Iwamatsu, H. Oda, S. Kamohara, N. Sugii, Y. Yamaguchi, T. Mizutani, T. Hiramoto, Ultralow-voltage operation of silicon-on-thin-box (SOTB) 2mbit SRAM down to 0.37 v utilizing adaptive back bias, in *2013 Symposium on VLSI Circuits*, June 2013, pp. T212–T213
22. R. Ranica, Enabling ultra-low VMIN SRAM operation in FDSOI technology, in *2018 IEEE SOI-3D-Subthreshold Microelectronics Technology Unified Conference (S3S) Short Course* (2018)
23. Y. Chen, W. Chan, W. Wu, H. Liao, K. Pan, J. Liaw, T. Chung, Q. Li, G.H. Chang, C. Lin, M. Chiang, S. Wu, S. Natarajan, J. Chang, 13.5 a 16nm 128mb SRAM in high-k metal-gate FinFET technology with write-assist circuitry for low-VMIN applications, in *2014 IEEE International Solid-State Circuits Conference Digest of Technical Papers (ISSCC)*, Feb 2014, pp. 238–239
24. S. Blonkowski, Filamentary model of dielectric breakdown (2010, April) [Online]. Available: https://doi.org/10.1063/1.3386517
25. F. Arnaud, S. Clerc, S. Haendler, R. Bingert, P. Flatresse, V. Huard, T. Poiroux, Enhanced design performance thanks to adaptative body biasing technique in FDSOI technologies, in *2017 IEEE SOI-3D-Subthreshold Microelectronics Technology Unified Conference (S3S)*, Oct 2017, pp. 1–5

Chapter 3
Body-Bias for Digital Designs

Sylvain Clerc and Ricardo Gomez Gomez

FD-SOI Body-Bias enables software defined V_T

3.1 Body-Bias for Digital Designs: Introduction

3.1.1 Body-Bias and the Digital Design Space

Process, voltage, temperature, and ageing lead to an I_{on} and I_{off} range defining a volume, the design space, where a circuit should meet its specifications. This chapter will detail how Body-Bias modifies the digital design space, either by tightening its volume, (as is the case of compensation), or by extending its limit towards higher performance.[1] This is illustrated by Fig. 3.1 where the frequency of a fan-out-of-4 logic cell critical path[2] is plotted against leakage with and without Body-Bias. All the values are normalized to typical process, voltage, and temperature. The optimization of design corners defining the design space volume limits is what matters to digital designers to meet their design specification, as will be described in Chap. 14, where we will review the relative merits of Adaptive Body-Bias and Adaptive Voltage Scaling.

In this chapter, we will focus on the benefits of Body-Bias for digital systems, but first we will define the qFO4 benchmark which will be used for data shown in this chapter and the Chap. 14. A key factor governing if Body-Bias can be beneficial or not is the position of the hosting system's supply voltage with respect to the voltage

---
[1] More on this design space evolution with Body-Bias in Chap. 16.
[2] Further abbreviated qFO4, see Fig. 3.2 and next section for an explanation of quasi-fan-out-of-4 logic path and why we call it 'quasi'.

S. Clerc (✉) · R. G. Gomez
STMicroelectronics, Crolles, France
e-mail: sylvain.clerc@st.com

© Springer Nature Switzerland AG 2020
S. Clerc et al. (eds.), *The Fourth Terminal*, Integrated Circuits and Systems,
https://doi.org/10.1007/978-3-030-39496-7_3

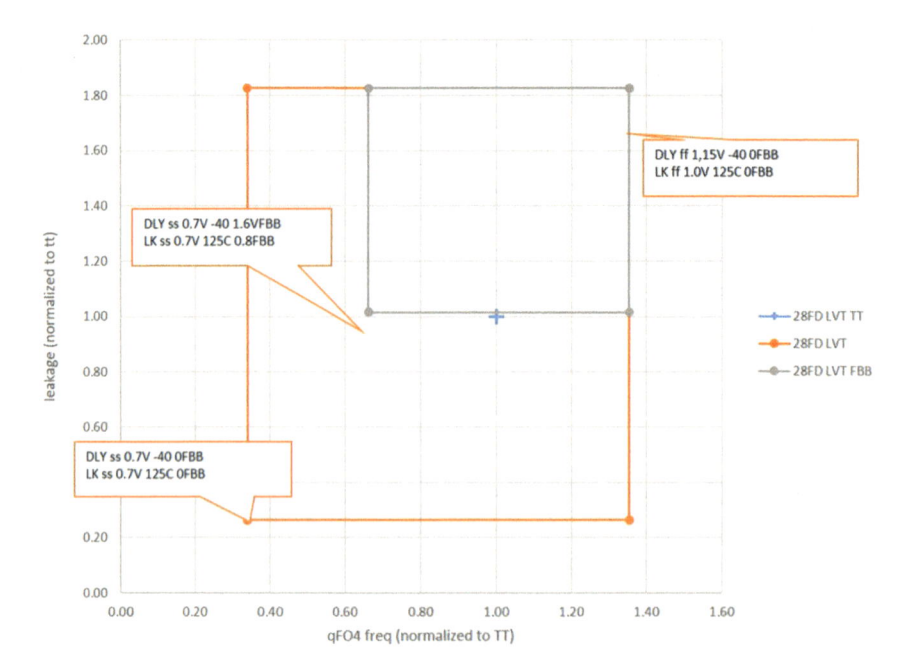

Fig. 3.1 qFO4 (explained further) logic cell critical path (leakage at 125 °C, frequency at −40 °C) plot with and without Forward Body-Bias, all values normalized to typical point (cross), the circuit is supplied below its V_{Tinv}. The extreme lower left and upper right corners define the design space

of temperature inversion (V_{Tinv}), see definition in Appendix B.1.3 and Chap. 10 (Sect. 10.1), it will be mentioned along this chapter.

3.1.2 Logic Performance Benchmark Method

Some of the facts exposed in this chapter are demonstrated through this Spice simulations benchmark. This design space exploration method for logic is based on qFO4 benchmark as reported in [1], the circuit schematic is shown in Fig. 3.2. A logic cone of 'self' loaded gates is assembled in a logic chain where multi input gates are all connected together and some load gates are added to have a stage load as close as possible to a fan-out-of-4 (FO4). The logic chain features 14 stages of logic connected to a Flip-Flop (FF) not shown on the figure and is architected as follows:

- inverter and buffers have loading cells connected to their output plus the cell of the next stage, hence each stage has a FO4 load
- 2 input gates have 1 loading cell connected plus the cell of the next stage, both inputs connected, hence each stage has a FO4 load

3 Body-Bias for Digital Designs

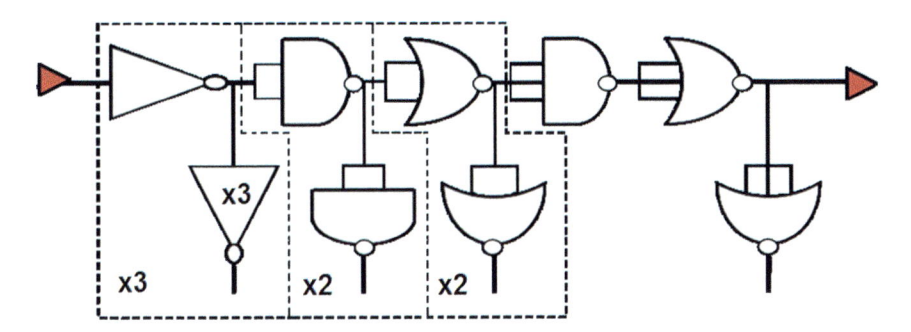

Fig. 3.2 qFO4 self-loaded logic cell delay structure

- 3 input gates do not have loading cells, but all three inputs of next stage are connected together, it has a FO3[3] load

The multi input connection averages out the active input dependent delay by filtering out the stacked MOS delay dependence. The self-loading quasi-FO4 filters out the transistor sizing effect of logic cells (i.e., height of cells or track number). FF inclusion ensures that setup/hold constraints calculation in delay is included, specifically the non-linear behaviour of latches resolution, giving a closer match with real design situation.

The key leakage power, delay, and internal energy parameters are extracted in the conditions detailed below:

- the logic delay is extracted by summing the data setup time, running across the logic cone to the Flip-Flop data pin, and the Flip-Flop's clock to output delay, loaded with intermediate capacitance with respect to its fanout drive capability. The setup limit is extracted at 10% clock to output variation compared to a clock to output delay reference extracted with very large setup.
- the internal energy[4] is extracted from 100 cycles integration average with data activity at 25% where data toggles every 4 cycles. The internal energy is translated back to dynamic power (for example, used in Chap. 14, Sect. 14.1) when appropriate according to clock frequency.
- the leakage power is averaged across input and clock states (i.e., four combinations), using Spice transient simulations with low pace change.

Both leakage power and internal energy are derived from the general definition of power and energy:

$$P = \frac{1}{T} \cdot \int_0^T v(t) \cdot i(t) dt \qquad (3.1)$$

---

[3]Hence the 'quasi' fan-out-of-4.

[4]The internal energy is the energy driven by the cell with input and output capacitance load energy excluded.

$$E = \int_0^T v(t) \cdot i(t) dt \qquad (3.2)$$

3.2 Digital Compensation Toolbox

Environmental, global manufacturing variation and ageing effect compensation is the domain of application where Body-Bias excels. We will detail in the following, how adjusting Forward Body-Bias can maintain iso-delay across all these variations.

3.2.1 Temperature Compensation

Here we have used the qFO4 benchmark and set a timing reference at the lowest voltage, slowest process, lowest temperature, maximum[5] Forward Body-Bias set at 0.9 V for fresh parts. In Fig. 3.3, we display the minimum Forward Body-Bias needed across temperature to reach the reference delay while varying V_T (LVT and HLVT) and transistor length for $L = 30$ nm, $L = 34$ nm, and $L = 40$ nm. From this simulation figure we can state that the Body-Bias curve is quasi-linear, rise and fall edge are not discriminant when considering a 14 gate stage path, and LVT transistor flavor needs more Body-Bias than their HLVT counterpart.

Figure 3.4 illustrates the `translation` in the (T, BB) space induced by process change. There, the same setup of simulation used for Fig. 3.3 has been used on the two devices flavors having the most demanding Body-Bias slope vs. temperature: SS process 0.75 V $-40\,°C$ 0.9 V FBB for the two upper curves, for the two lower curves the same delay reference is used but for TT process. It can be seen that the Body-Bias curve is shifted down by 500 mV.

The LVT flavor cells exhibit a poorer translation matching than the HLVT ones. This error can be estimated to 50 mV Body-Bias voltage which is 1 step of the embedded Body-Bias generator reported in Chap. 15.

3.2.2 Voltage Compensation or Body-Bias Modulation with Voltage

If your design still needs Body-Bias at high temperature low voltage, and slow process (illustrated in Fig. 16.4), because it is supplied close or above V_{Tinv}, the voltage compensation will keep the leakage of your slowest biased corners away from the worst leakage limit by lowering bias at higher supply.

---

[5] Implicit bias maximum applied at slowest RTL synthesis PVT, kindly refer to Sect. 16.3.

3 Body-Bias for Digital Designs

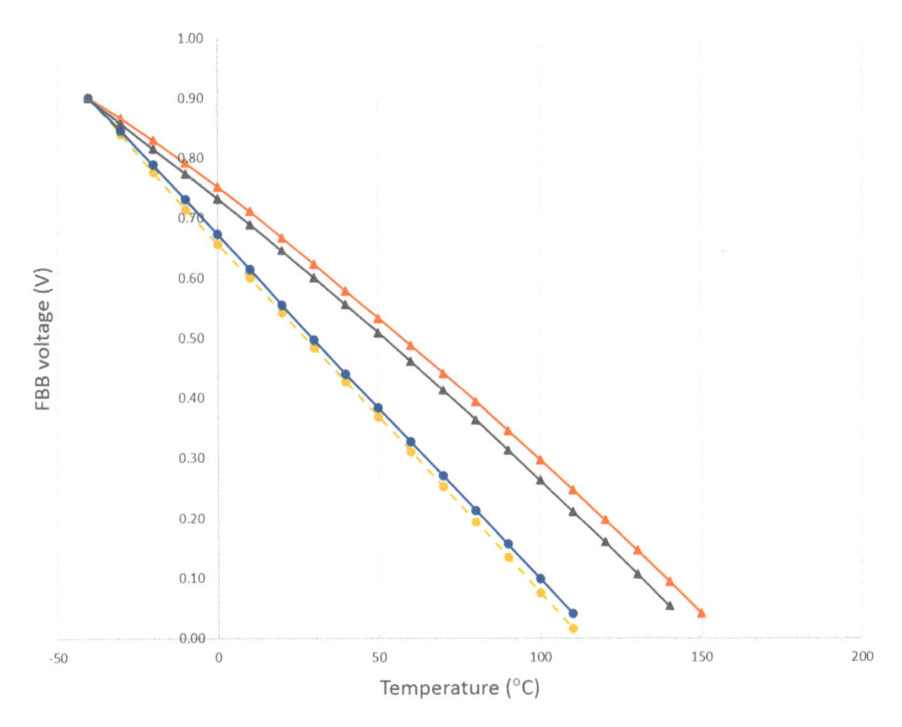

Fig. 3.3 28 nm FD-SOI qFO4 iso-delay Body-Bias Spice simulations across temperature, with varied Poly Bias (orange and blue), and V_T LVT (triangle symbols), HLVT (circle symbols). Supply voltage is below V_{Tinv}

> Voltage compensation enables to contain Forward Body-Bias leakage at maximum temperature in case your design is supplied above V_{Tinv}.

Furthermore, voltage compensation can also be used to adapt each design to on-PCB LDO tolerance enabling lower cost power management parts to be used, this case is very design specific and falls out of the scope of this general study.

Figure 3.6 displays iso-delay FBB with supply voltage for 2 V_T/PB flavors at −30 °C, and 0 °C, Above 0 °C temperature, Body-Bias is not needed anymore to match slowest timing reference. It can be seen that temperature generates a translation in the (V, BB) space. Further, Figs. 3.5 and 3.6 show that the voltage compensation bias rule can be approximated by a linear law whose slope is the lowest Body-Bias slope in (V, BB) space.

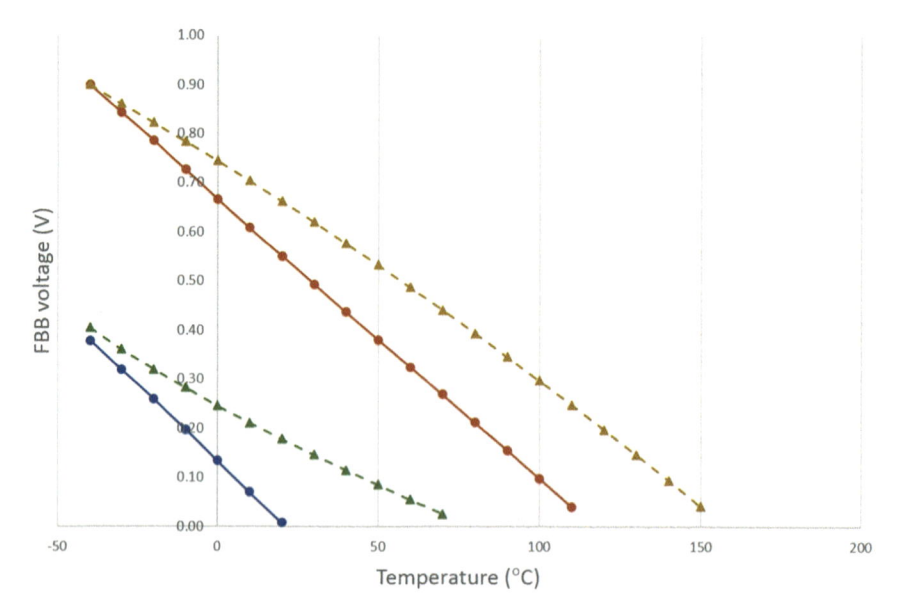

Fig. 3.4 28 nm FD-SOI qFO4 simulation of iso-delay Body-Bias vs. temperature with varied process, for two selected flavors of *PB/VT*. The two curves above display iso-delay FBB for slow dies, the two curves below the iso-delay FBB for typical dies. Supply voltage is below V_{Tinv}

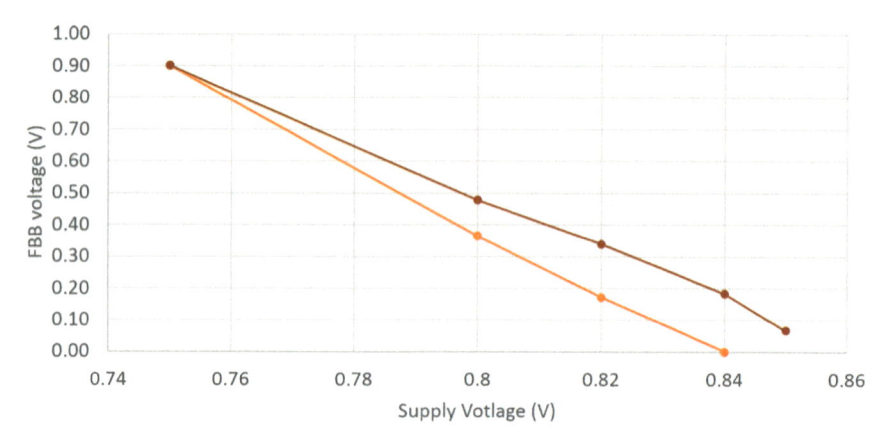

Fig. 3.5 28 nm FD-SOI qFO4 simulation of iso-delay Body-Bias vs. supply voltage for two PB/VT flavors

3.2.3 Process Compensation

The global variation induced by manufacturing tolerance can be compensated via Body-Bias as illustrated in Fig. 3.7, where the spread of delay and leakage is plotted without bias compensation (left side) and with bias compensation aiming a nominal delay target (right side). As it can be seen on the right side of the figure, the delay

3 Body-Bias for Digital Designs

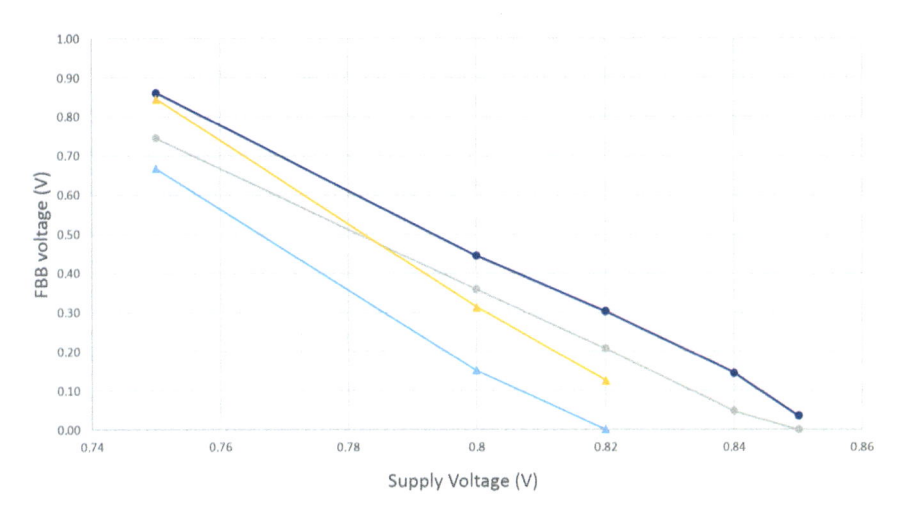

Fig. 3.6 28 nm FD-SOI qFO4 simulation of iso-delay Body-Bias vs. supply voltage on 2 V_T/PB flavors (LVT round symbols, HLVT triangle symbols) at $-30\,°C$ and $0\,°C$

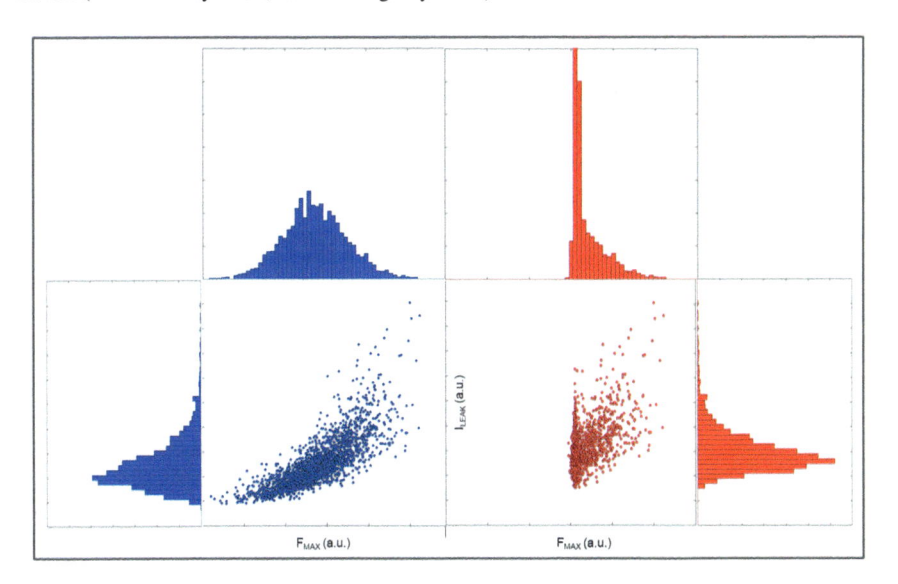

Fig. 3.7 Figure showing frequency and leakage before (left) and after (right) compensation with Body-Bias. qFO4 simulations, arbitrary units

spread can be halved without impacting the maximum leakage, which is a gating specification of product. However, this safe guard on leakage limit is only true in case the design voltage is away from V_{Tinv}, which will be revisited in Sect. 3.3 (illustrated in Fig. 3.13).

3.2.4 Ageing Compensation

Here we will not review the consequence of Body-Bias on device, as it has been exposed in Sect. 2.8, but illustrate how we can compensate the ageing effect by applying Body-Bias. For this demonstration we will use the ageing feature embedded in Spice models by proceeding as follows:

- the Spice model features an ageing mechanism which is extracted from elementary device characterization for NBTI, HCI, etc.
- this ageing mechanism is enabled by the designer by stating the duration, the conditions of stress, and which phenomenon is to be recorded. Here we have activated both HCI and NBTI (see Sect. 2.8).
- the age stress simulation data toggle activity has been set to 12.5%. The duration of the stress is varied for the purpose of the demonstration from 0.25 to 2 years.
- the final delay simulation is ran in a corner condition which differs from the ageing condition, but which takes into account the ageing effect on each device as recorded by the previous step.

Figure 3.8 illustrates the Body-Bias voltage needed to keep an iso-delay qFO4 simulation after ageing. The reference delay is extracted at fresh, 0.9 V Forward Body-Bias, the axis crossing point of the figure. It can be noted that 100 mV of bias is needed to maintain frequency for 2 years of stress (Y coordinate at 2 years for the upper curve), and 80% of this counter-ageing Body-Bias is needed after 0.5 year of stress.

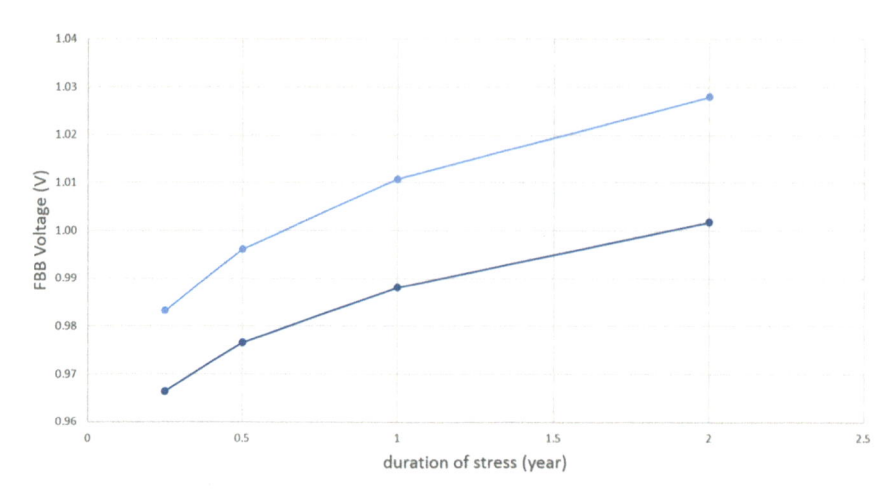

Fig. 3.8 28 nm FD-SOI qFO4 iso-delay Body-Bias in Volts across ageing in equivalent years, for two VT/PB flavors. The simulation is run at 0.75 V −40 °C, while ageing profile is extracted at High V, 165 °C. The reference delay is the point (0y, 0.9 V FBB)

3 Body-Bias for Digital Designs

Table 3.1 Asymmetric bias qFO4 Spice bench simulation in FD-SOI, leakage is simulated at 165 °C, all leakage normalized by slow-slow (SS) 0.75 V values at same temperature

	SS reference	SF		FS	
Bias symmetry	Symmetric	Symmetric	Asymmetric	Symmetric	Asymmetric
Bias NMOS	0.9	0.55	0.55	0.45	0.0
Bias PMOS	−0.9	−0.55	−0.15	−0.45	−0.45
Leakage	1.00	1.24	0.81	1.06	0.94

3.2.5 Asymmetric Body-Bias

If, instead of applying symmetric bias (same absolute value of Body-Bias voltage is applied for both NMOS and PMOS), an asymmetric bias is applied, then Body-Bias can compensate skewed process corners with lower leakage impact than symmetric bias.

Table 3.1 illustrates the relative leakage reduction which can be obtained from asymmetric bias with both slow NMOS–fast PMOS (SF) and fast NMOS–slow PMOS (FS) compared to slow NMOS–slow PMOS (SS) corner with symmetric bias.

The asymmetric-labelled column in each of the two (FS, SF) cases displays the leakage reduction when Body-Bias is applied at hot (this is a best case for leakage gain for asymmetric). As it can be seen, in this compensation case, instead of inducing extra leakage, Body-Bias enables 19% of leakage reduction at SF/hot while maintaining iso-delay vs. SS 0.75 V reference at −40 °C.

It must be noted that applying skewed corners in Digital SoC implies that these biased corners are covered by your Sign-Off, more on this in Chap. 16 (Sect. 16.3).

3.2.6 Compensation Costs and Gains

For this section, and in complement of qFO4 Spice simulation benchmarks, we will use the circuit reported in [2] and displayed in Fig. 3.9 to demonstrate the power gains which can be brought by Body-Bias.

Let us first detail what is inside the referenced circuit. The circuit embeds two replications of formally equivalent ARM A53 (single) cores designed to explore the effect of Body-Bias on power. The two CPUs have the same floorplan, include 500 kB AXI memory, 30k FFs, and both of them have their logic implemented in 12 tracks FD-SOI logic cells. Their worst case implementation corners[6] are both slow-slow process 0.9 V supply −40 °C but one of them is implemented with zero

---

[6] Kindly refer to Chap. 16 for implementation corners definition.

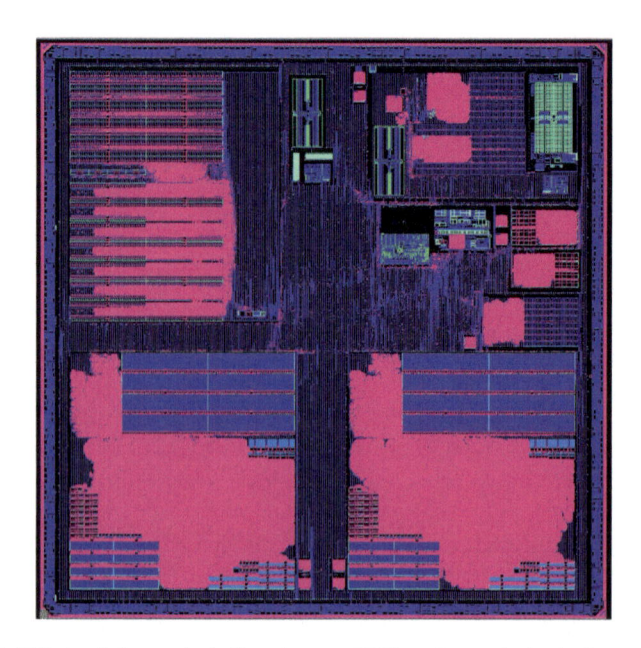

Fig. 3.9 FD-SOI circuit layout including the two CPUs under study in the lower half. ©2016 IEEE. Reprinted with permissions

bias and the other one with 0.6 V Forward Body-Bias (respectively, labelled BOC and BIC in Fig. 3.10, for, respectively, Bias-In-Corners and Bias-Out-of-Corners).

3.2.6.1 Body-Bias Leakage Reduction

Forward Body-Bias can lead to leakage reduction, which is counterintuitive as Forward Body-Bias brings down the transistor's V_T and hence increases the leakage at device level. However, by incorporating Forward Body-Bias corners, a global cell undersize during implementation is induced while matching the frequency constraint.[7] As long as this cell undersize aspect dominates over the increased device leakage with Forward Body-Bias, the leakage gain stands.

As an example, the static power measurement of the above CPUs shows a 30% leakage reduction when Body-Bias is integrated in the worst-case implementation corner. This is the result of worst-case process compensation: Body-Bias enhances the drive of transistors which shortens the logic cell delay. Then, the implementation tools in both synthesis and physical phases take profit of that increased drive to recover leakage by downsizing the W/L of the cells.

---

[7]Because there is a race between the global cell undersize and individual cell leakage, we denote this leakage reduction as 'indirect'.

3 Body-Bias for Digital Designs   69

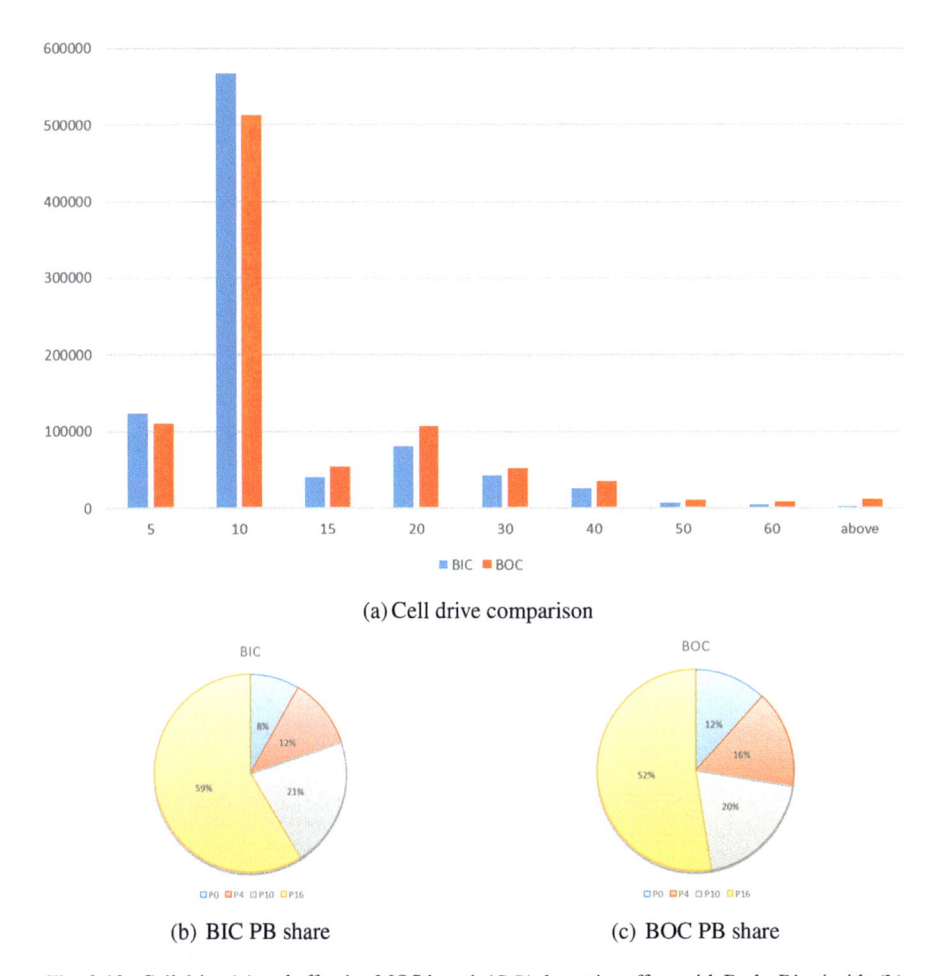

(a) Cell drive comparison

(b) BIC PB share

(c) BOC PB share

Fig. 3.10 Cell drive (**a**) and effective MOS length (PB) downsize effect with Body-Bias inside (**b**) vs. outside (**c**) implementation corners, the BIC and BOC labels stand for, respectively, Bias-In-Corners and Bias-Out-of-Corners, BIC implementation has higher low drive and large L cell count

The cell downsizing effect is displayed in Fig. 3.10 in which post-implementation gate netlist cell drive has been analysed. Figure 3.10a graphs should be read as follows:

- the X axes show normalized W/L of cells
- the normalized drive is defined as $norm(\frac{W}{L}) = \frac{100 \cdot 1.2 \cdot \text{drive}}{8 \cdot (PB+30\,\text{nm})}$
- the left-hand Y axes show the cell population spread sorted by cell drive

The small cell drive count is higher in the circuit with Forward Body-Bias inside its implementation corners.

As will be detailed in Sect. 3.3, this gain is valid for parts where downsizing is possible. This excludes, for example, the SRAM periphery, where sizing is fixed, and thus no gain should be expected. This said, Table 3.5 page 81 gives another example of the leakage reduction possible with Forward Body-Bias. There, a 40% gain is reported for an increase of 300 mV Forward Body-Bias in a design supplied close to V_T.

> Body-Bias can lower design leakage by W/L downsizing during synthesis as long as downsize leakage effect dominates over individual cell leakage increase.

3.2.6.2 Body-Bias Dynamic Power Gains

As for leakage, the dynamic power gain induced by Forward Body-Bias is indirect. By itself Forward Body-Bias does not change the dynamic power: Body-Bias has very little effect on gates cross-conduction current internal power and the total wiring capacitance which governs the circuit's dynamic power does not change with Forward Body-Bias. This is the case of the two designs shown in Fig. 3.9, which have similar floorplan hence similar total wire capacitance. These two designs, when supplied at the same voltage, will consume the same dynamic power. However, if a circuit needs to operate at iso-performance, dynamic power gain can be obtained by lowering the design supply voltage while applying Forward Body-Bias.

3.2.6.3 Yield or Minimum Operational Voltage Gain

Yield gains which can be obtained from Forward Body-Bias depend on each circuit and fabrication. However, we can give the reader several qualitative trends by joining the device level analysis on FF setup/hold[8] constraints and the digital flow corners definition covered in Sect. 16.3.

With respect to setup timing constraints, and providing that extra Body-Bias can be applied on slow dies, frequency can be recovered directly, from the speedup brought by Body-Bias.

> In case the yield or minimum voltage is limited by FF setup constraints a V_T adjust with Body-Bias can recover dies frequency.

---

[8]Kindly refer to Appendix B in case you are not familiar with digital constraints.

3 Body-Bias for Digital Designs

Table 3.2 28 nm FD-SOI 3-σ hold constraint at 0.7 V supply 25 °C under varied Forward Body-Bias, Silicon data extracted on 131k FF instances

FBB	0 V	0.4 V	0.7 V
Hold	113 ps	87 ps	74 ps

Hold constraints are dominated by clock skew at fast corners and by transitions and variability at slow corners. While at fast corners Body-Bias cannot help to recover yield, it can indirectly sharpen the signal transitions and V_{GT}[9] at slow corners, hence improving the slow corner's yield. This is illustrated in Table 3.2, where 3-σ hold constraint of 131k instances FF array is reported at 0.7 V 25 °C with various Forward Body-Bias values.

We denote this gain as indirect because the Body-Bias speedup may increase the hold constraint in absolute, but the spread due to variability will be lowered, making the hold fixing task of implementation tool indirectly easier.

> In case the yield or minimum voltage is limited by FF hold constraints in slow corners, Body-Bias can help recover yield indirectly by lowering variability and accelerating transitions.

3.2.6.4 Body-Bias Engineering and Deployment Costs

Body-Bias has associated costs listed below. Each item will be detailed further in this book in specific referenced chapters and sections:

- Extended corners coverage, refer to Chap. 16, Sect. 16.3.
- Dedicated power grid, and increased spacing for supply voltage power routing, refer to Chap. 16, Sect. 16.4.
- Intrinsic power consumption of Body-Bias Generator (BBGEN) which tops during bias transition, refer to Chap. 15.
- Latency of bias settling time which induces specific validation scenarios or bias controller latency, see Chap. 13.

As a circuit architect you will have to study and arbitrate between ABB or AVS or mix the two techniques, this subject is covered in Chap. 14 (Sect. 14.1).

---

[9] Gate overdrive voltage.

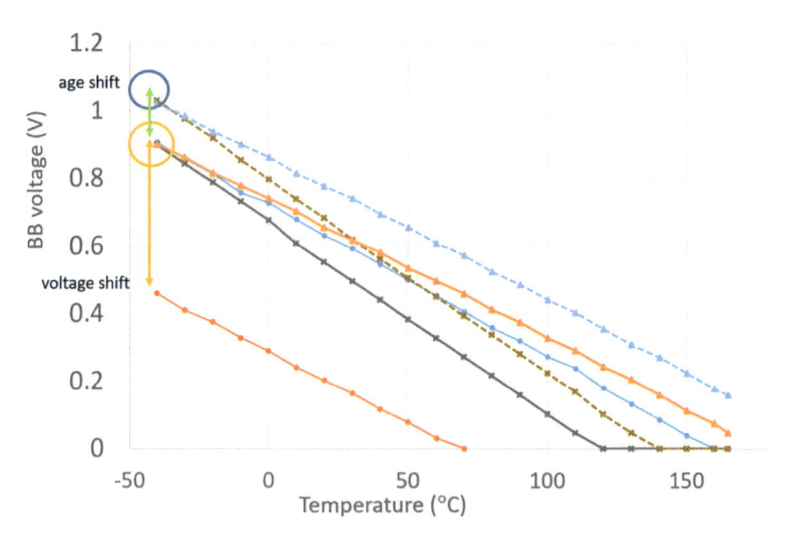

Fig. 3.11 Temperature compensation simulations across various V_T and PB with ageing and voltage variation, refer to text for explanations

3.2.7 Open Loop Bias Law

This section will cover how the circuit architect in charge of compensation should tune the circuit's bias as a function of the various parameters to compensate in an open loop Body-Bias control circuit, an example of such a design is detailed in Chap. 13.

In Fig. 3.11 we have plotted qFO4 simulations to cover together voltage, temperature, and ageing. The figure curves exhibit 3 groups of bias evolution curves:

- The middle set of three curves in triangle orange, cross dark grey, and round blue solid lines for, respectively, LVT Poly Bias 10, HLVT Poly Bias 10, LVT Poly Bias 4 show the iso-delay body-bias evolution across temperature for fresh parts. The three different V_T and MOS length logic cells curves have different bias slopes vs. temperature, all converging at $-40\ °C$ which is the normalized reference point (orange circle).
- The upper dashed lines show the iso-delay body-bias evolution across temperature range for aged parts. Two VT/PB flavors, HLVTPB10 and LVTPB10, were simulated, both converging at $-40\ °C$ which is the normalized reference point (dashed blue circle). The green arrow shows the bias translation needed to compensate end-of-life ageing.

- The lower curve shows the LVTPB4 iso-delay bias for $VDD = 0.8$ V. The orange arrow shows the bias translation caused by 50 mV supply voltage difference.

From the translation in the (P, V, T, a, BB) space illustrated by Figs. 3.4, 3.6, and 3.11 we can assume that each (P, V, T, a) parameter affects independently the Forward Body-Bias needed to maintain iso-delay, as denoted by Eq. (3.3).

> The bias law of the compensation system can be expressed as a summation of the different (V, T, a) individual parameters contributions, each one modulated by process.

$$BB = BB_{PT}(T) + BB_{PV}(V) + BB_{Pa}(a) + BB_{\text{Offset}} \qquad (3.3)$$

With $BB_{PT}(T)$, $BB_{PV}(V)$, and $BB_{Pa}(a)$ being the single parameter functions compensating, respectively, voltage, temperature, and ageing, themselves modulated by process. Furthermore, the bias evolution with either voltage or temperature can be considered linear in first order approximation or, if the system needs more precision, polynomial. We will see in Chap. 16 how this links to engineering test requirements. The BB_{Offset} term can cover the compensation system's uncertainty or some fabrication process shift. This process shift can be deliberate to target some application specific centering such as yield recovery or N/P balancing. This latter case is illustrated in Fig. 3.12.

Last, because the device response to Body-Bias is not uniform across V_T and PB flavors, the bias law will have to include the highest demanding bias value to maintain the frequency target. As an example, in Fig. 3.11 the LVTPB10 flavor is the most demanding Body-Bias cell flavor, said differently, has the flattest slope in the (T, BB) space.

3.3 Body-Bias Design Limits

3.3.1 The Design Leakage Ceiling

We have detailed in Sect. 3.2.6.1 the leakage reduction induced by lower $\frac{W}{L}$ used in synthesis. However, design leakage bound specification can limit the amount of Body-Bias which can be applied to your circuit in two ways:

- First, non-biased islands' leakage and non-synthesized blocks' sizing will not change with Body-Bias, nor their leakage will, limiting the effect of synthesis downsizing.

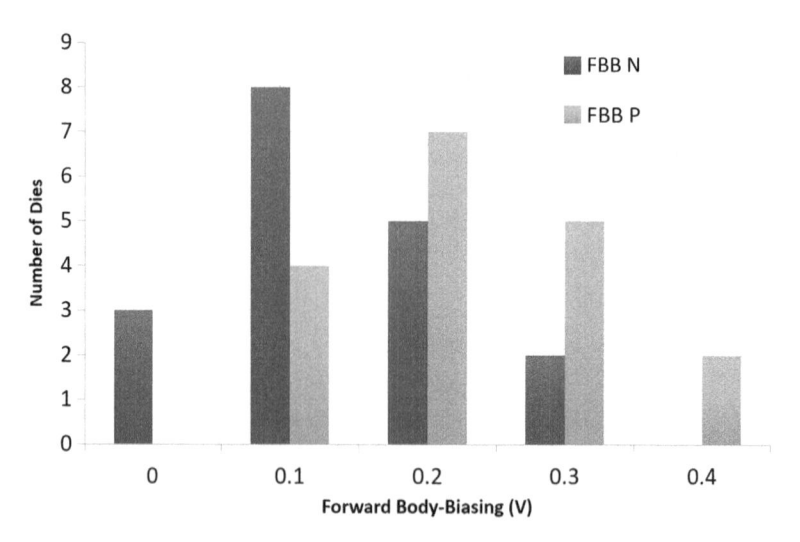

Fig. 3.12 Asymmetric Body-Bias offset applied on 18 dies of design reported in [3] at 25 °C 0.33 V iso-frequency of 2 MHz

- Second, the amount of Forward Body-Bias needed at maximum temperature may lead your slow parts to reach the leakage specification bound. This latter case is illustrated in Fig. 3.13, where a qFO4 compound V_T and MOS length logic path iso-delay Body-Bias vs. temperature is plotted for various supply values relative to V_{Tinv} (a). It can be seen that, as the supply voltage approaches V_{Tinv}, the minimum Forward Body-Bias needed at maximum temperature increases. In Fig. 3.13b, the relative position of Body-Bias compensated slow parts' leakage versus fast parts' leakage specification bound is reported. At 0.9 V Forward Body-Bias the slow parts already reach the leakage limit.

It should be noted here that the reference points at (−40 °C, 0.9 V) are not iso-delay. Instead, the iso-delay reference is represented with the dashed line in Fig. 3.13b, illustrating the Body-Bias down shift induced by a 50 mV supply increase.

To summarize, if your frequency specification does not change, you will not need to keep the initial bias of 0.9 V and your system may not be harmed by residual leakage at maximum temperature. On the contrary, if you are chasing for all-out maximum frequency, then the residual Forward Body-Bias leakage at hot may be a problem and the system should modulate Forward Body-Bias with supply voltage as exposed in Sect. 3.2.2.[10]

---

[10]Or else consider AVS, see Chap. 14 (Sect. 14.1) to review which option is the best for your design.

3 Body-Bias for Digital Designs

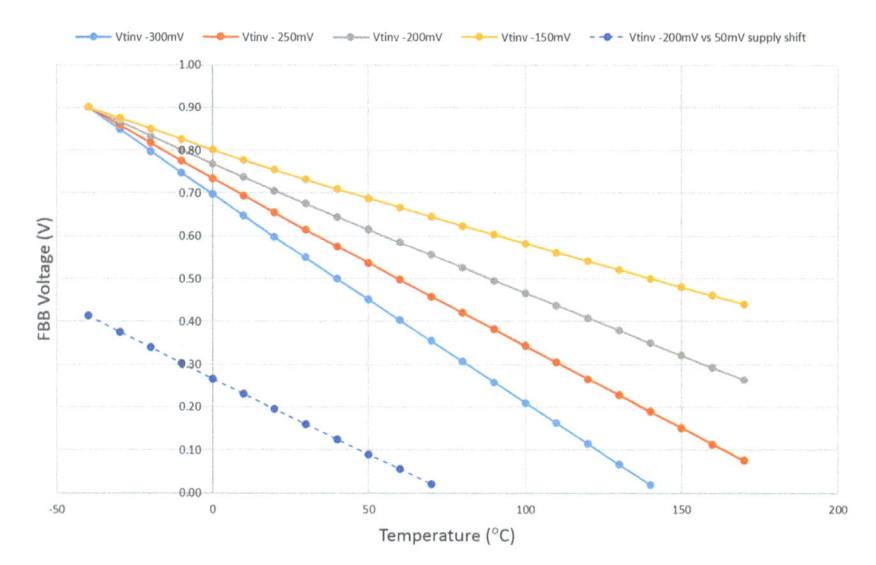

(a) Body-Bias as a function of T for various supply to V_{Tinv} distance (continuous line), residual bias at maximum temperature rises with supply. Bias Voltage shift induced by voltage modulation Body-Bias (dashed lines)

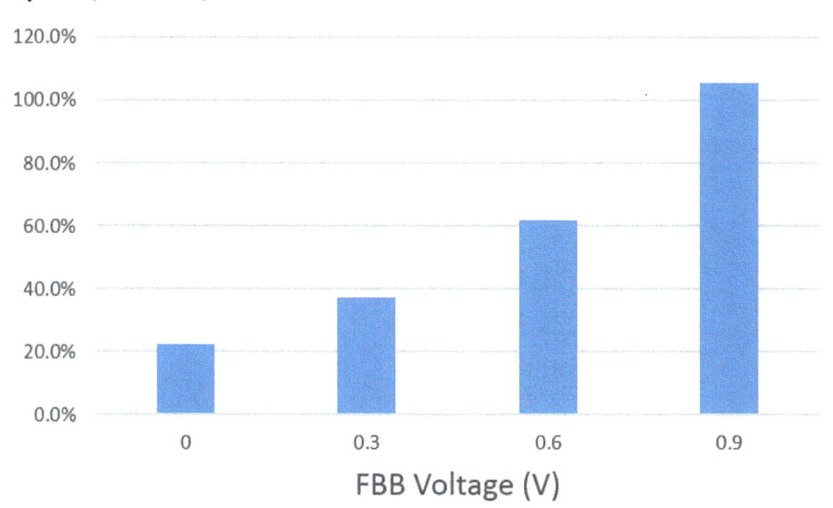

(b) Relative compensated slow parts leakage vs FBB value at maximum temperature

Fig. 3.13 Minimum BB at maximum temperature vs. supply distance to V_{Tinv} (**a**) and distance to leakage limit vs. BB at maximum temperature (**b**). In case of non-voltage modulated bias, the leakage on compensated parts increases with the residual Body-Bias at maximum temperature. In case voltage modulation bias is enabled, the leakage of compensated slow parts gets back to non-biased ratio of 20% (dashed blue line on (**a**) needs 0 bias at maximum temperature)

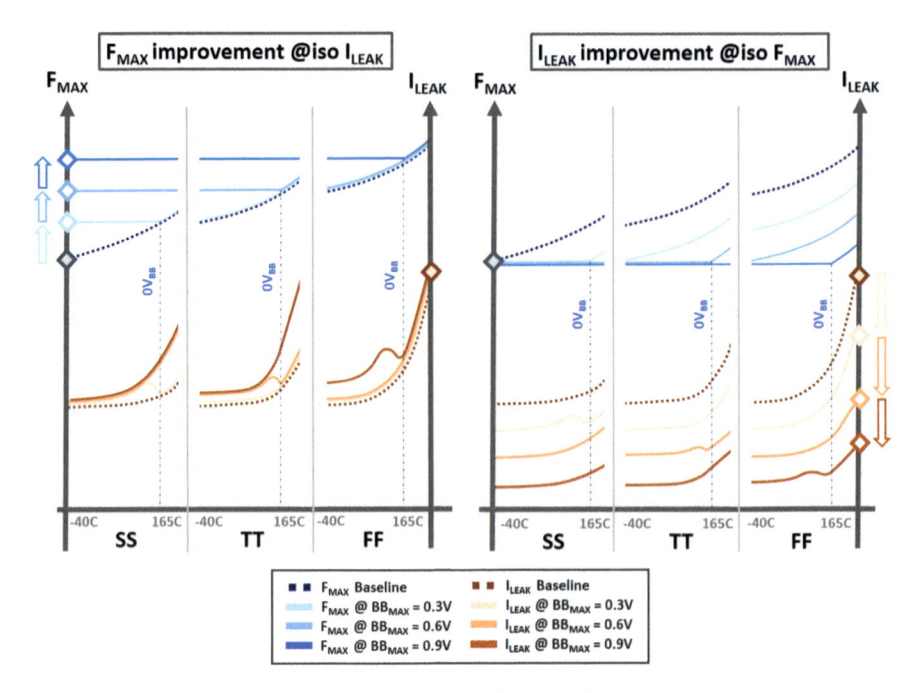

Fig. 3.14 F_{MAX} and I_{LEAK} improvement strategies in case of process and temperature compensation. The zero bias vertical line ($0V_{BB}$) is the limit where bias is not needed anymore to match F_{MAX} target (i.e., without bias). The leakage curves of baseline and compensated ones touch at $0V_{BB}$ and compensated leakages curves start with higher leakage at minimum temperature

> Non-biased parts' or non-synthesized block's leakage will set a leakage ceiling that low voltage synthesis with Body-Bias cannot improve. Have special attention to memory leakage.

The consequence leakage frequency trade-off is illustrated in Fig. 3.14, where we depict opposing strategies: an F_{MAX} target enlargement (left), or an I_{LEAK} improvement at iso-F_{MAX} (right), both brought by an increase on the maximum Body-Bias applied to slowest parts (tBB_{MAX}).

When chasing a speed enhancement (left figure), an increase on BB_{MAX} allows the designer to target a higher F_{MAX} specification (depicted as a vertical shift on the blue marks at the F_{MAX} axis). As described throughout this chapter, the body-bias voltage will then be modulated, achieving an iso-delay profile across P and T until the zero bias condition is reached ($0V_{BB}$). However, the increase on the BB_{MAX} can cause some residual V_{BB} to remain at maximum temperature/fast parts to sustain the iso-delay requirement, ultimately leading to a situation where the slow biased corner's leakage exceeds the worst-case I_{LEAK} ceiling (highlighted with an orange mark on the I_{LEAK} axis).

3 Body-Bias for Digital Designs

On the other hand, the circuit architect may decide to not increase the F_{MAX} constraint while increasing BB_{MAX}, as represented on the right side of Fig. 3.14. In this case, the cell's downsizing effect caused by the higher I_{on} with Body-Bias enables a reduction on the worst-case leakage specifications, depicted as a shift on the orange-coloured mark on the I_{LEAK} axis. This leakage gain with body-bias is, however, limited by two aspects. On the one hand, the leakage contribution of the non-synthesized or non-bias parts cannot be reduced with bias. This imposes a limit on the maximum achievable leakage reduction, saturating the gain with Forward Body-Bias similarly as what happens with timing (refer to Sect. 3.3.3). On the other hand, the leakage gain induced by the downsizing effect is limited by the minimum $\frac{W}{L}$ cells in the technology offer. Once a cell has reached the minimum $\frac{W}{L}$ ratio, more Forward Body-Bias at synthesis will induce a leakage increase as the cell cannot be further downsized.

> Residual BB needed at maximum temperature may lead the slow compensated parts to reach maximum leakage rating. Consider voltage modulation of Body-Bias to contain leakage at maximum temperature.

3.3.2 Thermal Bound

The potential gain offered by Body-Bias is bounded by iso-delay power budget or by thermal runaway limit. The thermal runaway is the practical consequence of transistor leakage increase: as leakage has a positive temperature coefficient, heat generated by the extra leakage of Body-Bias can cause a positive feedback situation where more leakage causes more heat, etc. until the system is powered off or damaged. The workaround is to use a package with a thermal resistance from device junction to ambient $R_{\Theta jA}$ which enables to dissipate a larger power than the power dissipated by circuit operation with Body-Bias at max voltage and temperature.

This phenomenon is illustrated by the graphs of Fig. 3.15, which display the package dissipated power in three cases of thermal resistance and an example circuit's total power evolution with temperature.

Assuming V_T drops of few mV/°C and recalling that I_{off} follows Eq. (3.4):

$$I_{\text{off}} = I_{TH} \cdot \exp\left(\frac{-V_T}{S_{VT}} \cdot \ln 10\right) \quad (3.4)$$

with S_{VT} the subthreshold slope in millivolt per decade denoting the amount of gate voltage drop needed to divide off current by a factor 10 and I_{TH} denoting a fixed current factor.

On the circuit side, the total power is the sum of both static and dynamic power, for a given activity P_{Dynamic} will remain constant, setting a fixed offset to total power evolution with temperature.

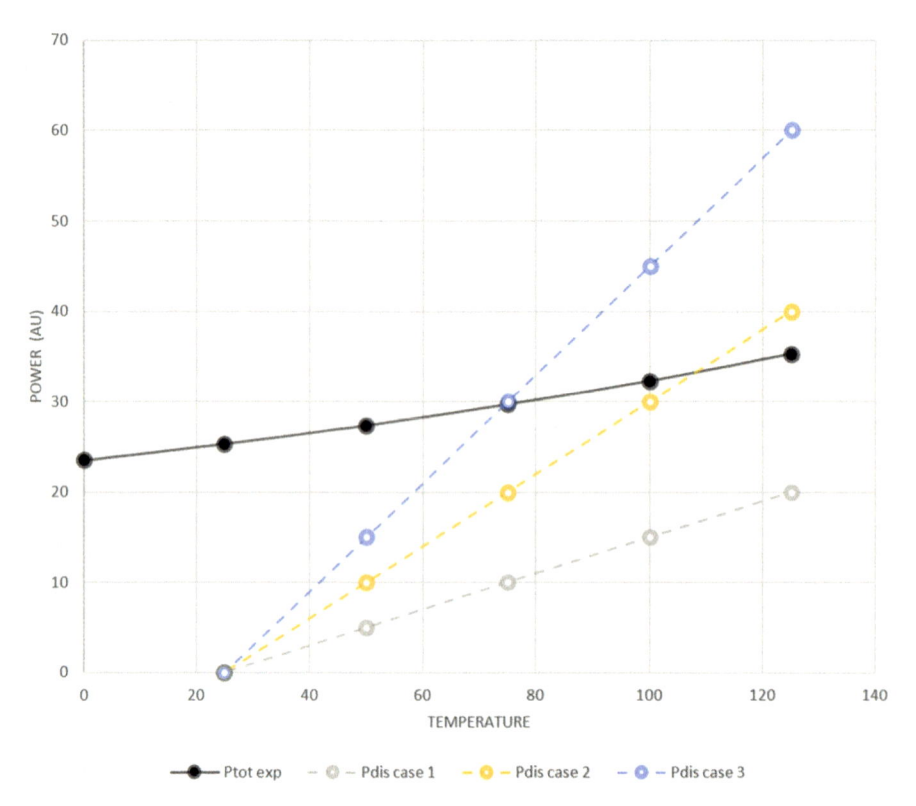

Fig. 3.15 Total (black solid line) and dissipated power (coloured dashed lines of three package thermal resistance cases), trend graphs as a function of junction temperature. When circuit total power does not intersect dissipation capability of the system, it goes in thermal runaway. The ambient temperature is set at 25 °C

$$P_{\text{Total}} = P_{\text{Dynamic}} + P_{\text{Static}} \tag{3.5}$$

$$P_{\text{Total}} = P_{\text{Dynamic}} + k \cdot \exp(-V_T) \tag{3.6}$$

On the package side, the junction temperature depends on the package thermal resistance times the power of the circuit, rewriting this power as $P_{\text{Dissipation}}$ to differentiate it from circuit total power, the package dissipation capability is given by Eq. (3.8).

$$T_j = T_A + R_{\Theta jA} \cdot P_{\text{Total}} \tag{3.7}$$

$$P_{\text{Dissipation}} = \frac{T_j - T_A}{R_{\Theta jA}} \tag{3.8}$$

The circuit ability to maintain steady temperature is a race between the system's dissipation power capability (Eq. (3.8)) and total power (Eq. (3.6)). When $P_{\text{Dissipation}}$

3 Body-Bias for Digital Designs

line intersects the system total power curve the system is stable with temperature (the case of the 2 upper dashed curves), when it does not, your system goes into thermal runaway (case of the lower grey coloured dashed line).

> As an architect, you should verify that your system is maintained into leakage/power/thermal specification bounds when applying Body-Bias.

3.3.3 The Unbiased Parts Timing Ceiling

Non-biased islands, like usually SRAM arrays,[11] are not affected by bias, this can lead to bias speedup saturation effect as displayed in Table 3.3 below where the frequency speedup on a CPU with non-biased SRAM culminates at 0.6 V. As will be seen in comparison with AVS technique section, this SRAM limitation effect is specific to ABB, AVS indeed accelerates SRAM arrays as well as periphery. Beyond this case of SRAM arrays, the reader should retain that non-modulated islands either in source voltage (AVS) or body voltage (ABB) limit the effect of modulation if it is not worked around at architecture level like introducing some memory wait-states or transfer buffering.

> Bias response saturation of non-biased island like SRAM array or non-biased scenario will set your Body-Bias timing ceiling.

3.3.4 Biasing the Other Way

What if the user wanted to apply a limited amount of Reverse Body-Bias in LVT/flip-Well domain?[12] The idea here would be to lower leakage of fast parts,

Table 3.3 28 nm FD-SOI speedup factor of biased logic vs. CPU with biased logic and non-biased SRAM bitcell/array at 25 °C, Silicon data

Forward Body-Bias	0 V	0.3 V	0.6 V	0.9 V	1.2 V
Biased logic reference	1	1.3	1.7	2.1	2.5
Non-biased bitcell CPU	1	1.3	1.4	1.45	1.5

---

[11] In situations of high temp skewed corners, SRAM bitcell has their stability weakened by V_T shift induced by Forward Body-Bias, see Chap. 5.
[12] In FD-SOI V_T modulation can be done by transistors WELLs, LVT transistors have their NMOS in NWELL and PMOS is PWELL. Refer to Chap. 2.

Table 3.4 qFO4 LVT flipped WELL devices (i.e., Forward Body-Biasable) leakage Spice simulation variation with 100 mV or 200 mV Reverse Body-Bias compared to Zero Body-Bias reference at 1.00 V 125 °C

Variation (%)	100 mV RBB	200 mV RBB
HLVTPB10	−12	+766
HLVTPB4	−13	+204
LVTPB4	−14	+64
LVTPB10	−13	+7

this time it would be `direct` leakage gain. Table 3.4 shows the leakage penalty in that case, the leakage evolution between 100 and 200 mV of Reverse Body-Bias is a tremendous 766% in worst case, putting a high precision constraint on bias voltage for timid gains (12–14%), the gain is clearly not worth the pain here.

> Biasing the other way (i.e., RBB for flipped WELL devices) gives a benefit lower than bias voltage generation cost and direct current of WELL diodes. Do not bias the other way.

3.4 Digital Performance Boost

Beside process and environmental compensation, the Body-Bias induced lower V_T can be traded off to boost your circuit performance provided that it remains within specifications (leakage, total power, heatsink budget or sink capability) with the extra leakage, as reviewed in the previous sections. Body-Bias performance boost can either be used to:

- implement On-demand boost for specific circuit mode, triggered by Operating System code, which we summarized in the chapter's motto as 'software defined V_T'. As for analog applications covered in Chap. 4 coming next, this performance modulation does not interact with signal path, no extra series transistor like in LDOs or DC-DCs is needed to enable this feature.
- trade-off a fraction of leakage gains induced by $\frac{W}{L}$ downsizing for extra speed
- push your circuit's speed beyond source voltage reliability limit.

A leakage-speed trade-off is illustrated in Table 3.5 where two implementations of a CPU were done at two Forward Body-Bias voltage, 600 and 900 mV, at 900 mV the clock frequency can either improve by 25% at iso-leakage `or else`[13] the leakage can be lowered by 40% at iso-frequency.

---

[13] Emphasized for the reader to understand that there is a trade-off to make here.

3 Body-Bias for Digital Designs

Table 3.5 Two 28 nm FD-SOI CPU implementation at 600 and 900 mV Forward Body-Bias

Forward Body-Bias	0.6 V	0.9 V freq	0.9 V leakage
Area	Baseline	−1.5%	−1.5%
Leakage	Baseline	Baseline	−40%
Frequency	Baseline	+25%	Baseline

Above V_{Tinv} and below reliability maximum voltage (VMAX), AVS gives more MegaHertz per Watt if your systems targets pure performance. Body-Bias boost regains optimality in case you seek multiple targets, like energy efficiency added to performance boost, continuum of operating point or ease of embedded voltage generation, this will be detailed in Sect. 14.1. To push performance further than VMAX Body-Bias can be used as a performance extra lever, similarly, some supply voltage stress relief can be traded off against increased Body-Bias to gain life time.

3.5 Ultra-Low Voltage Designs

The ever longing search for low power has led to extensive design activity in the ultra-low voltage domain in recent years. However the low supply voltage causes the circuits designed at ultra-low voltage to face several challenges:

- very low performance, even at ambient temperature, further...
- I_{on} is adversely impacted at low temperature dragging the performance even lower
- variability of device depends exponentially on gate overdrive voltage, $V_G = (V_{GS} - V_T)$

Forward Body-Bias can advantageously compensate the three above items via the natural increase of gate overdrive voltage it provides, this comes at no dynamic power increase (see [4]):

- performance boost provides quick gain as Ultra-low voltage devices are in a regime where a small increase in overdrive voltage provide a huge performance gain
- temperature compensation at cold is natural as the device is supplied well below V_{Tinv}
- variability is lowered exponentially with gate overdrive voltage headroom gained by Forward Body-Bias

Last, asymmetric transistor I_{on} induced by low voltage can be compensated selectively.

Further refer to Chaps. 9, 12 and 13 with design examples illustrating the ultra-low voltage ease of use Body-Bias can bring. In contrast to the previous statement of high performance design with supply voltage well above V_{Tinv} from Sect. 3.4, the design reported in [5] is an illustration of ultra-low voltage operation enablement in the context of wide operation range. Similarly, the design detailed in Sect. 13.2 demonstrates ultra-low voltage designs Forward Body-Bias sweet spot in a spectacular manner (see Fig. 12.13).

3.6 Body-Bias for Digital Designs: Conclusion

Leaving apart the ultra-low voltage situation where bias speedup is exponential, at super threshold, Forward Body-Bias delivers a performance increase in the range of 10–100% depending on device V_T, supply voltage, bias range. This device acceleration can either be used:

- to speedup circuits as an alternative to usual AVS
- to gain leakage induced by synthesis downsize from faster devices
- to gain dynamic power from lower operating voltage
- to reclaim yield from faster frequency or lower variability at low voltage

A trade-off between all of these aspects in the hand of circuit architect.

The Body-Bias tuning range is limited either by technology isolation ratings between WELLs or metal lines, not by diode breakdown at the usual bias voltage. Another limit comes from the amount of Body-Bias which is acceptable at maximum temperature not to degrade circuit total power specification from increased leakage. This latter limitation can be worked around by modulating bias as a function of temperature if device is supplied below V_{Tinv} or else as a function of supply voltage.

Compensation of device fabrication and environmental conditions is another way to utilize the Body-Bias acceleration: temperature, voltage, process, and ageing can all be compensated with Body-Bias at no reliability cost. Provided that adequate parameter monitoring is made available, bias modulation as a function of the parameter to compensate can be made linear or polynomial at very limited hardware cost, in the order of few k-gates (Chap. 13). The choice between linear and polynomial bias regulation engines and the amount of parameters to compensate for, depends on precision specifications and on-chip hardware budget.

Last, Body-Bias effect is magnified for circuits supplied close to V_T, at ultra-low voltage, the increase of gate overdrive voltage brought by bias gives exponential improvement in frequency, low temperature behaviour, and local variability without the temperature induced leakage limitation.

The Part II of this book will illustrate some design examples of general aspects studied in this section, then in the Part III Body-Bias deployment will be covered detailing the embedded bias voltage generation, bias control solutions, arbitration between AVS and ABB, and digital implementation aspects.

References

1. F. Abouzeid, S. Clerc, F. Firmin, M. Renaudin, G. Sicard, A 45nm CMOS 0.35V-optimized standard cell library for ultra-low power applications, in *International Symposium On Low Power Electronics and Design* (2009)
2. F. Abouzeid, C. Bernicot, S. Clerc, J.M. Daveau, G. Gasiot, D. Noblet, D. Soussan, P. Roche, 30% static power improvement on ARM Cortex®-A53 using static Biasing-Anticipation, in *2016 46th European Solid-State Device Research Conference (ESSDERC)* (2016), pp. 29–32
3. S. Clerc, M. Saligane, F. Abouzeid, M. Cochet, J.M. Daveau, C. Bottoni, D. Bol, J. DeVos, D. Zamora, B. Coeffic, D. Soussan, D. Croain, M. Naceur, P. Schamberger, P. Roche, D. Sylvester, 8.4 A 0.33V/−40 °C process/temperature closed-loop compensation SoC embedding all-digital clock multiplier and DC-DC converter exploiting FDSOI 28nm back-gate biasing, in *2015 IEEE International Solid-State Circuits Conference-(ISSCC) Digest of Technical Papers* (2015), pp. 1–3
4. R.I.M.P. Meijer, Body bias aware digital design, PhD manuscript, Eindhoven University of Technology, 2011
5. D. Jacquet, F. Hasbani, P. Flatresse, R. Wilson, F. Arnaud, G. Cesana, T.D. Gilio, C. Lecocq, T. Roy, A. Chhabra, C. Grover, O. Minez, J. Uginet, G. Durieu, C. Adobati, D. Casalotto, F. Nyer, P. Menut, A. Cathelin, I. Vongsavady, P. Magarshack, A 3 GHz dual core processor ARM cortex TM-A9 in 28 nm UTBB FD-SOI CMOS with ultra-wide voltage range and energy efficiency optimization. IEEE J. Solid-State Circuits **49**(4), 812–826 (2014)

Chapter 4
Body-Biasing in FD-SOI for Analog, RF, and Millimeter-Wave Designs

Andreia Cathelin

4.1 Introduction

Many unique advantages for designing in FD-SOI come from the fact that a very large V_T tuning range is available for all types of transistors in this technology. The body tie voltage can now be considered as a very efficient tuning knob, permitting to span any transistor's threshold voltage over more than 200 mV variation. Figure 4.1 presents such numbers, for both the regular-VT (RVT) and low-VT (LVT) transistors (N- and P-MOS) in the 28 nm FD-SOI CMOS technology from STMicroelectronics.

As a comparison, Figs. 4.1 and 4.2 present the numbers for LVT NMOS transistors in 28 nm CMOS technologies, bulk and FD-SOI (same fab). In bulk the body-bias tuning range is few tens of mV, whereas in FD-SOI is over 250 mV.

Any designer seeing such numbers immediately thinks of new opportunities opened by significant transistor's parameters performing such big ranges of variation. There are several blocks inside an electronics system that can be observed under this new perspective: either system level blocks in a transceiver's signal path or tuning/trimming elements. For the blocks in the signal path, it can be useful to obtain, for example, variable gain amplifiers, phase shifters, or variable delay elements by varying transistors' V_T. For the tuning or trimming schemes, here comes the opportunity of a new tuning knob in the landscape.

Figure 4.3 presents a cross-section of LVT devices in a 28 nm FD-SOI technology. For such devices, we apply what is called forward body-biasing (FBB) as follows: VBBN from 0 (regular state) up to possible 3 V for the NMOS transistors, and, respectively, VBBP from 0 (regular state) down to possible −3 V for the PMOS

A. Cathelin (✉)
STMicroelectronics, Crolles, France
e-mail: andreia.cathelin@st.com

© Springer Nature Switzerland AG 2020
S. Clerc et al. (eds.), *The Fourth Terminal*, Integrated Circuits and Systems,
https://doi.org/10.1007/978-3-030-39496-7_4

Fig. 4.1 Threshold voltage (V_T) variation for RVT and LVT transistors, N- and P-MOS in 28 nm FD-SOI technology, with respect to the respective applied body-bias voltage (VB). © 2017 IEEE. Reprinted with permission

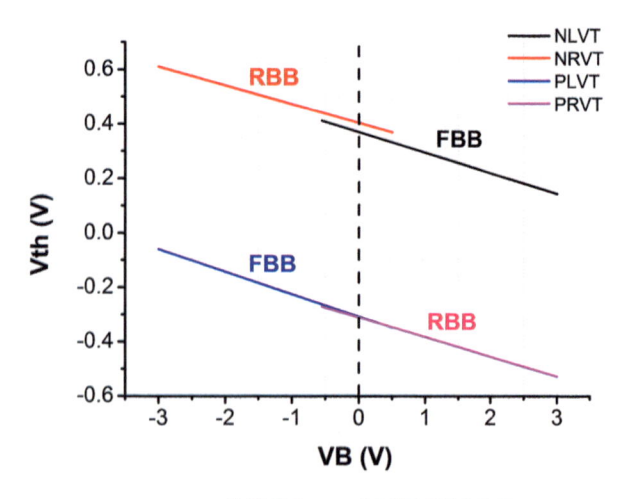

Fig. 4.2 Comparison of V_T variation through body-biasing effect for similar size transistors in 28 nm FD-SOI (blue curve) and bulk (red curve) from STMicroelectronics

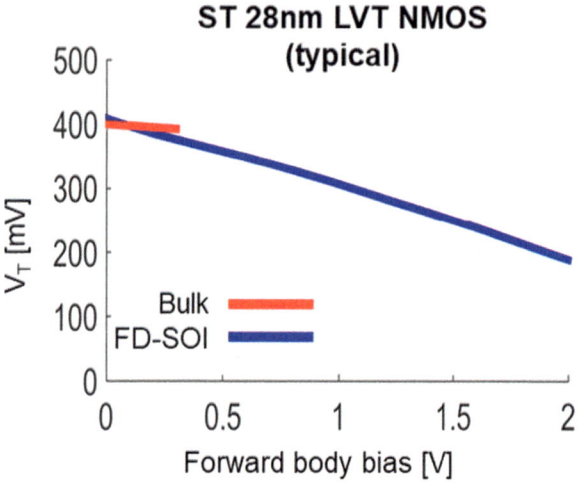

transistors [1]. One can observe as well from the cross-section sketch that given that the source and drain zones are raised and lay atop the thin buried oxide (BOX) and the conduction film, there are no parasitic diodes from there that would limit the body-biasing region, like was the case in bulk technology.

Another observation may be drawn from Fig. 4.3. In the case of tuning or trimming loops in an integrated system, there is an electrical value sensing (current, voltage, or power) generally somewhere on the signal path, which is then compared with a reference signal (most of the time external), then a command is processed on chip and the result of that command is the loop feedback signal. That loop feedback signal is, in regular CMOS implementations, a current or a voltage which is fed back somewhere else in the signal path to close the regulation loop. That signal, in bulk CMOS, is doing its desired action, but also comes in with parasitics, such as parasitic capacitors or resistances, or some non-linear elements. In the case of FD-SOI, the

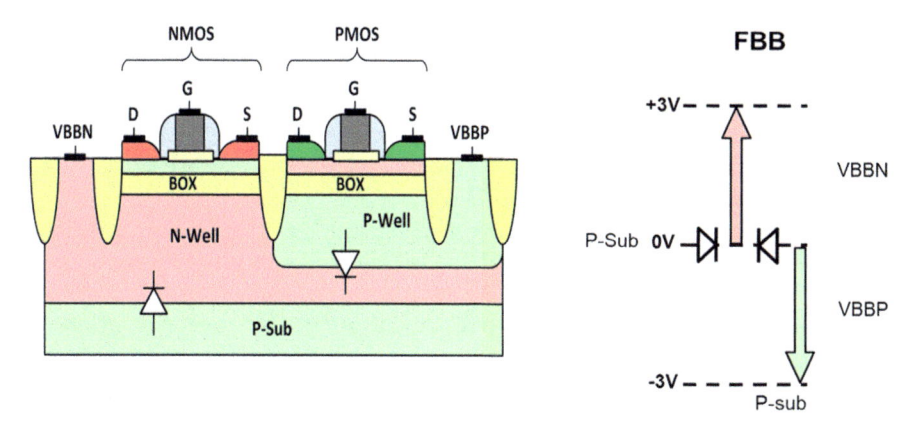

Fig. 4.3 Cross-section of LVT transistors in 28 nm FD-SOI and body-biasing opportunity through VBBN and VBBP voltages

loop feedback will be produced generally through a (pair of) body-bias voltage(s) (generally acting on N- and P-MOS), which arrive on transistors themselves on the signal path, but these signals are applied underneath the isolating BOX. In the case of FD-SOI these feedback signals do not bring in any more parasitic signals on the signal path, hence resulting in easier to control tuning or trimming schemes. And finally, tuning the body-biasing zones drains negligible static current, as these are isolated zones.

The following two sections will provide a discussion on two different methods of using body-biasing in analog, RF, and millimeter-wave designs. Generally, a circuit (bloc, or larger system on chip) will need body-biasing on both NMOS and PMOS families of transistors. Nevertheless, for expression simplification, we will call "body-biasing" or "body-biasing voltage" a group of body-biases, for N- and, respectively, P-MOS transistors (VBBN and VBBP).

4.2 On the Usage of Variable Body-Biasing Voltage on Chip

A first category of electrical schematics will use a variable body-biasing voltage over time and process, voltage, temperature, and aging (PVTA).

The very first usage will be in order to cancel the PVTA variations of system level parameters over time. It is of common knowledge that such system level parameters (like a filter cut-off frequency, receiver chain linearity, local oscillator phase noise, and so on) are permitted to vary only with $\pm 5\%$ inside a system. Whereas the localized devices (active and passives) electrical parameters show variations of almost $\pm 50\%$. By controlling V_T's of transistors placed in the signal path through a body-biasing regulation loop, one can obtain this very fine tuning behavior. Several examples can be cited here: we can regulate analog Gm-C filter cut-off frequency and linearity like in [2] through simultaneous N- and P-MOS body-biasing control. We can control as well second-order input intercept point (IIP2) linearity in RF front-ends based on LNTA and passive down-conversion mixers designs through passive mixers body-biasing control [3]. We can as well act on oscillators phase noise, like in [4] detailed in Chap. 7, where 200 GHz distributed oscillators are controlled through body-biasing.

Another usage of body-biasing is for reconfiguration of circuit, bloc, or system upon the application operation mode. In power amplifiers as a bloc level example, one can gradually change the power gain and power operation class by gradually acting on the body-biasing of the power generating transistors (for example, differential pairs), see [5] and [6]. Massively digital RF transmitter chains (made of chains of inverters) can be tuned also in terms of emission frequency mask via body-biasing [7], detailed in Chap. 9. As well, any other kind of linear blocs such as variable gain amplifiers, phase shifters can be very fine tuned by using transistors' body-biasing.

And finally, many other design schemes can be proposed specifically for FD-SOI and the usage of body-biasing. As an example, [8] detailed in Chap. 6, presents a novel tunable delay element for high data rate receiver chains, where course and fine tuning can be obtained, respectively, through gate and body-biasing of the same pair of transistors.

Another example shows that basic building blocks can be drastically simplified when working in an FD-SOI technology. In [9], the classical class AB amplifier is elegantly revisited thanks to body-biased devices, as presented in Fig. 4.4. The newly proposed class AB amplifier, which is now only a two transistors CMOS inverter, can have all its system level parameters (like Gm, noise, linearity) controlled through all PVTA parameters variation through the two control voltages VBN and VBP. And of course, the variability of a two transistors scheme is way improved with respect to the one composed of a dozen transistors.

For all these examples, the full circuit integration has to contain also a new type of power management bloc on chip commonly called body-bias generator (BBGEN).

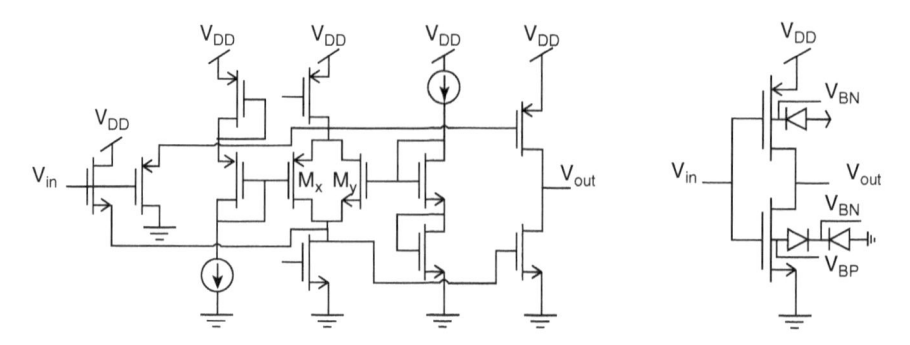

Fig. 4.4 Class AB amplifier. Left: typical schematic in bulk, and right: FD-SOI specific schematic. © 2017 IEEE. Reprinted with permission

4 Body-Biasing in FD-SOI for Analog, RF, and Millimeter-Wave Designs 89

This kind of BBGENs for an analog usage is much different in specifications with respect to those used for digital circuits compensation (see Chap. 15). Whereas for digital service BBGENs, the major features are the big number of transistors drive and the time agility, the analog service BBGENs have to perform a limited voltage ripple and to show a careful injected noise spectrum, while the number of served transistors can be much lower (tens to maybe hundreds, but not more).

4.3 On the Usage of Fixed Body-Bias Voltage on Chip

This second category of body-biased circuits will use fixed voltages (different from the standard operation ones) which are generated on chip, or may be trimmed by software from time to time.

The big category to be cited here is the data converters one, or more generally any kind of switched-capacitors designs. For data converters designers, one of the big features to care of is the quality of the sampling switches, made of CMOS pass-gates.

Figure 4.5 presents a numerical example for switches used in a multi-bit data converter. The quality of the CMOS switches is materialized by the value and variation of the switch resistance with respect to the voltage across its terminals. In the numerical example presented here [10], one can see that the switch resistance curve in red (28 nm CMOS bulk) is much flattened when getting to FD-SOI (green), and becomes almost flat in the case of FD-SOI with a fixed body-bias of ± 1.8 V (black curves left and right). The switch resistance variation improves by a factor of 40 in body-biased FD-SOI as compared to the standard bulk CMOS process, permitting to get smaller switches with smaller parasitic capacitances,

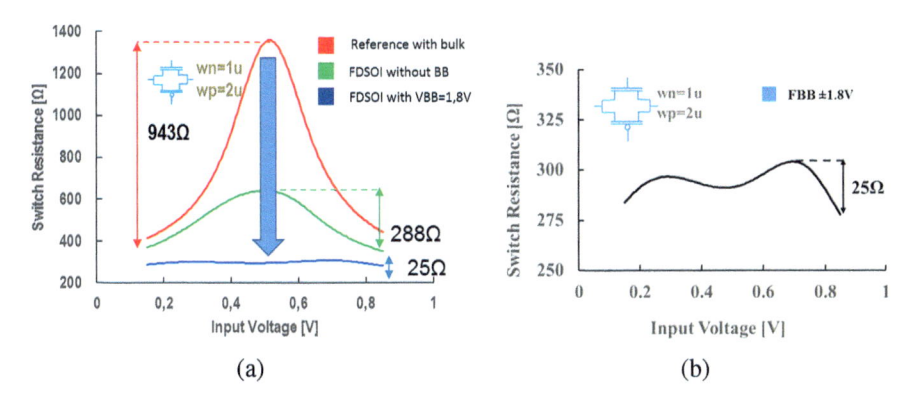

Fig. 4.5 Switch resistance behavior under body-biasing conditions, for data converters. (**a**) Comparison bulk vs FD-SOI and FD-SOI with body-biasing operation; (**b**) zoom in on the switch resistance for the FD-SOI with body basing case. All transistors have minimum technology length (Ldrawn 30 nm). © 2017 IEEE. Reprinted with permission

hence more compact layout, and then even better switches performances. In terms of system level performances, this leads either to unprecedented very high speed Gb/s data converters [11] with excellent energy efficiency or to multi-bit mid sampling rate data converters that can achieve excellent linearity without the need of any calibration schemes [10].

For the data converters designs, a new era has come when the FD-SOI has been introduced. Indeed, when checking the on-line survey[1] from professor B. Murmann [12], one can observe that each and every new FD-SOI data converter is pushing away the limits of the existing state-of-the-art perimeter, thanks to this outperforming quality of the CMOS switches (Fig. 4.6).

Another usage of fixed body-biasing serves for non-overlapping clock generation in massively digital RF Receivers [13]. Indeed, system level linearity can be well controlled and improved by body-biasing the chain of inverters generating the 180° reverted clocks needed in modern IQ architectures. This kind of body-biasing does not need on-line tuning, maybe in some cases a power up set-up or a slow temperature dependent adjustment.

Finally, another usage of fixed body-biasing is to enable operation at ultra-low voltages (for example, 0.5 V) and in the same time to increase the circuit speed, when referring to digital and mixed signal parts of a SoC [14].

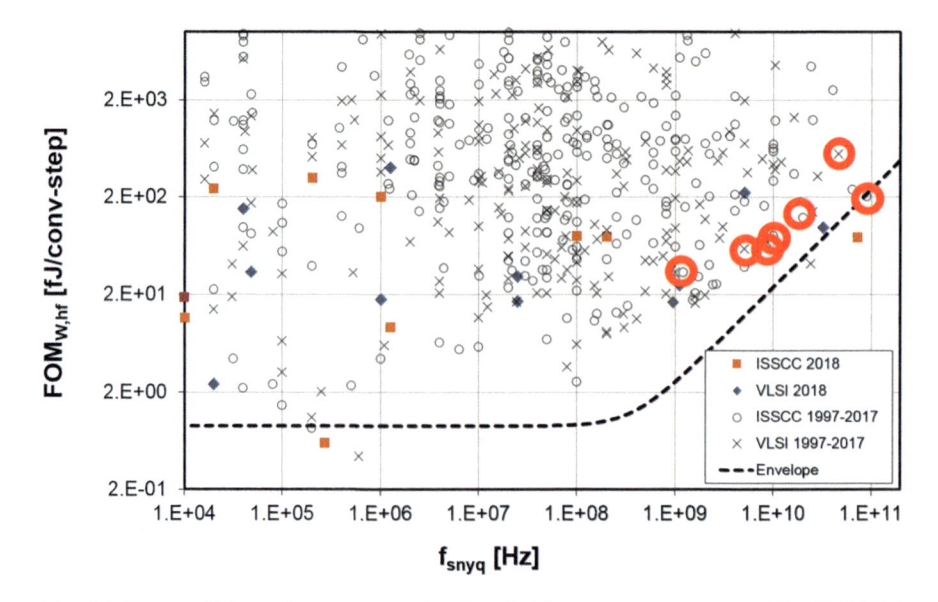

Fig. 4.6 Energy efficiency data converters data from B. Murmann survey, upon revision 20180716 [12]. The FD-SOI implementations are highlighted in red

---

[1]How high reputation among analog designers.

The power management unit needed in these cases for providing the fixed (or trimmable) value of the body-bias voltages is much easier to obtain. For the positive voltage, generally the thick oxide supply voltage already available on the SoC is used (see in the previous example 1.8 V). For the negative voltage (like -1.8 V in the previous example), a negative charge pump will be integrated on chip to this extent.

Chapters 6–10 will largely depict different types of integration, from building blocs to circuits and SoC's. They largely employ all these kinds of body-biasing techniques for analog, mixed signal, RF, and up to mmW types of circuits.

References

1. A. Cathelin, Fully depleted silicon on insulator devices CMOS: the 28-nm node is the perfect technology for analog, RF, mmW, and mixed-signal system-on-chip integration. IEEE Solid State Circuits Mag. **9**(4), 18–26 (2017)
2. J. Lechevallier, R. Struiksma, H. Sherry, A. Cathelin, E. Klumperink, B. Nauta, A forward-body-bias tuned 450 MHz Gm-C 3rd-order low-pass filter in 28 nm UTBB FD-SOI with >1dBVp IIP3 over a 0.7-to-1 V supply, in *2015 IEEE International Solid-State Circuits Conference (ISSCC) Digest of Technical Papers* (2015), pp. 1–3
3. D. Danilovic, V. Milovanovic, A. Cathelin, A. Vladimirescu, B. Nikolic, Low-power inductorless RF receiver front-end with IIP2 calibration through body bias control in 28 nm UTBB FDSOI, in *2016 IEEE Radio Frequency Integrated Circuits Symposium (RFIC)* (2016), pp. 87–90
4. R. Guillaume, F. Rivet, A. Cathelin, Y. Deval, Energy efficient distributed-oscillators at 134 and 202 GHz with phase-noise optimization through body-bias control in 28nm CMOS FDSOI technology, in *RFIC 2017* (2017)
5. A. Larie, E. Kerhervé, B. Martineau, L. Vogt, D. Belot, A 60GHz 28nm UTBB FD-SOI CMOS reconfigurable power amplifier with 21% PAE, 18.2 dBm P1dB and 74 mW PDC, in *2015 IEEE International Solid-State Circuits Conference (ISSCC) Digest of Technical Papers* (2015), pp. 1–3
6. F. Torres, M. De Matos, A. Cathelin, E. Kerhervé, A 31 GHz 2-stages reconfigurable balanced power amplifier with 32.6 dB power gain, 25.5% PAEmax and 17.9 dBm Psat in 28 nm FD-SOI CMOS, in *RFIC 2018*, Philadelphia (2018)
7. G. de Streel, F. Stas, T. Gurné, F. Durant, C. Frenkel, A. Cathelin, D. Bol, SleepTalker: a ULV 802.15.4a IR-UWB transmitter SoC in 28-nm FDSOI achieving 14 pJ/b at 27 Mb/s with channel selection based on adaptive FBB and digitally programmable pulse shaping. IEEE J. Solid State Circuits **52**(4), 1163–1177 (2017)
8. I. Sourikopoulos, A. Frappé, A. Cathelin, L. Clavier, A. Kaiser, A digital delay line with coarse/fine tuning through gate/body biasing in 28 nm FDSOI, in *ESSCIRC Conference 2016: 42nd European Solid-State Circuits Conference* (2016), pp. 145–148
9. M. Videnovic-Misic, P. Cathelin, A. Cathelin, B. Nikolic, Class AB base-band amplifier design with body biasing in 28 nm UTBB FD-SOI CMOS, in *IEEE S3S Conference*, San Francisco, October (2017)
10. A. Kumar, C. Debnath, P.N. Singh, V. Bhatia, S. Chaudhary, V. Jain, S. Le Tual, R. Malik, A 0.065 mm^2 19.8 mW single channel calibration-free 12b 600 MS/s ADC in 28 nm UTBB FDSOI using FBB, in *ESSCIRC Conference 2016: 42nd European Solid-State Circuits Conference* (2016), pp. 165–168

11. S. Le Tual, P.N. Singh, C. Curis, P. Dautriche, A 20 GHz-BW 6b 10 GS/s 32 mW time-interleaved SAR ADC with Master T&H in 28 nm UTBB FDSOI technology, in *2014 IEEE International Solid-State Circuits Conference Digest of Technical Papers (ISSCC)* (2014), pp. 382–383
12. ADC performance survey 1997–2018. In: ISSCC & VLSI Symposium [online]. https://web.stanford.edu/~murmann/adcsurvey.html. Accessed Jan 2019
13. R. Kasri, E. Klumperink, P. Cathelin, E. Tournier, B. Nauta, A digital sine-weighted switched-Gm mixer for single-clock power-scalable parallel receiver, in *CICC 2017*, Austin, April (2017)
14. L. Fanori, A. Mahmoud, T. Mattsson, P. Caputa, S. Rämö, P. Andreani, A 2.8-to-5.8 GHz harmonic VCO in a 28 nm UTBB FD-SOI CMOS process, in *2015 IEEE Radio Frequency Integrated Circuits Symposium (RFIC)* (2015), pp. 195–198

Chapter 5
SRAM Bitcell Functionality Under Body-Bias

Lorenzo Ciampolini

5.1 Silicon Product Yield in a Highly Competitive Market

As discussed all along this book, the device fourth terminal, or the body terminal, is a powerful mean to alter at run time the behavior of a circuit manufactured in FD-SOI technology. When dealing with SRAM, however, there is a diffused sense of panic with regard to the "behavior alteration" that could be obtained, and every discussion with the process engineers invariably ended up with the question: what will be the effect on yield? Let us then attack directly this chapter from this *hot* subject.

The yield of a silicon product is equal to the total number of working dies, over all manufactured dies. Yield influences the cost of the silicon product in different ways.

1. The manufacturing cost of a working die is influenced directly by yield. If your yield is 25%, then your silicon products cost four times higher than if your yield is 99%, and this is a very important matter if you are selling a low-price product. But a 99% yield design gives you practically the same die cost with respect to a 1 ppm yield-loss design (1 ppm yield loss = 999,999 dies working over one million dies produced).

2. If you are designing expensive microprocessor units, you might be interested do a secondary die screening based on their failing operating frequency; in this way, you could sell the fastest devices at a price higher than those working only at lower frequencies by calling them with different fancy names like i7-8700 or i7-8700K [1]. Even if the commercial products cited might not be directly obtained through such an electrical die sorting of the same design, their prices give a

---

L. Ciampolini (✉)
STMicroelectronics, Crolles, France
e-mail: lorenzo.ciampolini@st.com

© Springer Nature Switzerland AG 2020
S. Clerc et al. (eds.), *The Fourth Terminal*, Integrated Circuits and Systems,
https://doi.org/10.1007/978-3-030-39496-7_5

reasonable idea of how cheaper a slower die will be negotiated in nowadays, highly competitive silicon market. For expensive silicon products, the die cost might be irrelevant, but how the yield drops as a function of the maximum clock frequency will determine how much cash the foundry will be able to obtain from the market.
3. One should also consider the total chain of effects caused by a failure occurring after the silicon product has been integrated into a third-party product and has been shipped to the consumer. For example, the manufacturing cost of a silicon product implementing an audio processor nearly does not change between a 99% yield and a 1 ppm yield loss. Performances of such a system are also not critical, since audio processing implies output frequencies below the hundredth of KHz. Nevertheless, an unexpected memory failure occurring regularly during a summer vacation every time that the car is well-heated might lead to a replacement request at the car-maker expenses. Even neglecting any legal expenses that the car-maker could face in such a case, it makes an enormous difference between the case in which the car-maker has to replace only one audio board over 1,000,000 sold cars (1 ppm yield) or to replace an audio board every 100 sold cars (99% yield). The need to recall all of the sold cars for checks would be a real nightmare scenario, which could occur in case the failure affects circuits tied to security of human beings.

In conclusion, yield directly or indirectly influences the cost of the silicon products, and thus affects in a very important way the revenues and ultimately the gross margin of the manufacturer, which is fundamental to survive in a highly competitive market like the semiconductor market.

As we will see in the following text, yield varies along the supported operating conditions, that is, at the technology nominal voltage $V_{DD} \pm 5\%$ (or maybe $\pm 10\%$) and over a given temperature range. Across all these operating conditions, there will be a worst-case operating condition where yield is minimum. Designing with typical devices, i.e., using MOS devices that stick perfectly to the typical model characteristics, only provides a go/no go information about the functionality of the intellectual property (IP) over the supported conditions. As devices become smaller, however, they are more and more subject to uncontrollable manufacturing variability that scales up inversely with the device sizes. Small real-world devices thus follow a "distribution" centered around the typical model characteristics, and the functionality of a given IP might then fail for some *particularly bad* combination of variations of the devices composing it. In general, one can evaluate at design time by means of the Monte Carlo (MC) method (refer to Sect. 5.7) the worst-case IP failure rate F. For any IP instantiated a single time in a silicon product, and if the IP under consideration is the only IP that can fail within the silicon product, then the silicon product yield loss coincides with the worst-case IP failure rate F.

If an IP failing at a rate F is instantiated N times in the same silicon product, simple probability theory allows to calculate the expected chip yield. For any kinds of IPs, if the number N of samples of the same instance increases, then the chip failure rate F_N increases, since the N different instances, which are affected by variability, will represent N different and independent opportunities to fail. Whether

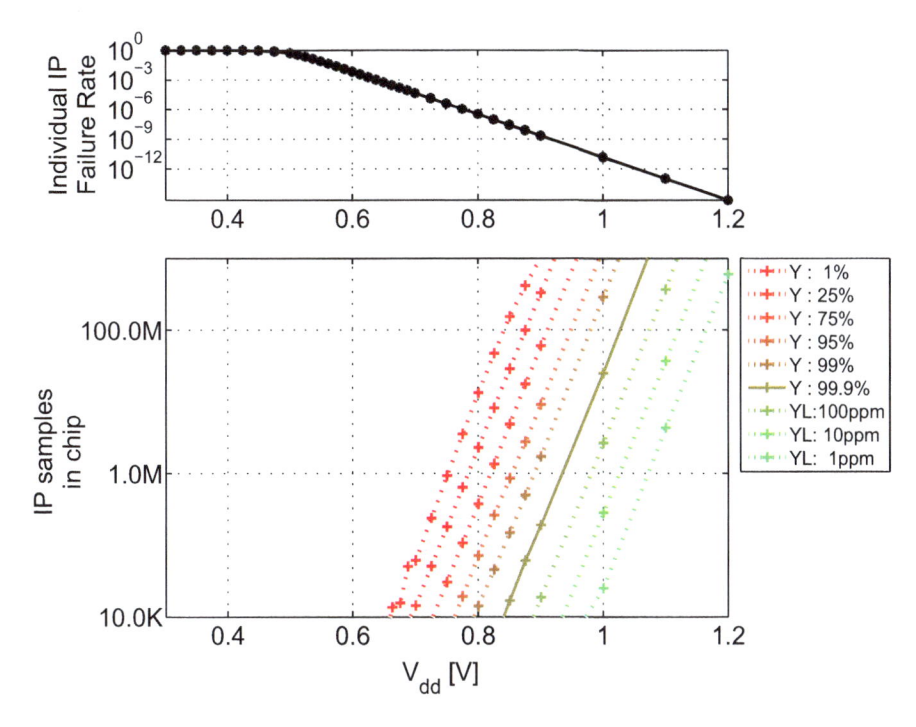

Fig. 5.1 Chip yield isolines as a function of the supply voltage and of the total number of samples of the IP in the silicon product (*yieldogram*, below), assuming that the single-IP failure rate (above) decreases monotonically with the supply voltage. High-yield isolines appear in green, while low-yield isolines appear in red

the design will be considered or not as safe depends thus on the worst-case IP failure rate F, on the number of repetitions N, but also on the targeted yield level. For large IPs, long-range variability might introduce long-range variability in the electrical characteristics of the IP, while for very small-sized IPs like SRAM bitcells, variability will be dominated by local mismatch.

Figure 5.1, above, shows an arbitrary failure rate versus voltage obtained from an analytical curve, which in the present case is an analytical Fermi–Dirac distribution at $-40\,°C$ with $\mu = 0.5$ eV. The first red curve of Fig. 5.1, below, shows the maximum number of times that this IP can be instantiated to obtain a 5% yield, as a function of the operating voltage. We can define the *safe design space* as the accessible amount of allowed repetitions of the IP as a function of both the operating voltage and the given target yield. Since the failure rate shown in Fig. 5.1 monotonically drops with increasing operating voltage, for a constant number of samples the yield increases with voltage, while for a constant level of yield, decreasing the number of samples allows to reduce the operating voltage. One should care that a simple linear plot of the failure rate says nothing about the yield for a large number of samples, it is required to observe the failure rate with a logarithmic scale.

5.2 The SRAM Expand-and-Shrink Trend

Commercial SRAMs embedded in silicon products must operate safely, that is, with neither write nor read fails, over all supported operating conditions. It can be trivial to support a 10% range around the nominal voltage V_{DD}, since devices operating in saturation do not change significantly their driving current when submitted to a small change in their source-to-drain voltage.

However, the ever-increasing amount of memory required by our everyday applications—streaming, listening, connecting, sensing the world—impose to use small-sized devices, to avoid wasting most of the product surface with SRAM. Unfortunately, small-sized devices manufactured in advanced technology nodes are known to be affected severely by uncontrollable variability that scales up inversely with the device sizes. Not only the SRAM bitcell devices are amongst the smallest devices of a product, one must also take in serious consideration the fact that there are more and more of these bitcells, i.e., that N tends to increase at each technology node. This means that the worst-case bitcell, or the bitcell displaying the worse metrics (performances or stability) amongst all bitcells present in the product, will be *very far* from a typical bitcell.

Since the total content of memory must operate safely, a given SRAM architecture must be calibrated on this worst-case bitcell, yielding performances that will be significantly lower than those that could be obtained by calibrating the design on a typical bitcell. So, in conclusion, the global trend between expand (the amount of memory) and shrink (the bitcell surface) leaves the SRAM surface cost unaffected, but creates an unprecedented trend to badly affect the performances of SRAM manufactured with deep-submicron technologies.

Figure 5.1, below, shows how fantastically low levels of failure rate a bitcell should achieve (under worst-case conditions) to allow high-yield production of a chip with 100 Mb SRAM content, around 10^{-12}! Since SRAM features in absolute the largest populations of on-die IPs, it is considered as the most dangerous yield detractor for a given technology.

In practice, if the SRAM population is exceedingly large, redundancy can be used to recover yield. If a given yield level is obtained at a given population size with no redundancy, the same yield level will be achieved at a larger population size if redundancy is used. For the sake of simplicity, the present text does not explicitly treat redundancy.

Self-test capabilities can be easily introduced in silicon products, this allows to self-test under worst-case conditions the memories in a silicon product. Some self-test facilities can also program by means of fuse-like devices the use of redundant columns or rows to substitute those featuring failing bits. This kind of hardware redundancy is also used to recover yield when the surfaces occupied by SRAM are so large that the occurrence of a *manufacturing defect*, i.e., an usually unpredictable local alteration of the die that might affect the performance or the functionality of some devices or of their interconnection, becomes probable.

5.3 Should We Measure or Calculate the SRAM Yield?

We can directly measure yield of a memory only by designing it (a few months of workload of a design team), producing it on silicon (workload of the fab teams and processing time usually equal to some months), and measuring it over all supported conditions (some weeks of workload of a characterization team or more, depending on the number of different design architectures you want to test).

Unfortunately, the yield of a particular process technology increases with the technology maturity: at the beginning of the development of a process technology, there might be a significant amount of manufacturing defects. The measured yield is due to the concurrent occurrence of such defects, of functionality fails due to the mismatch between the performances expected by designers and those really achieved by the manufactured devices all along the test vehicle (not only due to the bitcells), of functionality issues due to the design architecture (e.g., exceedingly large banks inducing too much parasitics, or insufficient assist circuitry) and, finally, of the fails arising from the bitcells themselves. In practice, the latter can be visible only when measuring a defect-free, well-centered die at typical process memory cut designed with an extremely robust architecture, where bitcells will start to fail below a certain level of supply voltage, as it will be shown in Fig. 5.3.

The measured yield achieves the intrinsic technology capabilities only at the development end of the technology, which is then said to achieve a *high maturity* level. It has to be noted that yield learning takes place with cycle times that are measurable in months: It is vital to use this time to solve any observed weaknesses related to the design.

For marketing reasons, it is mandatory to have an idea of the technology capabilities before it attains a high maturity level. Even if an experienced SRAM process engineer can correctly feel what the final yield of a new technology could or will be, sometimes customers demand more than feelings, particularly when they are expected to sign consistent cheques to access the technology, or when they are about to decide to start or not some very expensive, internal design projects. In these cases, it is better to have a method to calculate the SRAM yield, in order to be able to present to customers not only results based on current technology status, but also predictions based on technology specifications, i.e., extrapolations about the final status at high maturity level.

Since the yield of a design is the total number of working dies over all manufactured dies, yield is evidently related to the failure probability of a single die. This latter number is evidently related to the probability of occurrence of one or more bit fails within the die, if SRAM is the main yield detractor. Thus, probability theory allows to calculate yield, if one knows the probability of occurrence of one or more bit fails, or, in other words, if one knows the individual SRAM bitcell failure probability. We will discuss in the following sections how to calculate such a bitcell failure probability from spice-based electrical simulations.

5.4 The SRAM Circuit

Except when explicitly stated in the text, all results shown in this chapter are obtained on the single-port high-density (SPHD), single-p-well six-transistor (6T) bitcell described in [2]. Figure 5.2 shows the schematics of such a 6T bitcell, which is built around two cross-coupled inverters forming a latch element able to store either a logical "0" or a logical "1" in the bitcell *internal nodes*, which are connected to the so-called BitLines (BL and $\overline{BL}$) through transistors M_5 and M_6 (*pass-gates*). During a write operation, the BitLines are forced by external circuits to carry the data that must be written into the bitcell. The operation lasts as long as the pass-gates are turned on, in other words during the time the WordLine (WL) is high, or during the WL pulse. The transistors M_2 and M_4 (*pull-up*) are used to help writing and then keeping a logical "1," while transistors M_1 and M_3 (*pull-down*) are used to help writing and then keeping a logical "0."

Memory bitcells are typically assembled in a single IP that can be instantiated in a silicon product. This IP, named *memory cut*, contains in general a regular array of a number of bitcells equal to the *memory cut size*, where words can be stored and addressed by a logical address, which will not or will coincide with the physical row number, depending if the words are stored using interleaving or not, respectively. An arbitrary number of instances of the same cut or of cuts of different sizes might be present in the same silicon product that will feature a total memory capacity equal to the sum of sizes of all memory cuts. Since integrated circuit (IC) design of complex applications might require the use of a large amount of different sizes of the same memory type, memory compilers are provided by the founder or by an associate design company, to allow generating for each memory type (single-port, dual-port, Level-1 cache, etc.) several instances, as required by the silicon product.

Fail to write the bitcell within a single WL pulse is called in this text a **write ability** (WA) failure. A **read stability** (RS) failure consists in the bitcell content

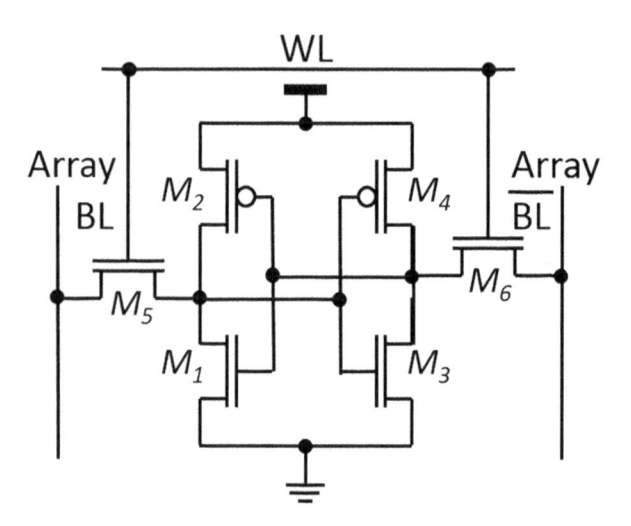

Fig. 5.2 Circuit of a 6-Transistor static random-access memory

toggling during the WL pulse, when both BitLines are set to logical "1" during a read operation. A write stability (WS) failure would consist in the bitcell being unable to toggle during any arbitrary WL pulse duration, even if BitLines are correctly set so as to make the bitcell content toggle; this kind of fail is called in the present text a static WA failure.

WA and RS failure modes can either be static, i.e., happening at low frequency (LF), or dynamic, i.e., at a given clock frequency. Dynamic failures are in general a function of the memory organization, since the number of rows and columns determines the parasitic capacitances acting on the various signals and the amount of leakage flowing in the system. A major dynamic failure mode is the **read ability** (RA) failure, i.e., when the SRAM readout circuit is unable to detect correctly which data is stored in the bitcell (refer to [2] for a more detailed description of the various failure modes).

Dynamic failures will be covered in the present chapter mainly by some measurement results on a particular circuit and finally presented only in the final yieldograms of Fig. 5.12 (refer to Sect. 5.10). In this figure, one considers for the sake of simplicity a single bitcell per column and per row, in order not to bias results on the particular architecture chosen. Those results obtained at very small device and parasitic capacitance and resistance values display a very small RC time constant of the circuit, with the advantage of a very short computing time. Results so collected display a correct behavior, if the signal time scale is not considered.

5.5 Can SRAM Spice Models Predict Yield?

Spice simulators are used widely throughout the world since many decades. After all this time, simulators manage to reproduce faithfully the electrical behavior of a lumped element circuit, if the correct numerical parameters are chosen for the analysis. Electrical models for MOSFETs are designed in such a way to reproduce all observable phenomena in MOSFET, and their parameters can be accurately extracted from silicon data and compiled into model cards to match with a great accuracy the experimentally observed characteristics.

Spice model cards can represent process-like variations by coupling model parameters to random variables, whose values vary according to pre-definite distributions. Verifying the statistical accuracy of all model cards, in other words: the model to hardware correlation (MHC), is a very complex matter, nevertheless required to guarantee that design of complex silicon products, involving billions of MOSFETs, using foundry model cards lead to working chips.

During the challenging path that brings manufactured devices on target, spice model cards represent the target characteristics of devices as they are expected to be when the technology reaches a high maturity level. The MHC evidently increases with the technology maturity: at the beginning of the development of a process technology, there might be (usually unexpected) process effects that might badly affect the performances or the leakage of devices with very small channel length.

As already stated, MHC is expected to be very high only at the development end of the technology, devices being then said to be *on target*.

The progress in MHC during technology development is monitored by a form of statistical quality control of the manufacturing process. A device characteristic is measured systematically on each wafer, on a small sub-sample of dedicated devices, which are designed in such a way to feature most physical effects occurring in devices used in silicon products. The mean μ and the standard deviation σ of the measurements on the most recent lots are thus available. The manufacturing process is in control if and only if all measured characteristics stay mostly between their lower specification level (LSL) and its upper specification level (USL).

A useful indicator of how often the LSL and USL are violated amongst the fabricated population is the process capability index $C_{pk} = \min(C_{plow}, C_{pup})$, where $C_{plow} = (\mu - LSL)/3\sigma$ and $C_{pup} = (USL - \mu)/3\sigma$. A C_{pk} equal to at least 2 implies thus that there are at least 6 sigma between the measured process parameter mean and its closest specification limit. This is often enough to guarantee that the process is in control, but C_{pk} depends evidently on how one defines the LSL and USL. If these levels are correlated with the spice model results, the MHC is monitored in the same time as the process quality.

In particular, spice models can be used to generate the process specifications limits (LSL and USL) for all monitored electrical characteristics. Let us assume that symmetric LSL and USL around a target T are defined using $\pm 6\sigma$ statistical results obtained with the model card. In such conditions, then for devices on target (whose statistical mean lies on T, i.e., halfway between LSL and USL) we have the following relationships: $C_{pk} = 2$ means a perfect MHC, $C_{pk} < 2$ means that the models predict an optimistic process spread (monitored devices are too often out of specifications), while $C_{pk} > 2$ means that the model spread is pessimistic (monitored devices fall never out of specifications).

In the symmetric specification limits that we are discussing, a $C_{pk} > 2$ process can be considered as *safer* than a $C_{pk} = 2$ process. However, as soon as C_{pk} becomes larger than 2, the more significant becomes the misalignment between silicon and model-predicted variability. This misalignment implies that the performances of the design, which are in general tuned on the model worst-case process, will be inferior to what the technology is capable, or, in other words, that the system is over-designed.

In conclusion, if LSL and USL are derived as described above, the process yield engineer task is not only to bring devices on target, but also to make the fab operates at $C_{pk} \geq 2$ and not exceedingly far from 2. If this is verified, then any design that has been validated using the statistical models will at high maturity level achieve at least the model-predicted performances and the model-predicted yield. In this sense, we could say that the real issue is not, as usually done in research *to fit models on measurements obtained on silicon*, but rather *to improve continuously the manufacturing process so that the manufactured silicon results fit the modeled performances*. In fact, the spice model cards may be seen as *technology specifications* that are coded in a very complex and highly non-linear form, instead of being detailed in some (probably endless) human-readable documentation.

Once the manufacturing process has reached a high maturity level, with devices on target and MHC well-verified, one still needs the continuous monitoring of electrical parameters, in order to be sure that no significant drifts appear in manufactured dies with time but, on the contrary, that the technology quality stays constant across all production time. And, in conclusion, the answers to the questions in last titles of Sects. 5.3 and 5.5 are that *we should both calculate and verify by measurements the SRAM yield* and that *SRAM spice models do not predict yield, but rather fix the specifications of high maturity memory yield*, respectively.

5.6 Supported Operating Voltage Range: The V_{min} Paradigm

Once an SRAM design is concluded and manufactured on silicon, the memory functionality must be verified over all supported conditions. Lowering the supply voltage drives the memory slowly outside of its specified operating range, until the worst-case bitcell of each die will cause the first single bit fail. At each voltage, the percentage of dies that still work (with no fails) over the total number of measured dies is the chip yield.

The blue dots in Fig. 5.3 show results obtained at $-40\,°C$ on a set of dies manufactured in a single lot during the FD-SOI 28 nm technology development [3]. Since the single measured lot might show some correlations between devices, i.e., it might be a fast or slow lot, one speaks of *local yield*, in opposition to the *mass-scale production yield*, that can be measured only on a statistical amount of different lots, and which is an important indicator of the technology capabilities.

The blue dots of Fig. 5.3 show a sharp fall of yield below 0.6 V. One defines for a given amount of memory the minimum operating voltage V_{min} as the minimum

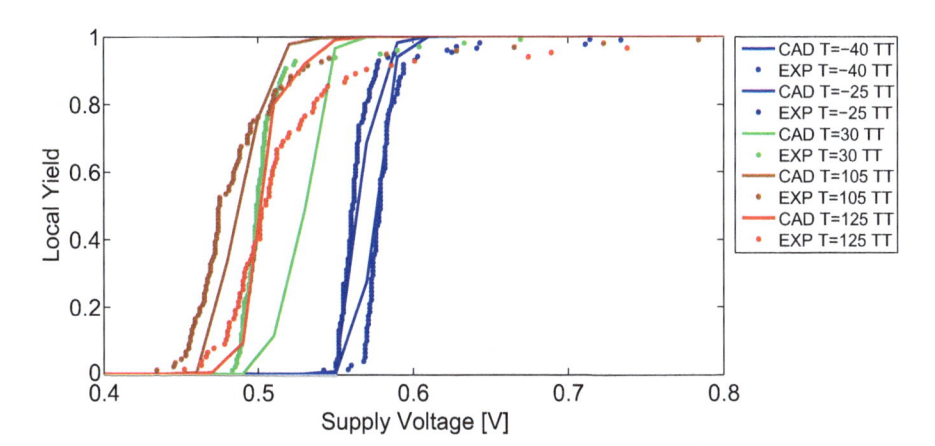

Fig. 5.3 Measured yield of 50 Mb of single-port high-density (SPHD) SRAM manufactured in a single lot during FD-SOI 28 nm technology development (symbols), compared to yield estimated with CAD models adjusted on the manufactured lot centering (lines). Different colors correspond to different temperatures

supply voltage that grants a targeted yield level, typically 95% during technology development. As the temperature changes, the yield curve changes significantly, both in its V_{min} and in its shape, becoming smoother at higher temperatures. Experimental uncertainties at high temperatures on such immature dies might explain some minor, erratic behavior (data around 0.7 V) due probably to manufacturing defects.

V_{min} at the maximum SRAM content without redundancy would be sufficient to define the safe design space at a technology node, if the SRAM failure rate dropped monotonically with the voltage supply, as in the example of Fig. 5.1, above. As we will see, this is not a correct assumption for nowadays SRAM, and one should find a better way to define the SRAM safe design space.

5.7 From Margins to Failure Rate: Validity of a LF Gaussian Model Across Temperatures

Spice foundry models can be used by simulating the bitcell response with different flavors of the MC method. For LF behavior, static RS and WA rates are calculated through the static noise margin (SNM) [4] and write margin (WM), respectively, the latter being also known as BitLine margin or BLM [5]. Margins are calculated with the Naïve MC method, which generates a set of circuit replica. The failure rate is then estimated assuming that single-side margins (for instance, one of the lobes of the butterfly curves for SNM) follow Gaussian distributions. The model data in Fig. 5.3 are calculated using this methodology and reproduce well the sharp fall of yield below 0.6 V and its temperature dependence. SNM and WM are said to describe the bitcell *stability* with respect to read and write operation, respectively, this word meaning in turn that a bitcell does not toggle under a read operation, but toggles as expected under a write operation, both operations being carried out at LF.

Gaussian distributions are based on the so-called law of errors. As the Nobel prize for physics G. Lippmann stated once, "Everybody believes in the exponential law of errors: the experimenters, because they think it can be proved by mathematics; and the mathematicians, because they believe it has been established by observation" [6]. While it has been demonstrated that some MOSFETs electrical parameters do not follow Gaussian law [7], the single-side margin distributions of SRAM bitcells are determined by many uncorrelated electrical parameters of different devices, and this might explain how Gaussian distributions reproduce well the experimental behavior.

Dynamic failures, on the contrary, like WA or RA, cannot be expected to follow a Gaussian distribution, but rather behave like gate delays [8]. In particular at low voltages, worst-case devices tend to operate in sub-threshold, and will produce abnormally slow responses that might fall significantly out of any Gaussian-estimated distribution tail. Following sections use therefore the importance sampling (IS) method for high-sigma MC estimation [9] when dealing with dynamical failures. Such a method directly delivers failure rates with no assumption about the margin distribution.

The extreme value theory (see, for example, Robert Aitken's chapter in [10] for an applied introduction to the matter) can be invoked to show that if we are only interested in the worst-case element over a large set of samples, then the shape of the original distribution of the elements is less important, leading finally only to Frechet, Weibull, or Gumbel distributions of worst-case samples. In other words, most of the information contained in the full distribution of samples is redundant or even irrelevant, if we are only interested in worst-case samples!

Considering now how margins change as a function of the temperature, one must first of all define which range is supported covering all the possible, future uses of the IC integrating the memory. For example, cache memories for home PCs might support a home-related [0 °C, 70 °C] temperature range, covering all possible junction temperatures ranging from turning on a cold PC in a cold winter morning up to the highest power consumption peaks during an hot summer day. Mobile applications might need to support a wider temperature range, for instance, descending down to −40 °C, in order to keep your device working during outdoor ski—or simply during your normal winter day-life if you are based in Alaska or in Siberia. Automotive or military ICs that are expected to work close to external sources of heat might require to support up to 165 °C or even more.

To understand the importance of the supported temperature range, one must remind a very well-known phenomenon: if MOSFETs performances are strongly dependent on the threshold voltage V_{th}, the latter depends strongly on the temperature, with a decrease in the order of a mV per each degree of temperature increase. If one takes exactly 1 mV [11], the MOSFETs of an SRAM that is expected to support a 200 °C temperature range will undergo a 200 mV change in their V_{th}, which is an impressive 20% for a nominal voltage around 1 V, but even becomes more significant at lower voltages.

Considering that the temperature dependence of V_T is tied to the temperature dependence of the energy band structure in the channel, it is easy to understand that this change is in general different between PMOS and NMOS, and will be affected by the device sizing and by its manufacturing process. This is bad news for devices like SRAM, that are built around inverters; as it is well-known, the switching voltage of an inverter is dependent on the ratio between its NMOS and its PMOS device. Having SRAMS with moving ratio between NMOS and PMOS means that their stability changes with the temperature. Finally, considering that temperature coefficients might differ significantly from one technology node to the other, and obviously depend on the type of manufacturing process, i.e., planar, FinFET, or FD-SOI, one understand how each technology node might involve reconsidering from the scratch how SRAM bitcell devices are actually sized.

The complex effects of the temperature-dependent balance between PMOS and NMOS are visible to an expert eye in the curves of Fig. 5.3, and the agreement between measurements and models is rather good, confirming that a Gaussian model is enough to predict at LF the safe design space for a particular memory capacity. But how can we manage adding to this fairly complex picture Body-Bias and supporting variable amounts of memory in the final, most important mass-scale production yield of a silicon product expected to work at HF?

5.8 Body-Bias Effects to SRAM

As said in the previous section, modulating the temperature influences the bitcell stability via the changes in the device V_{th}. Since, as discussed throughout the book, Body-Bias is a powerful method to modulate the device V_T (e.g., refer to Fig. 4.1), it can be expected to have a major impact on the bitcell stability. As discussed in Sect. 5.2, the bitcell stability is also a function of the manufacturing variability, which can be divided into uncorrelated or random device variability, also known as *mismatch*, which is the unavoidable component of variability that affects two devices placed as close as possible; and on correlated variability that affects all devices of a die, due to uncontrollable drifts of some processing conditions that affect all devices at the same time, like, for example, the thickness of the gate dielectrics. In summary, the way in which stability is affected by changes in PVT also depends on how the bitcell is centered, i.e., from the ratio between NMOS and PMOS and on the individual device sizing. This is evident in Fig. 5.4, which shows three dynamic failure rates [2] as obtained with the IS method (please refer to Sect. 5.7). Since the most important failure mechanism at the observed process corner, with no Body-Bias applied, is evidently RA, the bitcell is said to be read-limited under these conditions.

Since the same Fig. 5.4 also shows that WA is less critical than RA, one might decide sacrificing some margin in WA to recover RA. In a 6T SRAM bitcell with four NMOS devices with the pull-up being the smallest devices, making stronger this PMOS devices helps to maintain data and thus reduces the write margin, at the same time increasing the RS. RA can be improved instead by making stronger the NMOS devices that will be able to deliver more current during the read operation.

The particular design of the bitcell under test is single-p-well, so a common Body-Bias is applied by construction to both kinds of device at the same time,

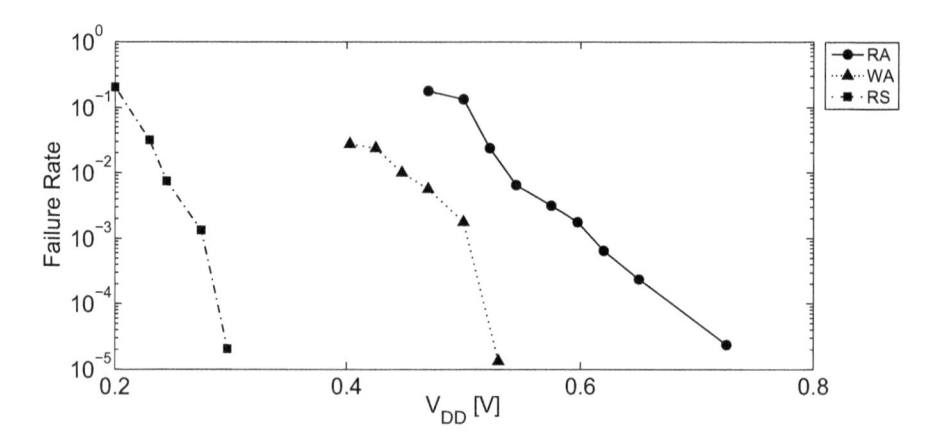

Fig. 5.4 Measured read ability (RA), write ability (WA), and read stability (RS) failure rates versus V_{DD} at a 2.5 ns WL pulse duration

5 SRAM Bitcell Functionality Under Body-Bias

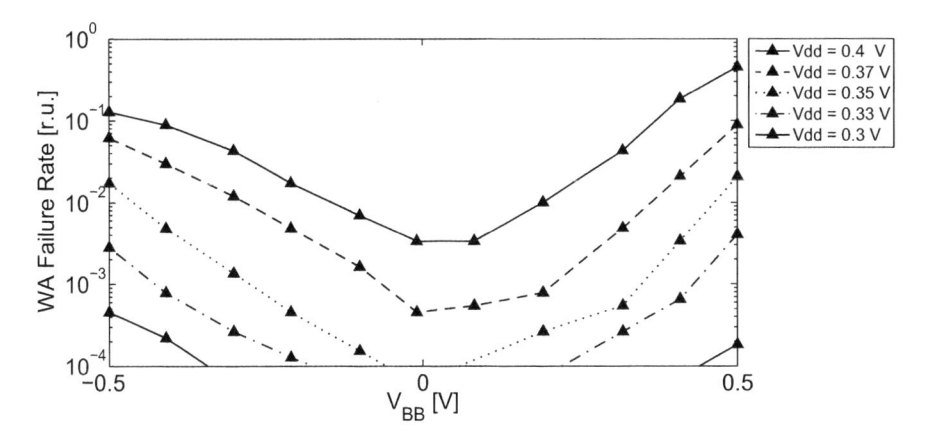

Fig. 5.5 Measured write ability (WA) failure rate versus Body-Bias at a 5 ns WL pulse duration

but has an opposite effect on the two different kinds of MOS. For instance, a positive Body-Bias will put the NMOS devices in Forward Body-Bias and PMOS devices in Reverse Body-Bias and in principle could be expected sacrificing RS while improving WA, while a negative Body-Bias will obtain the opposite behavior and in principle could be expected to improve RS while sacrificing WA.

The following Fig. 5.5 shows that in reality things are more complicated that this simple picture. Applying Body-Bias at ultra-low voltage sacrifices WA independently on the type of applied bias, since in the two different directions it triggers different write failure modes [2]. In particular, if for negative Body-Bias the WA is reduced as expected, because it becomes probable to meet a bitcell with a pull-up so strong that it fights efficiently against the external BitLine voltage trying to discharge the node carrying a logical "1" at the begin of the write operation, for positive Body-Bias the WA is also reduced, because it becomes probable to meet a bitcell with an opposite pull-up so weak that it is unable to pull up the node carrying a logical "0" at the begin of the write operation.

Another non-linear effect that can be observed in an SRAM under Body-Bias is the behavior of SRAM bitcell leakage, which is shown in Fig. 5.6. A general expected trend is that leakage increases when applying positive Body-Bias, since in general NMOS leakage dominates the SRAM bitcell leakage. Nevertheless, negative Body-Bias can also deteriorate leakage in some particular conditions [2], where the PMOS contribution becomes significant.

Figure 5.7 shows some results that were obtained on a particular amount of single-p-well bitcells of different surfaces [12]. There exists conditions in which applying a constant Body-Bias to a bitcell independently of the particular operating conditions (*Blind* dataset) produces only marginal improvements with respect to the default V_{min} behavior. Important improvements can be obtained only by modulating the amount of Body-Bias as a function of the particular operating conditions or of

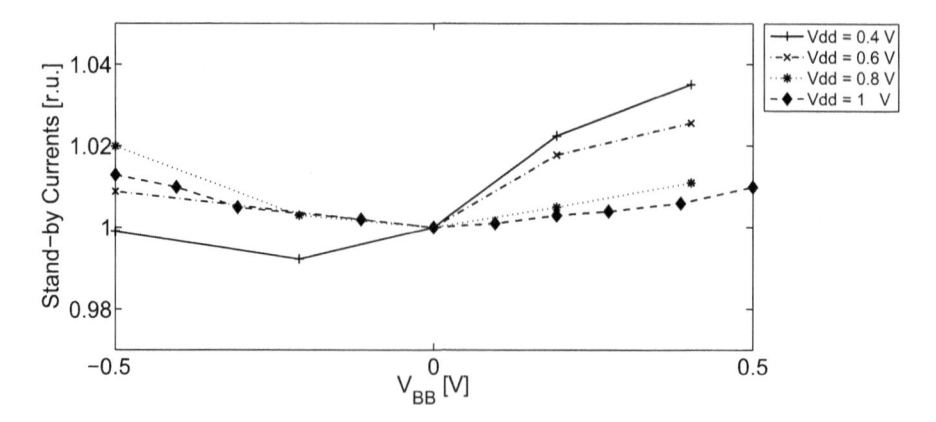

Fig. 5.6 Measured bitcell leakage versus Body-Bias at various supply voltage levels

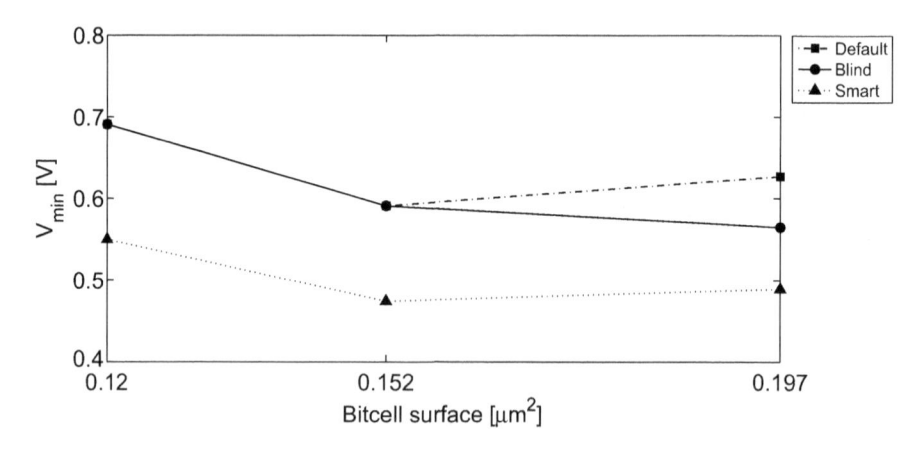

Fig. 5.7 The default V_{min} for a particular memory capacity (squares) for three different bitcells of increasing area can be marginally improved by applying a fixed Body-Bias level (circles). A smart usage of Body-Bias, with a variable Body-Bias level following the PVT, can improve much more V_{min} (triangles)

the particular process in which the memory has been manufactured, implementing thus a process compensation or a temperature Compensation solution (refer to Sect. 3.2).

5.9 Understanding HF Effects

Having discussed the dependence of the bitcell failure rate and, as a result, the dependence of mass-scale production yield on variability, on the PVT and on the applied Body-Bias, we now approach a final, fundamental factor that must be taken

5 SRAM Bitcell Functionality Under Body-Bias

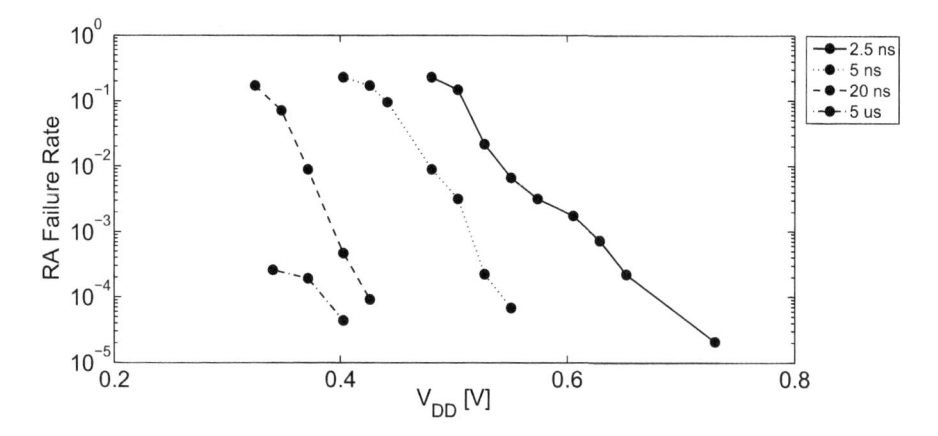

Fig. 5.8 Measured RA failure rate versus V_{DD} at variable WL pulse duration. At a given supply voltage level, shorter WL pulse duration imply higher failure rates

into consideration: the time that is available to perform operations, or, in design terms, the maximum clock frequency for which the product is targeted.

Figure 5.8 shows the failure rate relative to RA [2], as a function of the duration of the WL pulse (refer to Sect. 5.4). At a given WL pulse duration, failures increase by lowering the supply voltage.

However, failures at a given supply voltage can be reduced significantly by increasing the WL pulse duration. The failure rate relative to a dynamic operation depends critically on the equivalent RC time constants playing a role in the dynamic operation itself. These time constants will be in general longer for a wider bank architecture, and this dependence on the number of rows and columns adds a significant complexity to draw a picture of dynamic bitcell stability analysis.

Results obtained with a particular architecture and at a particular PVT might not be easily generalized, but nevertheless show some interesting properties, like the fact that each different margins gets affected in a different way by clock frequency. Figure 5.9 shows that in the architecture considered in [2], even if the bitcell RS does not look affected, the bitcell, which is write-limited at LF, becomes instead read-limited at HF, as shown in Fig. 5.4, where data were obtained at 2.5 ns (HF in this context).

Assist techniques [13] can be used to lower the failure rates under stringent conditions, either ultra-low voltage or HF [14]. The Negative BitLine (NBL) assist consists in writing a logical "0" by using a negative voltage value [15], in order to overdrive the pass-gate that shall drive the internal node to "0." Figure 5.10 shows the WA failure rate for a bitcell manufactured in a different technology than those presented in this chapter, at increasing NBL voltages, including data at no assist (no NBL), as obtained with different methods: Naïve MC (dot-dashed lines), IS MC (symbols), Naïve MC WM at applied NBL (solid lines), and extrapolations of WM at no applied NBL that assume perfect normality (dashed lines) [9]. For the latter data, a 1% accuracy band is materialized by dotted lines.

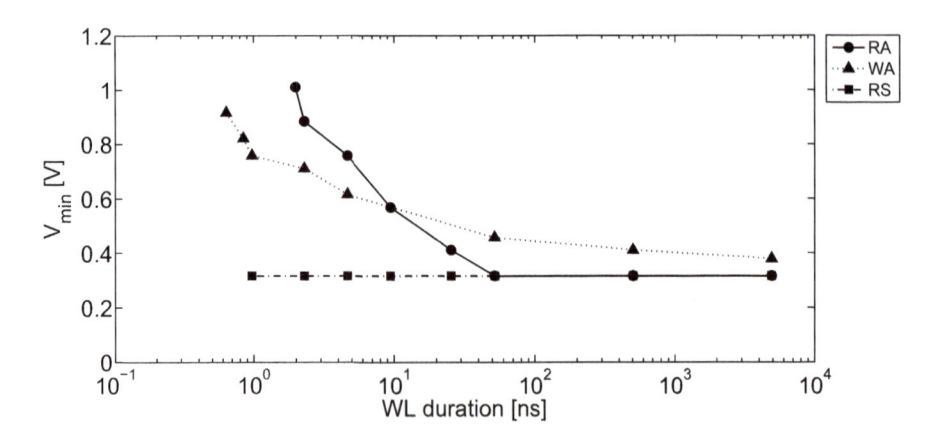

Fig. 5.9 Measured read ability (RA), write ability (WA), and read stability (RS) failure rates versus WL pulse duration at no applied Body-Bias

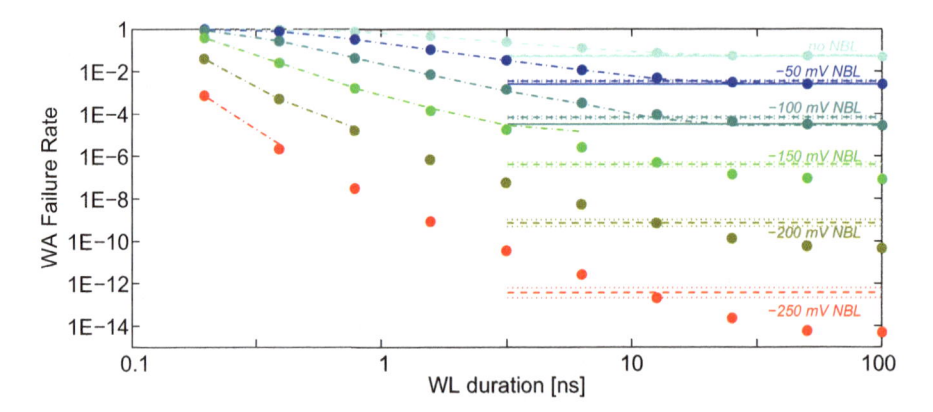

Fig. 5.10 WA failure rate vs. WL pulse duration at different levels of Negative BitLine (NBL) assist for write operation. Importance sampling MC data (symbols) is compared to Naïve MC data (lines) that can be obtained only for relatively high failure rates. Horizontal levels show WM-based data (solid lines) or predictions from WM-based Gaussian model (dashed lines), with materialized statistical uncertainty (dotted lines). Results obtained on a different technology than that used in all other figures

At HF (for WL duration exceedingly smaller than the circuit response time) the operation of writing a logical "0" fails with certainty (left part of the curve). When the WL duration becomes longer than the circuit response time (tends to LF), the failure rate decreases and saturates to the static failure value predicted by WM Naïve MC. For large levels of assist, WM Naïve MC becomes unfeasible. Unfortunately, the extrapolated levels look systematically pessimistic with respect to WM Naïve MC, when the latter are available (down to $V_{NBL} = -100$ mV). However, they give a zero-order approximation of the bitcell failure rate that can help in a preliminary circuit sizing.

5.10 Switching to the N_{max} Paradigm to Support Body-Bias

From the previous sections, one might understand the difficulty to support an industrial range of operating conditions for SRAM, in particular for HF applications operating at low voltage. Design companies may show some reticence to consider Body-Bias, since this may look simply as an additional parameter to control, while it is a powerful additional degree of freedom to extend the safe design space of memories. To understand how to determine from model-based estimations the supported operating condition range, one must abandon the V_{min} paradigm and make a step back to how yield is derived from the failure rate.

In the following text, we use the *yieldograms*, which are the plots shown in Fig. 5.11. These plots are similar to that shown in Fig. 5.1, but with swapped axis (voltage is on the vertical axis, while the number of IP samples is along the horizontal axis). All yieldograms are calculated for mass-scale production yield [16], and not for local yield as calculated at a single corner. The mass-scale production yield assumes a Gaussian distribution of the manufactured dies around

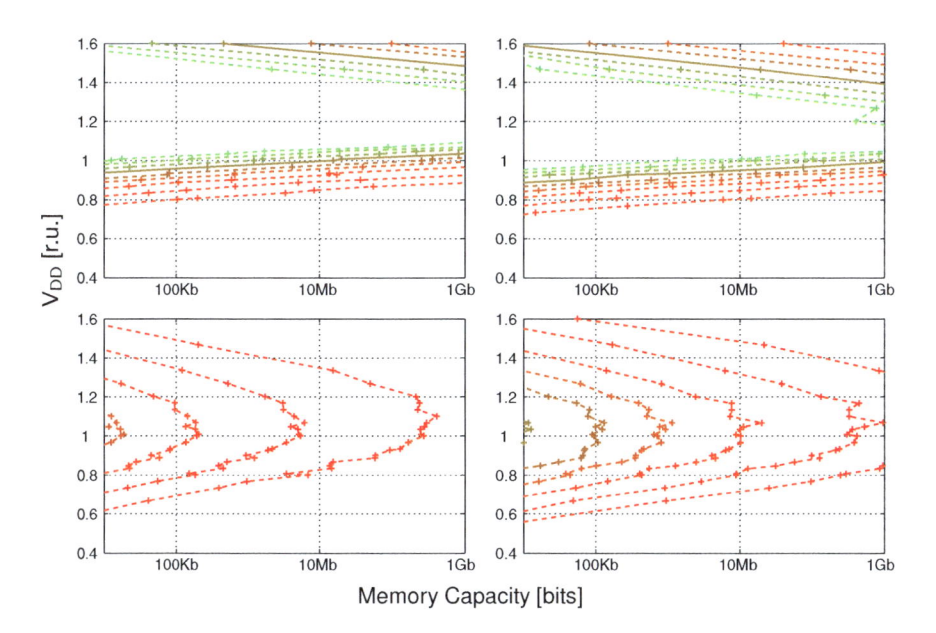

Fig. 5.11 Chip yieldograms at mass-scale production, as modeled using static margins calculated with CAD models of the FD-SOI 28 nm SPHD of Fig. 5.3 through the full temperature range [−40 °C, 165 °C]. Data presented on two pages shall be read clockwise, from the top row, as follows: yieldograms obtained (from the left to the right) at −0.75 V, at −0.5 V, at −0.25 V and at 0 V Body-Bias voltage; in the bottom row, yieldograms obtained (from the right to the left) at 0 V, at 0.25 V, at 0.5 V, and at 0.75 V Body-Bias voltage. Voltage is expressed in relative units, as in [15]

the typical conditions, which is what is currently assumed for the electrical models released by practically all foundries.

Figure 5.11 shows how the SPHD (please refer to Sect. 5.4) yieldograms evolve at LF with a variable Body-Bias. The figure, shared across two pages, shall be read clockwise, starting from the leftmost plot in the top row (obtained at a -0.75 V Body-Bias) and ending with the leftmost plot in the bottom row (obtained at a 0.75 V Body-Bias). The curves show the yield level as a function of the applied voltage and of the memory size over an impressive temperature range $[-40\,°C, 165\,°C]$, which was chosen so as to display the ultimate technology capabilities. In these figure, write-limited yield isolines can be identified with the very dense straight lines in the low-voltage parts of the top row yieldograms, while read-limited yield isolines can be identified with the less dense, rounded curves. Please refer to [15] to see an explicit decoupling of the two limitations, obtained in that case through the use of dedicated assist techniques.

Let us consider at first the yieldogram obtained at no Body-Bias, which is duplicated in each row at the rightmost position. In the range $[0.6, 0.8]$ (supply voltage is expressed here in relative units, as in [15]) the maximum capacity for a given yield level increases sharply with the voltage level. Since the yield isolines are very dense, the yield gradient is very large and oriented nearly parallel to the y-axis, i.e., yield increases sharply with the voltage for a given memory capacity. This is exactly what was visible for the blue dots in Fig. 5.3. For a low memory capacity, the voltage range $[0.8, 1.2]$ is a high-yield safe operating region. For larger capacity, a safe operating region can be defined only at low-yield level. Yield decreases smoothly at high voltage, this phenomenon has been observed elsewhere

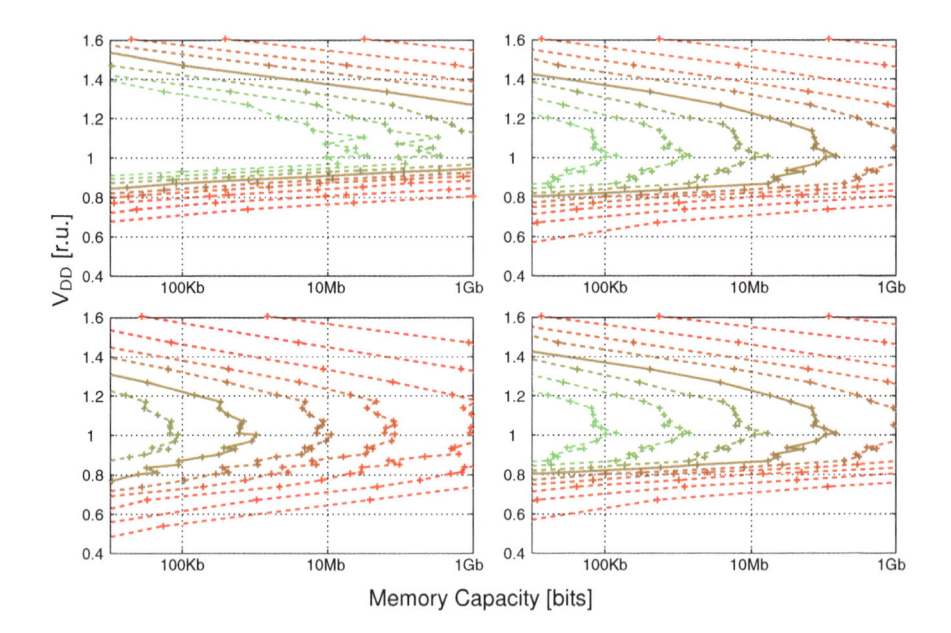

5 SRAM Bitcell Functionality Under Body-Bias

and it is known to be tied to read instabilities [17]. This definitely confirms that the V_{min} paradigm is insufficient to describe safely the technology capabilities: one must consider instead the maximum number of instantiation of an IP N_{max}, which is available at every PVT for every yield level, tracing an yieldogram to delineate the safe design space of the IP. It is a change of paradigm, but it has the advantage to apply to any arbitrary circuit, and not only to those failing only at low voltages.

The top row yieldograms are obtained at a negative Body-Bias. Going counter-clockwise, the third plot of the first row is obtained at a -0.25 V Body-Bias. Under this negative Body-Bias, NMOS devices enter Reverse Body-Bias (and are thus slowed), while PMOS devices enter Forward Body-Bias (and are thus accelerated). This Slow NMOS-Fast PMOS balancing makes it harder to write bitcells, especially at lower temperature, since in this technology transistors are slowed down at cold temperatures. The very dense, write-related yield isolines move upwards to higher voltage levels: write limitation has become more important and restricts the safe operating regions above $V_{DD} = 0.8$. Meanwhile, it becomes harder to toggle a bitcell while reading it: read-related yield isolines move rightwards, towards larger capacities.

The same movement continues in the second plot of the top row, obtained at -0.5 V Body-Bias, where write-related yield isolines continue to move upwards and restrict now high-yield operations above $V_{DD} = 0.9$. Meanwhile, read-related yield isolines move even more rightwards, towards larger capacities. And finally, at -0.75 V Body-Bias, in the first plot of the top row, read limitation has nearly disappeared, while write-related yield isolines continue to move upwards and restrict now high-yield operations above $V_{DD} = 1$. Negative Body-Bias allows thus to obtain significant LF yield for large memory capacities, even over the large temperature range chosen here.

Considering now the bottom row, starting from the rightmost yieldogram (obtained at no applied Body-Bias): moving clockwise to the left, to the third yieldogram of the row, obtained at 0.25 V Body-Bias, NMOS devices enter Forward Body-Bias (and are thus accelerated), while PMOS devices enter Reverse Body-Bias (and are thus slowed). Under these conditions, it becomes easier to write a bitcell: the dense yield isolines related to write limitations have now disappeared, because the bitcell appears to be read-limited over the full voltage range. The pattern related to read limitation has moved a bit leftwards (towards lower capacities): it seems impossible to use under these conditions the bitcell at high-yield, but one should remember that these results were obtained at LF.

Continuing to the second yieldogram of the bottom row, obtained at 0.5 V Body-Bias, the read-related pattern of yield isolines has moved leftwards, towards lower capacities, and the movement continue with the first yieldogram of the bottom row, obtained at 0.75 V Body-Bias. From these LF yieldograms, one can conclude that the bitcell is ideally centered for no applied Body-Bias, but that a negative Body-Bias allow to obtain high LF yield at high voltage.

5.11 HF Body-Bias Effects and Future SRAM Compensation Applications

Figure 5.12 shows, in its left part, data obtained from dynamical simulations of RS and WA using IS MC method at LF when applying a 0.75 V Body-Bias. Data looks very similar to that shown in the bottom-left yieldogram of Fig. 5.11, which was obtained at the same Body-Bias at LF. Failure rates calculated with the two method are very close, both methods indicating a RS-limited bitcell, with very low yield over all useful range of memory capacities. At this Body-Bias level, the NMOS are so strong, that it becomes probable to find bitcells where a simple data readout overwrites the stored values, if a LF clock is used, i.e., if the WL pulse duration is sufficiently long. The data are obtained indeed at a very long WL pulse duration that gives results similar (but not exactly equal) to those obtained by simple Gaussian extrapolation of static results.

The yieldogram shown in the right part of Fig. 5.12 has been obtained at HF, i.e., a very short WL pulse duration, otherwise in the same conditions as in the left part of Fig. 5.12. With such a short WL pulse duration, the WA ability failure rate has increased (refer to Fig. 5.9), but has not yet became so large to badly affect the chip yield.

Data in the right part of Fig. 5.12 looks very similar to that shown in the second graph in the top row of Fig. 5.11, which was obtained at -0.5 V Body-Bias voltage. The use of an HF causes the WA failures to introduce very dense, write-related yield isolines at low voltage, while reading with slow NMOS reduces significantly read-related instability and moves upwards the yield isolines. In summary, the read/write balance is modulated by a simple change in the clock frequency at which read/write operations are carried out.

The present results bring an important theoretical content since they proof:

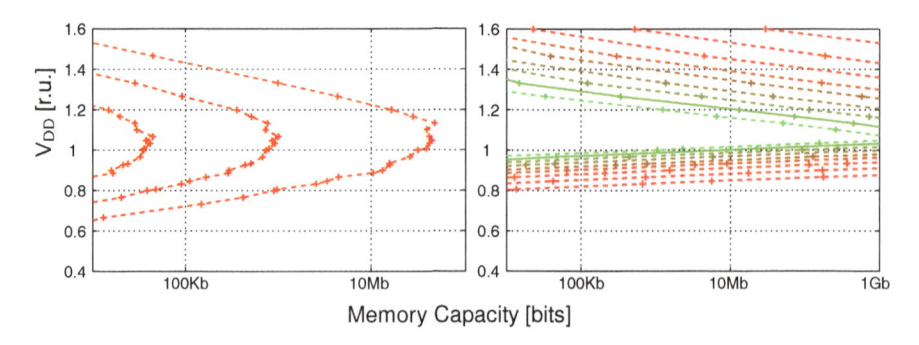

Fig. 5.12 Chip yieldograms obtained at 0.75 V Body-Bias voltage, as in the bottom-left yieldogram of Fig. 5.11, as modeled using dynamic margins only at worst-cases: $-40\,°C$ for write ability, and $165\,°C$ for RS. Data calculated at low frequency (left) and at high frequency (right). Voltage is expressed in relative units, as in [9]

1. At LF, the Gaussian model applied to static results produces data similar to LF dynamical stability analysis. If a memory cut is used at low frequency, the mean-over-sigma ratio of traditional margins like WM and SNM are sufficient to characterize the memory yield.
2. For each memory design, there exists a certain operating frequency, beyond which RS becomes negligible. Beyond this architecture-dependent frequency, SNM can be neglected and one must consider only dynamical margins like WA and RA, which are strongly non-Gaussian.

Concerning the second point, one must remind that the present yieldogram only shows the balance between RS and WA, neglecting RA and can therefore be applied to arbitrary designs, because, concerning RS and WA, only the time constant of the system changes by changing the number of rows and/or columns. It is ultimately the RA that fixes the range of frequencies allowed to the system, and the readout circuitry must be designed considering in the detail the memory cut architecture and the target yield level.

5.12 Conclusions

The results presented in this chapter show that the architecture definition and optimization of SRAMS designed in modern deep sub-micron technologies is a task that is strongly coupled to the choice of the supported voltage/temperature operating ranges and on the allowed amount of memory not protected by redundancy. Body-Bias is in this case an important, additional degree of freedom, that allows to influence the bitcell stability and obtain write performances at HF that approach those of LF. Yieldograms may help to delimit the safe design region, when so many different parameters play in the evaluation.

Even if Body-Bias introduces some complexity in a silicon product, once it is integrated in a system and made available to SRAM, it reveals to be a precious additional degree of freedom that can help in extending the safe design space. It acts as helping mean also with respect to another parameter that was not directly discussed in the present chapter: the circuit operating age. Results of previous section show that the current bitcell is designed in such a way to be ideally centered for no applied Body-Bias. Nevertheless, these results are obtained with models of fresh, as-produced silicon. It is known that aging of PMOS devices can increase their V_{th}, making them slower, and introducing instability in the SRAM bitcell, a phenomenon known as negative bias temperature instability (NBTI [18]).

Since the NBTI phenomenon can be described by an increase in the PMOS device V_{th}, it can also be mimicked by applying a Reverse Body-Bias on PMOS devices, i.e., a positive Body-Bias. Neglecting that yieldograms of Fig. 5.11 are obtained with Body-Bias being applied simultaneously on PMOS and NMOS devices, since both kinds of devices share the same well, the yieldograms in the bottom row show how yield evolves with circuit aging.

Age compensation is therefore a direct application of Body-Bias, since a negative Body-Bias will move back the bitcell balance towards less read limitations and more write limitations. In the present case, aging can happen in a much more rude way under much more rude operating conditions, like those taking place under strong high-energy particle irradiation. Whether in service electronics for high-energy experiments or in on-board electronics in space shuttles, aging will push the device V_T away from that of fresh conditions and thus decrease yield. Body-Bias is in this case the only way to recover yield, and it consists finally in a simple, remotely operated voltage change on the device fourth terminal!

References

1. Intel (2018). https://ark.intel.com/products/series/122593/8th-Generation-Intel-Core-i7-Processors
2. O. Thomas, B. Zimmer, S.O. Toh, L. Ciampolini, N. Planes, R. Ranica, P. Flatresse, B. Nikolić, *2014 IEEE International Electron Devices Meeting* (2014), pp. 3.4.1–3.4.4. https://doi.org/10.1109/IEDM.2014.7046973
3. R. Ranica, N. Planes, V. Huard, O. Weber, D. Noblet, D. Croain, F. Giner, S. Naudet, P. Mergault, S. Ibars, A. Villaret, M. Parra, S. Haendler, M. Quoirin, F. Cacho, C. Julien, F. Terrier, L. Ciampolini, D. Turgis, C. Lecocq, F. Arnaud, *2016 IEEE Symposium on VLSI Circuits (VLSI-Circuits)* (2016), pp. 1–2. https://doi.org/10.1109/VLSIC.2016.7573512
4. E. Seevinck, F.J. List, J. Lohstroh, IEEE J. Solid-State Circuits **22**(5), 748 (1987). https://doi.org/10.1109/JSSC.1987.1052809
5. H. Makino, S. Nakata, H. Suzuki, S. Mutoh, M. Miyama, T. Yoshimura, S. Iwade, Y. Matsuda, IEEE Trans. Circuits Syst. Express Briefs **58**(4), 230 (2011). https://doi.org/10.1109/TCSII.2011.2124531
6. H. Jeffreys, Mon. Not. R. Astron. Soc. **99**, 703 (1939). https://doi.org/10.1093/mnras/99.9.703
7. C. Millar, D. Reid, G. Roy, S. Roy, A. Asenov, IEEE Electron Device Lett. **29**(8), 946 (2008). https://doi.org/10.1109/LED.2008.2001030
8. K. Agarwal, S. Nassif, IEEE Trans. Very Large Scale Integr. VLSI Syst. **16**(1), 86 (2008). https://doi.org/10.1109/TVLSI.2007.909792
9. L. Ciampolini, J. Lafont, F.T. Drissi, J. Morin, D. Turgis, X. Jonsson, C. Desclèves, J. Nguyen, *2016 IEEE/ACM International Conference on Computer-Aided Design (ICCAD)* (2016), pp. 1–8. https://doi.org/10.1145/2966986.2967031
10. A. Singhee, R. Rutenbar, *Extreme Statistics in Nanoscale Memory Design*. Integrated Circuits and Systems (Springer, Berlin, 2010). https://books.google.fr/books?id=_-OypzRuaHsC
11. Y. Taur, T.H. Ning, *Fundamentals of Modern VLSI Devices* (Cambridge University Press, New York, 1998)
12. K.C. Akyel, Statistical methodologies for modeling the impact of process variability in ultra-deep-submicron SRAMs, PhD manuscript, Université de Grenoble, 2014
13. M. Qazi, M. Sinangil, A. Chandrakasan, IEEE Des. Test Comput. **28**(1), 32 (2011). https://doi.org/10.1109/MDT.2010.115
14. E. Karl, Z. Guo, Y. Ng, J. Keane, U. Bhattacharya, K. Zhang, *2012 International Electron Devices Meeting* (2012), pp. 25.1.1–24.1.4. https://doi.org/10.1109/IEDM.2012.6479099
15. K.J. Dhori, H. Chawla, A. Kumar, P. Pandey, P. Kumar, L. Ciampolini, F. Cacho, D. Croain, *2017 IEEE International Symposium on Defect and Fault Tolerance in VLSI and Nanotechnology Systems (DFT)* (2017), pp. 1–6. https://doi.org/10.1109/DFT.2017.8244429

16. L. Ciampolini, S. Gupta, O. Callen, A. Chabra, D. Dibti, S. Haendler, S. Kumar, D. Noblet, P. Malinge, N. Planes, D. Turgis, C. Lecocq, S. Azmi. J. Low Power Electron. **8**(1), 106 (2012). https://doi.org/10.1166/jolpe.2012.1175
17. D. Burnett, S. Balasubramanian, V. Joshi, S. Parihar, J. Higman, C. Weintraub, *2015 IEEE International Reliability Physics Symposium* (2015), pp. 6A.6.1–6A.6.5. https://doi.org/10.1109/IRPS.2015.7112760
18. G.L. Rosa, W.L. Ng, S. Rauch, R. Wong, J. Sudijono, *2006 IEEE International Reliability Physics Symposium Proceedings* (2006), pp. 274–282. https://doi.org/10.1109/RELPHY.2006.251228

Part II
Design Examples: From Analog RF and mmW to Digital. From Building Blocks and Circuits to SoCs

This part will describe Body-Bias usage design examples.

Chapter 6
Coarse/Fine Delay Element Design in 28 nm FD-SOI

Ilias Sourikopoulos, Andreia Cathelin, Andreas Kaiser, and Antoine Frappé

6.1 Delay Elements Review

Delay manipulation is a major concern for the reliable implementation of circuits whose purpose is timing. For example, clock phase generation, deskewing, data alignment, hazard mitigation are just some signal processing functions that could entail digital delays. However, as one acknowledges the fact that any sort of signal processing function intrinsically produces delays, then designing for delay as a performance metric is straightforward. It is exactly this intrinsic non-ideality, namely propagation delay, that is usually manipulated to build digital delay circuits (Fig. 6.1).

Depending on the application, the range and resolution when controlling a delay value differs, justifying numerous approaches in dealing with the issue. On the one hand, some sort of simple operation such as to avoid setup and hold violations in a datapath can be simply handled with some extra buffering. On the other hand, producing reliable clock signals comes along with designing elaborate systems with multiple control loops. The reason for the required complexity is high precision in implementing accurate timing operations. Common examples of systems that employ controlled delay lines include delay locked loops (DLL) [1], digitally controlled oscillators (DCO), phase locked loops (PLL) [2], asynchronous

---

I. Sourikopoulos
IRCICA, CNRS USR 3360, Villeneuve d'Ascq, France
e-mail: ilias.sourikopoulos@univ-lille.fr

A. Cathelin
STMicroelectronics, Crolles, France

A. Kaiser · A. Frappé (✉)
Univ. Lille, CNRS, Centrale Lille, Yncréa ISEN, Univ. Polytechnique Hauts-de-France, UMR 8520 - IEMN, Lille, France
e-mail: antoine.frappe@yncrea.fr

© Springer Nature Switzerland AG 2020
S. Clerc et al. (eds.), *The Fourth Terminal*, Integrated Circuits and Systems, https://doi.org/10.1007/978-3-030-39496-7_6

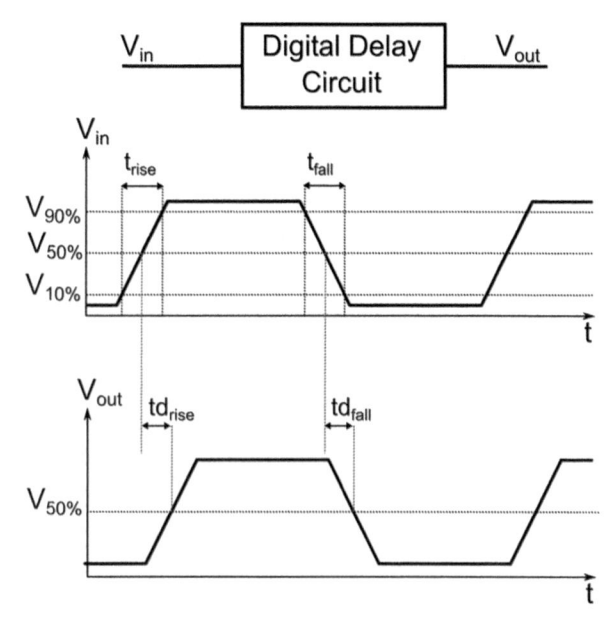

Fig. 6.1 Definition of rising and falling digital delay

pipelines [3], and time-to-digital converters (TDC) [4]. Delay circuits are also used to realize pulse-width control circuits with output programmable duty cycles used in ADCs or DACs [5] and more recently continuous-time digital filters [6]. Besides, it is not uncommon to employ dedicated delay elements for relatively simpler signal processing operations as well, like phase shifting, interpolation, or non-overlapping clock generation. Manipulating the mechanisms involved in the generation of delay has led to the introduction of multiple ideas throughout the years. Apart from the straightforward cascading of inverters, or any logic gate for that matter, there has been specific research activity targeting power-efficient, variable delay elements. Below, we present the prevalent techniques over the recent years: capacitive shunting, current starving, and thyristor-based design. After the presentation, a short discussion follows on the topology selected for further study.

6.1.1 Cascaded Inverters

One of the simplest ways to introduce a digital delay has been cascading inverters (Fig. 6.2). In this case, delay is created due to the finite slopes of charging and discharging the loaded inverter outputs. These charging and discharging slopes could be modeled based on the time constants set by the effective resistance of the switching transistors and the output capacitance. A scheme of cascaded inverters interleaved with multiplexers is a typical example implementation. A recent realization was proposed in [7], where high effort stages of cascaded inverters were used to produce delay values in the order of a nanosecond. Evidently, any

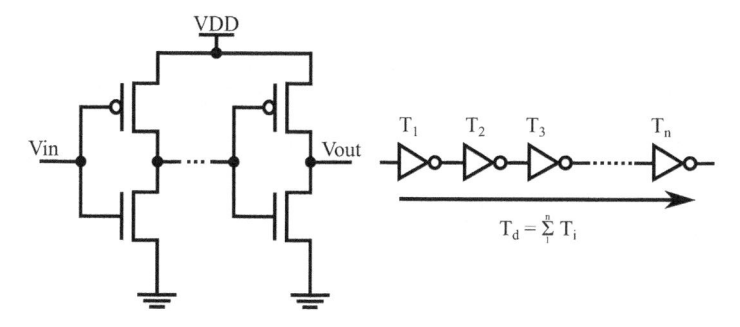

Fig. 6.2 Cascaded inverter delay

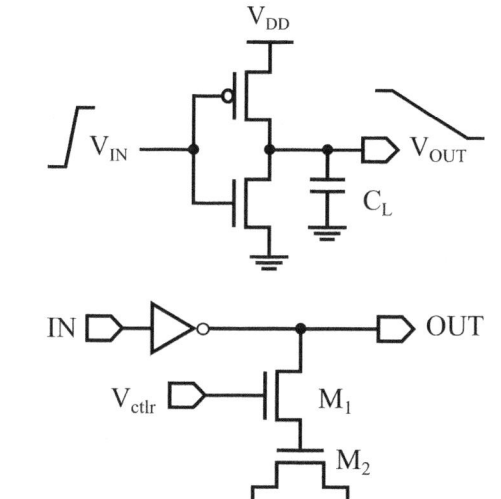

Fig. 6.3 Changing the slope due to propagation delay

Fig. 6.4 Capacitive shunting delay

sort of variation in process, supply, or temperature readily translates in variation in delay [8]. Moreover, the delay is inversely proportional to the switching slope. So, for bigger delay values there is more power consumption involved.

6.1.2 Capacitive Shunting

As seen above, for the general arrangement of cascades of inverters or in the case of inverters driving capacitive loads such as the one in Fig. 6.3, the output signal slope varies with the time constant. This represents the main mechanism for creating delay. Actually, delay can be varied by modifying output capacitance directly as in [9] or the charge flow to it, through control voltage V_{ctrl} as seen in Fig. 6.4 similarly to the proposed cell in [10].

Fig. 6.5 A semi-static approach for a delay circuit

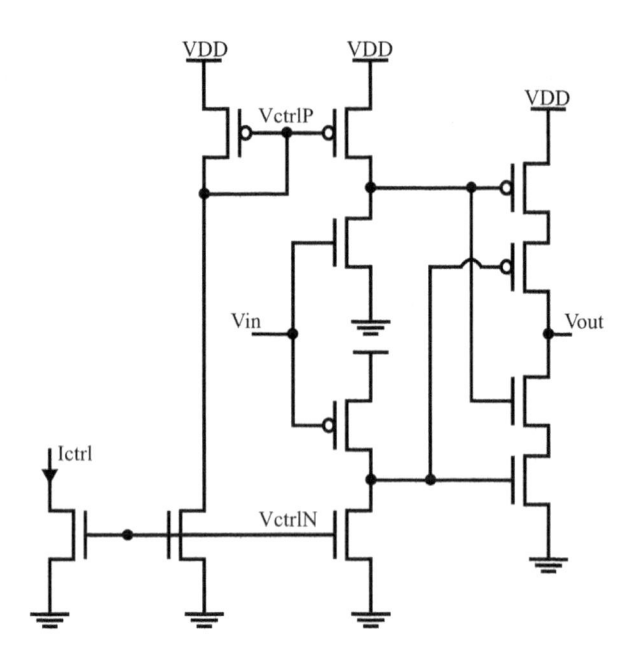

6.1.3 The Semi-static Approach

A semi-static approach to ensure minimum short-circuit current during transitions in a delay element is proposed by [11]. The topology is shown in Fig. 6.5 for producing rising-edge delays. Input is connected to two separate arrangements, where in each one, the complementary transistor is controlled by biasing a current mirror. Thus, a static consumption overhead is introduced. The two preliminary outputs are combined in a single stage by mitigating the short-circuit currents.

6.1.4 Current Starving

The current-starving technique is realized by adding extra MOS devices in series with the ones of the inverter. This effectively reduces (starves) the current associated with the switching events, which directly impacts the propagation delay.

As shown in Fig. 6.6, delay control can be established either by directly modulating the voltage on the gates of the "starving" transistors. A digital approach would be switching-in transistors parallel to the starving as shown in the work of [12] but this leads to a non-monotonic increase of the delay with the ascending binary input pattern. This is due to parasitic capacitance being added, which counteracts the reduction of effective resistance. In order to achieve a monotonic behavior, it is suggested to control only the current of the starving transistors. This could be done by mirroring the output of a current DAC at the cost of extra static consumption. This

6 Coarse/Fine Delay Element Design in 28 nm FD-SOI 123

Fig. 6.6 Current-starving technique

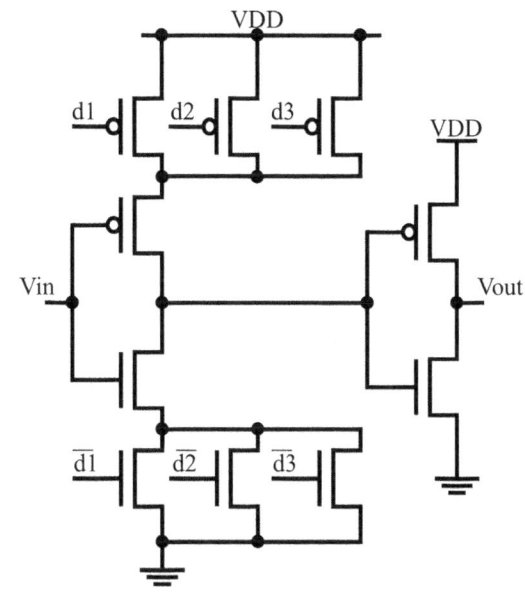

Fig. 6.7 Schmitt trigger concept for delay generation

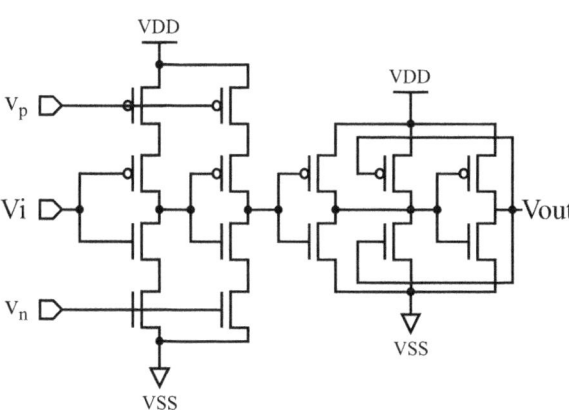

approach is proven to increase tolerance to PVT variations because the digital code reflects only the control current. On another note, the current-starving technique produces a less sharp transition slope, which actually means more short-circuit current. A way to limit this side-effect was proposed in [13] through the use of series diodes to limit the output swing, and effectively reduce power consumption.

An interesting modification to the current-starved inverter is a topology resembling the Schmitt trigger. The idea was proposed in [14] in the context of an SRAM cell. The slow transition of the starved inverter is remedied with a cascaded inverter. This idea was extended to Fig. 6.7 in the work of [15] where the Schmitt trigger is presented with positive feedback action from the output signal, therefore improving the output transitions. This renders this topology superior in terms of signal integrity and delay in comparison with current-starved, transmission-gate

Fig. 6.8 Thyristor-like switching technique

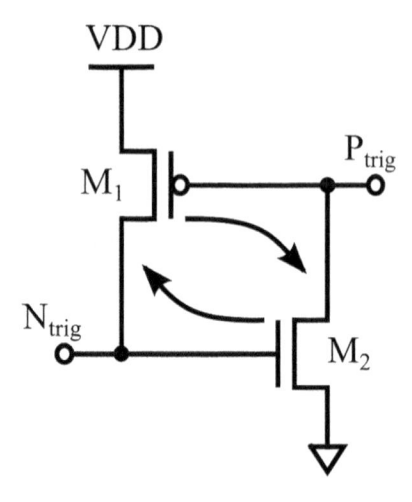

load based, cascaded inverter based set-ups. The idea of positive feedback for transition slope modification also draws from the thyristor-based delay elements, which are presented in the next section.

6.1.5 Thyristor Delay Elements

The basic principle of a thyristor device is the activation of the device when a certain conduction threshold is crossed. In a thyristor-based delay element this operation is replicated by a positive feedback mechanism that completes a delay event after crossing a threshold. The delay event could be a capacitor that is slowly charged or discharged and after a threshold voltage across its terminals is reached, the charging/discharging is forcefully accelerated through positive feedback action.

The concept is illustrated in Fig. 6.8, using a complementary transistor pair. In order to describe functionality, we will start from a steady state where both transistors are off and the gate voltages (V_{Gp} and V_{Gn}) are VDD for gate P and zero for gate N. As we lower the voltage on P, in the vicinity of VDD-VTHp, the PMOS begins to turn on. However, as the PMOS turns on, the voltage on its gate is kept driven towards the ground, not only because of the triggering action, but also because the NMOS is now beginning to turn on as well. Therefore, in addition to the initial triggering action on the PMOS gate, the NMOS introduces an additive force, which further accelerates the switch: a positive feedback loop. Obviously, to accommodate another switching cycle, the thyristor must be pre-charged again as there is no way to return to prior state. This can be realized by adding a pre-charge circuit. The complete topology of the cell, where sequencing of the delaying and pre-charging events take place is shown in Fig. 6.9.

It works with two similar parts for supporting rising and falling edges, where the delayed rising output pre-charges the part that delays the falling edge. For example, before any rising edge of D, $\bar{Q}$ has already reached high state through the preceding

Fig. 6.9 Thyristor-based delay element

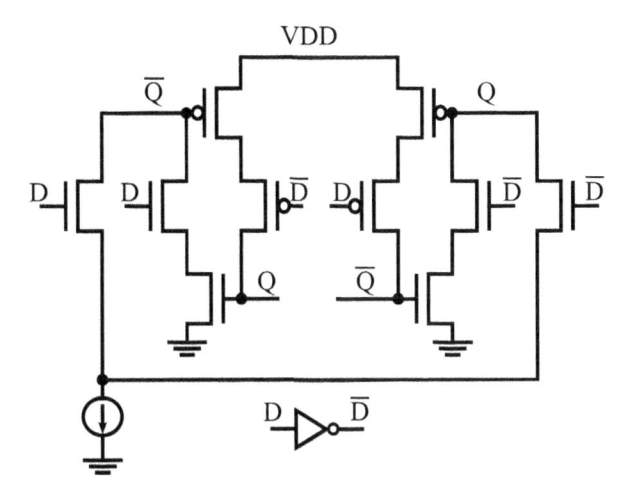

high to low transition of D. So, the left side thyristor is ready to delay the rising edge of D. The thyristor-based methodology exhibits attractive characteristics in terms of power consumption and supply sensitivity. Power is consumed during switching with a small shunt current. Primarily, there is no static consumption apart from any control current generation mechanism. Also, supply sensitivity is generally low because the delay is composed of two components: the controlled current part and the switching part. The switching part is the only part depending on the supply value, because this is when a charge cycle is completed through positive feedback. If this duration is negligible in terms of the desired delay value, sensitivity to the supply voltage is minimal. The thyristor-based delay element has attracted attention due to its special characteristics and various works have been proposed based on it. For instance, in [16], the fact that the thyristor-based topology can suffer from charge sharing is acknowledged. The parasitic capacitance tied to $\bar{Q}$ is shared with the source capacitance of the input transistors. The work proposed the addition of switches to pre-charge the output nodes, prior to switching the input.

In the work of [17], the thyristor-based topology was used to accommodate a delay cell that would support narrowly spaced bursts of asynchronous pulses. The delay element featured in this work produced delayed pulses of determined width. So, with respect to Fig. 6.9, only one side of the circuit was used. The circuit made use of an additional capacitor element, which can be slowly discharged through the current source. The same concept is followed for a similar circuit in the work of [6]. The design comprised two capacitors and two current sources with independent biases for the same delay cell, in order to support further configurability. The same general methodology is used for the thyristor delay topology in [18]. The work proposed two ideas on the basic scheme: (1) the addition of an extra current source to accommodate a sharper edge and reduce the transition shunt current and (2) a series diode connected with the output transistor pair in order to modify the thyristor activation point.

Table 6.1 Qualitative comparison of delay element types

	Casc. inverters	Cap. shunt	Semi-static	Cur. starved	Thyristor
Static power	No	No	Yes	Yes	No
Energy per toggle	∝ delay	∝ delay	∝ delay	∝ delay	Constant
Supply sensitivity	High	Average	Average	Average	Low
Temperature sensitivity	High	Average	Average	Average	Very low

6.1.6 Choosing a Delay Element

Table 6.1 presents a qualitative comparison of the characteristics of the delay element types that were detailed above. Evidently, there have been a lot of ideas to produce digital delay elements, but out of the main topology classes described above, the one that seems more attractive for advanced CMOS technology nodes is the thyristor-based one. This is because of its superiority against supply and temperature variations, while there is no static consumption. All the reviewed implementations of the thyristor-based delay element are based on a usual control mechanism during transitions by modulating the charge/discharge current. As different technology nodes are targeted and delay ranges differ, it is difficult to quantify the efficiency of the topologies. To overcome this issue, simulations have been reported for comparisons. But even so, it is not clear what amount of optimization has gone into the comparisons and yet, no variant has proposed a remarkable improvement over the main topology.

6.2 Coarse/Fine-Tuning Delay Element and Line Using Gate and Body-Biasing in 28 nm FD-SOI

It is clear that selecting the proper delay cell depends mainly on its assignment. For the circuit implementation of this section the target is the delay line specification described in [19]. The delay line is specified as a part of the feedback branch of a high-speed decision feedback equalizer, aiming at digital processing in the vicinity of 1–2 Gbps. This determines a regime for the delay range of a few hundreds of picoseconds. The delay line is also specified with a wide range of configurability, in the order of nanoseconds. Based on the aforementioned categories the proposed granular delay element is a thyristor-type and is based on an improvement of the one in [20]. The modifications effectively limit short-circuit current consumption and extend functionality in providing delay control through transistor body-biasing. Implemented in 28 nm UTBB FD-SOI (ultra-thin body and buried oxide fully depleted silicon on insulator) CMOS technology, the transistor's body terminal serves as a fine control knob, which complements the coarse gate control. This way an unprecedented coarse/fine control scheme is realized without any extra hardware [21].

6.2.1 Delay Element Design

The proposed delay cell topology is displayed in Fig. 6.10 and three stages can be readily distinguished. The first stage is a current-starved inverter, whose output (V_C) is connected to a second gated inverter stage. The second stage produces a feedback signal, V_F, which controls transistors placed in parallel with the ones implementing current-starving. Finally, a driving stage generates complementary output signals. The circuit follows the typical thyristor-type operation. For a rising input, operation can be briefly summarized as follows: As input V_{IN} rises, V_C discharges slowly due to the current-starving scheme and up to the point where the second stage inverter threshold is crossed. At that time, the parallel transistor is activated, through V_F, which effectively cancels the current-starving action. So, the discharge of V_C is then accelerated and the switch is completed. The element's consumption profile is dominated by the current-starved charging/discharging interval, which produces a short-circuit current up to the switching of the second stage. In an attempt to minimize the consumption of this action, the cell is designed with complementary inputs ensuring that the $V_{\overline{IN}}$ signal precedes its complement. This is realized by cascading delay cells. Arriving directly on the second stage inverter, the input's complement plays a preparatory role for the upcoming switch. As V_{IN} arrives in a high state, $V_{\overline{IN}}$ has been already driven low, shutting the path to the ground. This action enables the slow second stage output to be raised without short-circuit current loss. For the delay element at hand, substrate biasing contributes to current-starving. This is made possible thanks to FD-SOI technology. As seen in Fig. 6.11, the transistor channels are implemented with a buried oxide layer underneath, which prevents source-bulk junction leakage. This ensures a body-tie voltage range which is much wider than in bulk CMOS. Also, the body coefficient is much more important in FD-SOI (85 mV/V) than in the equivalent bulk node (40 mV/V).

Regarding control of the drain current, body-biasing is principally of the same nature as gate biasing. However, the electric field is applied through a thicker

Fig. 6.10 Coarse-fine delay element using transistor body-biasing

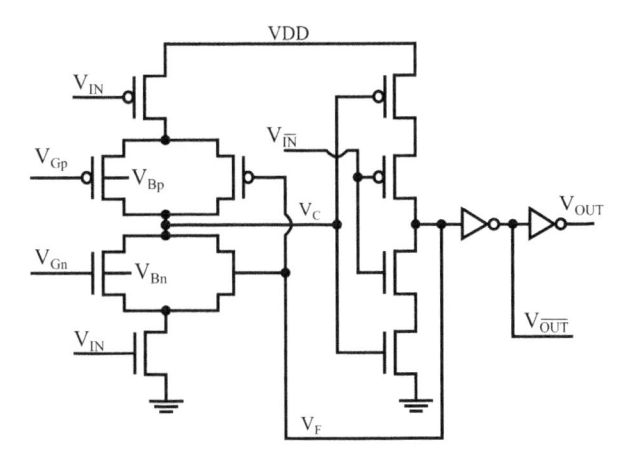

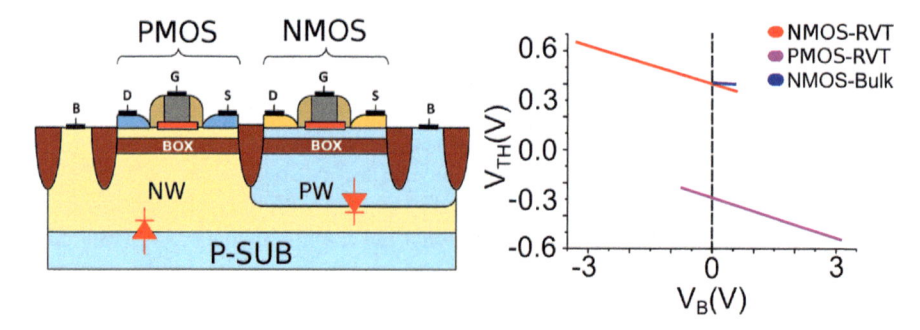

Fig. 6.11 FD-SOI transistor cross-section (left) and threshold voltage (V_{TH}) variation with body-biasing (V_B) for the technology offered RVT-type transistors (right)

oxide and a more distant contact. One could easily predict that the effect on current modulation would be less profound. This is exactly the key observation for establishing an extra control knob to fine-tune the delay.

The transistors of the circuit are sized to produce delays in the order of picoseconds. In Fig. 6.12, on the top, we simulate the falling-edge delay variation, T_d, against the control inputs. The NMOS starving transistor biases are fixed to $V_{Gn} = 500$ mV for the gate, and $V_{Bn} = 0$ mV for the body. The red curve refers to keeping the PMOS body as $V_{Bpn} = 1$V and varying V_{Gp}. It is a typical exponential variation with a steep ascent as current is starved under conduction threshold. In blue, we plot a family of curves, which refer to fixed values of $V_{Gp} = 500$ mV, 570 mV, etc. and varying V_{Bp}. The results confirm the above observation and reveal another important effect: careful combination of gate/body-biasing can lead to obtaining a segment of the delay range that refers to an almost linear part of the curve. To elaborate this, we can assume a gate voltage in the vicinity of 500 mV as seen in the bottom plot of Fig. 6.12. For this setting, the delay versus the body-bias variation provides a high correlation with a linear function, as well as a very fine-tuning capability in the order of fs/mV.

6.2.2 Delay Line Architecture

In an effort to fully investigate the topology and the control flexibility that body-biasing offers, a prototype delay line was designed and fabricated. The transistors of the first stage current-starved inverter were laid out in twin and triple-well arrangements in order to ensure isolation. The imposed constraints of well spacing lead to an overall unit delay element area of $3 \times 7\,\mu$m. The well arrangement is shown in Fig. 6.13.

The delay line design involves a cascade of granular cells organized in groups, as seen in Fig. 6.14. Group size was optimized based on the granular delay range in order to guarantee minimum overlap when programming a delay value. Each group

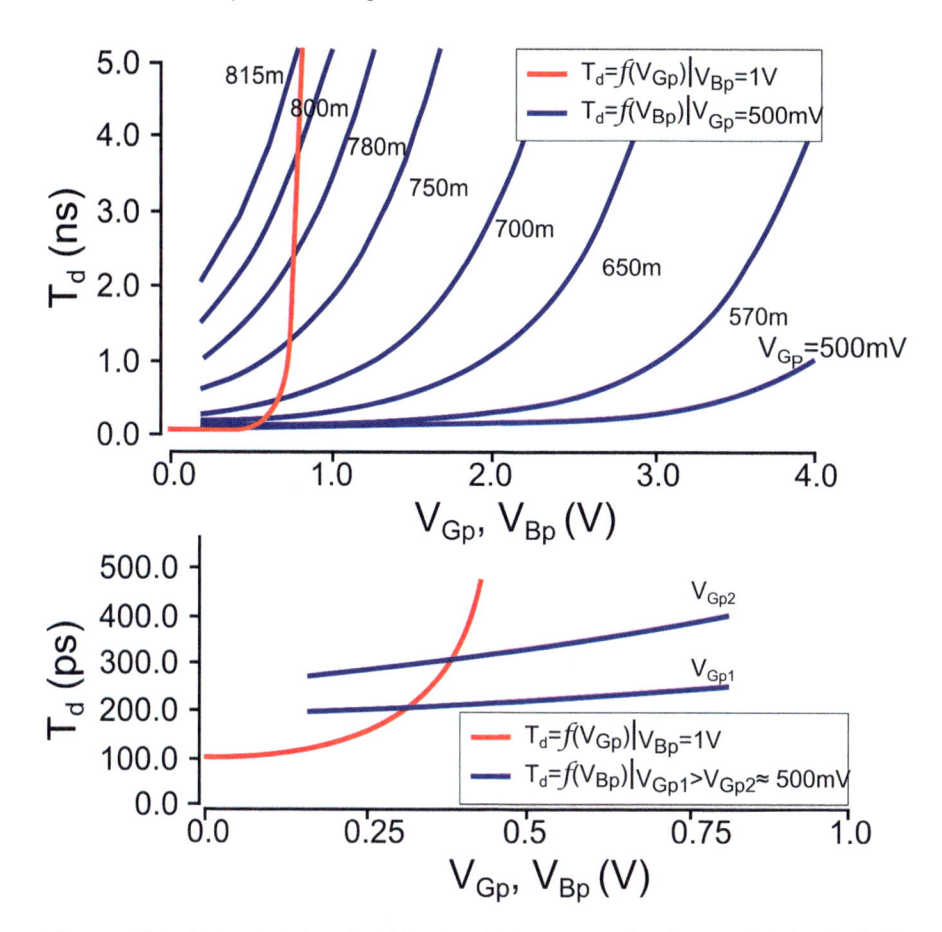

6 Coarse/Fine Delay Element Design in 28 nm FD-SOI

Fig. 6.12 Simulations of falling-edge delay time. Red curves: varying the gate while keeping body constant. Blue curves: varying the body while keeping the gate constant. Curves are presented on same x-axis to illustrate the evolution of delay, but are not coincident as they refer to different biasing conditions

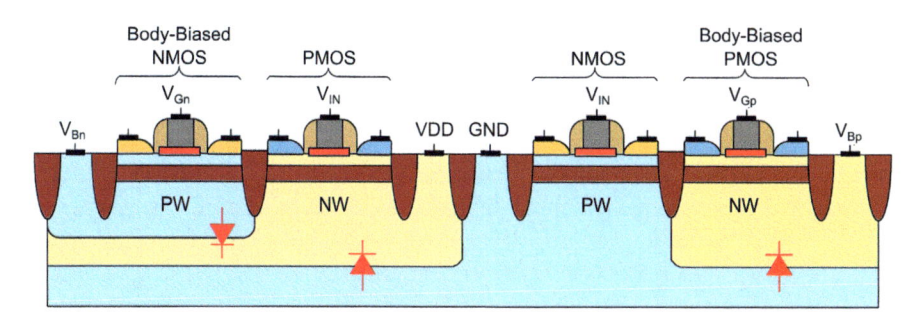

Fig. 6.13 Well arrangement accommodating the body-biasing stage

Fig. 6.14 (a) Delay line architecture. (b) Inter-group transition and delay line output are implemented with lead cells. (c) Lead cell logic

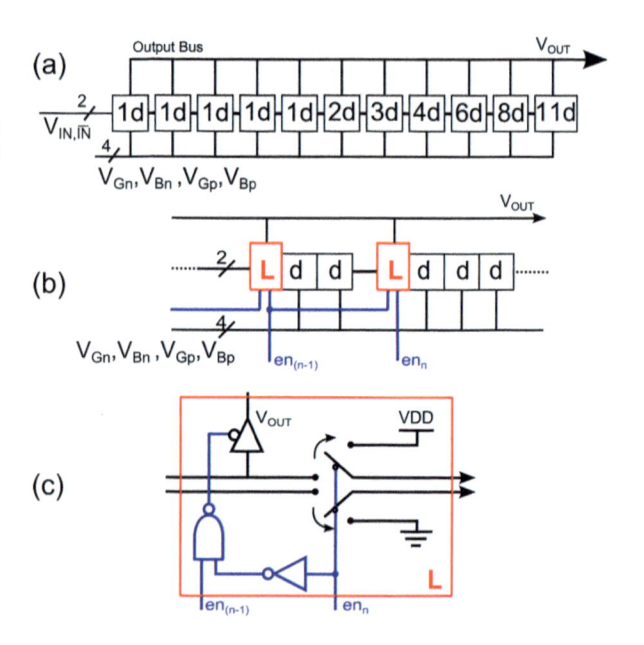

is associated with a tap output to facilitate programmability. For this purpose, all tap output nodes are connected to a common bus. Control of the bus is carried out with an 11-bit thermometer-coded word that activates the selected output.

In order to minimize power consumption, a local power-down scheme is established for the delay groups that are not active for a given delay value. This is realized through the use of lead cells (indicated as L-blocks in Fig. 6.14b), which are placed between groups. Their function is to enable the output of the last active group and propagate a steady state to the remaining ones. The logic is shown in Fig. 6.14c. The prototype chip was fabricated in 28 nm UTBB FD-SOI CMOS technology by STMicroelectronics. The chip photograph is shown in Fig. 6.15.

It comprises two delay lines: one with external analog control of body-biasing and one with digital control, through on-chip 8-bit R-string DACs. A dummy output was used to de-embed the delays coming from pad drivers and external routing. The delay lines were designed with independent power supplies to permit direct power consumption measurements. The delay line control words and on-chip DAC inputs were provided from custom on-chip control registers. In order to perform the measurements, a test-bench was built with the die being wire-bonded on a PCB. The on-chip control registers as well as on-board DACs used for generating gate and body-bias signals were programmed though the USB port of a PC, using a serial programming interface.

Fig. 6.15 Delay line chip detail

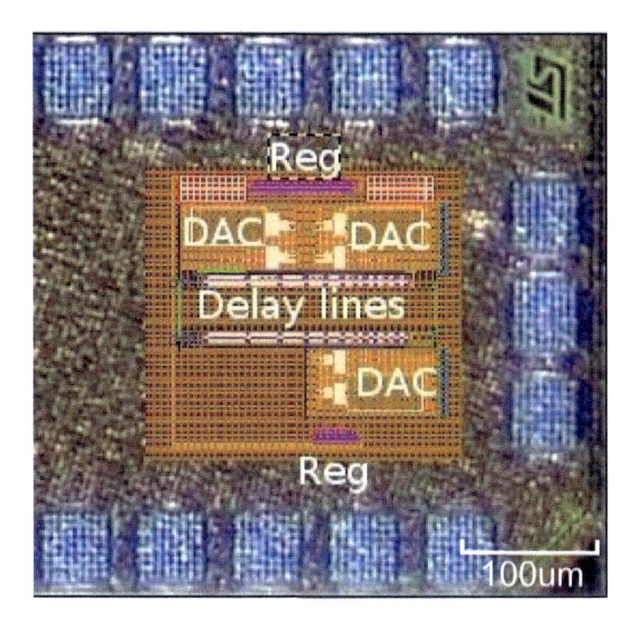

6.2.3 Delay Line Measurement Results

The good functionality of the delay line under different control scenarios was verified. Various measurements of rising and falling edges were performed in different rates up to 2 GHz. Figure 6.16 presents the variation of delay for a fully active line with body-biasing applied over a range of 1.6 V, both for rising and falling edges. The coarse/fine character is illustrated.

Another scenario entailed activating consecutively the delay groups to characterize the complete line programmability. Rising-edge delay range is measured under fixed gate biasing: $V_{Gn} = 400$ mV, $V_{Gp} = 600$ mV. Keeping $V_{Bp} = 1$ V, this scenario involved activating stages incrementally and varying V_{Bn} body-biasing between 0.2 and 1 V. These are the limits between inter-well junction diode inversion and nominal supply. The delay versus the programming vector, up to activating all stages, is displayed in Fig. 6.17.

The performance results are summarized in the second column of Table 6.2. "Minimum coarse sensitivity" refers to the minimum variation of delay achievable with enabling only the first group, while "minimum fine sensitivity" refers to varying body-biasing at this setting. The "maximum delay" field refers to the general gate and body-biasing extremes and verifies the known wide-spanning capabilities of a thyristor-type delay. Moreover, granular delay characterization was pursued through multiple measurements of stage by stage activation to de-embed lead cells and output drivers. The results refer to a scenario of rising-edge delays under $V_{Gn} = V_{Bn} = 500$ mV, $V_{Gp} = 0$ V and $V_{Bp} = 1$ V. The measured control sensitivity is 50 fs/mV and energy efficiency is 12.5 fJ/event. It is to note that delay

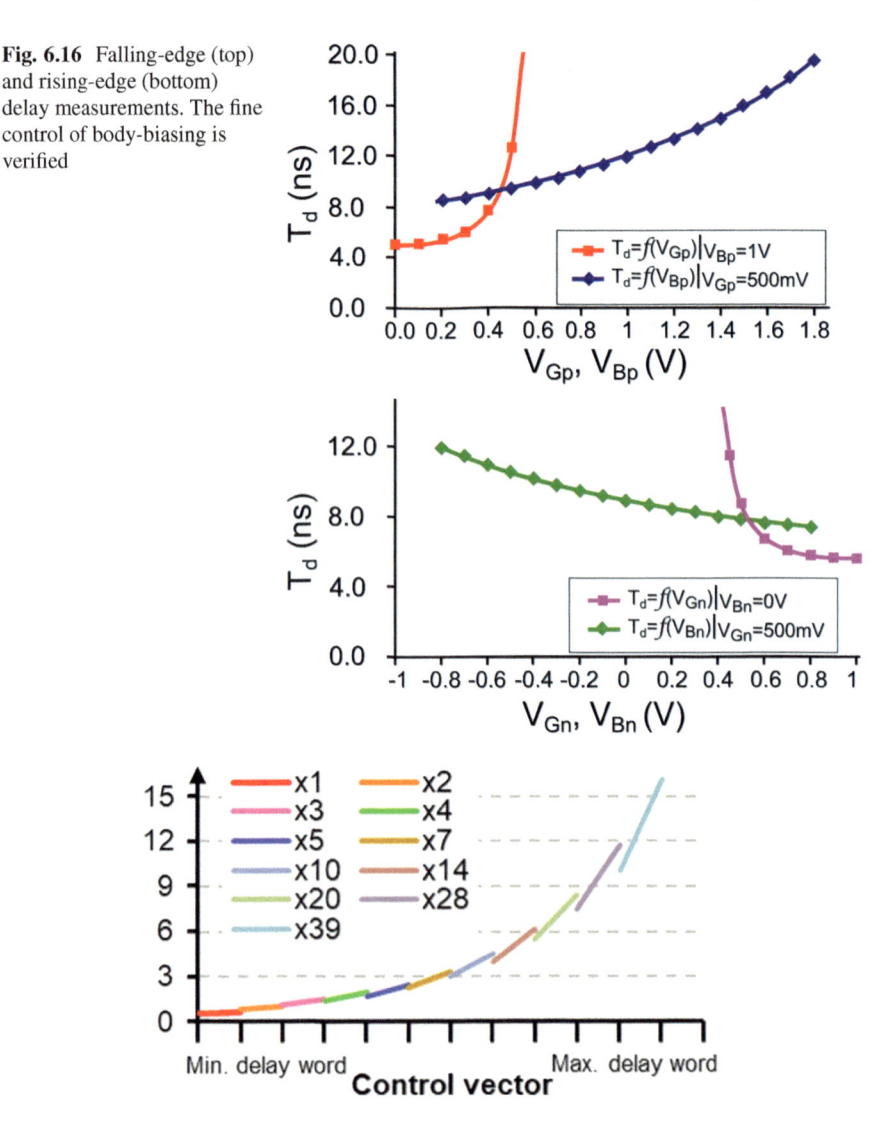

Fig. 6.16 Falling-edge (top) and rising-edge (bottom) delay measurements. The fine control of body-biasing is verified

Fig. 6.17 Rising delay vs. control vector: 1–39 delay elements at minimum and maximum values of body-bias. The curve shape reflects group sizing

range trades off with sensitivity. Therefore, depending on the application needs, obtaining a smaller sensitivity is possible while impacting the delay range.

Comparison with the current state of the art is done in Table 6.3. The advanced FD-SOI CMOS node and the unique body control scheme mitigate the use of extra hardware, such as current sinks, which impose a static consumption overhead. This way the lowest reported power consumption is achieved.

Table 6.2 Performance summary

This work	Delay line	Granular delay
Range (2 GHz input)	530 ps–16.13 ns	110–500 ps
Min. coarse sensitivity	120 ps/mV	600 fs/mV
Min. fine sensitivity	150 fs/mV	50 fs/mV
Max. delay	170 ns	4.5 ns
Efficiency	668 fJ/bit	12.5 fJ/bit
Area	980 μm^2	21 μm^2

Table 6.3 Delay element state of the art

Ref.	[18]	[17]	[20]	This work
Tech. node	0.35 μm	90 nm	65 nm	28 nm FD-SOI
Supply	3.8 V	1 V	1 V	1 V
Delay range	4 μs–22 ms	5 ns–1 μs	95–250 ps	110 ps–4.5 ns
Control	Current sink	Gate voltage	Current sink	Gate/Body (coarse/fine)
Sensitivity	100 kHz/μA (osc)	40 ps/mV (est.)	N/A	50 fs/mV
Power	120 pJ/event	50 fJ/event	40 fJ/event	12.5 fJ/event

[20] reported consumption as per tap in a CT-DSP filter
[18] reported a three-element oscillator consumption including a digital counter

References

1. M. Hossain, F. Aquil, P.S. Chau, B. Tsang, P. Le, J. Wei, T. Stone, et al., A fast-lock, jitter filtering all-digital DLL based burst-mode memory interface. IEEE J. Solid-State Circuits **49**(4), 1048–62 (2014). https://doi.org/10.1109/JSSC.2013.2297403
2. L. Xu, S. Lindfors, K. Stadius, J. Ryynanen, A 2.4-GHz low-power all-digital phase-locked loop. IEEE J. Solid-State Circuits **45**(8), 1513–21 (2010). https://doi.org/10.1109/JSSC.2010.2047453
3. I.J. Chang, S.P. Park, K. Roy, Exploring asynchronous design techniques for process-tolerant and energy-efficient subthreshold operation. IEEE J. Solid-State Circuits **45**(2), 401–10 (2010). https://doi.org/10.1109/JSSC.2009.2036764
4. J.-P. Jansson, A. Mäntyniemi, J. Kostamovaara, A CMOS time-to-digital converter with better than 10 ps single-shot precision. IEEE J. Solid-State Circuits **41**(6), 1286–96 (2006). https://doi.org/10.1109/JSSC.2006.874281
5. J.-R. Su, T.-W. Liao, C.-C. Hung, Delay-line based fast-locking all-digital pulsewidth-control circuit with programmable duty cycle, in *2012 IEEE Asian Solid State Circuits Conference (A-SSCC)* (IEEE, Piscataway, 2012), pp. 305–8. https://doi.org/10.1109/IPEC.2012.6522686
6. C. Vezyrtzis, W. Jiang, S.M. Nowick, Y. Tsividis, A flexible, clockless digital filter, in *2013 Proceedings of the ESSCIRC (ESSCIRC)* (2013), pp. 65–68. https://doi.org/10.1109/ESSCIRC.2013.6649073
7. Y.W. Li, K.L. Shepard, Y.P. Tsividis, A continuous-time programmable digital FIR filter. IEEE J. Solid-State Circuits **41**(11), 2512–20 (2006). https://doi.org/10.1109/JSSC.2006.883314
8. J.M. Rabaey, A.P. Chandrakasan, B. Nikolić, *Digital Integrated Circuits: A Design Perspective*, 2nd edn. Prentice Hall Electronics and VLSI Series (Prentice Hall, Upper Saddle River, 2003)
9. M. Bazes, A novel precision MOS synchronous delay line. IEEE J. Solid-State Circuits **20**(6), 1265–71 (1985). https://doi.org/10.1109/JSSC.1985.1052467

10. M. Maymandi-Nejad, M. Sachdev, A digitally programmable delay element: design and analysis. IEEE Trans. Very Large Scale Integr. VLSI Syst. **11**(5), 871–78 (2003). https://doi.org/10.1109/TVLSI.2003.810787
11. L.H. Jung, N. Shany, A. Emperle, T. Lehmann, P. Byrnes-Preston, N.H. Lovell, G.J. Suaning, Design of safe two-wire interface-driven chip-scale neurostimulator for visual prosthesis. IEEE J. Solid-State Circuits **48**(9), 2217–29 (2013). https://doi.org/10.1109/JSSC.2013.2264136
12. M. Maymandi-Nejad, M. Sachdev, A monotonic digitally controlled delay element. IEEE J. Solid-State Circuits **40**(11), 2212–19 (2005). https://doi.org/10.1109/JSSC.2005.857370
13. J.-L. Yang, C.-W. Chao, S.-M. Lin, Tunable delay element for low power VLSI circuit design, in *TENCON 2006 - 2006 IEEE Region 10 Conference* (2006), pp. 1–4. https://doi.org/10.1109/TENCON.2006.344092
14. A. Sekiyama, T. Seki, S. Nagai, A. Iwase, N. Suzuki, M. Hayasaka, A 1-V operating 256-kb full-CMOS SRAM. IEEE J. Solid-State Circuits **27**(5), 776–82 (1992). https://doi.org/10.1109/4.133168
15. N.R. Mahapatra, A. Tareen, S.V. Garimella, Comparison and analysis of delay elements, in *The 2002 45th Midwest Symposium on Circuits and Systems, 2002. MWSCAS-2002*, vol. 2 (2002), pp. II-473–II-476. https://doi.org/10.1109/MWSCAS.2002.1186901
16. J. Zhang, S.R. Cooper, A.R. LaPietra, M.W. Mattern, R.M. Guidash, E.G. Friedman, A low power thyristor-based CMOS programmable delay element, in *Proceedings of the 2004 International Symposium on Circuits and Systems ISCAS '04*, vol. 1 (2004), pp. I-769–72. https://doi.org/10.1109/ISCAS.2004.1328308
17. B. Schell, Y. Tsividis, A low power tunable delay element suitable for asynchronous delays of burst information. IEEE J. Solid-State Circuits **43**(5), 1227–34 (2008). https://doi.org/10.1109/JSSC.2008.920332
18. B. Saft, E. Schafer, A. Jager, A. Rolapp, E. Hennig, An improved low-power CMOS thyristor-based micro-to-millisecond delay element, in *ESSCIRC 2014 - 40th European Solid State Circuits Conference (ESSCIRC)* (IEEE, Piscataway, 2014), pp. 123–126. https://doi.org/10.1109/ESSCIRC.2014.6942037
19. I. Sourikopoulos, A. Frappé, A. Kaiser, L. Clavier, A decision feedback equalizer with channel-dependent power consumption for 60-GHz receivers, in *International Symposium in Circuits and Systems (ISCAS)*, Melbourne (IEEE, Piscataway, 2014), pp. 1484–87. https://doi.org/10.1109/ISCAS.2014.6865427
20. M. Kurchuk, C. Weltin-Wu, D. Morche, Y. Tsividis, Event-driven GHz-range continuous-time digital signal processor with activity-dependent power dissipation. IEEE J. Solid-State Circuits **47**(9), 2164–73 (2012). https://doi.org/10.1109/JSSC.2012.2203459
21. I. Sourikopoulos, A. Frappé, A. Cathelin, L. Clavier, A. Kaiser, A digital delay line with coarse/fine tuning through gate/body biasing in 28nm FDSOI, in *ESSCIRC Conference 2016: 42nd European Solid-State Circuits Conference* (2016), pp. 145–48. https://doi.org/10.1109/ESSCIRC.2016.7598263

Chapter 7
Millimeter-Wave Distributed Oscillators in 28 nm FD-SOI Technology

Raphaël Guillaume, Andreia Cathelin, and Yann Deval

7.1 Introduction

Millimeter-wave (mm-wave) and terahertz (THz) frequency bands have many attractive properties, such as the ability to penetrate through matter, non-ionizing radiation, and large bandwidth availability. These features make mm-waves and THz systems suitable for many applications such as non-ionizing biomedical imaging, non-destructive quality control, and very high data rate communication.

Because of the recent increasing demand on mm-wave/THz applications, many researchers have been interested in integrating signal sources in this frequency range, especially in cost-optimized silicon technology as SiGe and CMOS.

Several silicon-based mm-wave and THz oscillators have been recently reported. Nevertheless, most of them are integrated in SiGe technology and the targeted applications drive the need for high performances and very large scale integration that FD-SOI CMOS can offer. Moreover, all these mm-wave/THz systems require an efficient signal source with a maximized output power and, for some applications, the phase noise performances are critical. For these reasons we explore here the 28 nm FD-SOI CMOS technology for mm-wave/THz applications and demonstrate its capabilities for mmW design with industrial margin, through a well-defined design methodology for integrated oscillators.

R. Guillaume · A. Cathelin (✉)
STMicroelectronics, Crolles, France
e-mail: andreia.cathelin@st.com

Y. Deval
IMS-Bordeaux, Talence Cedex, France

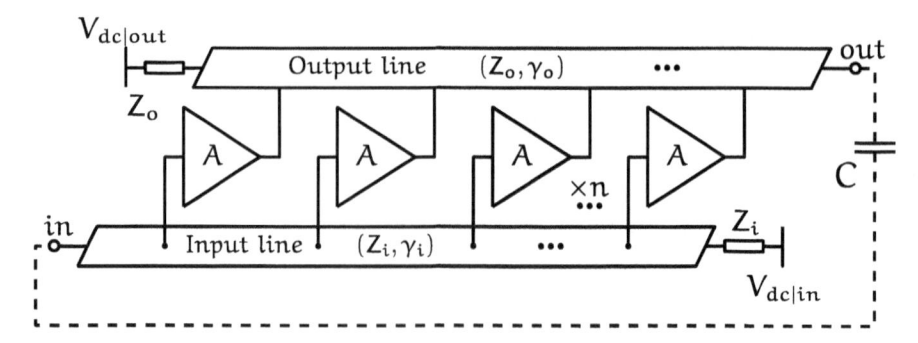

Fig. 7.1 Simplified n-stage distributed oscillator

We have chosen for this study the distributed oscillator (DO). As stated in [1], the distributed theoretical maximum oscillation frequency of such oscillators is $f_{osc|max} = \pi/2 \cdot f_T$, where f_T is the transition frequency of the used amplification stages.

As the studied technology (28 nm FD-SOI CMOS) has intrinsic f_T/f_{max} in the order of 300 GHz, we have chosen two oscillators test-cases: one around 135 GHz ($\approx f_T/2$) and one the closest possible to intrinsic f_T (intrinsic device + BEOL), at $\approx$200 GHz.

The DOs have many interesting properties mainly due to the distributed nature of the passive components (Fig. 7.1)

- First of all, the oscillation frequency is mainly function of the TL equivalent lumped component which depends only on the TL dimension [2, 3]. This means that the oscillation frequency has a very low dependence on the technological dispersion of active components.
- Due to the distributed structure, the transistors' parasitic capacitances are absorbed into the TLs [2]. This allows better open loop gain-bandwidth product and thus farther close loop operation frequency compared to a lumped harmonic oscillator.
- The TL exhibit better high-frequency quality factor than lumped components [4], thus the frequency phase noise of a mm-wave/THz DO will be better than a classical harmonic oscillator.
- By choosing the adequate number of amplification stages, it is possible to generate low-noise quadrature signals [2].

All these properties make the DO a good candidate for mm-wave and THz frequency generation on advanced CMOS technology. Nevertheless, a good understanding of DO behavior is essential. Consequently, a theoretical study is necessary.

7.2 Distributed Oscillator Theory for Operation Frequencies Close to f_{max}

Let us consider the DO presented in Fig. 7.2. The gates and drains of the n identical transistors are connected to TLs with a constant spacing of l. Each transistor has a large signal transconductance of G_m. The TLs have a characteristic impedance Z_c and a complex propagation constant $\gamma = \alpha + j\beta$, in which α and β are the attenuation and phase constant of the transmission lines, respectively. We assume here that the TLs are terminated to a matched load of Z_c. According to DO theory [2], the oscillation condition in an ideal behavior is given by

$$G_m n (Z_c/2) e^{\alpha n l} e^{j\beta n l} = 1. \tag{7.1}$$

Assuming purely real characteristic impedance Z_c, (7.1) gives the Barkhausen amplitude and phase criteria:

$$\left| G_m n (Z_c/2) e^{\alpha n l} \right| = 1, \tag{7.2}$$

$$e^{j\beta n l} = 1 \Leftrightarrow \beta n l = \pi. \tag{7.3}$$

Noting that $\beta = \frac{2\pi f}{v_\phi}$ and $v_\phi \approx \frac{1}{\sqrt{L_u C_u}}$ [4], (7.3) gives the oscillation frequency:

$$f_{osc} = \frac{1}{2 n l \sqrt{L_u C_u}}, \tag{7.4}$$

where L_u and C_u are the TLs inductance and capacitance per unit length, respectively. It is noteworthy that (7.4) can be written as $f_{osc} = \frac{1}{\sqrt{L_{tot} C_{tot}}}$, where $L_{tot} =$

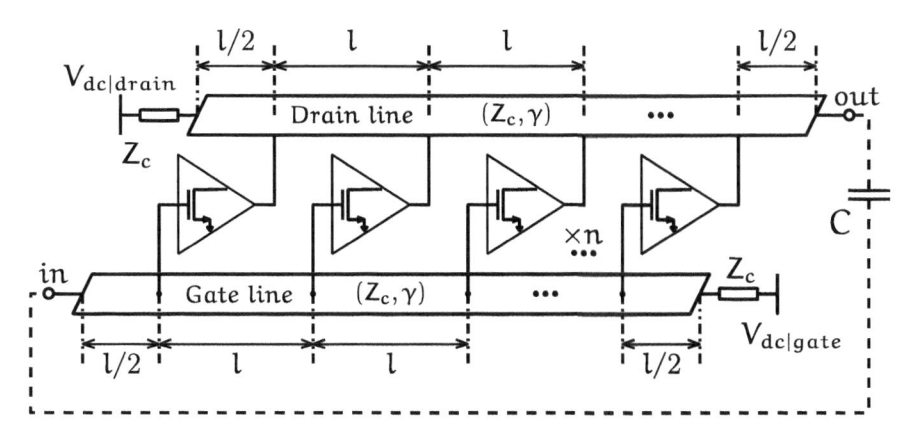

Fig. 7.2 Distributed oscillator architecture used in the theory

$2nlL_u$ and $C_{tot} = 2nlC_u$ are the total inductance and capacitance of the entire loaded line. Thus, the oscillation frequency given by (7.4) is approximately 2π times larger than that of a lumped oscillator with a tank inductance and capacitance of values L_{tot} and C_{tot}.

Equations (7.1)–(7.4) assume an ideal transistor phase-shift of π which is never the case at RF and mm-wave frequencies given the limited values for transistors' f_T/f_{max} [3, 5].

Indeed, when we apply a voltage on the gate to source junction of a standard CMOS transistor, as depicted in Fig. 7.3a, it induces a current i_{ch} in the channel with a delay (inducing a fist added phase-shift $\Delta\varphi_1$). This is because it takes time for the gate signal to travel along the gate RLC network. Moreover, due to parasitic capacitance between the gate and drain, the gate voltage also causes a feedforward current from the drain i_{dg} (inducing a second added phase-shift $\Delta\varphi_2$).

The transistor gate to drain phase relationship is illustrated in Fig. 7.3b. The total drain current i_d is the vector sum of i_{ch} and i_{dg}. Since the drain current is delayed, the drain voltage should also be delayed on top of the conventional inversion behavior of the CMOS transistor. Thus, a total transistor phase-shift correction $\Delta\varphi$ is introduced as $\Delta\varphi = \Delta\varphi_1 + \Delta\varphi_2$. Such effect is normally neglected in RF design, but in mm-wave and THz frequency ranges, it becomes very significant.

When applying this phase-shift correction to the DO theory, the oscillation condition for operation frequencies close to f_{max} becomes

$$G_m n \left(Z_c/2\right) e^{j\Delta\varphi} e^{\alpha nl} e^{j\beta nl} = 1. \tag{7.5}$$

Assuming purely real characteristic impedance Z_c in the oscillation frequency vicinity, (7.5) gives the Barkhausen amplitude and phase criteria:

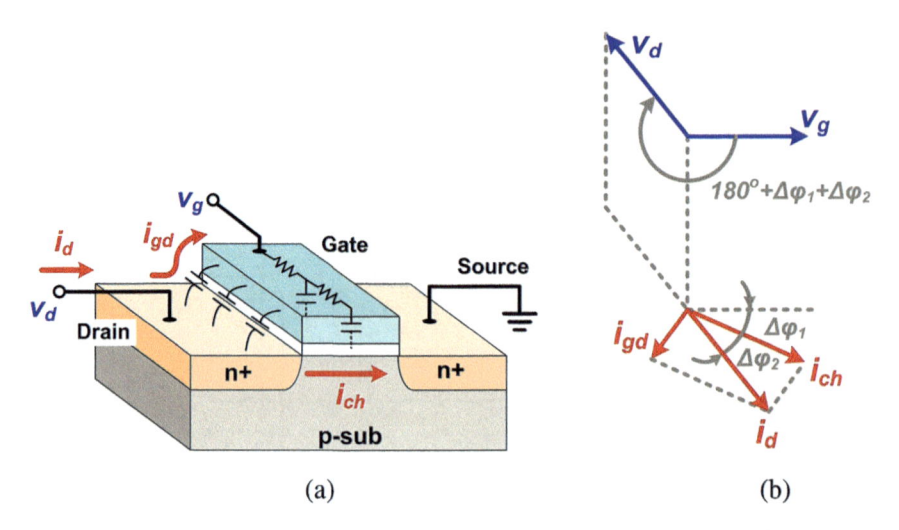

Fig. 7.3 Standard CMOS transistor and its gate/drain phase relationship. (Figures from [5].) (**a**) Standard CMOS transistor. (**b**) Gate/drain phase relationship

7 Millimeter-Wave Distributed Oscillators in 28 nm FD-SOI Technology

$$\left| G_m n \, (Z_c/2) \, e^{\alpha n l} \right| = 1, \tag{7.6}$$

$$e^{j(\beta n l + \Delta\varphi)} = 1 \Leftrightarrow \beta n l + \Delta\varphi = \pi. \tag{7.7}$$

As previously said, $\beta \approx 2\pi f \sqrt{L_u C_u}$; therefore, Eq. (7.7) gives the oscillation frequency:

$$f_{osc} = \frac{1 - \Delta\varphi/\pi}{2 n l \sqrt{L_u C_u}}. \tag{7.8}$$

It is noteworthy that, for a given TL length, the phase-shift correction involves a reduction of the oscillation frequency. In other words, to oscillate at a given frequency the phase-shift correction involves a reduction of the TL length and hence of the circuit surface.

The results of (7.8) can be found by considering Fig. 7.4. Indeed, intuitively the phase-shift induced by the TL between the same transistor drain and gate has to be equal to $\pi - \Delta\varphi$. As the TL length between one transistor drain and gate is always equal to nl, this gives $\beta nl = \pi - \Delta\varphi$ which is the result of (7.7).

Finally, in [6], a general theory of phase noise in oscillators is presented. The phase noise is given by

$$L(\Delta\omega) = 10 \cdot \log \left[\frac{kT}{V_{out}^2} \cdot \frac{1}{R_p(C\omega_0)^2} \cdot \left(\frac{\omega_0}{\Delta\omega}\right)^2 \right], \tag{7.9}$$

with k the Boltzmann's constant, T the absolute temperature, R_p and C the equivalent parallel resistance and capacitance of the phase-shift network, ω_0 the oscillation frequency, and $\Delta\omega$ the offset from the carrier.

Considering that V_{out} is proportional to $(V_{GS} - V_T)$, one can observe that the phase noise can be tuned using the threshold voltage V_T and hence using the body-bias.

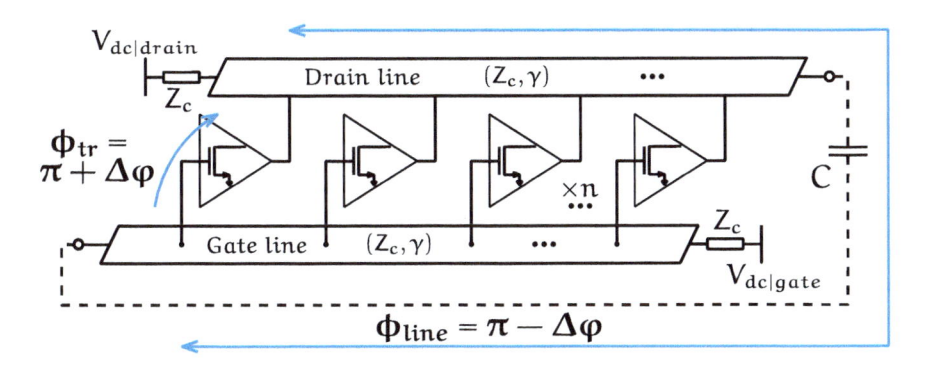

Fig. 7.4 Distributed oscillator for operation frequency close to f_{max}

7.3 Amplification Stage Design

The circuits have been implemented in 28 nm FD-SOI CMOS technology from ST Microelectronics. In these designs, only L-min LVT NMOS transistors have been used. Indeed, LVT transistors provide higher performances in terms of f_T/f_{max}.

To realize the amplification stages, a common source topology was chosen. Indeed, as said in the previous section, this topology allows high efficiency, high linearity, and a good output power.

A transistor with a total width of W_{tot} is composed of several fingers with a width W_f (Eq. (7.10)).

$$W_{tot} = N_f \cdot W_f . \tag{7.10}$$

The finger width W_f is critical for millimeter applications. It is indeed directly connected to the gate resistance responsible for the f_{max} degradation. Charts have been published in [7] (Fig. 7.5a) in order to select the appropriate gate width W_f according to the technological node.

Even if the 28 nm technology node is not reported, this chart allows making the projection that the optimum gate finger width is around 0.8 μm. This value has been confirmed as well by simulations, as shown in Fig. 7.5b.

A transistor with a total width W_{tot} can be composed of one or several cells, as presented in Fig. 7.6.

The layout in Fig. 7.6a is suitable for small transistors with only one gate access. The layout in Fig. 7.6b allows two gate accesses on each transistor sub-cells. This highly improves the reliability of the component by limiting its stress.

The total transistor width is estimated according to the desired output power and the current density in the transistor. The optimal current density $J_{DS|opt}$ is around 0.3 mA/μm. This value is independent of the technology as it has been demonstrated by [10].

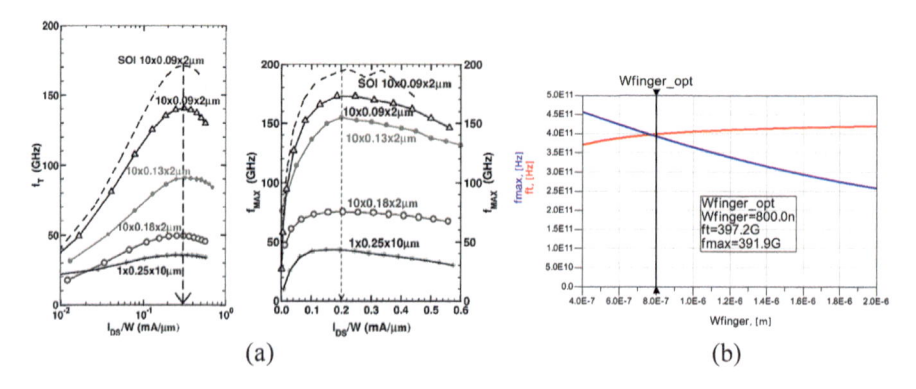

Fig. 7.5 Finger width determination: (**a**) charts proposed in [8] and (**b**) 28 nm FD-SOI NMOS f_T/f_{max} simulations as a function of finger length at a constant current density

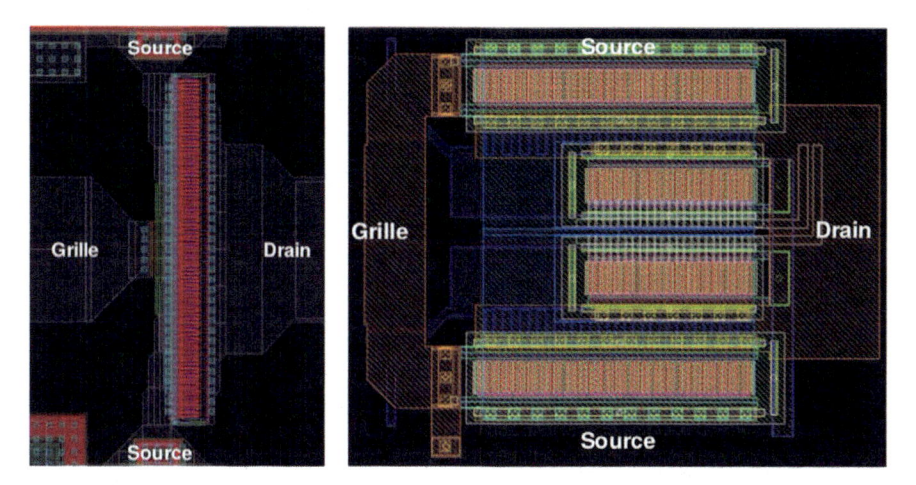

7 Millimeter-Wave Distributed Oscillators in 28 nm FD-SOI Technology 141

Fig. 7.6 Example of transistor layout from [9]: (**a**) single cell transistor and (**b**) multiple cells transistors (the topology on the right contains two different sizes transistors split each into two other ones)

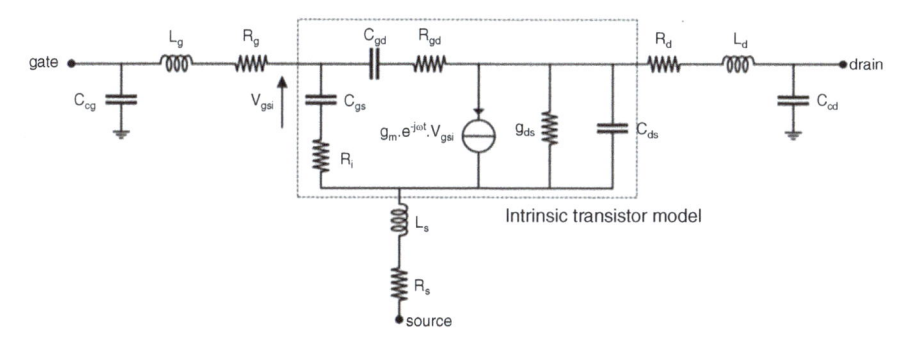

Fig. 7.7 Small-signal model of MOS transistor with its interconnections [11]

In our case, the specification is to have a total power consumption of 20 mA. Thus, assuming four amplification stages, each stage has to be biased at around 5 mA. From this and the optimal current density, the optimal total transistor width $W_{tot|opt}$ of each amplification stage is deduced using (7.11).

$$W_{tot|opt} = \frac{I_{DC|stage}}{J_{DS|opt}} \approx 16\,\mu m. \quad (7.11)$$

Our amplification stage transistor is finally sized. The total width is equal to 16 μm and the optimal finger width is 0.8 μm, with 20 fingers.

The next step is to layout specific access in order to minimize the effect of the full BEOL on the f_T/f_{max}. The small-signal model of a MOS transistor with its interconnection parasitic elements is depicted in Fig. 7.7.

Equations (7.12) and (7.13) present, respectively, the f_T and f_{max} equations deduced from the small-signal schematic.

$$f_T \approx \frac{g_m}{C_{gs}\sqrt{1+2\frac{C_{gd}}{C_{gs}}}}. \qquad (7.12)$$

$$f_{max} \approx \frac{g_m}{4\pi C_{gs}\sqrt{(R_g + R_s + R_i)\left(g_{ds} + g_m\frac{C_{gd}}{C_{gs}}\right)}}. \qquad (7.13)$$

From (7.12) and (7.13) it is obvious that f_T/f_{max} performances are degraded by the parasitic interconnections. Several techniques have been used here to reduce the effect of the BEOL interconnection and thus optimize the overall transistor performances.

First of all, to reduce parasitic capacitance, it is possible to use thin staircase accesses for the drain and the source as shown in Fig. 7.8. These accesses allow low drain-to-source parasitic capacitances, as well as the drain and source surface facing the gate is reduced. Using these techniques, [9, 11] report f_{max} amelioration of almost 20% comparing to classic layout method for the full BEOL liad out transistor.

A second optimization is the use of a dual gate access. As reported in [12], the use of dual gate accesses reduces the gate access resistance. In [9, 11], the author reports an f_{max} amelioration of approximately 8% comparing single gate access transistor, as shown in Fig. 7.9.

Figure 7.10 presents the optimized BEOL of the amplification stage transistor. A 3D view is also available in Fig. 7.11a, while Fig. 7.11b presents a cross sections through the plane A.

Fig. 7.8 Example of layout with thin staircase accesses for drain and source [9]

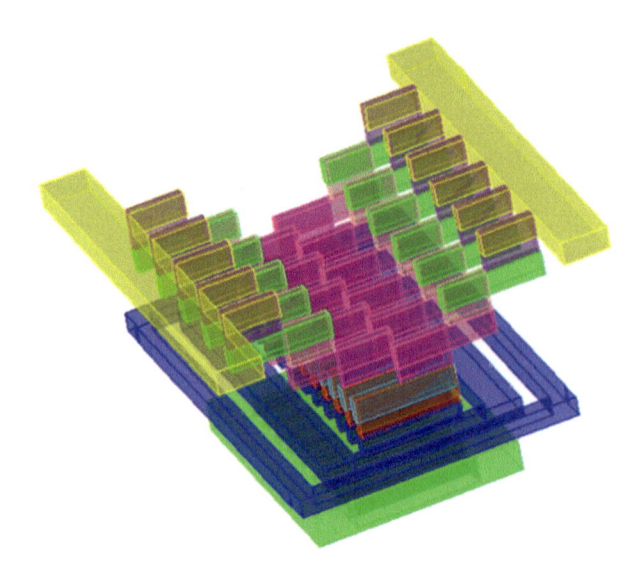

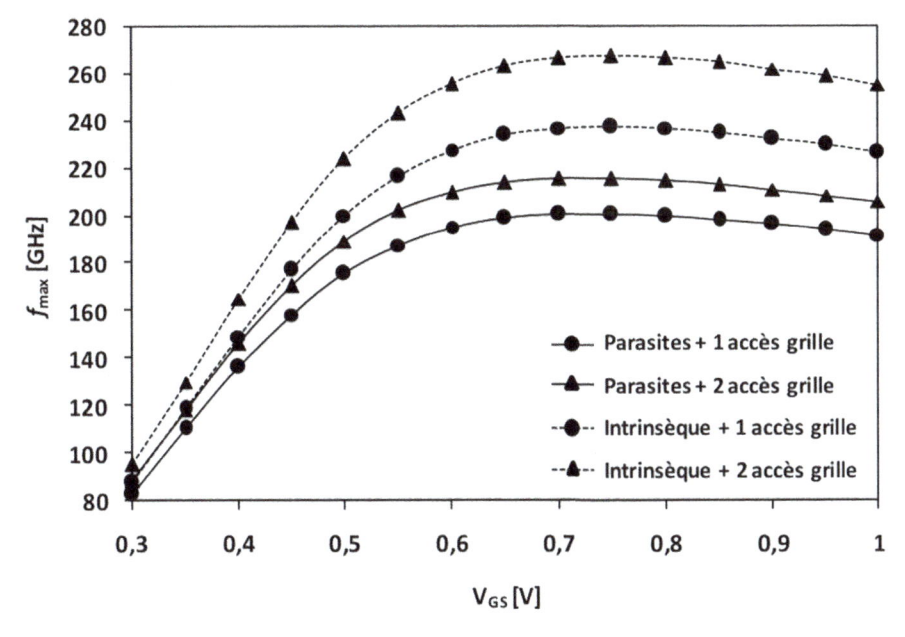

7 Millimeter-Wave Distributed Oscillators in 28 nm FD-SOI Technology 143

Fig. 7.9 Impact of dual gate access on f_{max} [9]

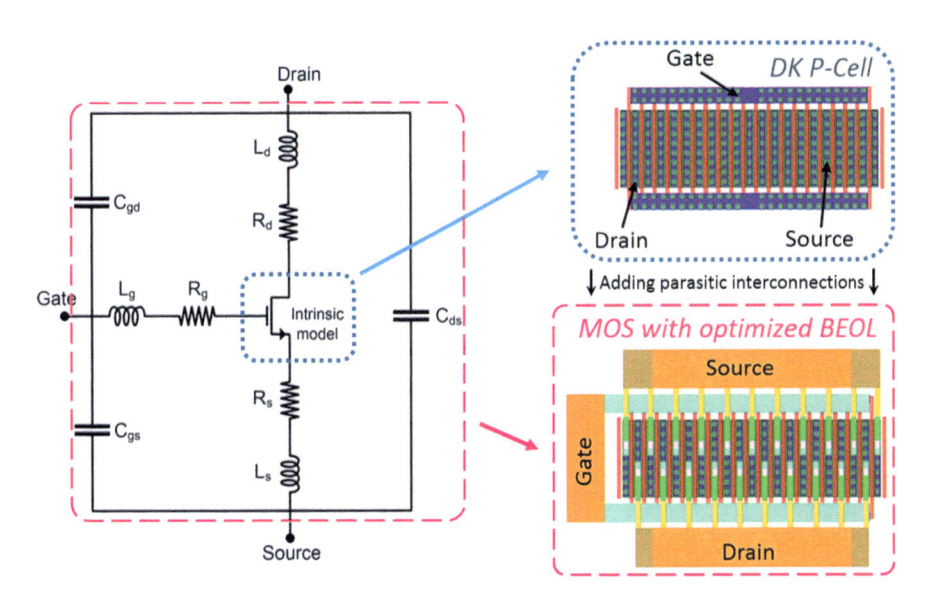

Fig. 7.10 BEOL layout optimization of the amplification stage transistor

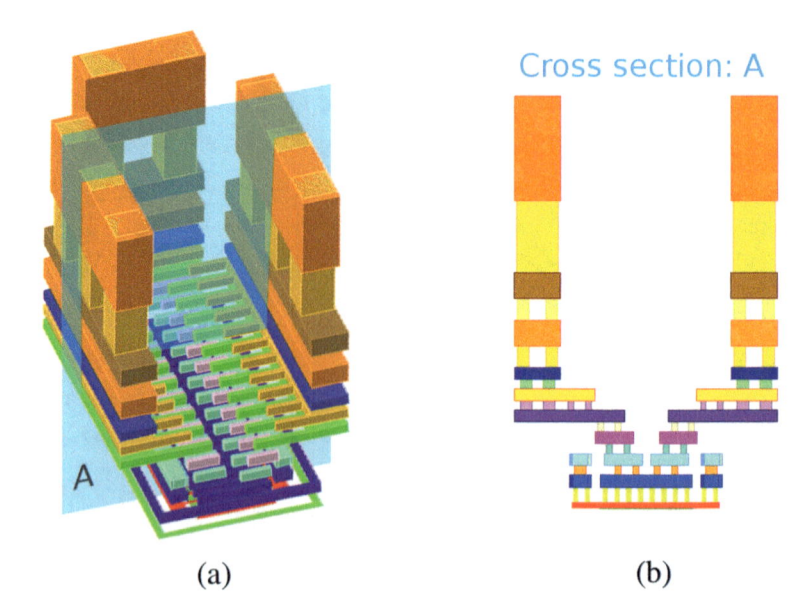

Fig. 7.11 Transistor layout with optimized BEOL. (**a**) 3D view and (**b**) cross section on the plane A

At this point, post-layout simulations must be made to evaluate the impact of the full BEOL on the f_T/f_{max} performances. To extract the parasitic capacitances and resistance a specific CAD tool has been used [13].

The cutoff frequency f_T is, by definition, the frequency at which the current gain is unity. The current gain $H21$ can be expressed from S parameters by

$$H21 = \frac{-2 \cdot S21}{(1 - S11)(1 + S22) + S12\, S21}. \tag{7.14}$$

The maximum oscillation frequency is defined as the frequency at which the maximum stable power gain is equal to unity. The maximum stable power gain can be expressed from S parameters by

$$G_{max} = \left(K - \sqrt{K^2 - 1}\right) \frac{|S21|}{|S12|}, \text{ for } K > 1. \tag{7.15}$$

The parameters f_T/f_{max} are calculated using (7.14) and (7.15) times the frequency. Figure 7.12 presents the f_T/f_{max} results with and without BEOL.

As expected, the gate access optimization permits a very small degradation of f_{max} with the full BEOL. The design kit P-Cell, which is up to metal 1, was simulated with an f_{max} of approximately 390 GHz while this value is near 370 GHz with the optimized BEOL.

The P-Cell f_T was simulated at 360 GHz. With the proposed BEOL, this value rushed approximately 280 GHz. Table 7.1 resumes theses values.

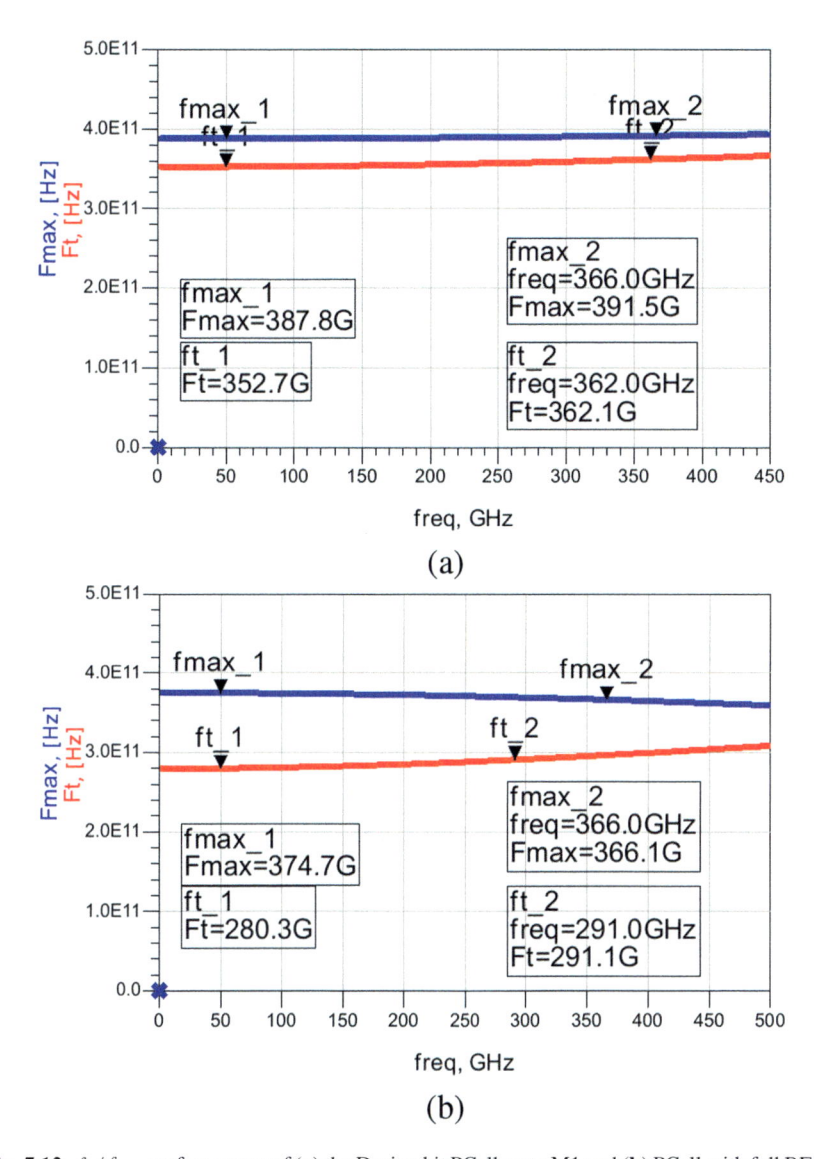

Fig. 7.12 f_T/f_{max} performances of (**a**) the Design kit PCell up to M1 and (**b**) PCell with full BEOL

It is possible to modulate the NMOS V_T using the transistor body-bias feature. To ensure that the transistor always works at its optimum, it is important to study the effect of body-bias on the f_T/f_{max}. Figure 7.13 presents the f_T/f_{max} as a function of gate voltage for several body voltages. The body voltage is here sweep from 0.5 to 2.5 V by 0.5 V steps.

It is noteworthy that the optimum bias point (i.e. the optimum gate voltage) is not necessary at maximum f_T, respectively, f_{max}. Indeed, as visible in Fig. 7.13,

Table 7.1 f_T and f_{max} simulations with and without BEOL parasitic interconnections

	Simulation with DK P-Cell	Simulation with P-Cell + BEOL
f_T	360 GHz	280 GHz
f_{max}	390 GHz	370 GHz

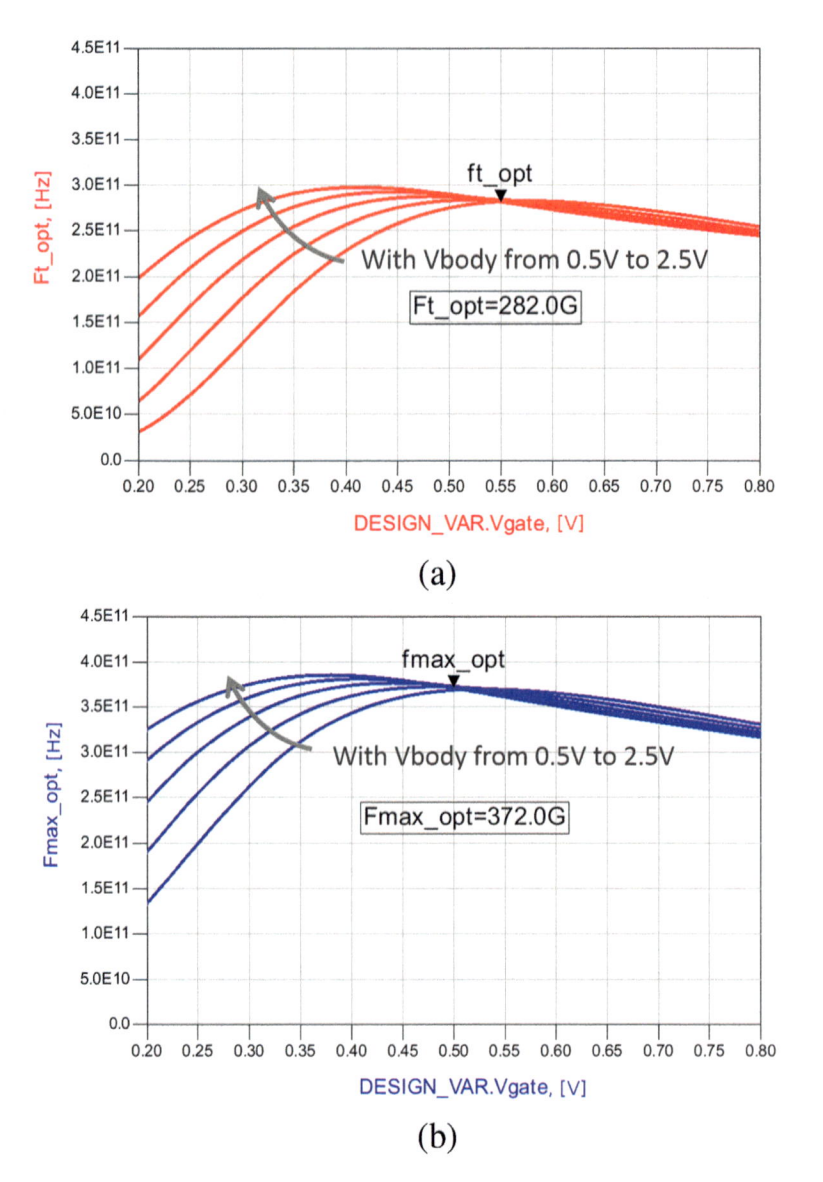

Fig. 7.13 Effect of the body-biasing on f_T/f_{max}. (**a**) f_T and (**b**) f_{max} as a function of Vgate for several Vbody

it provides the gate voltage at which f_T, respectively, f_{max}, does not change with body-biasing. This gate voltage is equal to 0.55 V for f_T and 0.5 V for f_{max}. Thus, it appears judicious to choose a gate voltage optimum point between 0.5 and 0.55 V.

7.4 Transmission Line Design

To choose the transmission line topology, it is important to take into account the distributed oscillator operating environment from the beginning. Indeed, a distributed oscillator transmission line must be loaded with a matched load. Thus, the transmission line characteristic impedance must be chosen accordingly.

To simplify our design, we choose to load the distributed oscillator with the measurement spectrum analyzer. As this kind of test equipment has usually 50 Ω ports, we choose to design a 50 Ω transmission line.

To reduce at maximum the substrate losses, we chose to use a microstrip (MS) line with a ground plane on metal 1 and metal 2. Nevertheless, due to layout density rules, it is almost impossible to have a perfect MS line. Thus, as presented in Fig. 7.14, the implemented structure was a microstrip with grounded all levels metal walls.

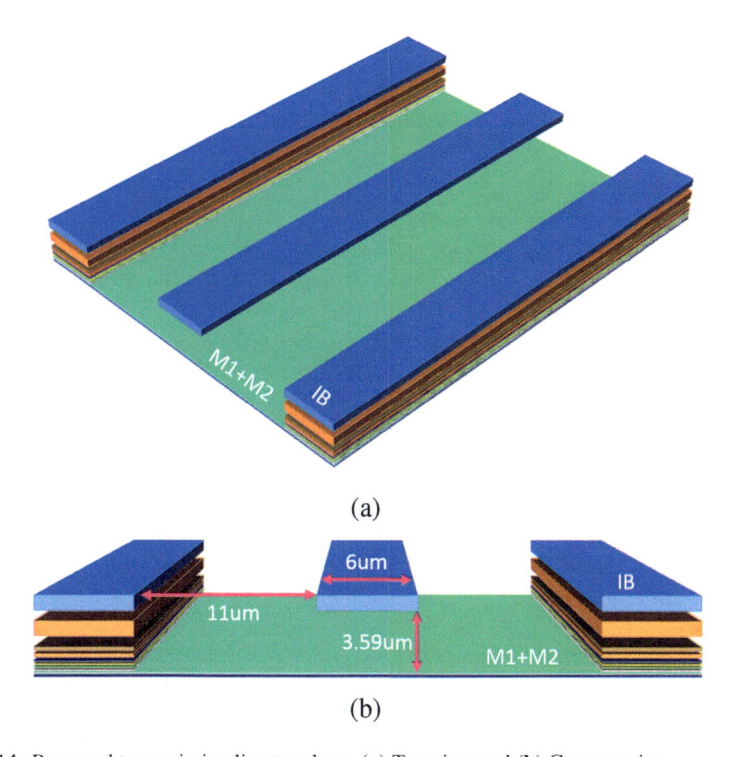

Fig. 7.14 Proposed transmission line topology. (**a**) Top view and (**b**) Cross section

This kind of topology is usually called grounded coplanar lines (CPW-G). Nevertheless, in our case, the grounded walls are far enough from the line to ensure that the major part of the electric field is still concentrated between the line and the ground plane.

The first step in transmission line characterization is to determine the characteristic impedance Z_c and the complex propagation constant γ. To do this, it is possible to exploit the S parameters coming from an electromagnetic (EM) simulation.

Indeed, the characteristic impedance can be written as:

$$Z_c = 50 \cdot \sqrt{\frac{(1+S11)(1+S22) - S12\,S21}{(1-S11)(1-S22) - S12\,S21}} . \tag{7.16}$$

And the complex propagation constant is obtained from:

$$\gamma = \frac{1}{l} \cdot acosh\left(\frac{(1+S11)(1-S22) + S12\,S21}{S12 + S21}\right), \tag{7.17}$$

with l the length of the transmission line. The complex propagation constant can be rewritten as:

$$\gamma = \alpha + j \cdot \beta . \tag{7.18}$$

The real part α represents the attenuation constant. It is usually expressed in dB/m. The imaginary part β is the phase constant. It represents the phase-shift per unit length in rad/m.

Using the Momentum EM tool, it is possible to extract the transmission line S parameters. Figure 7.15 presents the extracted Z_c and γ of the proposed transmission line over 1 mm by using Eqs. (7.16) and (7.17). It is noteworthy that our 50 Ω 1 mm length transmission line corresponds to a full wavelength at around 171 GHz.

From the real part of γ (see Eq. (7.18)), it is possible to extract the attenuation constant α, presented in Fig. 7.16a. From these results, it is possible to determine the losses per millimeter. For example, our transmission line exhibits around 1.7 dB/mm losses at 200 GHz.

The imaginary part of γ is the phase constant β, presented in Fig. 7.16a. It represents the phase-shift per unit length in rad/m. Thus, it increases until a maximum at which the wavelength is equal to $\lambda/2$ (modulus λ). Then, since the phase-shift has a modulus of 2π, the phase-shift will decrease to zero. At zero, it corresponds to a full wavelength phase-shift. Thus, if one looped back a transmission line of 1 mm and assuming no losses in the transmission line, oscillations must appear at around 171.2 GHz.

By considering a small part of the transmission line, typically smaller than $\lambda/20$, it is now possible to extract an RLCG model to use it to size our distributed

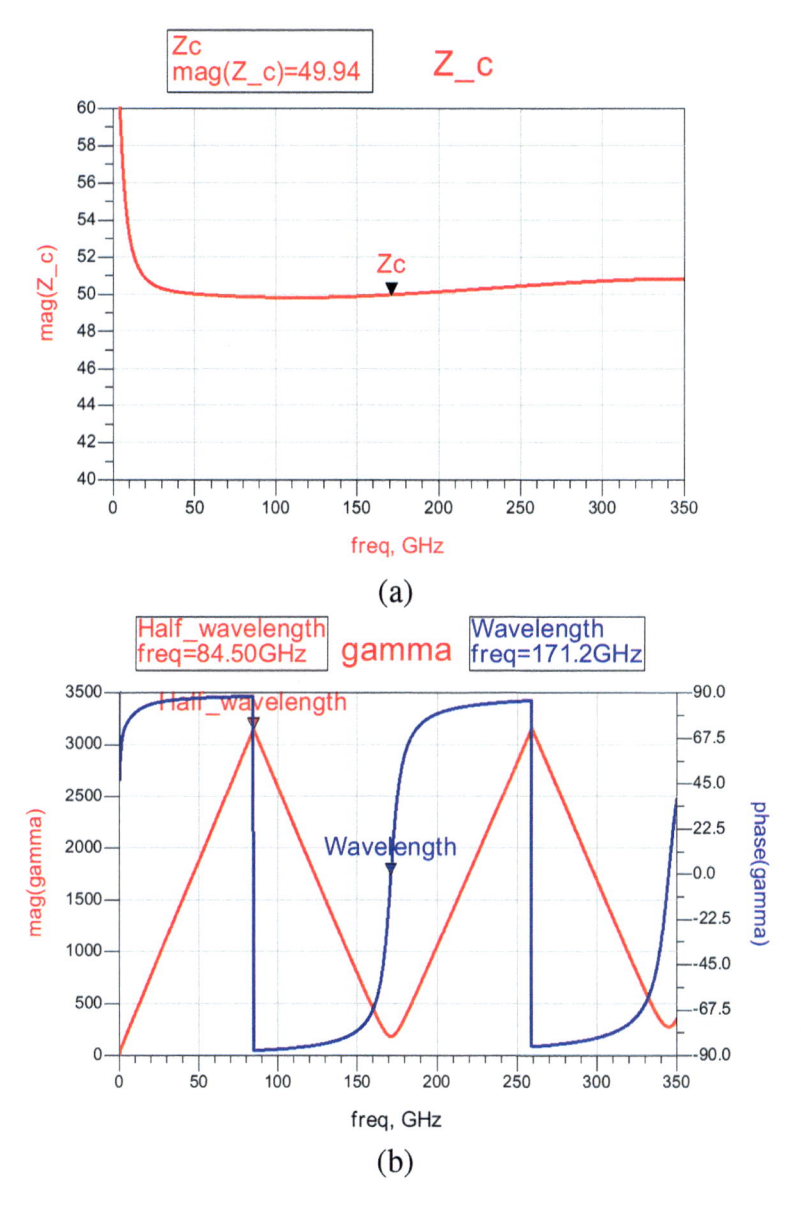

Fig. 7.15 Electromagnetic extraction of transmission line. (**a**) Characteristic impedance Z_c and (**b**) complex propagation constant γ over 1 mm

oscillator. Figure 7.17a–d presents, respectively, the resistance R_u, inductance L_u, capacitance C_u, and conductance G_u per unit length of our transmission line.

By substituting L_u and C_u in Eq. (7.8) and by rewriting this equation, it is now possible to calculate the total line length l_{tot} needed to oscillate at a given frequency.

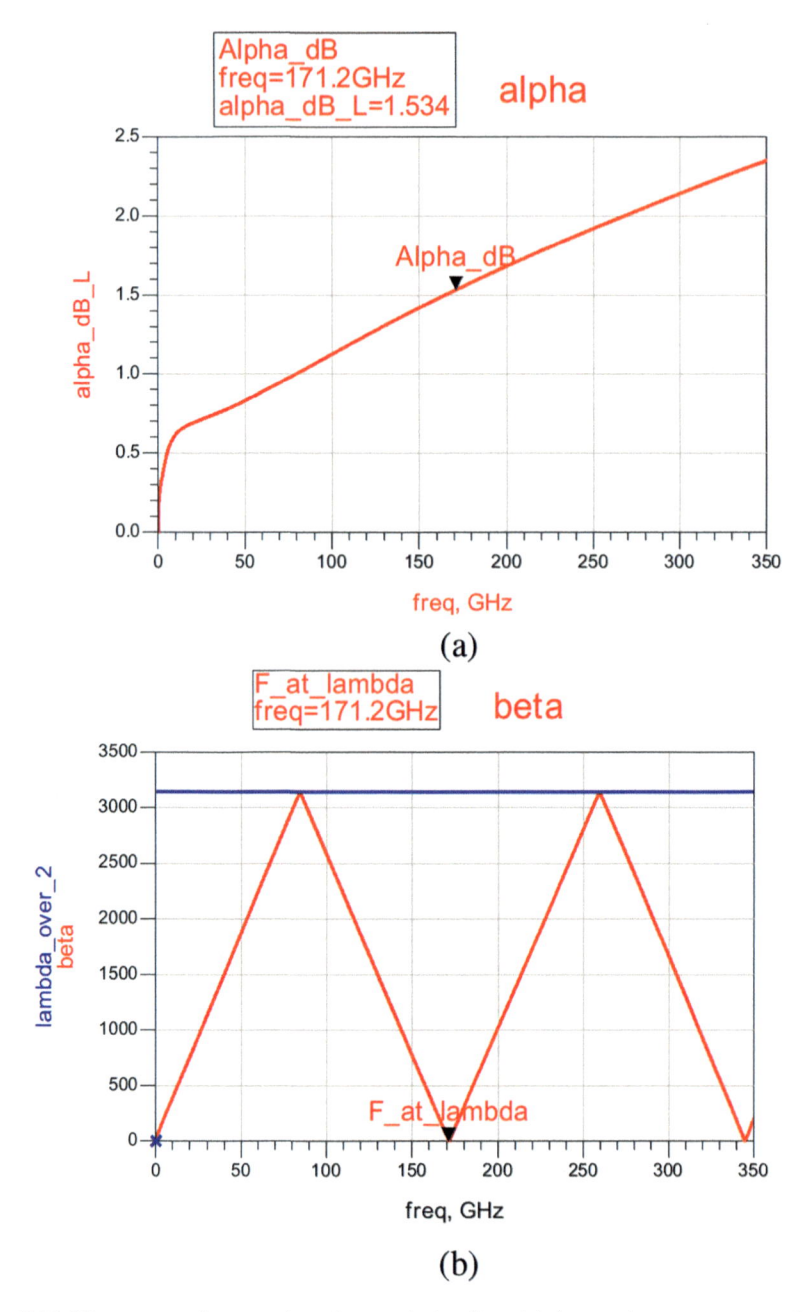

Fig. 7.16 Electromagnetic extraction of transmission line. (**a**) Attenuation constant α and (**b**) phase constant β over 1 mm

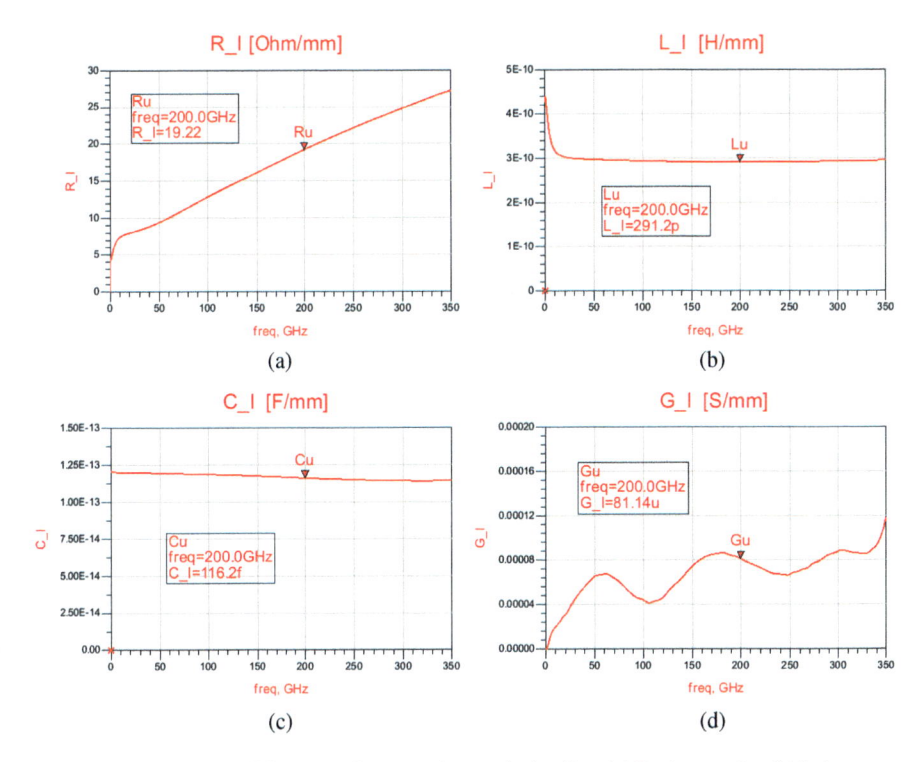

7 Millimeter-Wave Distributed Oscillators in 28 nm FD-SOI Technology

Fig. 7.17 Extracted RLCG model of proposed transmission line. (**a**) Resistance R_u, (**b**) inductance L_u, (**c**) capacitance C_u, and (**d**) conductance G_u per unit length

$$l_{tot} = 2nl = \frac{1 - \Delta\varphi/\pi}{f_{osc}\sqrt{L_u C_u}}. \tag{7.19}$$

At this point, the only unknown parameter is $\Delta\varphi$. As presented in previous section, this parameter represents the amplification stage additional phase-shift due to parasitic effect at mm-wave and THz frequencies.

To calculate this added phase-shift, it is necessary to know the distance between the drain and gate lines. Indeed, the connection accesses between the TL and the amplification stage will also influence this parameter.

A way to do this is to fix the distance between the drain and gate lines and calculate the added phase-shift in this configuration. Then calculate l_{tot} using Eq. (7.19) and check if it exists a form factor allowing this line spacing and this line length. Several iterations can be necessary to find a solution.

To fix the distance between the drain and gate lines, a trade-off appears. Indeed, the two lines have to be far enough to not significantly impact each other. But, they also have to be close enough to minimize the added phase-shift and to allow to close the loop. In our cases, the simulated added phase-shift is 0.37π at 135 GHz and 0.48π at 200 GHz. Corresponding to a total TL length of approximately 800 μm for the 135 GHz DO and 450 μm for the 200 GHz DO, as resumed in Table 7.2.

Table 7.2 Determination of the transmission line total length as a function of targeted oscillation frequency

Targeted frequency	135 GHz	200 GHz
Added phase-shift	0.37π	0.48π
Corresponding TL length	800 μm	440 μm

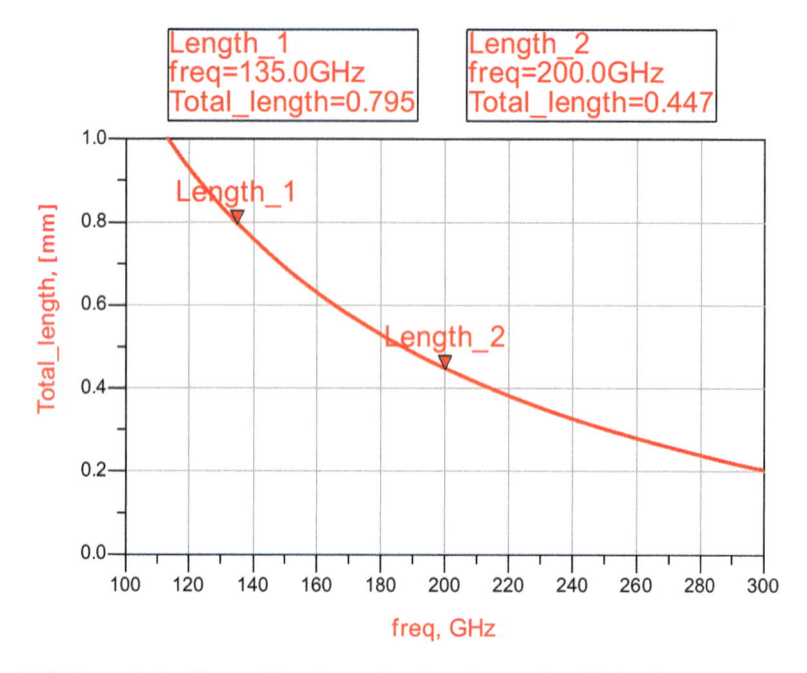

Fig. 7.18 Transmission line total length as a function of targeted oscillation frequency

Assuming a linear approximation for the added phase-shift, it is possible to draw the TL total length as a function of the targeted oscillation frequency, as depicted in Fig. 7.18.

7.5 Circuits Measurements

7.5.1 Standalone Transistor Measurements

In order to well understand the transistor behavior at mm-wave and THz frequencies, a standalone transistor and some test structures have been embedded on the side of our first test-chip.

First of all, in the following figures, some analog transistor parameters measurements up to 110 GHz will be presented. Figure 7.19a, b presents the RF transimpedance ($G_{m|RF}$) and the gate resistance (R_g), while the parasitic capacitances C_{gd} and C_{gg} are presented in Fig. 7.19c, d.

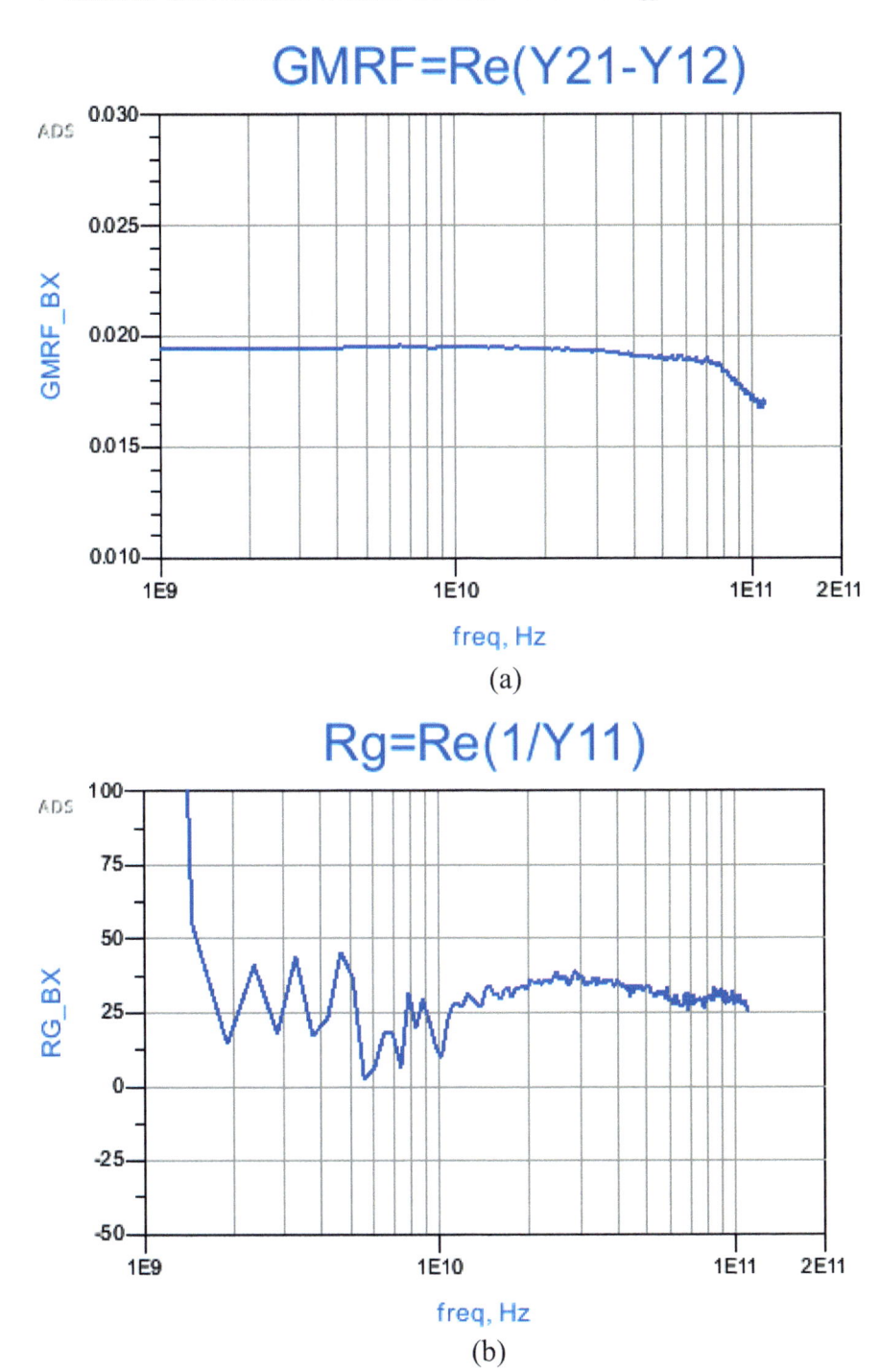

Fig. 7.19 Standalone transistor measurement up to 110 GHz: (**a**) RF transimpedance, (**b**) gate parasitic resistance, (**c**) gate to drain parasitic capacitance, and (**d**) gate to substrate parasitic capacitance

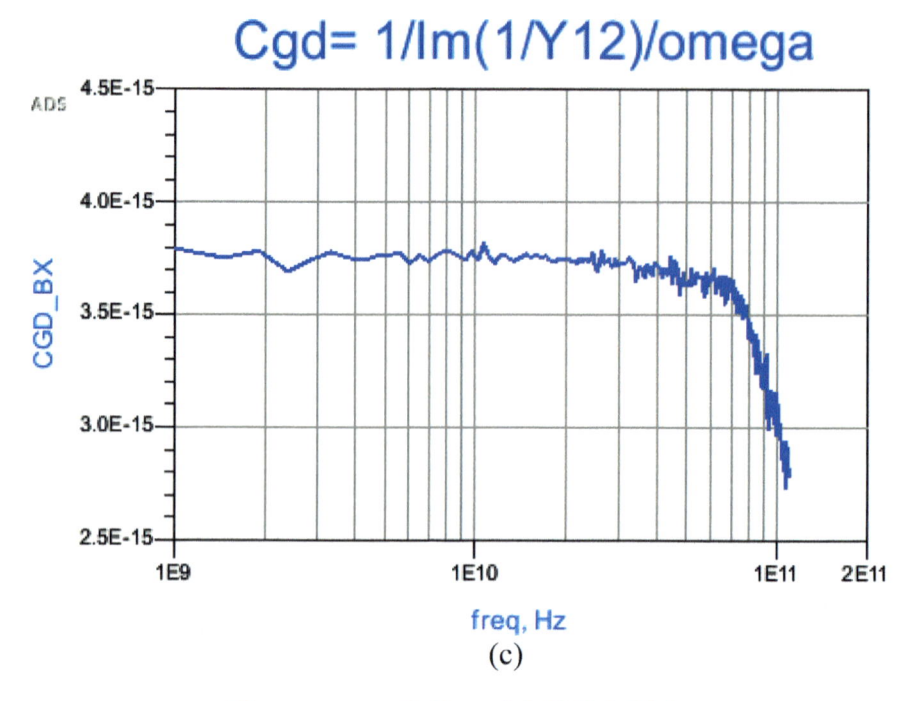

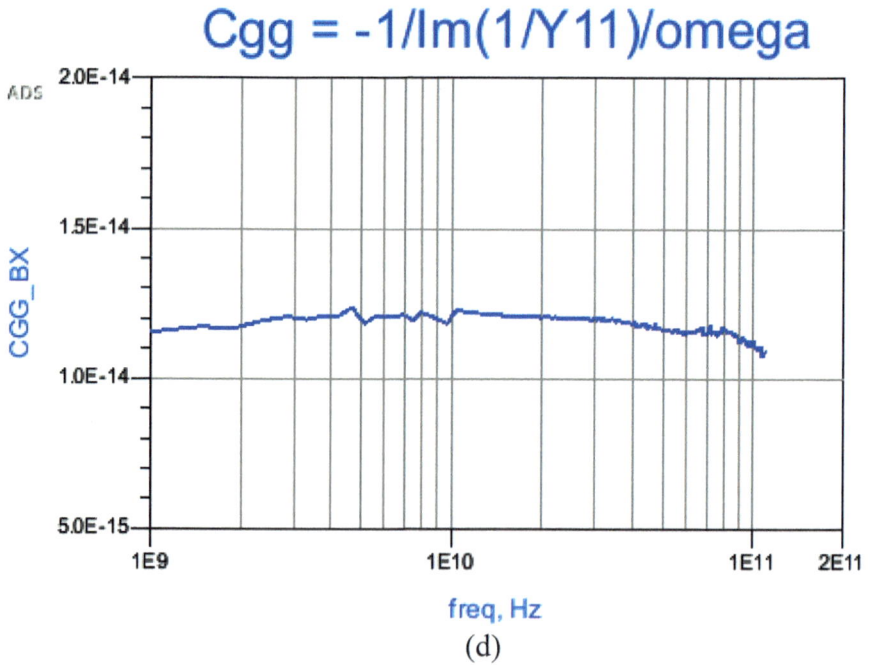

Fig. 7.19 (continued)

The cutoff frequency (f_T) is, by definition, the unity current gain frequency. To measure it, one can plot the H21 of the DUT and spot the zero crossing. Nevertheless, most of the time the measurement apparatus does not reach this frequency and H21 must be extrapolated. H21 is decreasing with a slope of 20 dB/dec so it is possible to extrapolate it by taking the asymptote of H21 and extending it to zero.

In the same way, the maximum oscillation frequency f_{max}, which is the frequency at which the maximum power gain is equal to 1, is measurable by taking the crossing of Mason's gain function U with zero. The only difference is that U is decreasing with 40 dB/dec.

Figure 7.20 shows in blue lines the H21, U, f_T, and f_{max} measured on the standalone transistor with the first bench, up to 110 GHz.

Thanks to our second bench, from 220 to 330 GHz, it becomes possible to plot H21 and U until the zero crossing. Figure 7.20 shows in pink lines the H21, U, f_T, and f_{max} measured on the standalone transistor with the second bench, from 220 to 330 GHz.

It is remarkable that for both f_T and f_{max} the continuity between the two frequency bands seems to be respected. Thus, for our transistor featuring a gate length of 30 nm and 20 fingers of 800 nm width and including an optimized BEOL, measurements give 263 GHz of f_T and 371 GHz of f_{max}.

Fig. 7.21 and Table 7.3 present the comparison between simulations and measurements for f_T and f_{max}. It can be seen that we obtain an excellent correlation. This confirms the very accurate models of the design kit P-Cell.

7.5.2 Measurement Setup

The microphotographs of the two distributed oscillators manufactured in 28 nm FD-SOI CMOS technology from ST Microelectronics are presented in Fig. 7.22. The core areas are, respectively, 0.034 mm^2 and 0.014 mm^2 without pads for the 134 GHz and the 202 GHz topologies.

The on-wafer measurement setup is depicted in Fig. 7.23. It consists of: a 100 μm Cascade I170 GSG probe followed by an S-bend waveguide, a SAM-A70 harmonic mixer, and a Rohde and Schwarz signal source analyzer on the drain pad. Another Cascade GSG probe is followed by a bend waveguide and a 50Ω load, connected to the gate pad. The DC gate and drain voltages are directly provided through the bias-tees included in the probes. The body-bias is provided through a DC-pad not visible in Fig. 7.22.

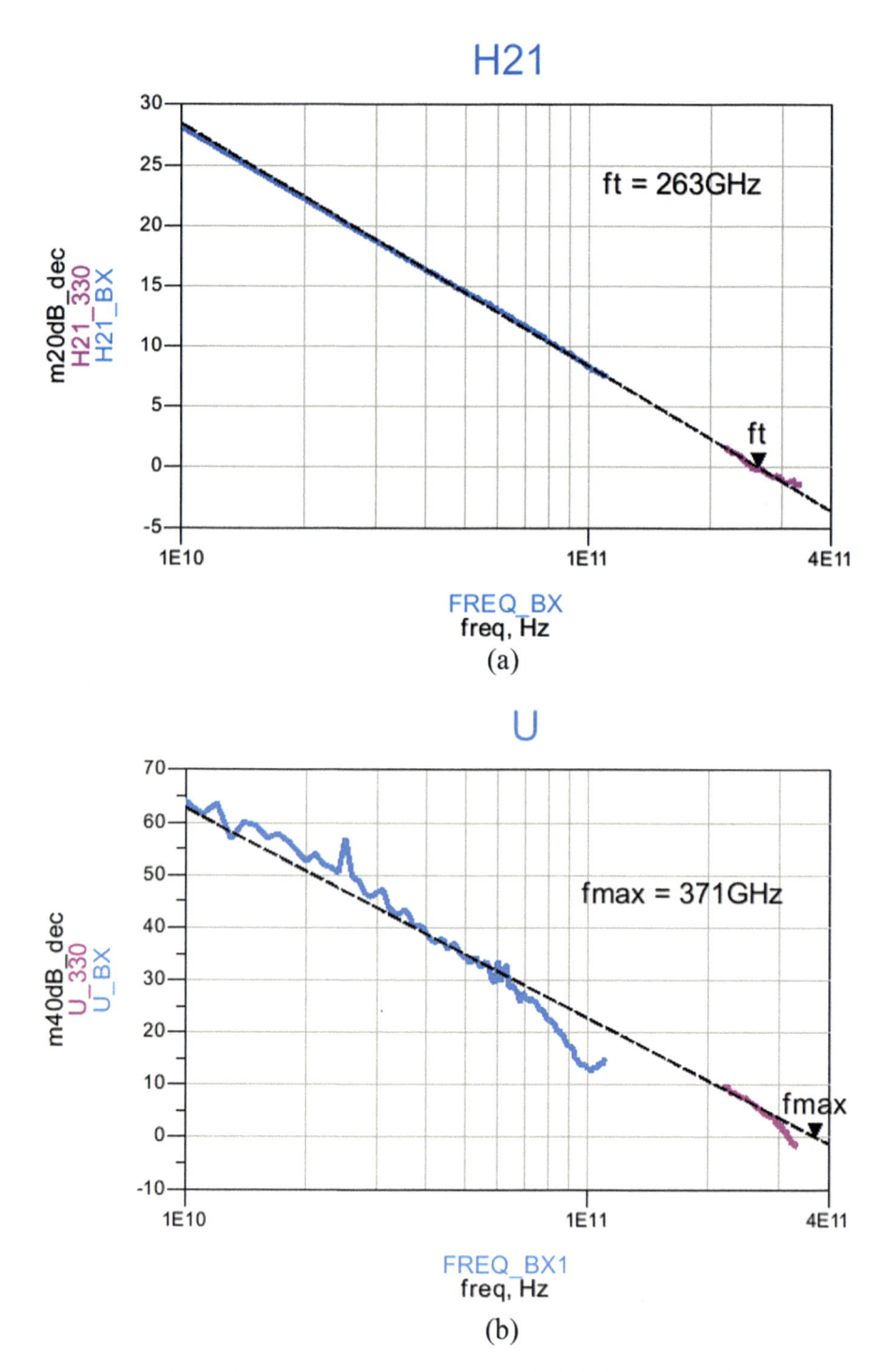

Fig. 7.20 Standalone transistor measurement up to 110 GHz in blue lines and from 220 to 330 GHz in pink lines: (**a**) current gain H21, (**b**) Mason's gain U, (**c**) cutoff frequency f_T, and (**d**) maximum oscillation frequency f_{max}

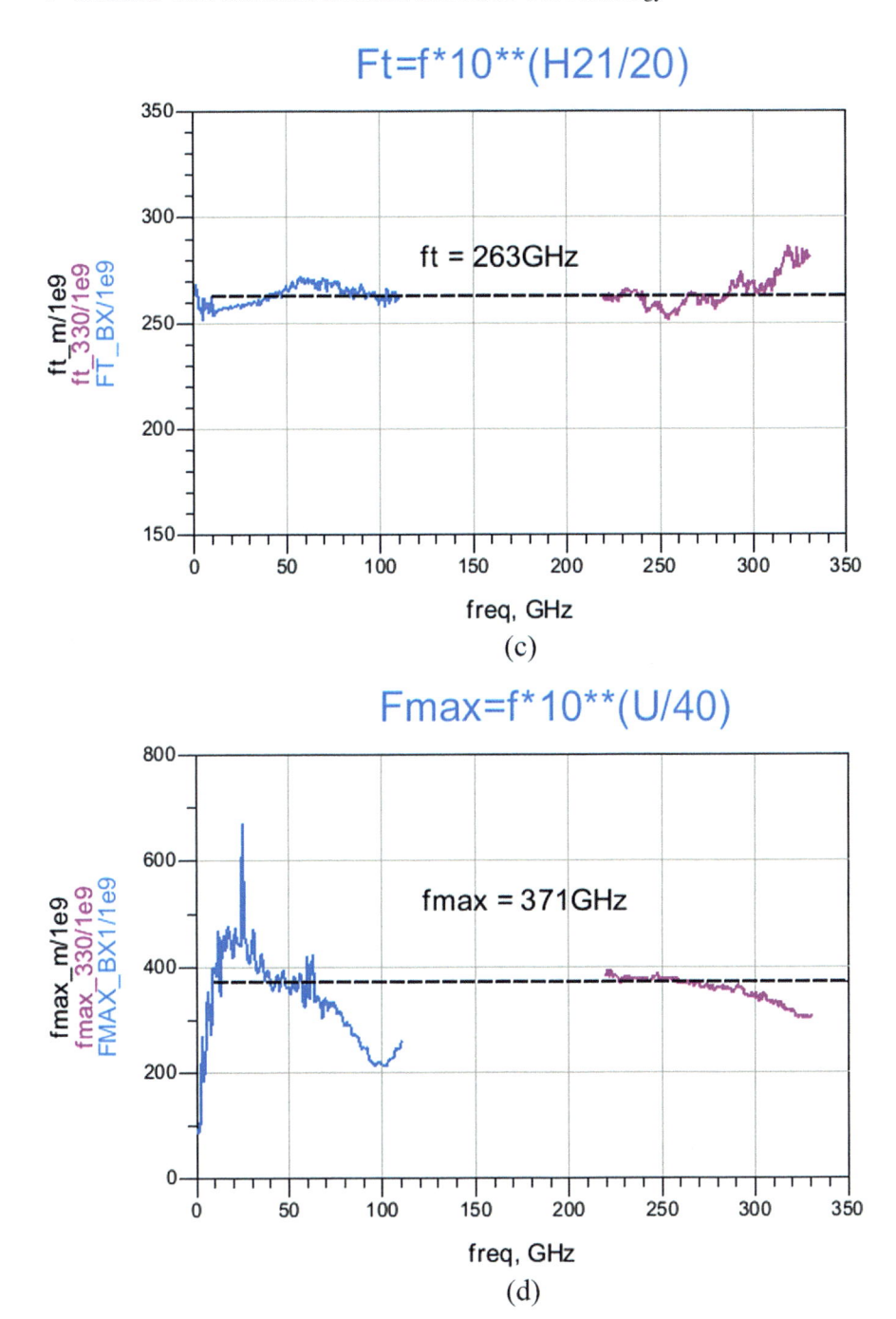

Fig. 7.20 (continued)

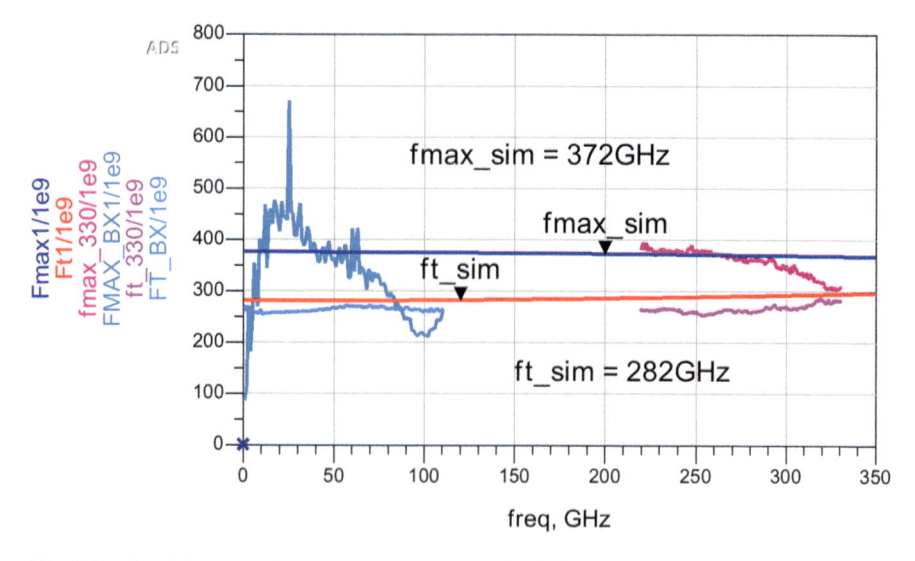

Fig. 7.21 Standalone transistor measurement versus simulation results for cutoff frequency f_T and maximum oscillation frequency f_{max}

Table 7.3 f_T and f_{max} simulations with and without BEOL parasitic interconnections versus measurement

	Simulation with DK P-Cell	Simulation with P-Cell + BEOL	Measurements
f_T	360 GHz	280 GHz	260 GHz
f_{max}	390 GHz	370 GHz	370 GHz

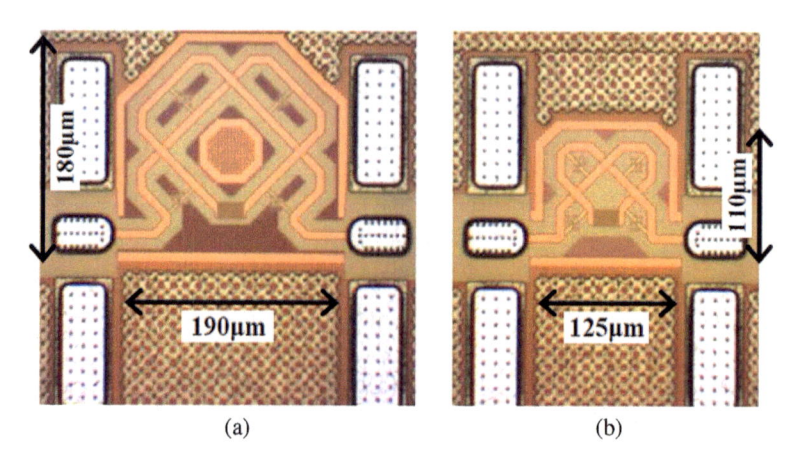

Fig. 7.22 Microphotograph of the distributed oscillators (same metric scale). (**a**) First 134 GHz distributed oscillator. (**b**) Second (202 GHz) distributed oscillator

7 Millimeter-Wave Distributed Oscillators in 28 nm FD-SOI Technology

Fig. 7.23 Measurement setup

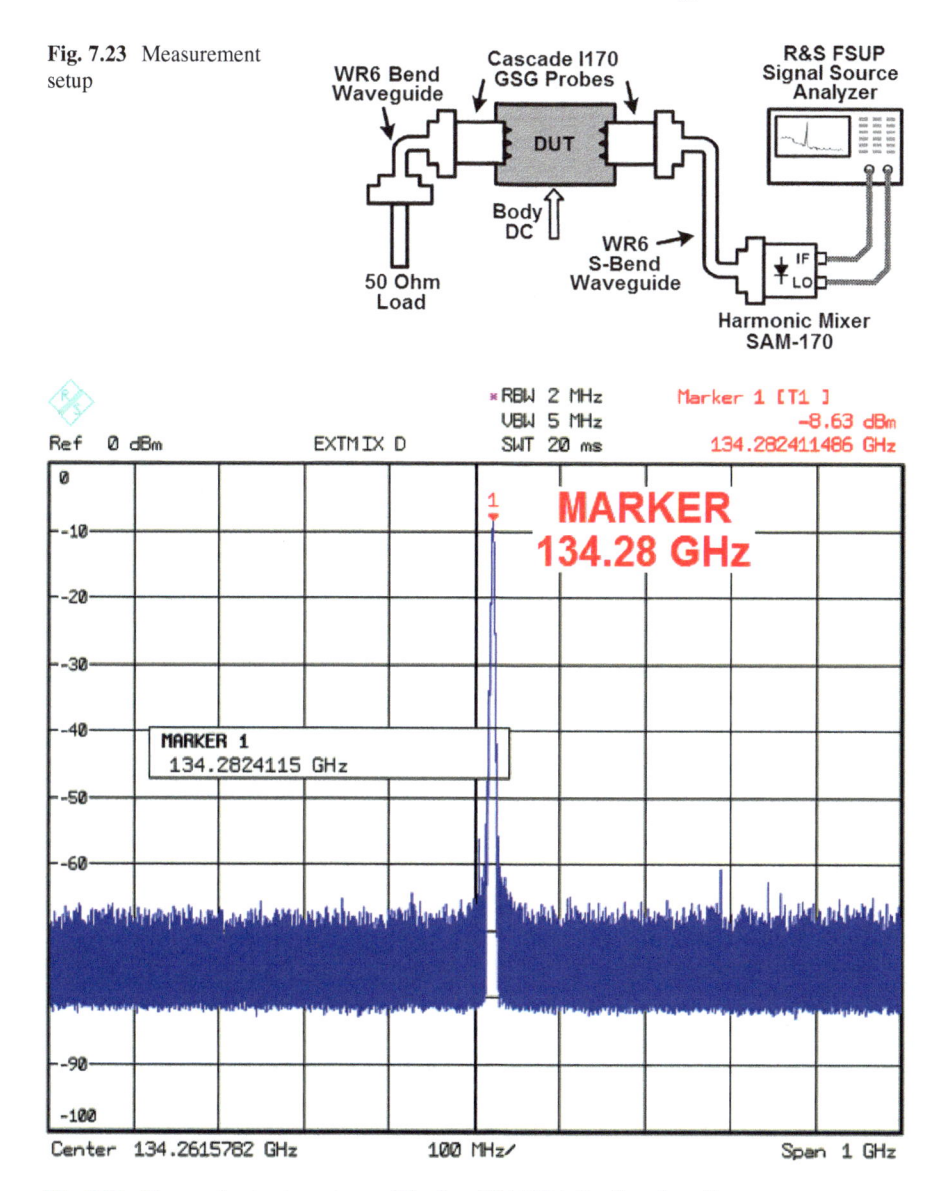

Fig. 7.24 Measured output spectrum of the first (134 GHz) distributed oscillator

7.5.3 Measurement Results

The first distributed oscillator was measured at a frequency of 134.3 GHz with a power consumption of 20 mW (Vdrain = 1 V), as shown in Fig. 7.24. The measured phase noise is depicted in Fig. 7.25 with −99.6 dBc/Hz at 1 MHz offset from the carrier.

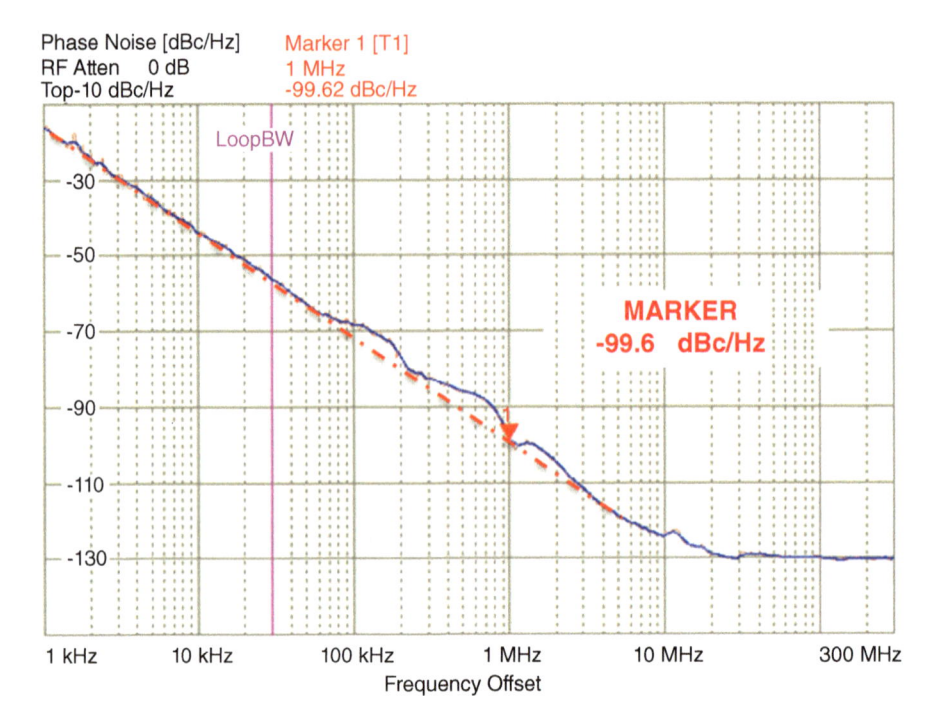

Fig. 7.25 Measured phase noise of the first (134 GHz) distributed oscillator

The de-embedded output power is 0.4 dBm (1.1 mW) corresponding to a remarkable DC-to-RF efficiency of 5.48%. The measured output power, DC-to-RF efficiency, and phase noise at 1 MHz offset are shown in Fig. 7.26 for different gate (Vgate) and body (Vbody) voltages.

As expected, increasing bias condition increase the output power until a compression. After this compression, increasing bias voltages result logically on a reduction of DC-to-RF efficiency. This behavior is easily understandable since the DO can be seen as a looped back distributed power amplifier. Indeed, increasing Vbody is equivalent to boost the transistors' gain (since the transconductance g_m is inversely proportional to V_T). Nevertheless, as for power amplifier, when operating in large signal the gain compresses and thus the efficiency decreases.

Moreover, as predicted by the theory the phase noise changes positively with the body-bias voltage change until the output power compression. A remarkable result here is that thanks to the FD-SOI unique wide range of body-bias control, a large range of phase noise values can be covered and hence a minimum phase noise operating point can be found.

Finally, the oscillation frequency variation over bias voltages' variation is only 60 MHz. This represents a very small frequency variation of 0.045% thanks to its low sensitivity to active component parameters.

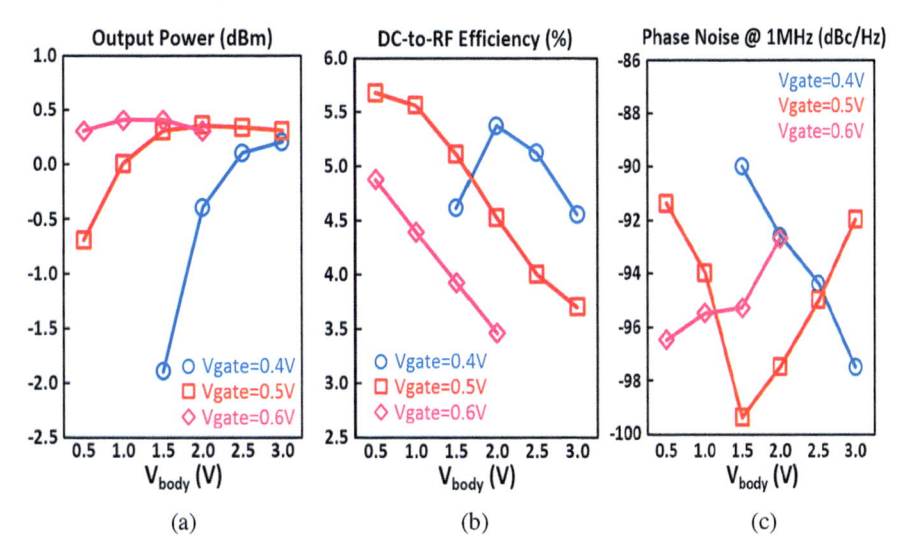

Fig. 7.26 Measured output power, DC-to-RF efficiency and phase noise at 1 MHz offset for different Vbody and Vgate, for the first (134 GHz) oscillator. (**a**) Output power. (**b**) DC-to-RF efficiency. (**c**) Phase noise at 1 MHz offset

As shown in Fig. 7.27, the second distributed oscillator was measured at a frequency of 202.2 GHz with a power consumption of 20 mW. The optimum phase noise at 1 MHz offset is −100.4 dBc/Hz, as presented in Fig. 7.28.

The oscillation output power is measured at 0.3 dBm (1.07 mW) corresponding to a DC-to-RF efficiency of 5.38%. The measured output power, DC-to-RF efficiency and phase noise at 1 MHz offset are presented in Fig. 7.29 for different gate (Vgate) and body (Vbody) voltages.

Similarly to precedent results, a phase noise optimum point can be reached by tuning the body-bias voltage. The oscillation frequency variation over bias change is 50 MHz, representing a variation of only 0.025%.

7.5.4 On-Wafer Mapping Measurement for Variability Study

For each circuit implementation, several circuits at different locations (#8) on the same wafer have been measured in order to provide information on the potential variability. Using the same Vgate and Vbody-biasing operation point, the oscillation frequency shows variations of only around 0.01%. Similarly, the output power level shows a standard deviation of less than 0.1 dBm. This demonstrates the low variability feature of such DO architectures, as shown in Fig. 7.30.

In order to have a better representative sample, for the first (134 GHz) oscillator, a second wafer from another tape-out has been tested and 24 other locations have been

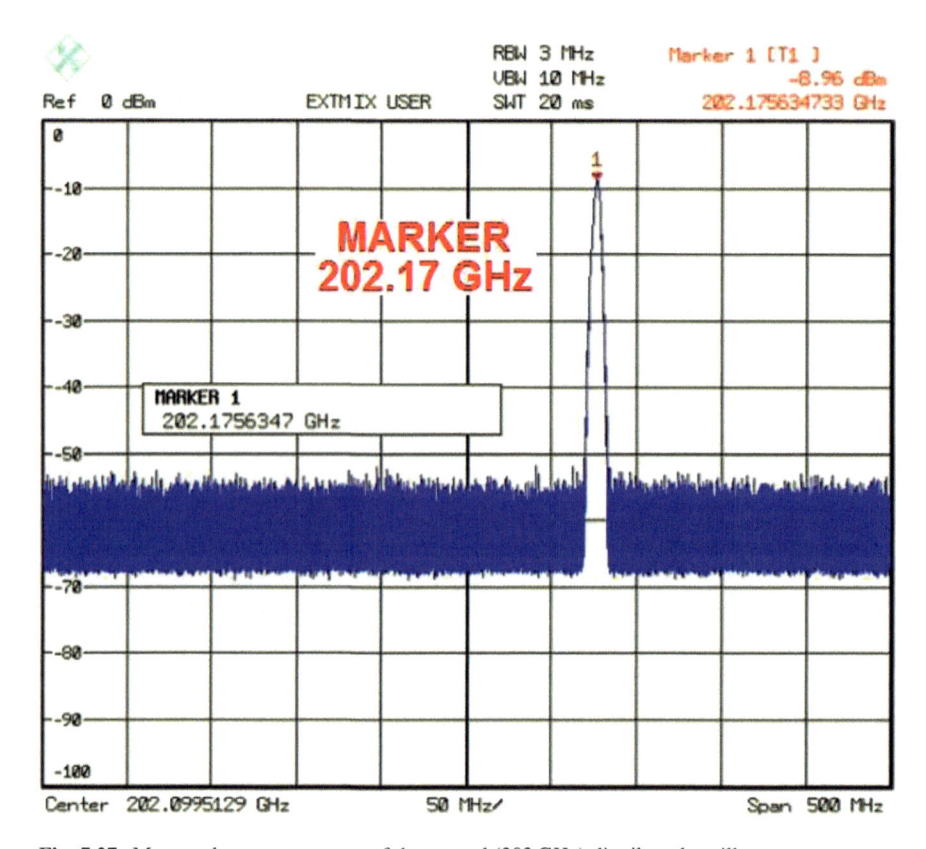

Fig. 7.27 Measured output spectrum of the second (202 GHz) distributed oscillator

measured. Slight layout modifications have been made to reduce losses in TL and in transistor accesses, and hence increase DC-to-RF efficiency. These modifications include metal dummies removal (one row of dummies was removed thanks to ground plan density optimization) and transistors' source access reduction (source access was reduced to dive more quickly to the ground plane).

Using the same Vgate and Vbody-biasing operation point, the oscillation frequency shows a standard deviation of around 0.015%. The output power level shows a standard deviation of 0.4 dBm. Distribution histograms for the oscillation frequency and output power are presented in Fig. 7.31 (in blue bar). Standard deviations are calculated only from the 24 new locations. Nevertheless, the 8 previous locations are also presented (in dark bar) in order to compare results. These results confirm the low variability feature of such DO architectures. Moreover, the very low variation of the center frequency confirms that the DO oscillation frequency is mainly defined by the TL total length and does not depend on active components variations.

7 Millimeter-Wave Distributed Oscillators in 28 nm FD-SOI Technology

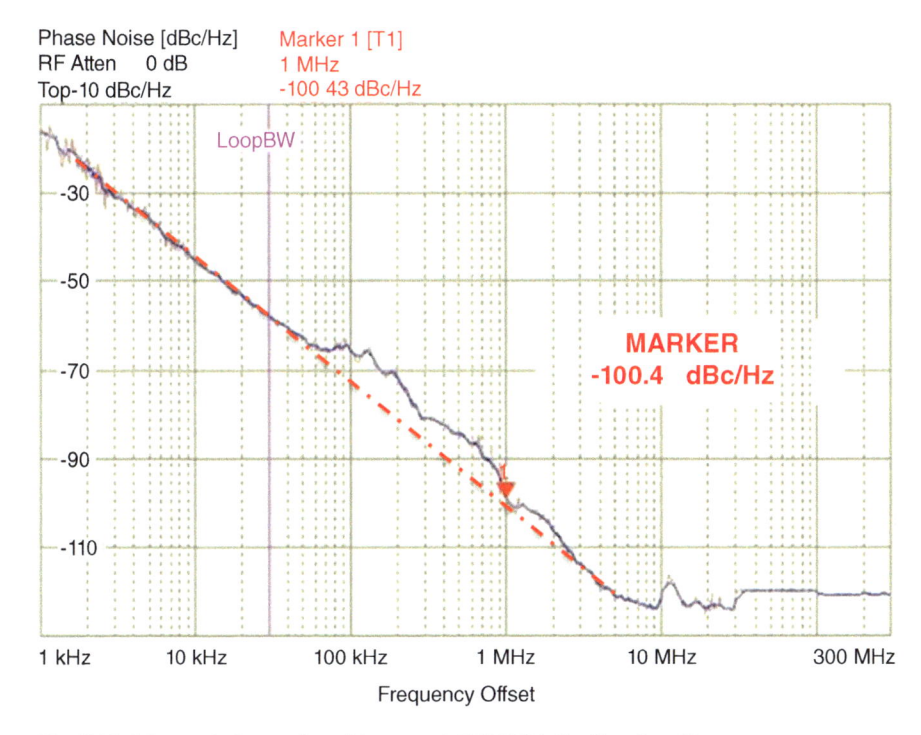

Fig. 7.28 Measured phase noise of the second (202 GHz) distributed oscillator

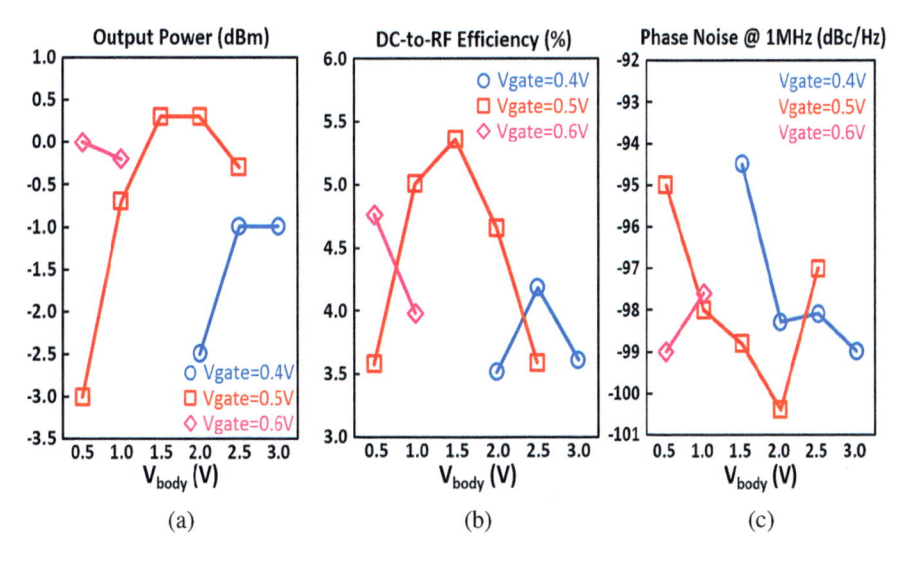

Fig. 7.29 Measured output power, DC-to-RF efficiency, and phase noise at 1 MHz offset for different Vbody and Vgate, for the second (202 GHz) oscillator. (**a**) Output power. (**b**) DC-to-RF efficiency. (**c**) Phase noise at 1 MHz offset

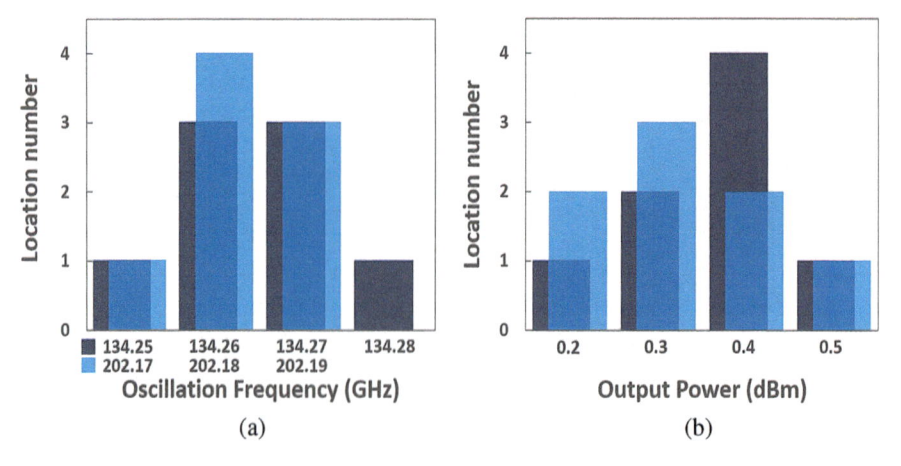

Fig. 7.30 Distribution histograms for the oscillation frequency and output power as a function of on-wafer location. (Eight different locations measured on a same wafer for each topology. Dark bar for the first (134 GHz) and blue bar for the second (202 GHz) oscillator.) (**a**) Oscillation frequency. (**b**) Output power

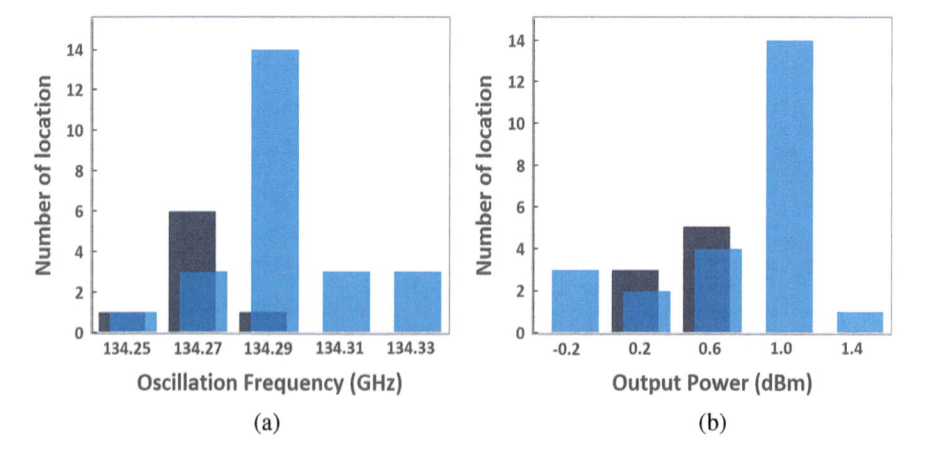

Fig. 7.31 Distribution histograms for the (**a**) oscillation frequency and (**b**) output power as a function of on-wafer location. 32 different locations on two different wafers (dark bar for the first wafer and blue bar for the second wafer)

Distribution histograms for the DC power consumption and DC-to-RF efficiency are depicted in Fig. 7.32 (Dark bar for the first wafer and blue bar for the second wafer). As predicted, the reduction of parasitic losses allows to rushed better output power with lower power consumption on the second wafer. Thus, the DC-to-RF efficiency is logically improved as shown in Fig. 7.32b. Power consumption presents a standard deviation of 0.57 mW and DC-to-RF efficiency exhibits a standard deviation of only less than 0.5%.

7 Millimeter-Wave Distributed Oscillators in 28 nm FD-SOI Technology

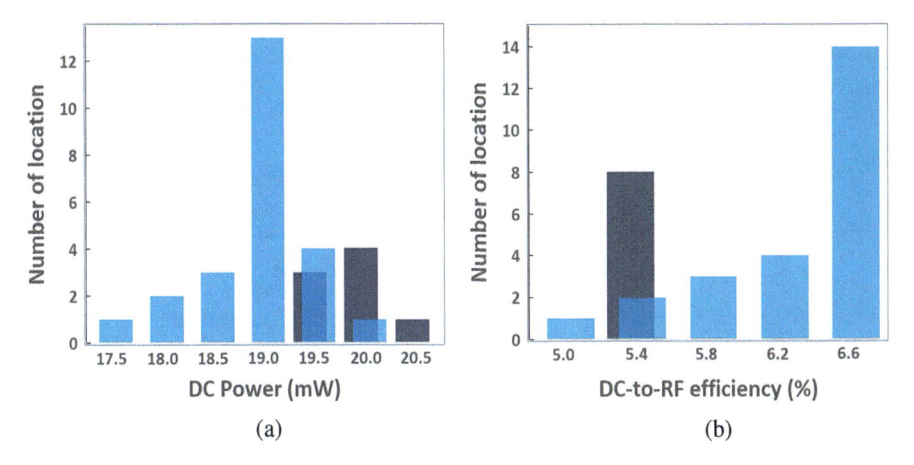

Fig. 7.32 Distribution histograms for the DC power consumption and DC-to-RF efficiency as a function of on-wafer location. 32 different locations on two different wafers. (Dark bar for the first wafer and blue bar for the second wafer.) (**a**) DC power consumption. (**b**) DC-to-RF efficiency

Fig. 7.33 Distribution histograms for the phase noise at 1 MHz offset as a function of on-wafer location. 32 different locations on two different wafers (dark bar for the first wafer and blue bar for the second wafer)

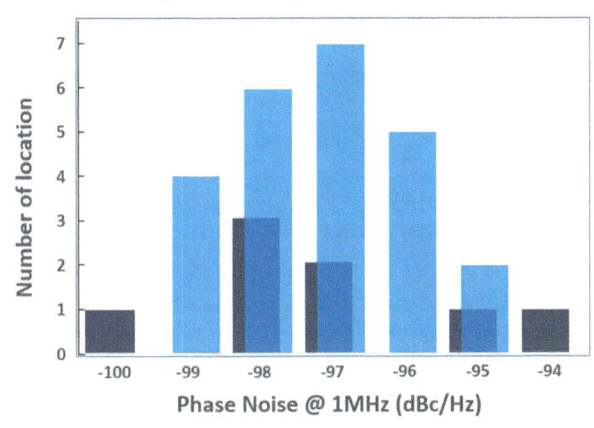

All these low process variability results enabled the use of such distributed oscillators as unlocked frequency reference for a certain category of sub-THz applications.

7.5.5 Phase Noise Optimization Through Body-Bias Control

For the phase noise, the measured variation for a constant Vbody and Vgate is around 6% for a standard deviation of approximately 1.4 dBc/Hz, as shown in the distribution histogram Fig. 7.33.

Nevertheless, phase noise is inversely proportional to output voltage level [6] and in our topology output voltage is proportional to overdrive voltage and hence to the

Fig. 7.34 Phase noise optimization by body-biasing; measurement over eight locations on wafer, for the first (134 GHz) oscillator

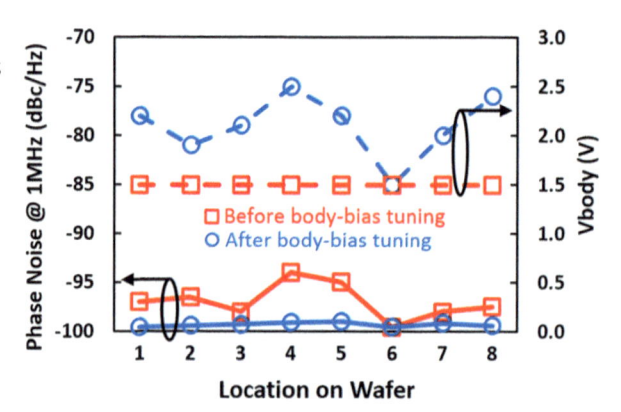

threshold voltage. Thus, the FD-SOI large V_T variation range allows a wide range of phase noise tuning and hence permits to find a minimum phase noise operating point.

This feature is depicted in Fig. 7.34, where the phase noise at 1 MHz offset of the eight locations on the first wafer was measured for a constant Vbody in square symbol lines. The circle symbol lines present the phase noise after optimization through body-biasing.

Finally, the phase noise of all location can be reduced between −99 and −100 dBc/Hz. This phase noise optimization is made at the expense of a power consumption variation between 20 and 25 mW and thus DC-to-RF efficiency reduction down to around 4%.

Moreover, for some applications which require low power consumption but less optimized phase noise, the body-biasing can be used to reduce the consumption (as the transistors are biased at constant Vgate condition, body-bias changes involved V_T changes and hence bias and g_m variation) at the expense of less performant phase noise.

7.6 State-of-the-Art Comparison and Conclusion

Measurement results are summarized and compared to state of the art in Table 7.4 using the classic oscillator FoM:

$$FoM = PN(\Delta f) - 20\log\left(\frac{f_0}{\Delta f}\right) + 10\log\left(\frac{P_{DC}}{1\,\text{mW}}\right). \quad (7.20)$$

To our knowledge, this work presents for the very first time oscillators at mm-wave and THz frequencies integrated in a 28 nm CMOS technology. The advantages coming from the high $f_T/f_{\max}$ performances for the active devices are, for these processes, counter-balanced by the limited back-end performance in such very dense nodes. This work compares positively with the state of the art, showing good DC-

7 Millimeter-Wave Distributed Oscillators in 28 nm FD-SOI Technology

Table 7.4 Comparison with previous state of the art

Ref.	Freq. (GHz)	P_{out} (dBm)	DC-to-RF efficiency	$PN_{@1MHz}$ (dBc/Hz)	P_{DC} (mW)	|FoM|	Techno.
[14]	121	−3.5	2.07%	−88	21.6	176.3	130 nm CMOS
[15]	195	6.5	15.2%	−98.6	29.2[a]	189.7[a]	55 nm SiGe
[16]	210	1.4	2.44%	−87.5	56.5[a]	176.4[a]	130 nm SiGe
[17]	212	−7.1	0.65%	−92	30	183.8[a]	130 nm SiGe
[18]	219	−3	2.09%	−77.4	24	170.4	65 nm CMOS
This work	134	0.4	5.48%	−99.6	20	188.9	28 nm FD-SOI CMOS
This work	202	0.3	5.38%	−100.4	20	193.5	28 nm FD-SOI CMOS

[a]Calculated from provided data. Work published in [3]

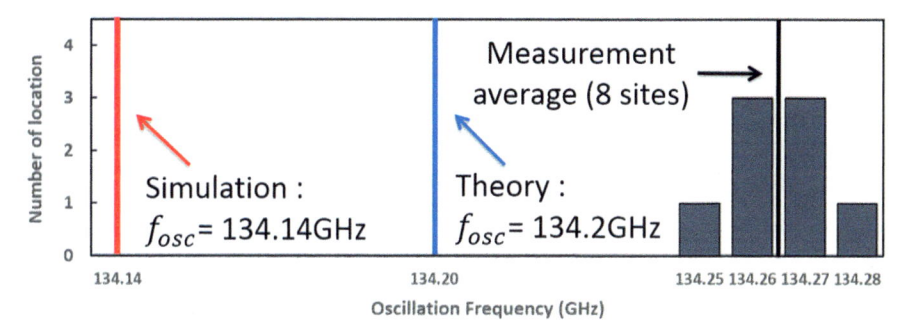

Fig. 7.35 Oscillation frequency comparison between measurements over the eight locations of the first wafer, post-layout simulation, and theoretical value (precision less than 0.1%)

to-RF efficiency for a very compact solution, despite the potential VLSI integration style 10 ML BEOL. Performant phase noise values have been measured taking advantage of the body-bias enabled phase noise optimization. And, the numerous locations measurements on wafer demonstrate the robustness of such integrated DO solution.

Another remarkable result appears when comparing these values with respect to post-layout simulated values and with theoretical value coming out from the DO theory developed in this chapter. Indeed, as depicted in Fig. 7.35, an excellent correspondence is obtained demonstrating the accuracy of the proposed mm-wave design flow and confirming the good precision of the DO theory. Indeed, all the presented numbers (simulation, theory, statistical measurement) are within less than 0.1% precision.

References

1. N. Seller, Contribution à l'Étude, au Développement et à la Réalisation d'Oscillateurs à Controle Numérique en Technologie Silicium Avancée, Theses, Université de Bordeaux, 2008. Available: http://www.theses.fr/2008BOR13648
2. H. Wu, A. Hajimiri, Silicon-based distributed voltage-controlled oscillators. IEEE J. Solid-State Circuits **36**(3), 493–502 (2001)
3. R. Guillaume, F. Rivet, A. Cathelin, Y. Deval, Energy efficient distributed-oscillators at 134 and 202 GHz with phase-noise optimization through body-bias control in 28 nm CMOS FDSOI technology, in *2017 IEEE Radio Frequency Integrated Circuits Symposium (RFIC)* (2017), pp. 156–159
4. D.M. Pozar, *Microwave Engineering*, 4th edn. (Wiley, Hoboken, 2011)
5. R. Han, E. Afshari, A CMOS high-power broadband 260-GHz radiator array for spectroscopy. IEEE J. Solid-State Circuits **48**(12), 3090–3104 (2013)
6. A. Hajimiri, T.H. Lee, A general theory of phase noise in electrical oscillators. IEEE J. Solid-State Circuits **33**(2), 179–194 (1998)
7. E. Morifuji, H.S. Momose, T. Ohguro, T. Yoshitomi, H. Kimijima, F. Matsuoka, M. Kinugawa, Y. Katsumata, H. Iwai, Future perspective and scaling down roadmap for RF CMOS, in *1999 Symposium on VLSI Technology. Digest of Technical Papers (IEEE Cat. No.99CH36325)* (1999), pp. 163–164
8. A. Cathelin, B. Martineau, N. Seller, S. Douyere, J. Gorisse, S. Pruvost, C. Raynaud, F. Gianesello, S. Montusclat, S.P. Voinigescu, A.M. Niknejad, D. Belot, J.P. Schoellkopf, Design for millimeter-wave applications in silicon technologies, in *ESSCIRC 2007 - 33rd European Solid-State Circuits Conference* (2007), pp. 464–471
9. A. Larie, Design of highly linear 60 GHz power amplifiers in nanoscale CMOS technologies, Theses, Université de Bordeaux, 2014. Available: https://tel.archives-ouvertes.fr/tel-01142532
10. T.O. Dickson, K.H.K. Yau, T. Chalvatzis, A.M. Mangan, E. Laskin, R. Beerkens, P. Westergaard, M. Tazlauanu, M. Yang, S.P. Voinigescu, The invariance of characteristic current densities in nanoscale MOSFETs and its impact on algorithmic design methodologies and design porting of Si (Ge)(Bi) CMOS high-speed building blocks. IEEE J. Solid-State Circuits **41**(8), 1830–1845 (2006)
11. B. Martineau, Potentialités de la technologie CMOS 65 nm SOI pour des applications sans fils en bande millimétrique, Thesis, Lille 1, 2008. Available: http://www.theses.fr/2008LIL10041
12. B. Martineau, A. Cathelin, F. Danneville, A. Kaiser, G. Dambrine, S. Lepilliet, F. Gianesello, D. Belot, 80 GHz low noise amplifiers in 65nm CMOS SOI, in *ESSCIRC 2007 - 33rd European Solid-State Circuits Conference* (2007), pp. 348–351
13. Parasitic Extraction - StarRC. Available: https://www.synopsys.com/implementation-and-signoff/signoff/starrc.html
14. O. Momeni, E. Afshari, High power terahertz and millimeter-wave oscillator design: a systematic approach. IEEE J. Solid-State Circuits **46**(3), 583–597 (2011)
15. H. Khatibi, S. Khiyabani, A. Cathelin, E. Afshari, A 195 GHz single-transistor fundamental VCO with 15.3% DC-to-RF efficiency, 4.5 mW output power, phase noise FoM of −197 dBc/Hz and 1.1% tuning range in a 55 nm SiGe process, in *IEEE Radio Frequency Integrated Circuits Symposium RFIC* (2017), pp. 152–155
16. C. Jiang, A. Cathelin, E. Afshari, An efficient 210 GHz compact harmonic oscillator with 1.4 dBm peak output power and 10.6% tuning range in 130nm BiCMOS, in *IEEE Radio Frequency Integrated Circuits Symposium RFIC* (2016), pp. 194–197
17. P.Y. Chiang, O. Momeni, P. Heydari, A 200-GHz inductively tuned VCO with -7-dBm output power in 130-nm SiGe BiCMOS. IEEE Trans. Microw. Theory Techn. **61**(10), 3666–3673 (2013).
18. H.-T. Kwon, D. Nguyen, J.-P. Hong, A 219-GHz fundamental oscillator with 0.5 mW peak output power and 2.08% DC-to-RF efficiency in a 65 nm CMOS, in *2016 IEEE MTT-S International Microwave Symposium (IMS)* (2016), pp. 1–3

Chapter 8
Millimeter-Wave Power Amplifiers for 5G Applications in 28 nm FD-SOI Technology

Florent Torres, Andreia Cathelin, and Eric Kervé

8.1 Introduction

The 5G future mobile network is planned to be deployed from 2020, in a context of exponential mobile market and exchanged data volume evolution. The 5G will leverage revolutionary applications for the advent of the connected world. For this purpose, several network specifications are expected, notably low latency, reduced power consumption, and high data-rates even if no standard is yet defined. The frequency bands traditionally used for mobile networks will not permit the needed performances and several mmW frequency bands are under study to create a complementary frequency spectrum. However, these mmW frequency bands suffer from large attenuation in building material and in free-space. Therefore, several techniques will be implemented to tackle these limitations in dense urban areas like backhauling, FD-MIMO, and beamforming phased array. This is leading to a large number of transceivers for base stations and end-user devices. CMOS technology offers undeniable advantages for this mass market, while FD-SOI technology offers additional features and performances. The power amplifier is the most critical block to design in a transceiver and is also the most power consuming. In order to address the 5G challenges, several specifications concerning power consumption,

---

F. Torres (✉)
STMicroelectronics, Crolles, France

IMS-Bordeaux, Talence Cedex, France

Ericsson, Lund, Sweden
e-mail: florent.torres@ericsson.com

A. Cathelin
STMicroelectronics, Crolles, France

E. Kervé
IMS-Bordeaux, Talence Cedex, France

Table 8.1 mmW PA specifications for 5G application in the 30 GHz band

	Targeted specifications at ≈30 GHz (31.8–33.4 GHz 5G band)
P_{sat} (dBm)	>17
PAE_{max} (%)	>30
PAE @ 6 dB P_{sat} back-off (%)	>10
Power gain (dB)	>16
$P_{1\,dB}$ (dBm)	>15
Configurability	Yes

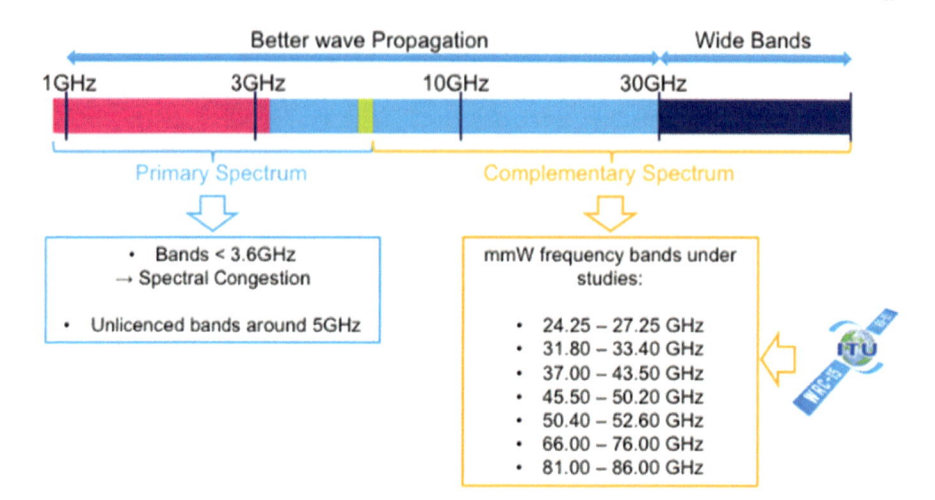

Fig. 8.1 5G expected frequency spectrum

linearity, and efficiency are expected. The environment variations in beamforming phased array and the industrial context drive the need for robust topologies, while power amplifier configurability is benefic in a context of adaptive circuits. This section addresses these challenges by exploring the conception of a robust and reconfigurable power amplifier targeting 5G applications while integrating specific design techniques and taking advantage of 28 nm FD-SOI CMOS technology features for configurability purposes. The targeted specifications for this work are summarized in Table 8.1.

At the time of writing of this book, standard for 5G is expected to emerge during the World Radiocommunication Conference (WRC-19) organized by the International Communications Union (ITU) and so no mmW frequency spectrum is yet allocated for 5G at this day.[1] However, several mmW frequency bands are under studies by the 5G industrial actors since ITU WRC-15 conference [6]. One of the expected 5G frequency spectra is summarized in Fig. 8.1.

---

[1] World Radiocommunication Conference 2019 (WRC-19) took place in Sharm El-Sheikh, Egypt, from 28 October to 22 November 2019.

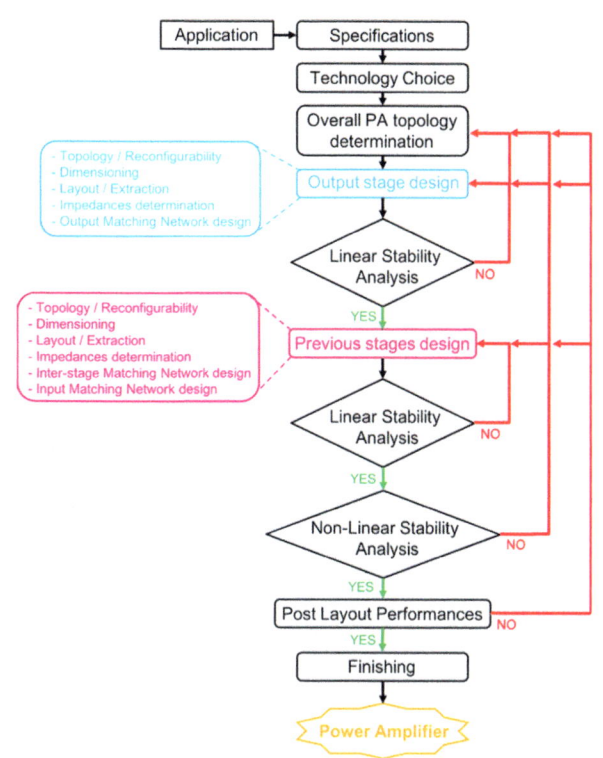

Fig. 8.2 Complete design flow of a mmW power amplifier in deep sub-micron CMOS technology

8.1.1 Design Flow for Integrated mmW PA Design

Figure 8.2 synthetizes a proposed integrated mmW PA design flow. This methodology and the techniques presented here have been used to design the power amplifier presented in the following sections.

The first step in the design flow is to define the overall PA topology: single or multi-stage, single-ended, differential, or balanced. This has to be decided regarding the specifications and targeted applications. In deep sub-micron technology nodes, the parasitics at high frequency necessitate to choose topologies offering the best stability and reducing their impact. This differs from the classical less advanced technology nodes design flow or low frequency designs as the parasitics have a less critical impact. Expected performances at high-level can be evaluated with an early technology exploration. This will allow the designer to know the possibilities and the limits of the technology to better choose topology, dimensions, and design techniques to achieve expected performances.

Then the power amplifier stages are designed successively from the output stage, that is the most critical for overall power amplifier performances, to the input stage. Each stage is designed following the quasi-similar scheme.

A topology is selected to fit the specifications and configurability needs if necessary. The transistors and different elements composing the stage are then dimensioned. Several design techniques as presented in this chapter are implemented. The active part of the stage is then designed, with a special care for interconnections induced parasitic reduction. These parasitic elements must be extracted early in the process with EM RLC extraction tools, in order to be taken into account during early post-layout simulations because they are critical at mmW frequencies. Optimal impedances at the output and input of the stage can then be determined using load-pull and source impedance sweep. The corresponding output matching network can be designed, with a special care for induced losses limitation. For this purpose, only thick top metal layers in 28 nm FD-SOI technology should be considered for passive devices implementation. Optimized fast and accurate EM simulations can then be conducted to design and refine the layout of matching networks.

Performances of the output stage can then be simulated and estimated while a linear stability analysis has to be performed. If the stage is unstable, the designer should reconsider the design and go back to the previous step. If the linear analysis is conclusive, the next step is to design the previous stages.

As for the output stage, the topology, dimensioning, layout, and parasitics extraction have to be conducted first. The optimum impedances are then determined. Inter-stage matching network between the output of this stage and the input of the next stage is then designed.

Input matching network can also be implemented when the input stage has been designed. It corresponds to the last step of the design process before layout finishing involving ground planes, bias and supply path and pad integration. Each stage must satisfy the linear stability analysis conditions or need to be re-designed.

Then when full power amplifier is designed, post-layout simulations have to be performed. Non-linear stability has to be explored; if the conditions are satisfied, the performances under post-layout simulations can be estimated. If these stability conditions are not satisfied, the design should be modified.

Finally, if the performances obtained during post-layout simulations fit with the specifications, the power amplifier can be fully finished and is ready to be manufactured. If the performances are not sufficient, design must be corrected.

8.1.2 Power Amplifier Configurability Discussion in the Context of FD-SOI Technologies

In the introduction, we pointed at the necessity and advantages of power amplifier configurability for 5G applications and SoC control implementation. Several configurability types are possible depending on the fixed objectives. In [4], varactor loaded lines allow the power amplifier reconfiguration for frequency band switching. In

[11], switched PA configuration allows several saturated output power modes and proposes lower power consumption for lower power modes.

In this section we focus on the configurability that can be implemented by using 28 nm FD-SOI body-biasing node. A back-gate enabled configurability is interesting to explore because it does not add supplementary stage or complexity to an existing design, on the signal path.

Power amplifier configurability through body-biasing node has been explored in [10]. In this reference, the body-bias is used to implement pseudo-Doherty operations. In fact, this is possible because of the body-biasing effect over V_T. The V_T variation allows dynamic current density modification for a fixed gate bias condition.

To illustrate the operating class choice by gate biasing, Fig. 8.3 presents $I_D = f(V_{GS})$ curves of a LVT NMOS transistor with 30 nm gate length and 50 fingers of 1 μm under 1 V supply voltage, showing class A and class AB corresponding bias.

In [10] and [13], this ability allows to bias all the transistors for class C operations, while the V_T variation through body-biasing over "main" transistors enables class AB operation. Then, body-biasing over "auxiliary" transistors allows a high-gain mode, enabling class A operation for all devices [10]. This pseudo-Doherty operation is illustrated in Fig. 8.4.

A second method of using body-biasing for power amplification stages configurability is described in the following, and has been deployed in the PA design presented in this chapter.

In other context than pseudo-Doherty implementation, the V_T variation presents advantages for configurability thanks to dynamical operating class modulation.

Let us focus on a common-source topology to illustrate this behavior. For a common-source transistor in saturation loaded by R_L and fed by an ideal voltage source, the small-signal voltage gain is given by Eq. (8.1).

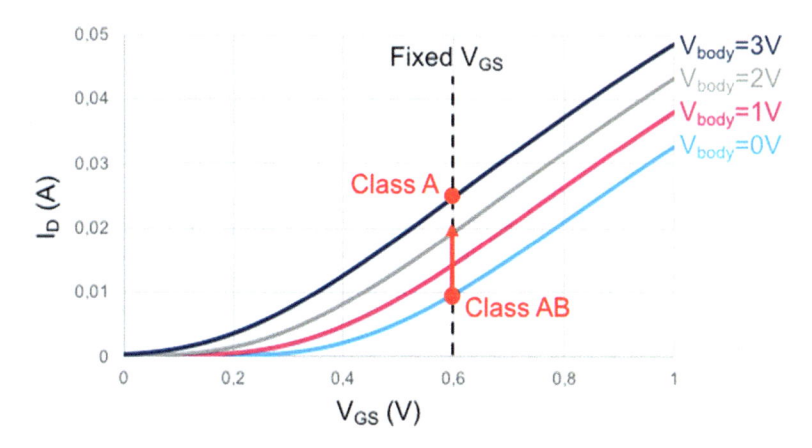

Fig. 8.3 Operating class shifting induced by body-biasing

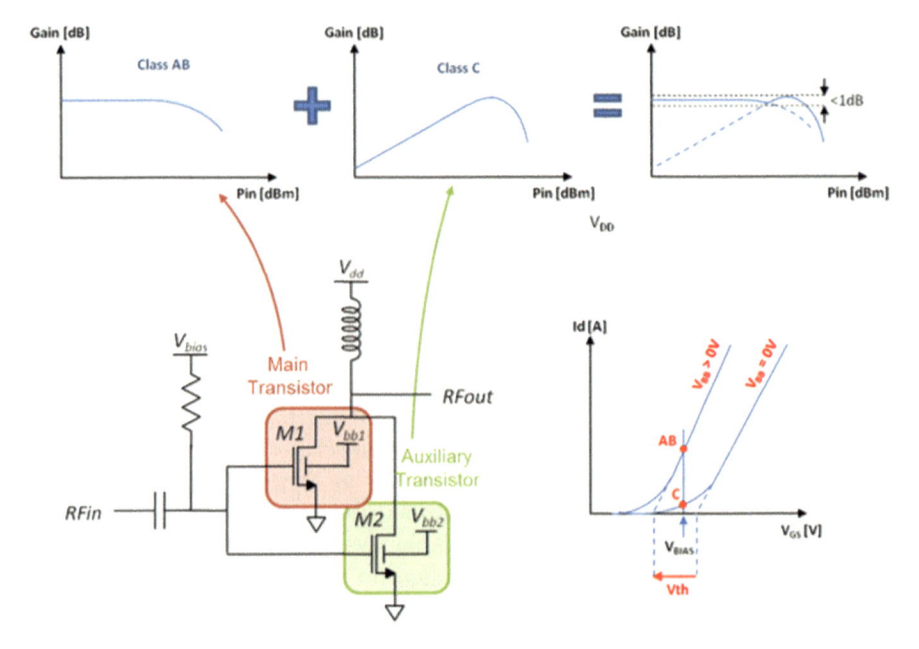

Fig. 8.4 Body-biasing enabled pseudo-Doherty operation [12]

$$A_V = -g_m \cdot \left(\frac{r_0}{R_L}\right) \quad (8.1)$$

where r_0 is the MOS device output resistance, R_L the load resistance, and g_m the transconductance with λ the channel length modulation parameter (Eq. (8.2)).

$$g_m = \frac{W}{L} \cdot \mu_m \cdot C_{OX} \cdot (V_{GS} - V_T) \cdot (1 + \lambda V_{DS}) \quad (8.2)$$

Therefore, the gain amplitude is determined by g_m parameter. Two solutions easily appear to obtain higher gain for fixed V_{DS} and transistor geometrical parameters:

- The classical solution is to use higher gate bias V_{GS} in order to reach higher gain in higher operation class.
- The body-biasing enabled V_T diminution that provides dynamical operation class shifting.

As both solutions induce the same effects over g_m, it is interesting to highlight the advantage of using body-biasing for gain enhancement instead of classical V_{GS} improvement.

In Fig. 8.5, $I_D = f(V_{GS})$ curves of a LVT NMOS transistor with 30 nm gate length and 50 fingers of 1 μm for a V_{DS} of 1 V are plotted to highlight the V_{GS} and Vbody variations needed to achieve a fixed I_D level.

8 Millimeter-Wave Power Amplifiers for 5G Applications in 28 nm FD-SOI... 175

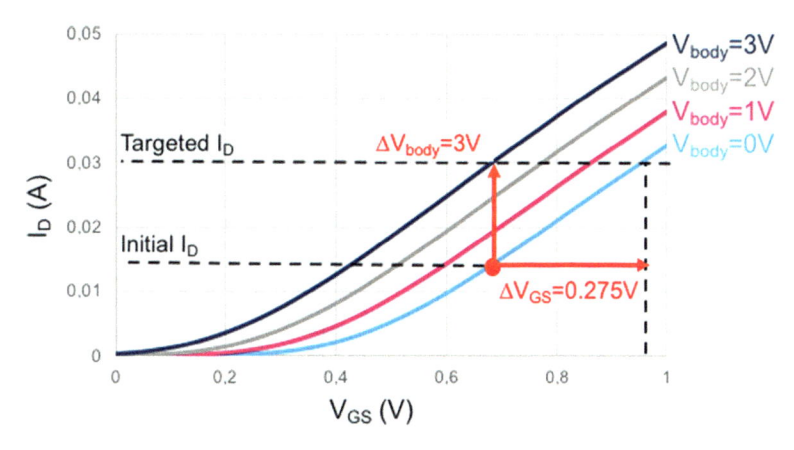

Fig. 8.5 V_{GS} and V_{body} dynamic comparison for fixed I_D level target

If we want to achieve a higher fixed I_D level from an I_D and V_{GS} reference, two solutions are possible. The V_{GS} value can be improved by 275 mV, while the body-biasing can be set to 3 V to reach the same I_D level. This means that a small variation of V_{GS} is leading to a large drain current variation while body-biasing dynamic is higher. Therefore, the wide dynamic possible through body-biasing node allows a very fine tuning of operating class. This is enabling continuous class shifting with easily reachable intermediary modes. Simulations have been conducted for the power amplifier presented in this chapter to illustrate this effect. For similar operating conditions, a 100 mV V_{GS} increase is leading to a 6.7 dB gain variation, while 100 mV increase of body-biasing is leading to a 0.8 dB power gain improvement. This is illustrating the fact that the body-biasing node in this technology enables both dynamic fine continuous class switching and fine grain control with no additional stage or additional design complexity.

In addition, we saw in the previous section that multi-stage topology enables interesting performance enhancement possibilities. In the following paragraph, configurability in multi-stage amplifier is discussed.

In multi-stage PA, for reconfigurable operation in 28 nm FD-SOI technology using body-biasing, a question may arise: In which stage the body-biasing should be implemented? In a multi-stage amplifier each stage must be linear enough to not degrade the linearity of the next stage. Therefore, acceptable overall linearity performances can be achieved. Practically, this means that the 1 dB output compression point of each stage must be higher than the 1 dB input compression point of the following stage. However, the use of body-biasing to increase the gain has a negative impact on linearity and has been observed in measurements provided later on. Thus, it is more benefic to implement the configurability through body-biasing on the last stage of the amplifier, generally the power stage, to limit the induced linearity deterioration caused by the saturation of previous stages.

8.2 Reconfigurable Balanced mmW PA Implementation in 28 nm FD-SOI Technology

In this section, we focus on a power amplifier implementation targeting the mmW 5G applications challenges highlighted in the introduction. High performances are aimed in terms of efficiency, gain, and linearity while gain configurability for system level control in SoC is expected. For this purpose, the design flow depicted in the previous section has been applied and several specific design techniques have been used in order to provide both high performances with robust integration.

First, active devices layout optimization and dimensioning are presented. Then the choice of the overall power amplifier topology and its implementation is discussed. Moreover, the design of each amplification stage and associated matching networks are depicted. We also discuss the points to check for a robust implementation targeting industrial production.

8.2.1 Active Devices

In the implemented circuit only LVT NMOS transistors are used. The design of active devices is crucial. While several parameters can be optimized directly from the Metal1-Pcell available in the design kit with a wise dimensioning, the layout optimization is also important as it adds unneglectable parasitic elements. Both optimizations are discussed in this section. The reader of this Chapter is kindly directed also to refer the Chap. 7, Sect. 7.3, where similar design considerations are presented.

8.2.1.1 Dimensioning

In order to design a transistor with a fixed total width W_{tot} from a transistor unitary cell several topologies are available. In fact, a transistor with a total width W_{tot} is composed by N_f number of unitary single-finger transistors exhibiting a width $W_{tot} = Nf \cdot W_f$ with W_f being the unitary finger width.

Therefore, it is possible to directly implement this W_{tot} transistor with N_f fingers or it is possible to parallelize several elementary transistors cells (Fig. 8.6).

The sub-division advantage of a W_{tot} transistor in W_{el} elements is to multiply the gate accesses and so distribute the stress between gates. Therefore, both higher reliability and lower gate resistance are provided. Reducing the gate resistance has been an important part of the layout optimization strategy depicted in the next section. The use of multiple elementary transistors also allows a more compact and adaptive layout. Moreover, it limits the non-quasi-static effects that can occur in large transistor, corresponding to a non-null difference of potential between the two extremities. Furthermore, large transistors also introduce higher intrinsic parasitic elements that make the impedance matching harder to implement.

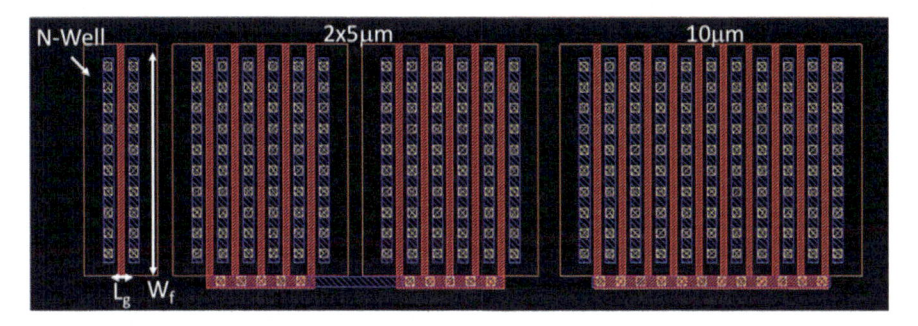

8 Millimeter-Wave Power Amplifiers for 5G Applications in 28 nm FD-SOI... 177

Fig. 8.6 Different transistors topologies

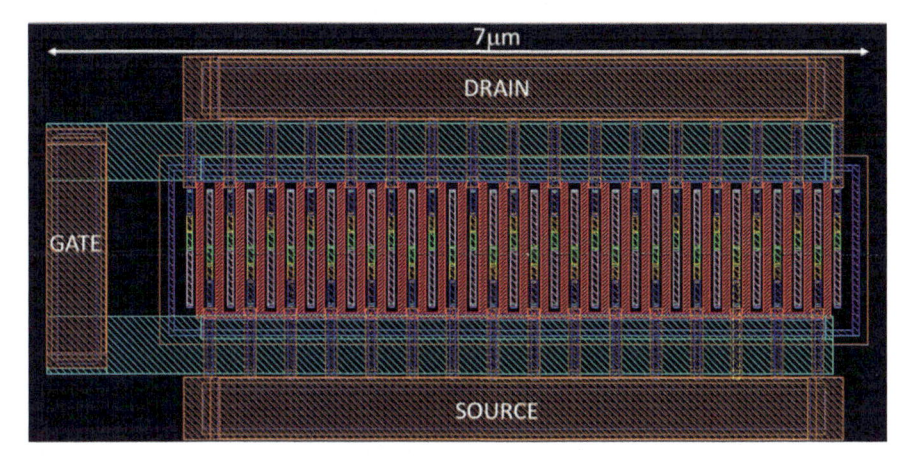

Fig. 8.7 Layout view of designed elementary cell

The maximum W_{tot} of a transistor can be calculated with (8.3) from [9]. At 30 GHz the $W_{tot_{max}}$ is around 330 μm. In this work we have designed an elementary transistor cell with 32 μm W_{el} for easy integration (Fig. 8.7).

$$W_{tot_{max}}[\mu m] = \frac{10^4}{f[GHz]} \quad (8.3)$$

It is possible to define several adapted values of W_{el} because it has no significant impact over performances as long as the physical dimension remains small to avoid the non-quasi-static effects. The most important dimensions in the elementary transistor cell are the unitary transistor gate width and length. Concerning the gate width W_f, please do refer to Chap. 7, Sect. 7.3 of this book.

It is possible to make the projection that the optimal gate finger width will be around 1 μm. In order to verify this value, we determined the f_T and f_{max} of a fixed Metal1-Pcell with a total width of 400 μm, where the finger width is varied. The number of finger is adaptive at each iteration in order to always provide a 400 μm fixed W_{tot}. The results are reported in Fig. 8.8.

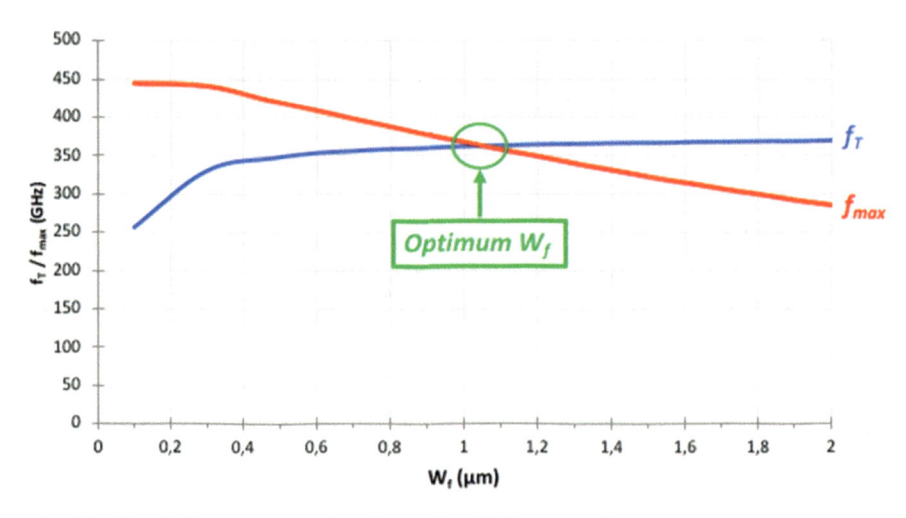

Fig. 8.8 f_T and f_{max} versus gate length for a fixed 400 μm transistor

The evolutions reported in Fig. 8.8 are linked to the intrinsic gate resistance and capacitance. Indeed, with a small value of W_f, the gate resistance is limited and higher f_{max} is achieved, while lower f_T is reported due to higher gate capacitances.

Concerning the gate length, in this design we implement a non-minimum gate length of 60 nm. There are several reasons for this gate length value justification related to target both performances and robustness of integration. Indeed, a non-minimum gate length reduces the local process variability and therefore the performance dispersion between chips. Higher gate length also reduces the current per gate finger at high power and thus limits the stress in respect with electromigration. This phenomenon is developed in a section further on. Furthermore, higher gate length is leading to lower C_{DS} parasitic capacitance and thus eases the output matching. Parasitic gate resistance and C_{GD}/C_{GS} ratio are also limited. In addition, 60 nm gate length demonstrates great large-signal performances during early simulations. This choice is also ensuring comfortable transconductance at the desired frequency band. Figure 8.9 presents simulated performances for L_g 30 nm and respectively 60 nm (drawn gate length).

8.2.1.2 Layout Optimization Strategy

Specific layout optimization strategy is mandatory in these frequency bands in deep sub-micron technologies, in order to maximize f_T and f_{max} while reducing the parasitic interconnects. The general consideration is presented in Chap. 7. The back end of line (BEOL) cross section used in this design is depicted in Fig. 8.10.

For this specific design, in order to reduce the gate access resistance, a double gate access on M2 has been implemented and the gate access is routed from M2 to IA thick top metal layer. In addition to f_{max} enhancement due to gate resistance

Fig. 8.9 f_T and f_{max} comparison between gate length of 60 and 30 nm

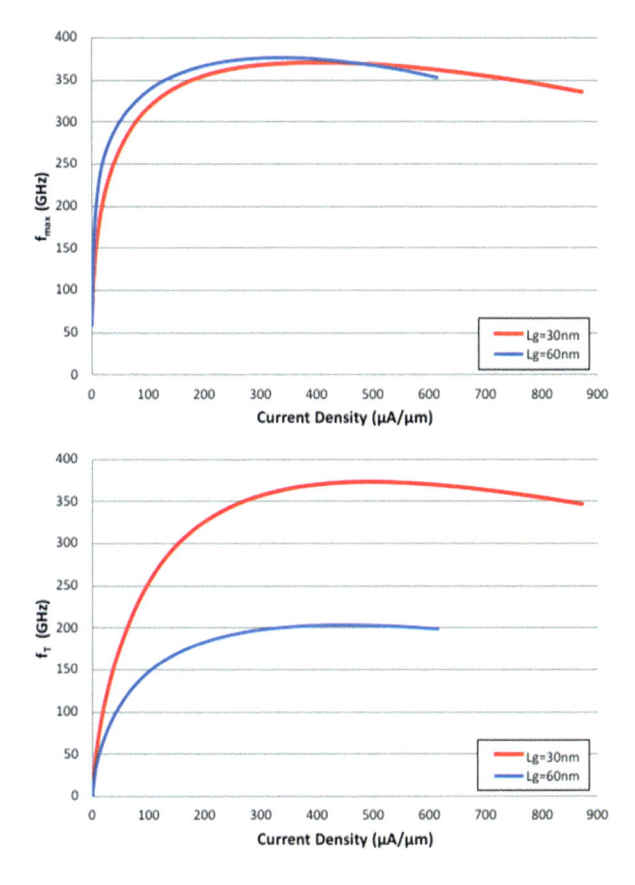

Fig. 8.10 Back end of line example for the 28 nm FD-SOI technology used in this design, 10 metal layers + AluCap. AluCap as last metal level: LB, 2 thick top metal levels: IA and IB, 2 medium metal levels: B1 and B2 and 6 thin metal levels: M1 to M6

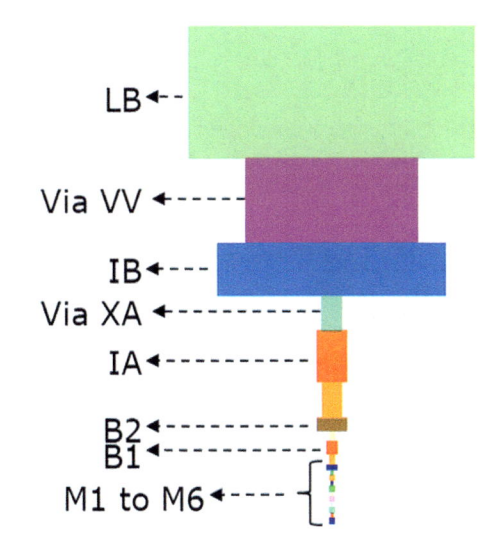

reduction, this double gate access also allows to reduce the stress over gate access at high power and so provides higher robustness than a classical single gate access. As double gate access at poly-silicon level was not available in the design kit at the time this circuit has been designed, it had to be carefully implemented in order to fulfill the DRM rules. In later design kits, this option is now available directly for the RF transistors due to its inherent advantages. It is also noticeable that the poly-silicon gate access can be widened for higher resistance reduction.

Source and drain accesses can also be optimized in order to reduce the associated parasitic resistances and the C_{DS} capacitance. The classic layout method is to implement the full surface of each drain and source finger, from M1 to the desired top metal layer. However, while this solution allows source and drain access resistance reduction, it is leading to a large C_{DS} value due to the fringe and parallel plate capacitance between drain and source fingers.

Therefore, in this work we propose a staggered structure. From M1 to M2, the whole finger surface is implemented. Then external accesses are implemented from M3 to IA metal level, with iterative reduction of finger length in each metal layer. Therefore, less surfaces are fringing between interdigitated source and drain fingers while accesses resistances are reduced thanks to metal layer stacking. Corresponding fringe capacitances to the ground are also reduced for f_{max} and f_T improvement. Accesses width and the number of vias on each finger and metal layers have been optimized in respect with electromigration. DRM and density rules have also been fulfilled. During the design optimization, all the capacitances and access resistances have been extracted with RCC extraction.

In these conditions, effective f_T of 220 GHz for the Metal1-Pcell and 190 GHz for the full optimized BEOL elementary transistor cell have been simulated. A 3D view of this elementary transistor cell is available in Fig. 8.11, for better visibility only vias between IA and B2 are visible.

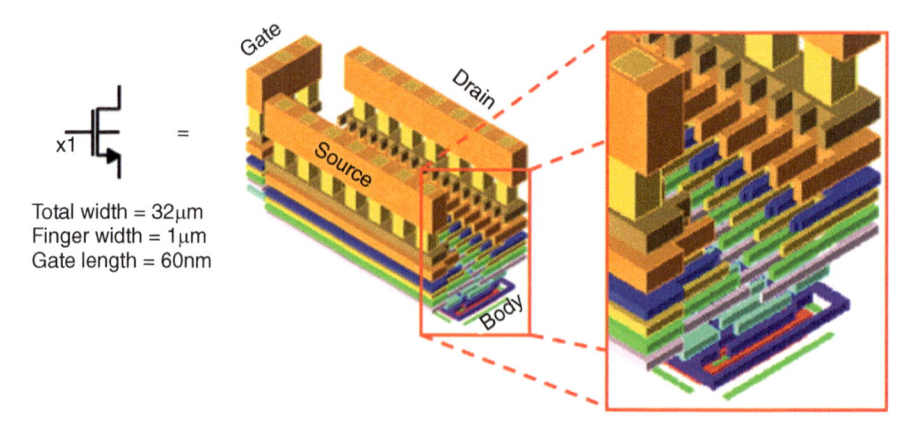

Total width = 32μm
Finger width = 1μm
Gate length = 60nm

Fig. 8.11 3D view of elementary transistor cell optimal layout

8.2.2 Power Amplifier Topology

8.2.2.1 Choice of Overall Topology

The specifications in terms of performances for 5G mmW power amplifiers can be summarized as follows:

- Power amplifier is the most power consuming block in a transceiver. As a major 5G specification is to reduce the network power consumption, highly efficient PAs are expected with reduced power consumption.
- The complex modulations and waveforms that will be used for 5G will present high PAPR and are leading to stringent requirements concerning linearity.
- Gain configurability is benefic for PA as it enhances system level control for SoC. Several modes also allow to modulate performances and power consumption depending on the use case. Large gain needs for 5G transceivers have also been identified.

In beamforming phased array, the environment difference of each circuit can lead to antenna mismatch and to harmful voltage overshoot at the output of individual power amplifiers.

Therefore, the overall power amplifier structure must be chosen wisely to target robust integration in the 5G and SoC implementation context, targeting the previously depicted challenges.

Balanced amplifier topology provides several advantages. Indeed, this topology offers robustness to input and output impedance mismatch and thus is ideal for beamforming phased array implementation. The use of 90° hybrid couplers at the output performs power combining and so improves the $P_{1\,dB}$ and maximum achievable output power by 3 dB. Moreover, the 90° recombination of amplification paths with 90° phase imbalance reduces the IM3 by 6 dB and achieves ACPR improvement. Therefore, impedance mismatch insensitivity, output power levels, and linearity improvements are performed by balanced topology. For these reasons, we decided to choose an overall balanced topology for the power amplifier implemented in this design.

After the overall power amplifier topology determination, the next step is to define the amplification paths architecture. Both amplification paths of the balanced amplifier are identical. Multi-stage topology allows higher gain than single-stage and is used in the designed power amplifier targeting high-gain level. The number of stages is important to define as we highlight that PAE decreases and power consumption increases with the number of cascaded stage. For this purpose, as we target high PAE and low power consumption simultaneously with high gain, we decide to limit the number of stages to two.

We have decided to use differential topology for both amplification stages of each amplification path as it performs supplementary power combining and stability improvements thanks to virtual dynamic ground. The use of baluns and transformers for impedance matching and input splitting/output combining

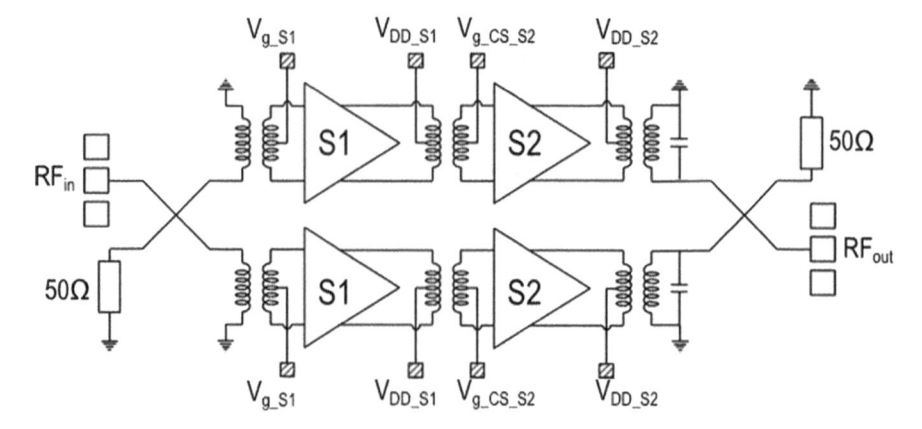

Fig. 8.12 Designed balanced power amplifier overall topology targeting 5G and SoC integration challenges

also provides advantages over robustness and reliability of the amplifier. It also eliminates the need for DC blocking capacitors on RF paths. Furthermore, the center taps of baluns/transformers are used to apply bias and supply voltages and eliminate the need for choke inductors with large area footprint.

The overall power amplifier topology is shown in Fig. 8.12 and all separate sections are described over the following paragraphs.

8.2.2.2 Balanced Topology Implementation

The design of quadrature hybrid coupler is challenging as it has to be compact to limit the area footprint while limited losses are expected to avoid critical overall amplifier efficiency reduction. Several distributed 90° hybrid coupler architectures are available in the literature and have been extensively used in MMIC like branch-lines coupler or coupled-lines coupler (Fig. 8.13).

However, integration of these couplers over expected 5G mmW frequency bands is challenging as they are based on $\lambda/4$ transmission lines that have a high area footprint around 30 GHz.

Recently, a 90° hybrid coupler distributed design featuring twisted layout and compact size has been demonstrated in [8], and has been validated in 28 nm FD-SOI balanced design [13], showing measured robustness to at least 3:1 VSWR conditions. Another advantage of this coupler that has been explored in [8] is the possibility to adapt the size and design of the coupler, and therefore to propose a pragmatic design.

In this work we used this promising coupler design to perform balanced topology. We provide in this section the design procedure developed in [12] and used in this work.

In order to design the 90° hybrid coupler, it is first possible to identify the coupler as a simple lumped elements model as presented in Fig. 8.14

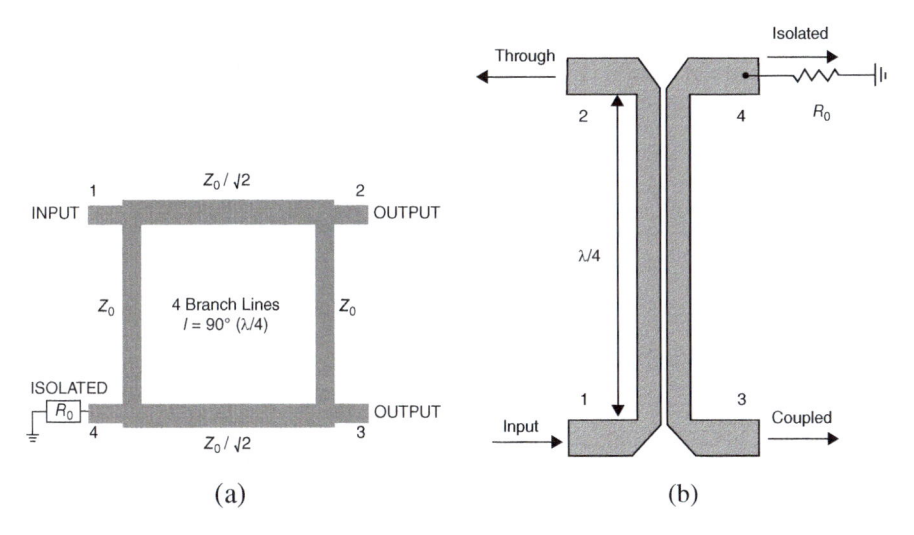

8 Millimeter-Wave Power Amplifiers for 5G Applications in 28 nm FD-SOI... 183

Fig. 8.13 Branch-lines (**a**) and coupled-lines coupler (**b**) illustration [2]

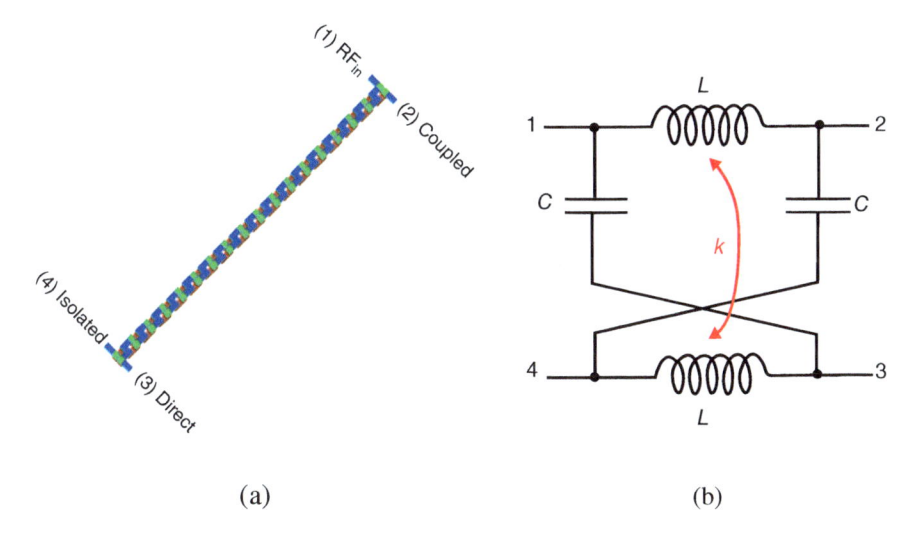

Fig. 8.14 Quadrature hybrid coupler layout (**a**) and corresponding simplified lumped element model (**b**)

It is then possible to determine the values of L and C lumped elements from Eqs. (8.4) and (8.5), where f_0 is the coupler central frequency, Z_0 its characteristic impedance, and k the coupling coefficient.

$$L = \frac{Z_0 \cdot (2-k)}{2\pi \cdot f_0} \tag{8.4}$$

$$C = \frac{(2-k)}{Z_0 \cdot 2\pi \cdot f_0} \tag{8.5}$$

Fig. 8.15 Unitary twisted cell

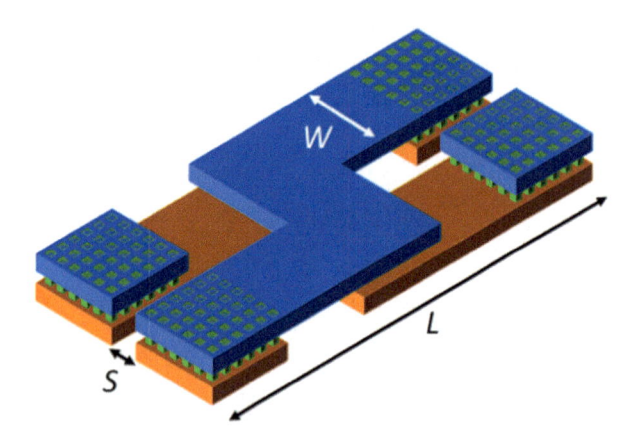

Therefore, the coupler dimensioning is dependent on the frequency, targeted performances, and coupling coefficient. To synthetize these values, the starting point is a unitary twisted cell design (Fig. 8.15). The inductances and capacitances are distributed along the two tracks. This cell is implemented over IA and IB thick top metal layers in order to reduce the parasitic routing resistances. Its dimensioning is made to respect the DRC rules while showing sufficient inductance, capacitance, and limited resistance on each track. The implemented geometrical values are 25, 5, and 1.6 μm for L, W, and S, respectively.

The unitary inductance L_u, capacitance C_u, and resistance R_u over each track can be extracted with EM tools.

Therefore, it is possible to design the hybrid coupler by cascading several unitary twisted cells in order to achieve the targeted inductance and capacitance estimated previously. The total inductance, capacitance, and resistance of N cascaded unitary cells are

$$L_{tot} = N \cdot L_u \quad (8.6)$$

$$C_{tot} = N \cdot C_u \quad (8.7)$$

$$R_{tot} = N \cdot R_u \quad (8.8)$$

Several parameters are important to verify in order to ensure optimal quadrature operation. The amplitude between both coupled and direct ports should be the same at the frequency of operation. In other words, the difference of amplitude ΔAmplitude should be 0 dBm at central frequency. A phase difference, ΔPhase of 90° between outputs is also required. Losses in the coupler also have to be estimated and must be limited to maximize the efficiency from a system level. Frequency bandwidth is a key parameter, 1 dB ΔAmplitude bandwidth will be taken as reference. These parameters are obtained by conducting small-signal analysis study with all the coupler terminations loaded with Z_0.

Fig. 8.16 40 GHz 90° hybrid coupler performance

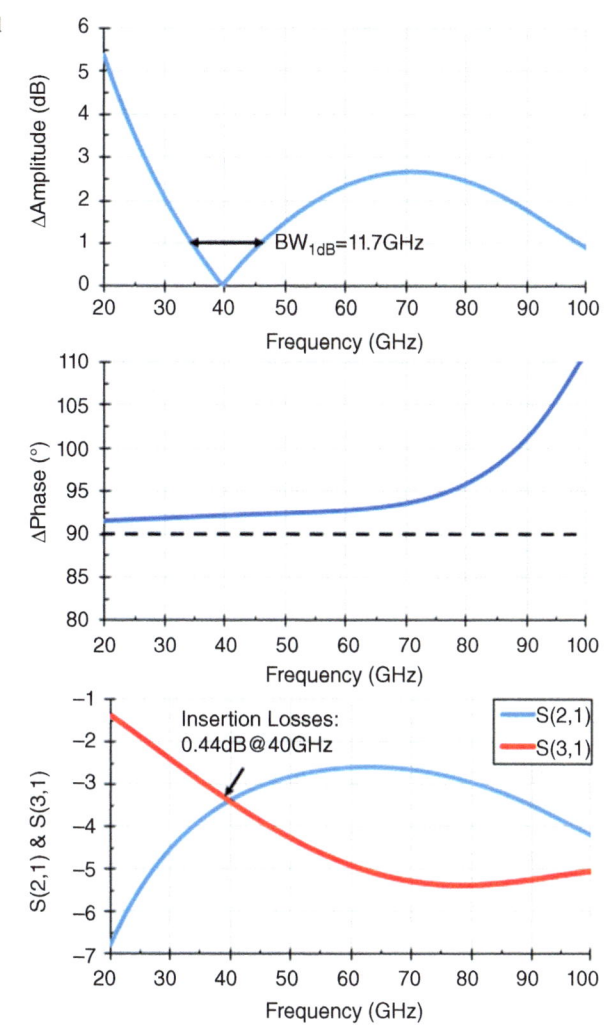

$$\Delta\text{Amplitude}[dB] = |S_{21}[dB] - S_{31}[dB]| \qquad (8.9)$$

$$\Delta\text{Phase}[°] = \text{phase}(S_{21}) - \text{phase}(S_{31}) \qquad (8.10)$$

If we take the example of a 50 Ω 40 GHz hybrid coupler, 17 cascaded twisted elementary cells are needed, leading to track lengths of 452 μm and to total L and C value of 219 pH and 43.8 fF, respectively, for a k of 0.9. The performances of this coupler after post-layout simulations are available in Fig. 8.16. From these results it is possible to determine losses of 0.44 dB at a frequency of operation of 40 GHz with a ΔPhase of 92°. A 1 dB ΔAmplitude bandwidth of 11.7 GHz is observed, with a ΔPhase of 92° ± 0.2° over the bandwidth.

8.2.3 Power Stages Design

After the implementation of overall balanced topology elements targeting robust integration with enhanced performances, a specific stages design strategy is adopted to provide high performances, configurability, and robustness to industrial margin for SoC oriented integration. As depicted in the specific mmW integrated PA design flow, we first focus on the output power stage S2 and then on the driver stage S1 design.

8.2.3.1 Design and Implementation of S2 Power Amplification Stage

In an amplification chain, the output power stage is expected to deliver power and gain. For this purpose, we decide to implement a cascode topology as it enhances gain, reverse isolation, output power, and bandwidth compared to a common-source topology. Differential topology has been chosen in order to perform power combining and so improves P_{sat} and $P_{1\,dB}$ by 3 dB and enhances stability compared to a single-ended topology.

As discussed in Gain configurability section in the beginning of this chapter, the configurability through body-biasing node should be implemented in the last stage of the amplification chain for linearity considerations. Therefore, we implemented body-biasing configurability on S2 stage. The overall S2 topology is available in Fig. 8.17.

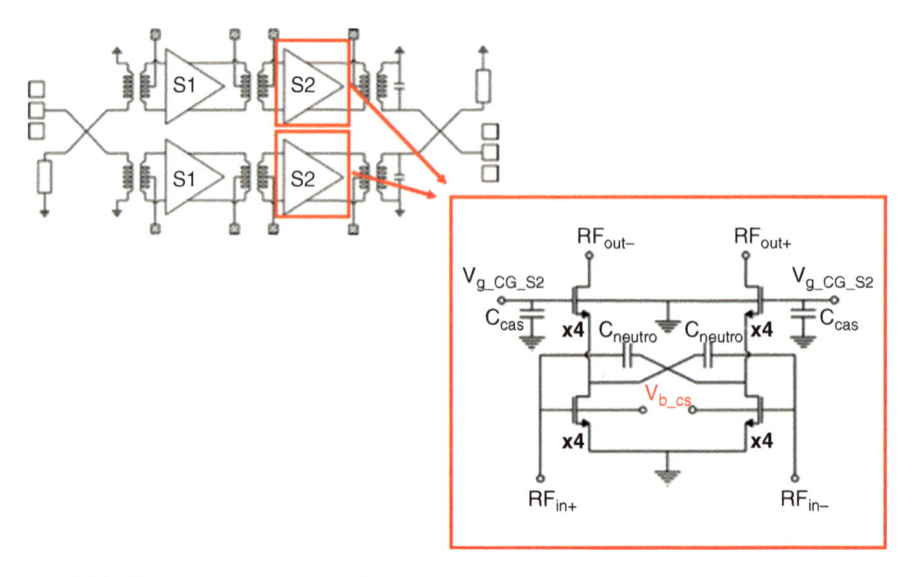

Fig. 8.17 S2 output power stage topology

In a cascode stage, the operating class of the overall structure is defined by the gate bias of the common-source (CS) stage. The common-gate (CG) stage is implemented to improve the output impedance of the structure, compared to a standalone common-source stage, in order to enhance the gain. Therefore, to implement body-biasing enabled gain configurability over a cascode stage and perform dynamic and continuous operating class switching, the varying body-biasing node must be located on common-source stage. This power stage (S2) is biased in class AB under nominal conditions and can be dynamically switched to class A with body-biasing. To our knowledge, this stage is the first body-biasing enabled reconfigurable cascode structure implemented in 28 nm FD-SOI technology in the literature.

Both CS and CG stages are implemented with 4x elementary transistor cells in parallel, leading to a W_{tot} of 128 μm for both CS and CG transistors. This value W_{tot} has been determined as a compromise between gain, P_{sat}, peak PAE, and power consumption performances. The W_{tot} maximum size is also limited by the input matching. Indeed, higher transistor length is leading to a lower input impedance. Therefore, input matching can be difficult to achieve if the optimal input impedance is low as the impedance transformation ratio will be large. In this technology, the input impedance real part limit is around 10 Ω, under this value the matching network will be very difficult to synthetize.

The cascode capacitance C_{cas} dimension of 200 fF has been determined in order to distribute the nominal supply voltage equally between CS and CG transistors. This capacitance has been implemented with a MOM capacitor, with interdigitated fingers from M2 to M6 and is connected to the common-gate stage gate access on M2 metal layer. Capacitive neutralization (C_{neutro}) has been realized over the common-source stage in order to improve reverse isolation, gain, and stability. A C_{neutro} value of 33 fF is optimal and has been estimated by Rollett factor stability considerations. C_{neutro} is implemented with MOM capacitor from M2 to M5.

The layout of S2 stage is available in Fig. 8.18

Several layout optimizations have been realized in order to limit the parasitic elements induced by layout and routing and also the local variability, for a better robustness to industrial margins. Dummies transistors with full identical BEOL have been implemented aside all useful transistor stages. The role of these transistor dummies is to reduce the local process variability that could occur during manufacturing at such deep sub-micron node. In general, variations are occurring at the edge of an element array. Therefore, if a variability occurs during process, the dummies transistor will be impacted, while the useful transistors in the array core will be "shielded."

Concerning the differential common-source stage, gate access implemented on IA level is enlarged in order to reduce the gate access resistance at routing level. Antenna diodes have been inserted at each gate access on the differential structure and are needed due to the AluCap routing. Antenna diodes are necessary as a charge accumulation can occur over AluCap metal layer during the manufacturing. Therefore, these extra charges are flowing to the ground through the diodes instead of penetrating active devices by the gate and induce damage. The size of these

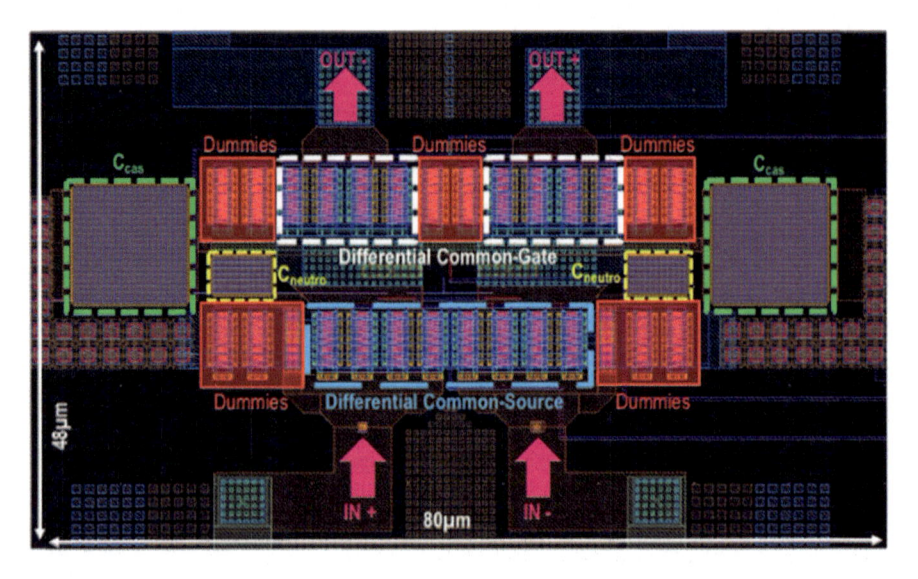

Fig. 8.18 S2 output power stage layout

diodes has been chosen to be as small as possible while fulfilling the design rules in order to limit the gate-to-ground added capacitance. An IB top thick metal layer rail is implemented over all the differential common-source in order to connect their source access to the lateral ground planes. The drain contacts are implemented on IA metal layer. The common-source stage gate bias is applied through input transformer center tap. All the distances between transistors are reduced to the minimum allowed by the design rules.

The common-gate stage design is also optimized for parasitic reduction. The gate bias of this stage is the only supply or bias applied directly to the node and not using any center tap of balun or transformer. The gate access is routed on M2 metal layer as no RF signal is flowing through this node. Therefore, a different transistor BEOL topology has been used for common-gate transistors. It only differs from the common-source at the gate contact node, implemented only in M2 and not routed up to IA metal layer. This also eases the inter-stage routing that we explain later in the section. Figure 8.19 shows the differences between common-source and common-gate transistors. The transistors drain accesses on this stage are implemented over IA metal layers and are wide in order to reduce drain resistance and to offer better thermal resistance at high output power. Connections to output matching network are realized with a wide 10×7 via array between IA and IB for the same reasons.

The inter-stage is a critical node in cascode topology. While the capacitance-to-substrate is naturally reduced by the SOI technology, parasitic resistances and inductances have to be reduced through layout. Therefore, wide inter-stage connection over IA and IB metal layers has been designed, with a large 16×5 via array. This node is also designed in respect with DRM density rules and electromigration

8 Millimeter-Wave Power Amplifiers for 5G Applications in 28 nm FD-SOI... 189

Fig. 8.19 Difference of BEOL between CS and CG elementary transistor cells

Fig. 8.20 Interconnections in the cascode stage

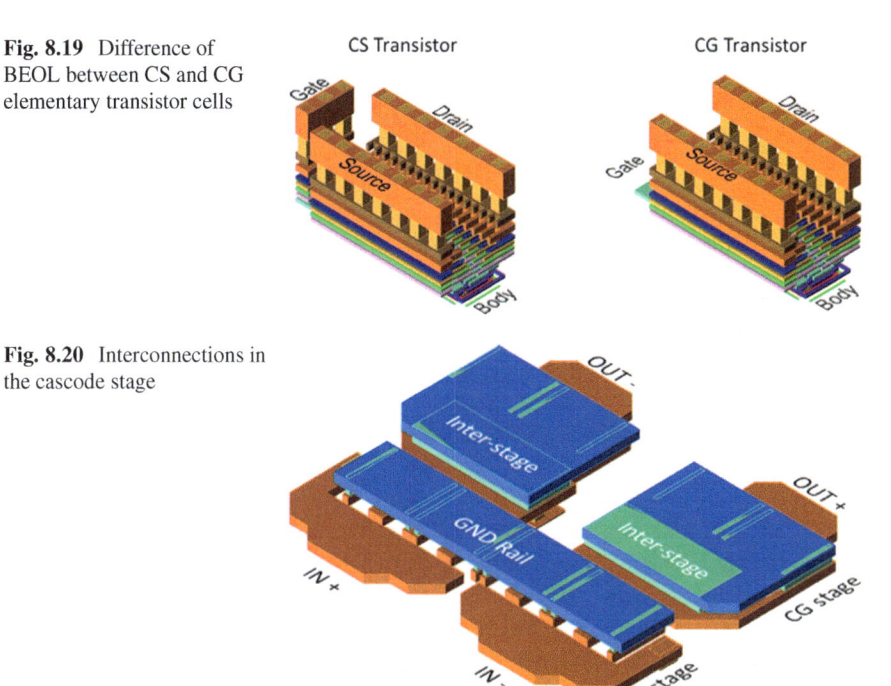

purposes. Figure 8.20 shows the interconnections between common-source and common-gate stages in the cascode topology.

The C_{neutro} placement between gate and drain of the differential branches has to be integrated in design process very early as the routing induced parasitic on this node can affect the necessary C_{neutro} value. Therefore, routing should not be implemented on thin metal layers and routing length should be limited. For this purpose, we implement the C_{neutro} routing over intermediate metal layers B1 and B2 available in the technology. The accesses at gate side are connected through via array to IA. The routing is then integrated over dummy transistor (which BEOL layers differ from the others dummies), as near as possible from the stage input gate. Then the C_{neutro} MOM capacitor is placed at the inter-stage, as near as possible from the two stages for a compact design. Then the crossing occurs over drain accesses, one C_{neutro} is using a B1 routing, while the other is using a B2 routing. They are then connected to the drain of common-source stage, at the inter-stage node, with a large via array. Even if the routing has been designed in order to reduce the parasitics, an EM extraction is mandatory before fixing definitely the C_{neutro} value.

Finally, the back-gate accesses for both common-source and common-gate stages are routed on M1 metal layer. 10 kΩ poly-resistors have been implemented over all body and gate bias accesses in order to avoid any RF signal leakage through these nodes. During the design optimization, all the interconnections induced parasitics have been extracted using 2.5D EM simulator.

8.2.3.2 Design and Implementation of S1 Power Amplification Stage

In a multi-stage amplifier, the role of the driver stage is to provide additional gain while limiting the power consumption. For this purpose, a common-source configuration has been implemented as it provides enough gain for a limited power consumption compared with a cascode architecture. As the driver stage does not need to provide a high gain nor high power, the driver transistors width is generally chosen as half the power stage transistors width. Therefore, we choose to use transistors with 64 µm total width and composed by two elementary transistor cells. As for the output stage, differential topology has been used to provide higher stability while performing power combining. The S1 stage topology is available in Fig. 8.21.

Differential topology has also been implemented in order to allow the use of neutralization capacitances to eliminate the Miller effect induced by C_{GD} over a common-source stage and so enhances stability, bandwidth, and gain. In this stage, the C_{neutro} optimal value is 16 fF It has been synthetized with a MOM capacitor from M2 to M5. The realized layout of S1 stage is available in Fig. 8.22.

As for the output stage, the interconnections parasitics have been reduced in the design by using large gate and drain accesses over IA thick metal level. An IB thick metal rail is implemented to connect the sources to lateral ground planes. Transistor dummies have been integrated to reduce the local process variability on this stage. The gate bias and drain supply voltage are applied through center taps of input/interstage matching networks. Body-biasing is applied through a 10 kΩ resistor to avoid any RF signal leakage. C_{neutro} routing is following the same strategy than in the previous section for the output stage. All the interconnections have been extracted

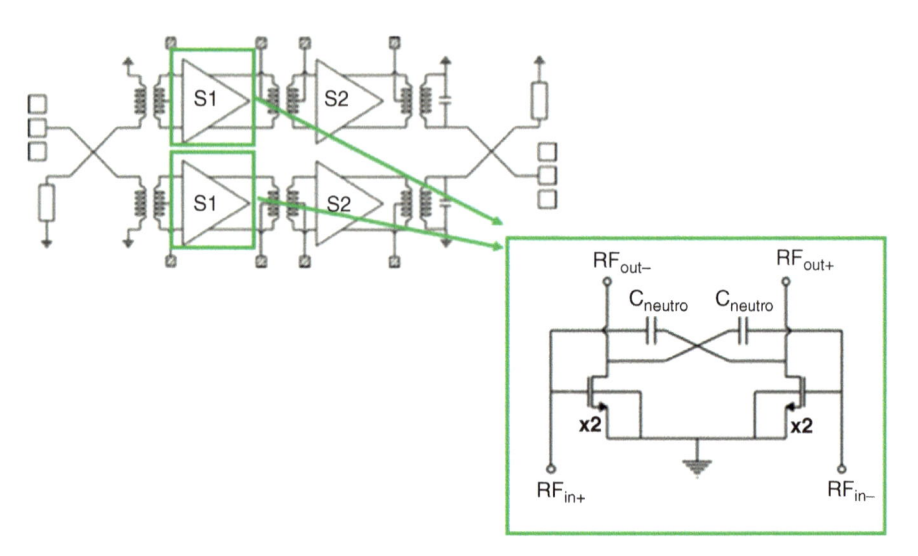

Fig. 8.21 S1 driver power stage topology

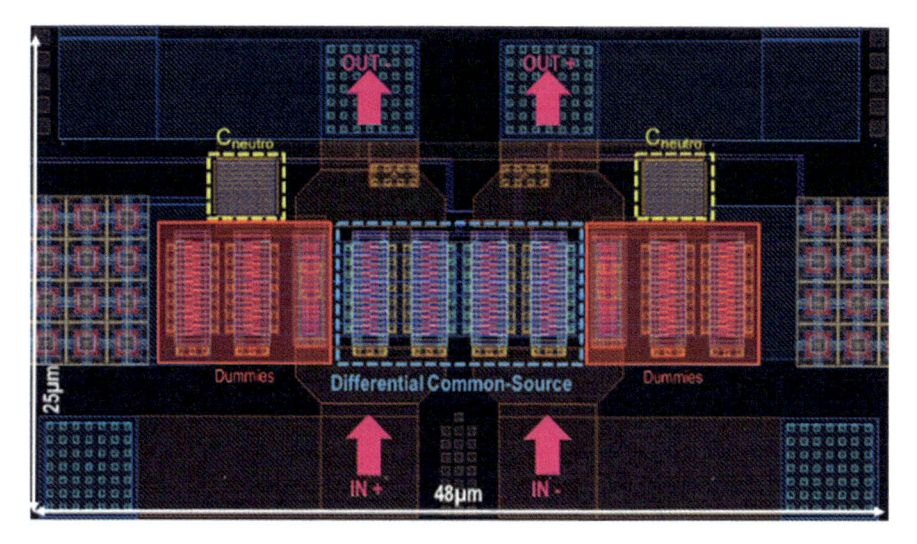

Fig. 8.22 S1 driver power stage layout

with momentum EM simulations during the design process. S1 stage is biased in class A in order to enhance the gain and linearity.

8.2.4 Impedance Matching Network Implementation

While specific stage design is important to provide the best operations under robust reliability conditions for targeted applications, the design of matching networks is also very important to perform the right impedance transformations and thus enhances performances while limiting the losses to achieve a maximum of efficiency.

8.2.4.1 Output Matching Network Optimization Strategy

The output matching network is the most critical matching network design of all the amplification chain and must be designed carefully to achieve the best performances for targeted applications. In this section, we propose a novel approach for output impedance matching of reconfigurable power amplifiers.

Traditionally, the ideal load impedance is determined with a load-pull simulation in nominal operating conditions. Then the matching network is designed and implemented to match the determined impedance. Even in the case of configurability, the impedance is matched for the nominal operation. This is the case in reconfigurable circuits implemented in 28 nm FD-SOI [10] where the output load is chosen for the maximum linearity conditions.

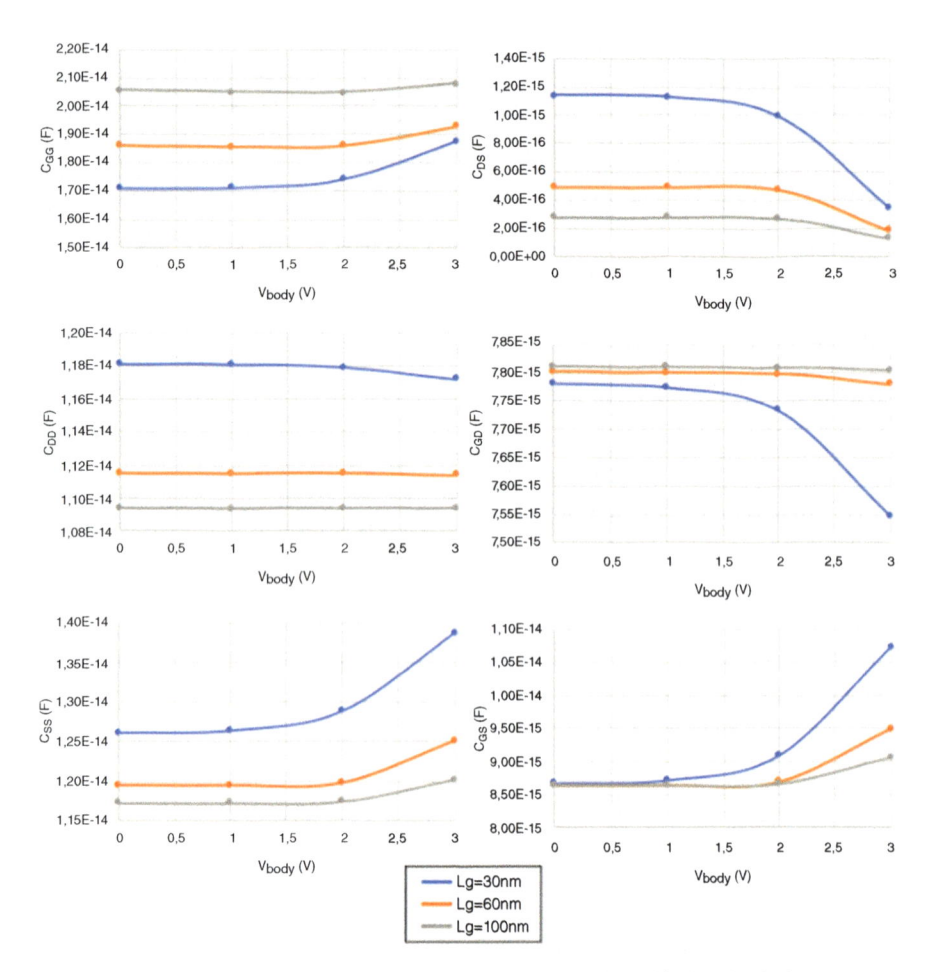

Fig. 8.23 Active device intrinsic capacitances evolution versus body-biasing simulations

However, with the body-biasing, the intrinsic parameters of the active devices are modified. This is illustrated in Fig. 8.23, where C_{GG}, C_{DD}, C_{SS}, C_{GD}, C_{DS}, and C_{GD} intrinsic capacitances are plotted versus body-biasing for several gate lengths for a 128 μm transistor Metal1-Pcell.

Therefore, the output optimal impedance of S2 stage varies depending on the body-biasing. This effect is noticeable by conducting load-pull for the extreme body-biasing conditions of 0 and 3 V. The results of these simulations are available in Fig. 8.24.

As it is noticeable, for a body-biasing of 0 V, the differential optimal impedance to enhance both PAE and gain is $Zload_{Class-AB} = 41 + j67$. In addition, for a 3 V body-biasing, the optimal impedance is $Zload_{Class-A} = 54 + j45$. Therefore, the choice of one or the other optimal impedance will lead to privilege an operating mode among the others. Therefore, in this work as we want an optimal matching for

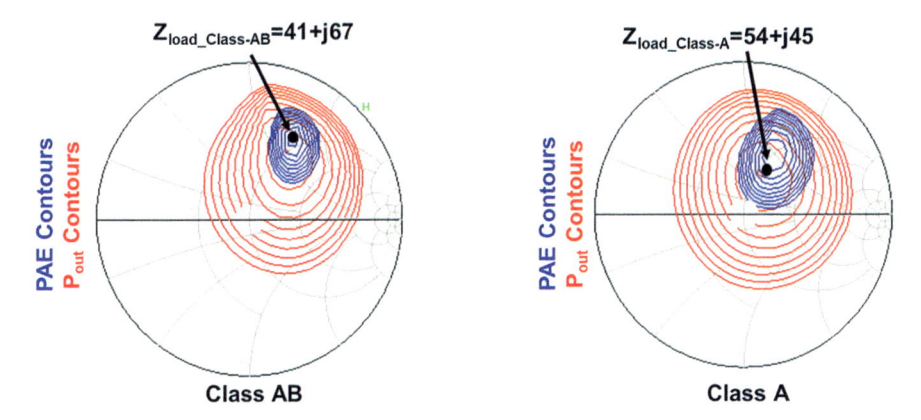

Fig. 8.24 Load-pull for extreme body-biasing conditions of 0 and 3 V and associated optimal differential impedances

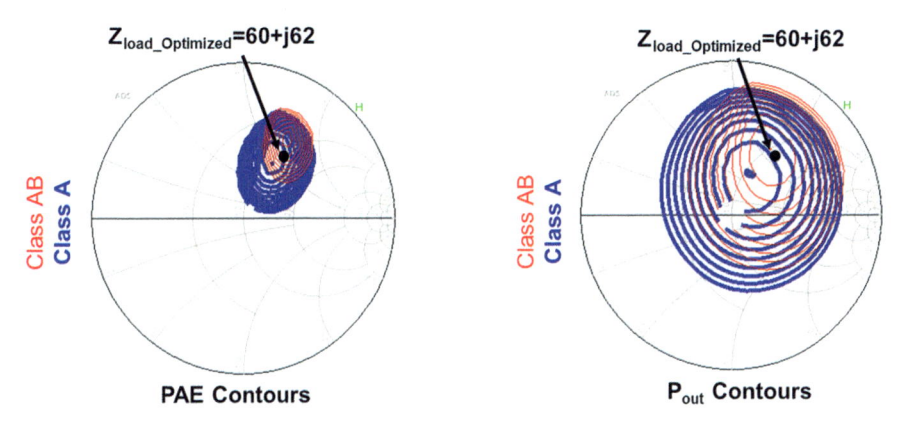

Fig. 8.25 Optimum output load determination strategy illustration

all operating modes, the choice of output load is made as a trade-off between both extreme body-biasing modes instead of choosing to privilege the maximum linearity mode. This choice is made to ensure quasi-constant PAE and output power among all operating classes and so avoid the efficiency decrease traditionally observed with this kind of configurability [10, 12]. The chosen differential output load impedance is thus $Zload_{optimized} = 60 + j62$. This output load impedance strategy choice is illustrated in Fig. 8.25.

This output load matching is synthetized with a flipped balun in stacked configuration, transforming the optimum load to 50 Ω and performing the output differential-to-single conversion. The primary winding is featuring two turns and is implemented over IB metal layer with a track width of 6 μm. The crossing between the two turns is made in IA; therefore, no losses are added due to a crossing implemented in thinner metal layers. The secondary consists in a single turn implemented on LB AluCap with 6 μm width. The secondary balun winding is

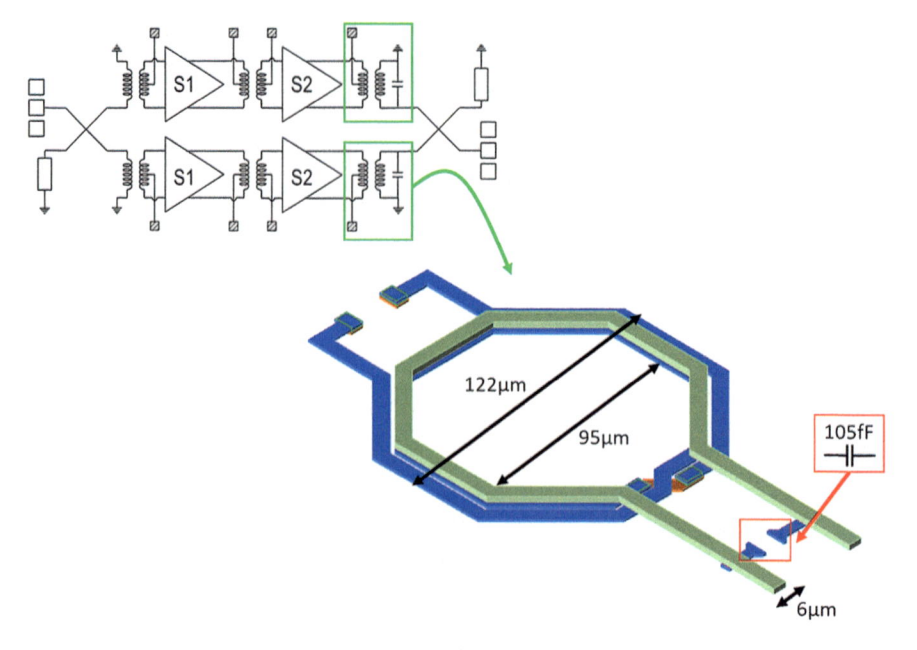

Fig. 8.26 Output matching network balun with associated dimensions

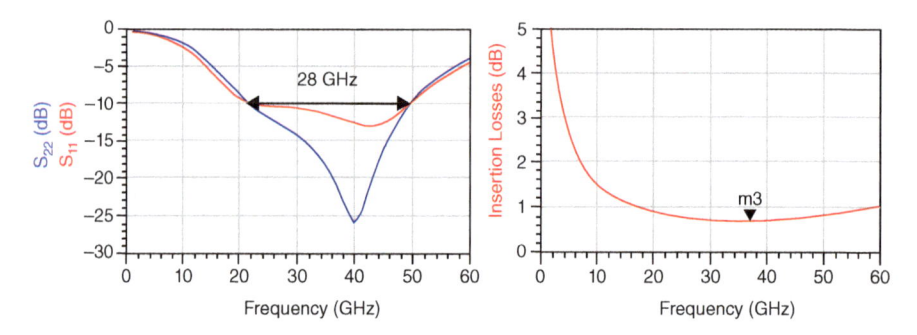

Fig. 8.27 Output balun performances

positioned over the second turn of the primary in order to maximize the coupling. A 3D view and the associated dimensions are available in Fig. 8.26. A capacitance of 105 fF is introduced between the two secondary terminations in order to refine the balun central frequency. This capacitance also broadens the balun bandwidth.

The performances of this balun are available in Fig. 8.27. A bandwidth of 28 GHz is achieved from 22 to 50 GHz. Insertion losses under 1 dB are achieved from 14 GHz.

8.2.4.2 Inter-Stage and Input Matching

The inter-stage and input matching networks do not need to be adapted to body-biasing conditions as the body-biasing only affects the output impedance of S2 and no configurability is implemented on S1. The input and output optimal impedances of S1 stage and the input optimal impedance of S2 stage have been determined with load-pull and source impedance sweep simulations. These identified differential impedance values are available in Table 8.2.

As explained in the design flow section, inter-stage matching network is first designed. S1 and S2 stages are both differential. Therefore, a transformer is used to perform differential impedance transformation. Both primary and secondary windings are featuring a single-turn configuration. The primary is implemented over IB metal layer, while the secondary is implemented over LB AluCap. Primary center tap is used to apply the driver supply voltage, while the secondary center tap is used to apply common-source gate bias of the output stage cascode topology. A 3D view is available in Fig. 8.28, while the S-parameters issued from post-layout simulations

Table 8.2 S1 and S2 stages optimum impedance values

Z_{in} opt S2	$10 + j52$
Z_{out} opt S1	$58 + j106$
Z_{in} opt S1	$15 + j102$

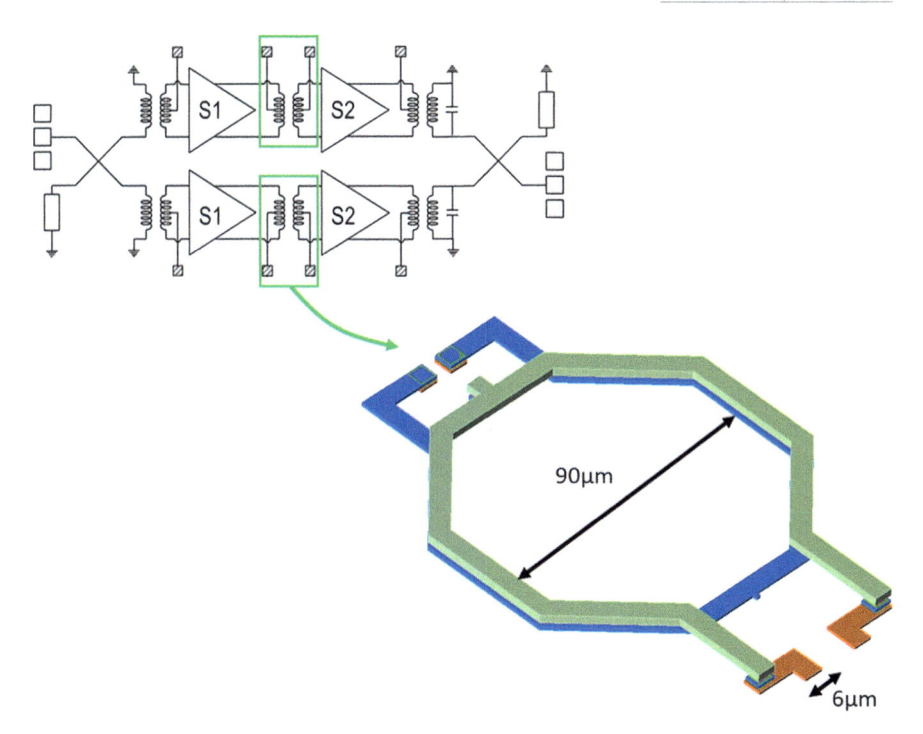

Fig. 8.28 Inter-stage matching network 3D view and dimensions

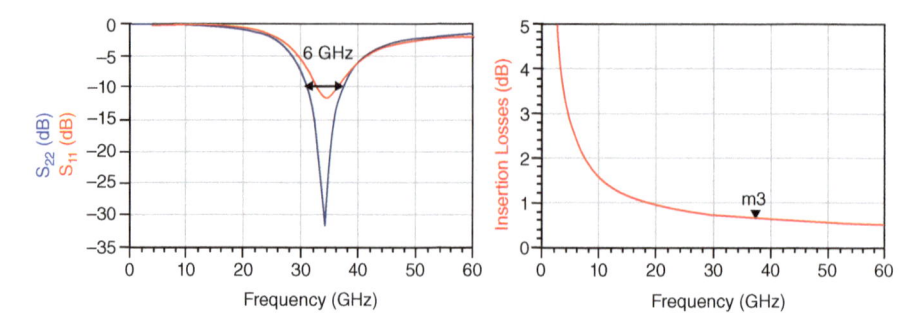

Fig. 8.29 Inter-stage matching network post-layout performances

and insertion losses are plotted in Fig. 8.29. S22 and S11 parameters are centered at 34 and 35 GHz, respectively, with a 6 GHz bandwidth. This transformer presents low insertion losses under 1 dB from 17 GHz.

Regarding the input matching network, it has to perform impedance transformation from 50 Ω to Zin_{opt_S1} and single-to-differential conversion. Therefore, a stacked flipped balun has been integrated. The primary is featuring a one-turn configuration implemented on AluCap. The secondary winding presents a 3-turn topology over IB. The crossings between turns are implemented on IA in order to avoid excessive resistive losses. Part of the second and third turns is implemented over IA metal layer due to DRM design constraints. The primary winding is stacked over the secondary turn of the secondary winding in order to maximize the coupling. A 3D view is available in Fig. 8.30, while the S-parameters and insertion losses performances can be found in Fig. 8.31. The S11 and S22 present a central frequency at 34 and 35 GHz, respectively, with a 7 GHz bandwidth while insertion losses under 1 dB are reported from 30 GHz. The center tap of the secondary is used to apply the driver stage gate bias voltage.

8.2.5 Robust Integration and Reliability

In this work we targeted high performances, configurability, and robust integration for industrial applications. For this purpose, several reliability conditions have been checked like electromigration and safe operating area while ESD protections and optimized ground path have been implemented. These reliability purposes, mandatory for the targeted applications, are discussed in the following section.

8.2.5.1 ESD Protection

Electrostatic discharges, ESD, can occur during manufacturing, measurement process, and during circuit operation. This phenomenon is due to the environment and can be dramatic for integrated circuits as it can induce critical damages and failures.

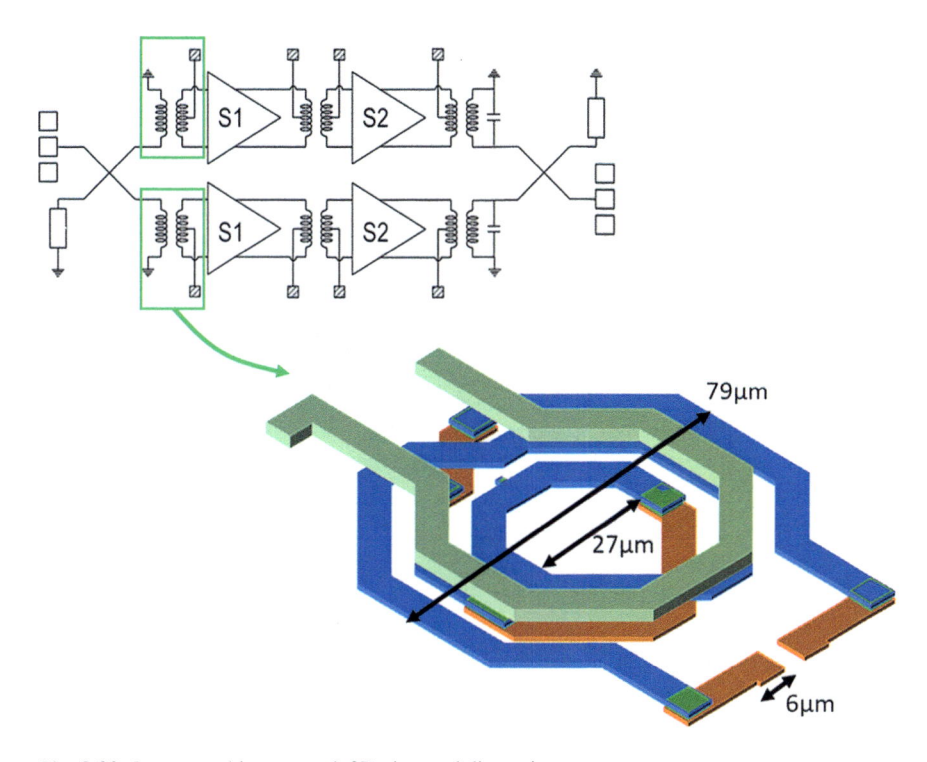

8 Millimeter-Wave Power Amplifiers for 5G Applications in 28 nm FD-SOI... 197

Fig. 8.30 Input matching network 3D view and dimensions

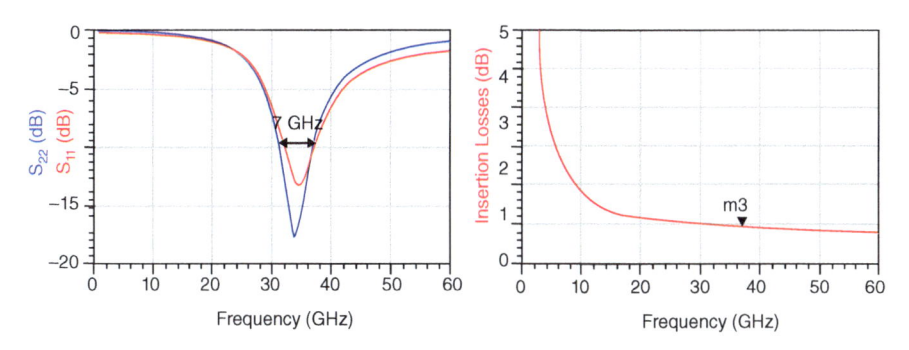

Fig. 8.31 Input matching network post-layout performances

Indeed, these discharges are critical for MOS transistors. When ESD is located on MOS gate, the gate oxide voltage is increasing and its destruction is occurring when breakdown voltage is reached. This phenomenon is amplified with CMOS technology downscaling. Indeed, a thinner gate oxide is leading to increased ESD sensitivity. When ESD is located on other transistor terminals, the associated PN junctions can suffer from avalanche breakdown phenomenon, increasing the device local temperature and causing permanent damages to the integrated circuit.

Fig. 8.32 Diode-based ESD protection

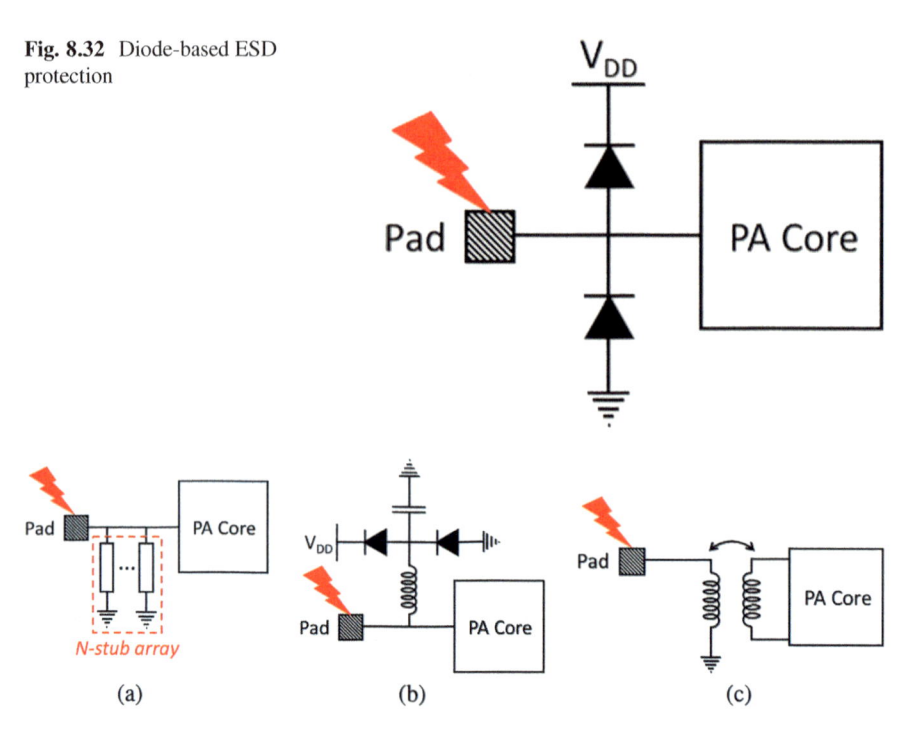

Fig. 8.33 ESD protections usable at mmW frequencies on RF paths. (**a**) Stubs protection. (**b**) Clamping diodes with RF leakage blocking protection. (**c**) Transformers protection

Therefore, for a robust integration, it is mandatory to implement ESD protections in the circuit. Several ESD solutions are provided in the literature, depending on the frequency of use and the placement on the circuit nodes. At low frequency, clamping diode-based circuit protections can be implemented easily (Fig. 8.32). This avoids the excessive current penetration in the circuit core in case of ESD while diodes are in blocking state during normal operation. However, at mmW frequency, the diode parasitic capacitance presents a low impedance. Thus, it can short the RF input/output signals. Therefore, this solution is not adapted to RF paths ESD protection at mmW frequencies.

Other solutions are available to overcome this issue, based on stubs, clamping diodes with RF signal leakage blocking or transformers (Fig. 8.33). In this circuit, we decide to use baluns and transformers as input/output and inter-stage matching networks not only for impedance transformation purpose but also for the robustness they confer to the circuit. In fact, transformers primary and secondary are magnetically and electrically coupled but physically isolated, providing ESD protection for each stage of the PA on the RF paths. The use of transformers also provides galvanic isolation between stages. Therefore, the PA design is sectored and robust to local ESD or stage temperature rise. However, it is not a sufficient condition to completely protect the circuit from ESD. Indeed, bias and supply voltages are applied on center taps of the baluns/transformers, and so are physically and electrically connected to the corresponding stages. To protect these nodes from ESD, we integrate ESD protection on each pad as described in Fig. 8.32, as no RF signal is flowing at these nodes.

8.2.5.2 Electromigration

In order to provide a power amplifier robust integration, it is important to fulfill electromigration rules available in the DRM for several years of operation at fixed temperature depending on the targeted applications. These rules allow to calculate at which temperature the power amplifier can operate for 10 years with no electromigration induced damages. These calculations can be done by considering the maximum current flowing at different circuit nodes.

In this design, we check carefully the electromigration conditions fulfillment. In practice, the temperature at which 10 years of operations are possible is limited by the number of vias in power cell transistor fingers due to the staggered structure presented previously. For the most extreme encountered measured conditions, corresponding to the maximum back-gate biasing mode operating at saturated output power (see measured results in the following sections), electromigration rules are fulfilled for 10 years of operation at 100 °C. By considering a lifetime derating available in the DRM, a 125 °C electromigration reliability is achievable for 2 years in these extreme operating conditions.

8.2.5.3 Safe Operating Area

The safe operating area is the ensemble of voltage and current conditions and dynamics in which any circuit element can operate during a certain time without self-harm. In fact, an operation outside these conditions could lead to an instant destruction of the active device due to junction or gate oxide breakdown. In cascode topology, this check is also important as high voltage swing can occur in the inter-stage node between common-source and common-gate even if the DC voltage supply is distributed between both stages. Therefore, a SOA check is mandatory before fixing the operating conditions and can be done through Spectre simulator during a transient simulation. In this work, the SOA conditions have been carefully explored and fulfilled for all operating conditions, ensuring safe operations and a maximum of reliability even under maximum body-biasing conditions. All the measurement conditions performed and depicted later in this chapter strictly fulfill the SOA conditions.

8.2.5.4 Ground Return Path Optimization

Ground path optimization is mandatory in deep sub-micron CMOS technologies as the thin metal layers available introduce parasitic resistance and inductance over

Fig. 8.34 Elementary ground and supply planes cell 3D view

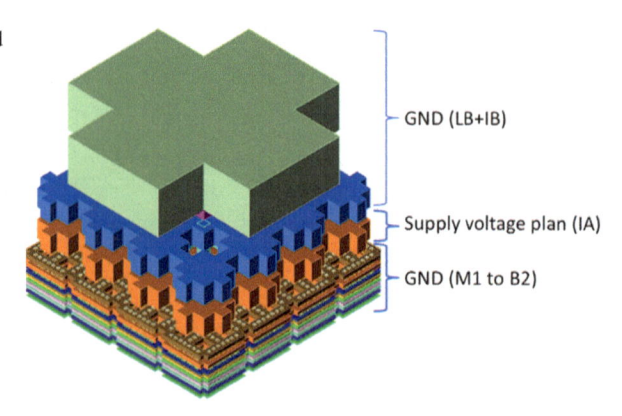

the ground path. These parasitic elements impact can be critical as it can induce oscillations. This issue is critical in single-ended power amplifiers where a signal return path from the output to the input through the ground plane is possible and leading to oscillations. In this work, even if a differential architecture is used allowing higher stability, ground plane has been optimized to reduce parasitics in order to avoid unexpected performances degradation. Furthermore, it is also important to optimize the voltage supply plan for the same reasons.

For this purpose, in this work, the ground plane and supply voltage planes are composed by a large array of elementary cells. The thin and intermediary metal layers from M1 to B2 are stacked to create a wider ground. IB metal layer is also connected to the ground. The IA metal layer is used to implement the supply voltage plane. Therefore, the supply voltage plane is implemented between two ground planes. This creates a distributed decoupling capacitance between supply and ground planes, enhancing the stability of the overall structure.

The elementary ground and supply plane cell has been carefully designed to fulfill density rules while enhancing the number of via and the metal width to minimize parasitics. A 3D view of this structure is available in Fig. 8.34.

It is also noticeable that two different supply planes have been implemented independently, one for the driver cell and the other for the power cell. Large MOM decoupling capacitors of 1 pF are also implemented as near as it is possible from the center taps where the supply voltages are applied to the stages to enhance overall stability.

8.3 mmW Power Amplifier Measurement Results

The following sections focus on on-wafer measurements of the reconfigurable balanced power amplifier implemented in 28 nm FD-SOI from STMicroelectronics presented previously. Figure 8.35 shows a photomicrograph of the manufactured power amplifier.

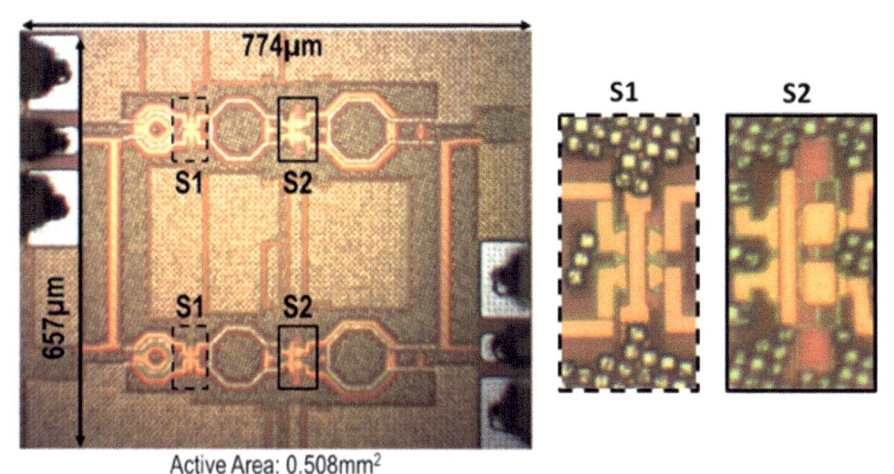

8 Millimeter-Wave Power Amplifiers for 5G Applications in 28 nm FD-SOI... 201

Fig. 8.35 Manufactured power amplifier photomicrograph

The configurability abilities enabled by the extensive use of body-biasing voltage tuning, a 28 nm FD-SOI technology flavor described in the previous chapters, are also highlighted.

8.3.1 Measurements at Optimal Operating Point

In this work, the optimal operating point in measurements is obtained at 31 GHz for a S2 supply voltage VDD_{S2} of 1.98 V, corresponding to the GO2 nominal voltage at maximum rating and made possible by the cascode architecture. S1 supply voltage VDD_{S1} is set to 0.7 V in order to enhance overall PA efficiency by limiting the power consumption. Gate bias values are 800 mV, 422 mV, and 1.5 V for Vg_{S1}, $Vg_{CS_{S2}}$, and $Vg_{CG_{S2}}$, respectively. The following results are corresponding to small-signal and large-signal measurements at this optimal operating point. In the following sections, the body-biasing value Vb_{CS} is sweeping between 0 and 1.65 V to illustrate the performances enhancement and the different effects enabled by the use of body-biasing node for PA configurability purpose for both small-signal and large-signal measurements. This choice of operating conditions for supply, bias, and body-bias voltages ensures reliable and stable power amplifier behavior while maximizing reached performances and is summarized in Table 8.3. In the following sections, if no different values are stated, the temperature of measurements is assumed to be 25 °C.

Table 8.3 Optimal operating point

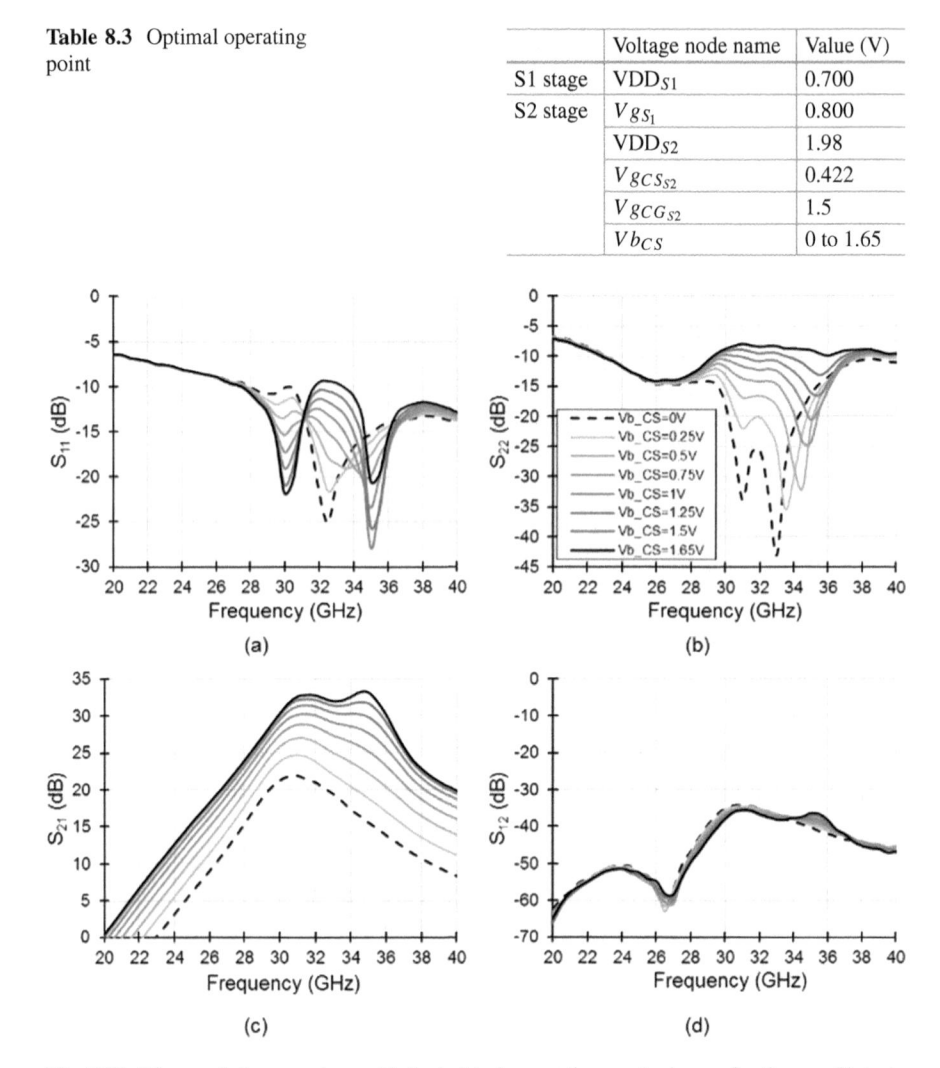

	Voltage node name	Value (V)
S1 stage	VDD_{S1}	0.700
S2 stage	Vg_{S1}	0.800
	VDD_{S2}	1.98
	$Vg_{CS_{S2}}$	0.422
	$Vg_{CG_{S2}}$	1.5
	Vb_{CS}	0 to 1.65

Fig. 8.36 Measured S-parameters with body-biasing continuous tuning: reflection coefficients S11 (**a**) and S22 (**b**), transmission coefficients S21 (**c**) and S12 (**d**)

8.3.2 Small-Signal Measurements with Body-Biasing Tuning

Small-signal analysis is performed for several values of Vb_{CS} between 0 and 1.65 V from 20 to 40 GHz. The graphs exhibiting plotted values of $S11$, $S22$, $S21$, and $S12$ are available in Fig. 8.36a–d, respectively. As it is noticeable on the $S21$ graph, the small-signal response in gain evolves with the body-bias variation. When $Vb_{CS} = 0$ V, the maximum small-signal gain value is 21.9 dB with a 3 dB bandwidth $BW_{-3\,dB}$ of 4.5 GHz from 29 to 33.5 GHz. The value of gain is incrementally improved to reach a maximum level of 32.6 dB for $Vb_{CS} = 1.65$ V. A second gain peak is measured at 35 GHz with higher Vb_{CS} values. It is residual from the high gain achievable for higher Vb_{CS} values that, with mismatched C_{neutro} and C_{GD}, is leading to oscillating conditions at driver stage. This is also corresponding to the S12 peak encountered at the same frequency. The gain achievable at this higher frequency allows BW_{-3dB} broadening and explains the achieved bandwidth. A maximum BW_{-3dB} of 6 GHz, from 30 to 36 GHz, is observed. These results demonstrate that the proposed power amplifier bandwidth covers the potential 5G band from 31.8 to 33.4 GHz for all operating modes. Power amplifier S21 measured behavior illustrates that the body-biasing allows dynamic gain level control over a >10 dB wide range of values all over the targeted frequency bandwidth.

The S11 curve at nominal operation presents an attenuation pole at 32 GHz and a lower one at 28 GHz. These notches evolve with the body-biasing and gradually shift the attenuation poles at 30 and 35 GHz for $Vb_{CS} = 1.65$ V. For all modes S11 illustrates that the power amplifier is well matched at the input. S22 curve at nominal operation shows two attenuation poles at 31 and 33 GHz. The curve is incrementally flattened with the Vb_{CS} improvement. This phenomenon is induced by the gain enhancement achieved through body-biasing. Indeed, the output power for the same small-signal input power level is enhanced. Therefore, the output is matched in power for higher Vb_{CS} values, while at nominal operation with $Vb_{CS} = 0$ V it is matched at small-signal. S22 remains low enough to ensure good output matching and stable operations over all Vb_{CS} conditions.

Regarding the S12, corresponding to the output-to-input isolation, the curves are very slightly affected by body-biasing variations. The value remains under −35 dB, demonstrating a good isolation under all body-biasing conditions over all the measured frequencies. This excellent value of output-to-input isolation is reached thanks to the overall balanced topology, the S2 stage cascode design, and the extensive use of cross-coupling C_{GD} capacitance neutralization in all stages.

8.3.3 Large-Signal Measurements with Body-Biasing Tuning

Large-signal measurements have been performed using the same operating conditions than small-signal analysis. The power gain, PAE, and power consumption results for a 31 GHz frequency of operation are available in Fig. 8.37a–c, respectively, with Vb_{CS} value varying from 0 to 1.65 V.

The power gain graph confirms the wide gain span, from 21.9 dB at $Vb_{CS} = 0$ V to an extreme value of 32.6 dB at $Vb_{CS} = 1.65$ V with incremental levels. To the 21.9 dB of power gain value is corresponding a P_{1dB} of 15.3 dBm, obtained with a gain expansion of 0.9 dB, typical to class AB operation, allowing to push the 1 dB output compression point P_{1dB} to higher output power level. To the 32.6 dB gain curve, with typical class A behavior, is corresponding a P_{1dB} value of 11.6 dBm. Therefore, two extremes modes can be identified as it follows:

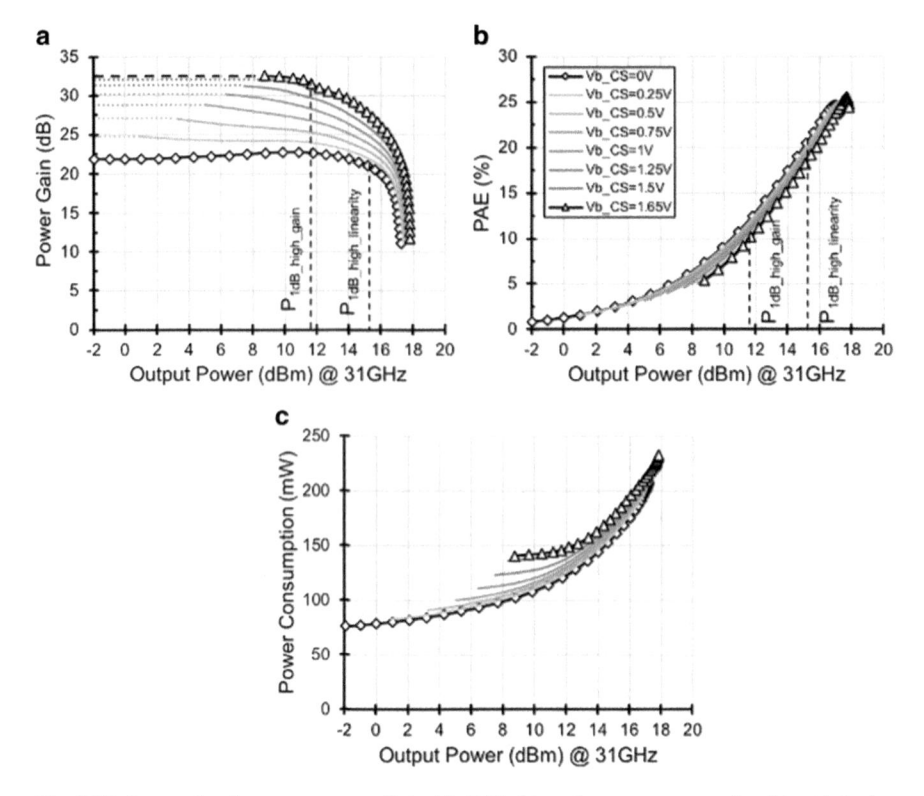

Fig. 8.37 Large-signal measurements: Gain (**a**), PAE (**b**), and power consumption (**c**) with body-biasing tuning variation from 0 to 1.65 V

- A high-linearity mode when $Vb_{CS} = 0\,\text{V}$.
- A high-gain mode when $Vb_{CS} = 1.65\,\text{V}$.

A saturated output power of 17.3 dBm is obtained in high-linearity mode and reaches 17.9 dBm in high-gain mode. There is only a slight change in P_{sat} value regarding the operation mode. On the PAE curves, it is noticeable that for all values of Vb_{CS}, the peak efficiency is quasi-constant from 24.7% in high-linearity mode to 25.5% in high-gain mode. The PAE at 1 dB output compression point, $\text{PAE}_{-1\,\text{dB}}$, is reduced from 21 to 10% in high-linearity and high-gain modes, respectively, as $P_{1\,\text{dB}}$ decreases with Vb_{CS}.

The PAE at back-off is an important parameter to take into account as it is in this condition that the PA likely has most chances to operate in advanced modulation schemes. The PAE at back-off is a challenging parameter to enhance for future transceivers. In this work, PAE at 6 dB P_{sat} back-off is very slightly affected with only 1.7% by Vb_{CS} variation and remains over 10% in all operating modes. This value is improved compared to other power amplifiers realized in the same technology at higher frequencies [10]. The quasi-constant values of PAE_{max} and P_{sat} for all the body-biasing conditions confirm the output load optimization strategy exposed in the previous section.

As one challenging specification for future wireless network is to reduce the overall system power consumption and as we expose the fact that the power amplifier is the most power consuming block in the overall transmitter architecture, it is important to limit the power consumption of the power amplifier. Power consumption curves, for this work, are plotted in Fig. 8.37c. It is noticeable that depending on the operating mode, the power consumption differs and justifies the performances difference. In high-linearity mode, the DC power consumption is 76.1 mW, while in high-gain mode the overall power consumption reaches 140.2 mW. The power consumption at saturated output power is 229 and 204 mW for high-gain and high-linearity modes, respectively. Indeed, it is possible to operate the power amplifier in high-linearity mode most of the time to lower the system power consumption and then to increase the gain level only when necessary to limit the overall average power consumption.

These large-signal measurements are highlighting that the body-biasing tuning allows a fine grain wide range gain control. This permits to provide precisely any necessary gain level located in the achievable range for performances enhancement, contrary to classical gain control stages with fixed gain steps [17]. This tuning node is also usable to compensate any manufacturing, environment, or time induced performance drift in the context of robust adaptive systems. The body-biasing voltage control also enables operating continuous class shifting. This continuous class shifting is achieved with quasi no efficiency degradation thanks to the output impedance matching strategy and demonstrates the possibilities offered by the technology to implement efficient reconfigurable PA with no additional stage requirement.

8.3.4 AM–PM Measurements with Body-Biasing Tuning

Linearity is a key performance for future wireless systems as complex modulation schemes are leading to high PAPR levels. The $P_{1\,dB}$, corresponding to 1 dB gain compression output power level is a parameter that describes linearity but an additional information regarding the phase is also necessary. For this purpose, it is possible to conduct AM–PM measurements. The AM–PM parameter is the output signal phase modification induced by an amplitude variation at the input of the power amplifier.

To measure this parameter, we performed large-signal S21 measurement with an input power sweep from -37 dBm to 1 dBm at 31 GHz. For each input power point, the real and imaginary parts of S21 are measured. From these S21 measurements, it is possible to estimate the output power P_{out} using (8.11) to (8.14).

$$P_{out}[dBm] = P_{in}[dBm] + \text{Gain}[dB] \quad (8.11)$$

$$\text{Gain}[dB] = 20\log\left(\sqrt{Re(S_{21})^2 + Im(S_{21})^2}\right) \quad (8.12)$$

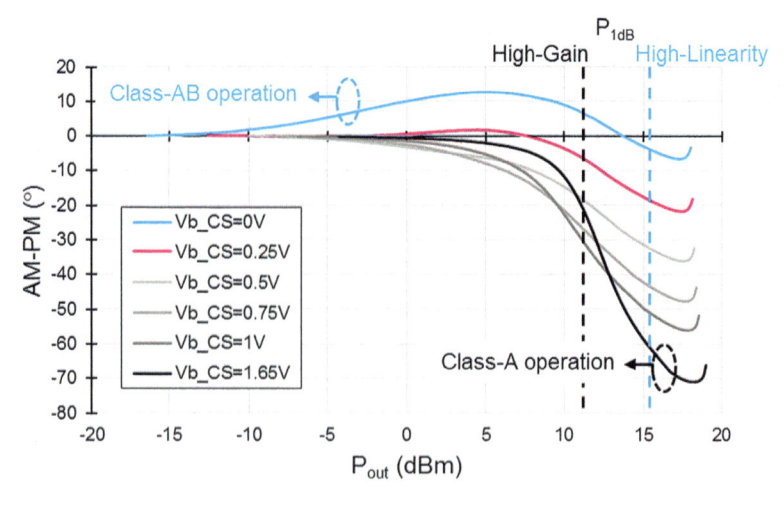

Fig. 8.38 AM–PM versus P_{out} measurements at 31 GHz with body-biasing tuning from 0 to 1.65 V

$$P_{out}[\text{dBm}] = P_{in}[\text{dBm}] + 20 \log \left(\sqrt{Re(S_{21})^2 + Im(S_{21})^2} \right) \quad (8.13)$$

$$\text{Phase}_{S_{21}}[°] = \tan^{-1} \left(\frac{Im(S_{21})}{Re(S_{21})} \cdot \frac{180}{\pi} \right) \quad (8.14)$$

Finally, normalized S21 phase is plotted in function of P_{out} for several Vb_{CS} values from 0 to 1.65 V like in previous sections. The AM–PM measurements results are available in Fig. 8.38.

Several remarks can be made regarding these results. The curve for $Vb_{CS} = 0$ V is corresponding to typical class AB operation. Indeed, a phase expansion with a maximum of 12.7° deviation from normalized origin value is reached. This phase expansion is typical from gain expansion in class AB mode. The AM–PM value at $P_{1\,\text{dB}}$ of 15.3 dBm in this mode is 3.6°. The curve for $Vb_{CS} = 1.65$ V shows typical class A operation. No phase variation occurs at low input power when the gain is flat and then the phase decreases when the output power approaches $P_{1\,\text{dB}}$ to finally present a high deviation during compression and up to saturation. In this mode, the AM–PM value at $P_{1\,\text{dB}}$ of 11.6 dBm is $-21.6°$. Intermediary modes are leading to intermediary AM–PM curves. $Vb_{CS} = 0.25$ V curve is corresponding to class AB operation with a lower phase expansion of 1.6°, while Vb_{CS} from 0.5 to 1 V curves shows class A type of operations. From these behaviors, it is noticeable that at high output power levels, from 11.2 dBm to P_{sat}, the $Vb_{CS} = 0$ V curve gives the lowest AM–PM variation. Then, in the output power interval from 4.5 to 11.2 dBm, the curve corresponding to $Vb_{CS} = 0.25$ V intermediary mode is leading to the lowest AM–PM variations compared to all other modes. Finally, under 4.5 dBm output power level, both $Vb_{CS} = 0.25$ V and 1.65 V modes are leading to a very low phase variation depending on output power level.

8 Millimeter-Wave Power Amplifiers for 5G Applications in 28 nm FD-SOI... 207

Therefore, there is an optimal mode regarding phase linearity depending on the output power level. It is then possible to dynamically modulate the AM–PM value with the body-biasing node voltage depending on the desired trade-off between gain, power consumption, output power level, and linearity performances. This dynamic modulation could be done with an adaptive scheme implementation in the context of SoC. Linearization techniques like digital pre-distortion could also be implemented in this context, linked to body-bias and targeted performances levels to compensate the phase deviation dynamically.

8.3.5 Measurements Over Frequency Range

As it is exposed in Sect. 8.3.2, the gain BW_{-3dB} in high-linearity mode, corresponding to $Vb_{CS} = 0$ V, is covering the band from 29 to 33.5 GHz. In high-gain mode, for $Vb_{CS} = 1.65$ V, it is covering the band from 30 to 36 GHz. In this section we expose the large-signal measurements in both extreme modes at several frequencies from 28 to 35 GHz. For this purpose, PAE_{max} and PAE_{-1dB} versus frequency are plotted in Fig. 8.39a, while P_{sat} and P_{1dB} versus frequency curves are presented in Fig. 8.39b.

It is noticeable in PAE_{max} curves that for both modes from 32 to 35 GHz, there is no difference on the values, both curves are superimposed. This difference starts slightly at 31 GHz and is higher for lower frequencies. At 28 GHz, PAE_{max} is 18.7% and 22.6% at $Vb_{CS} = 0$ V and $Vb_{CS} = 1.65$ V, respectively. This difference of PAE_{max} values can be explained by the fact that the output of the power amplifier is better matched at P_{in} corresponding to PAE_{max} value for $Vb_{CS} = 1.65$ V at these frequencies. P_{sat} values are always higher for $Vb_{CS} = 1.65$ V over all the measured frequency range and always above 15 dBm for both modes. Maximum values are obtained at 31 GHz and have been detailed in the previous section. A

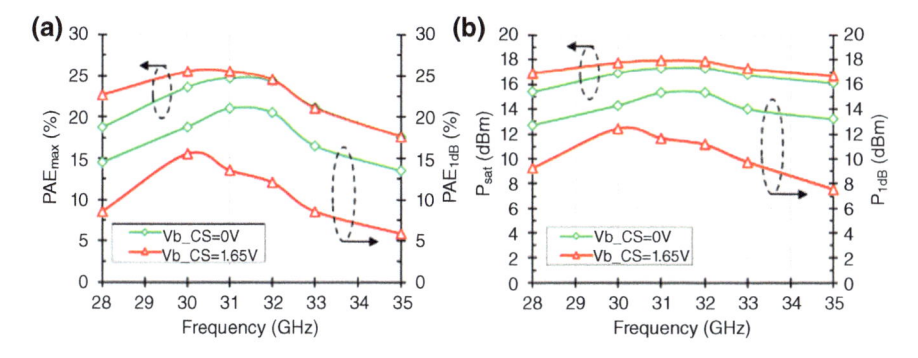

Fig. 8.39 Large-signal measurements: PAE_{max}/PAE_{-1dB} (**a**) and P_{sat}/P_{1dB} (**b**) in extreme modes from 28 to 35 GHz

maximum difference of 1.4 dBm between the two modes is obtained at 28 GHz and can confirm the hypothesis of the non-optimal matching at 28 GHz.

The 1 dB output compression point curves are exhibiting several noticeable behaviors. First, $P_{1\,dB}$ values are higher for $Vb_{CS} = 0$ V, for all frequency values for the same reason detailed in Sect. 8.3.3 at 31 GHz as $P_{1\,dB}$ is reduced with Vb_{CS} tuning. Maximum value of 15.3 dBm in high-linearity mode is obtained at 31 and 32 GHz. Minimum values are obtained at 28 and 35 GHz with 12.7 and 13.2 dBm, respectively. In high-gain mode, a maximum $P_{1\,dB}$ level of 12.4 dBm is reached at 30 GHz, while the lowest value is obtained at 35 GHz with 7.5 dBm. Then, we can remark that the $P_{1\,dB}$ difference between both extreme modes is minimized at 30 GHz with 1.9 dBm of difference, while at 35 GHz this value is maximum with 5.7 dBm.

On the $PAE_{-1\,dB}$ versus frequency curves, it is noticeable that the values for $Vb_{CS} = 0$ V are higher. This is naturally explained because of the higher level of $P_{1\,dB}$ in this configuration, leading to higher corresponding efficiency value. Maximum values of 21 and 20.5% are reached at 31 and 32 GHz, respectively, while minimum values of 14.5 and 13.5% are obtained at 28 and 35 GHz, respectively. For high-gain mode, a maximum of 15.5% is obtained at 30 GHz, while minimum value of 5.8% is obtained at 35 GHz. These two curves exhibit the same variations than $P_{1\,dB}$ curves depicted previously.

These frequency measurements for the same operating conditions confirm the PA operation and performances suitable for 5G applications on the 31.8–33.4 GHz 5G band. The large-signal measurements over frequency show that at 35 GHz second gain peak identified during S21 measurements for $Vb_{CS} = 1.65$ V, the achieved linearity and efficiency levels are low compared to the targeted values. No further measurements have been conducted at this frequency.

8.3.6 Power Amplifier Behavior for Temperature Variations

Small-signal and large-signal measurements have been performed for a temperature of operation from 25 to 125 °C, in order to evaluate the design robustness to temperature variations. The same operating conditions than in the previous sections have been applied. We check carefully that that the temperature is stabilized between two measurements.

8.3.6.1 Large-Signal Measurements from 25 to 125 °C with Body-Biasing Tuning

In these measurements, we report the values of the power gain, P_{sat}, PAE_{max}, and P_{DC} for Vb_{CS} varying from 0 to 1.65 V at temperatures comprised between 25 and 125 °C. Results are reported in Fig. 8.40.

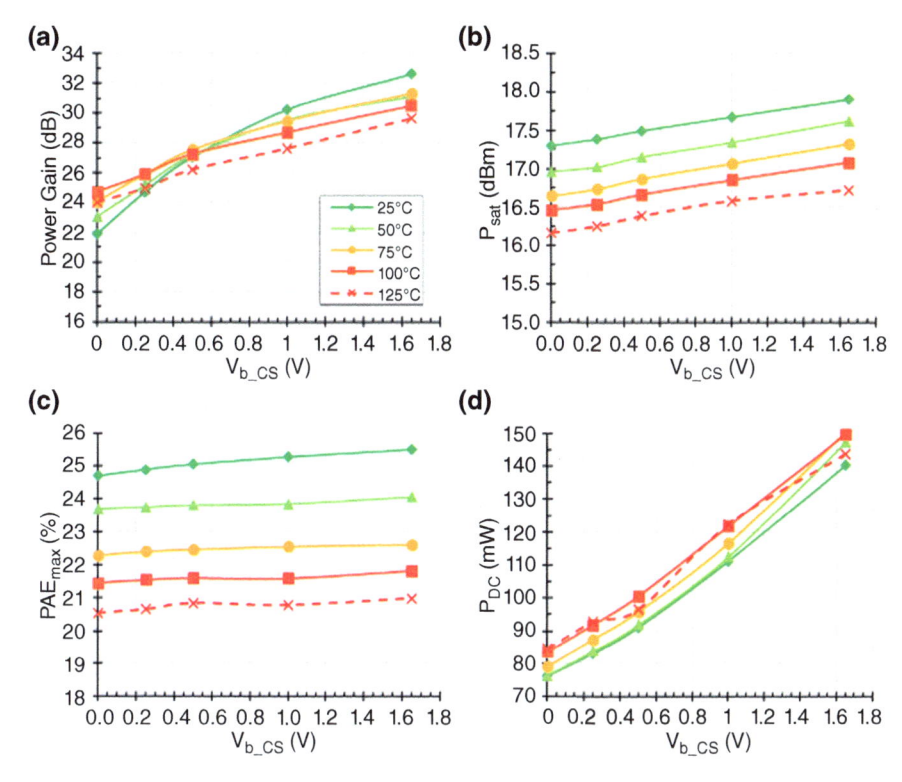

Fig. 8.40 Large-signal measurements at 31 GHz from 25 to 125 °C with Vb_{CS} tuning from 0 to 1.65 V: power gain (**a**), saturated output power (**b**), maximum power added efficiency (**c**) and DC power consumption (**d**)

First, we focus on P_{sat} and PAE_{max}. The effect of temperature over P_{sat} is an incremental reduction with a maximum total decrease of 1.2 dBm (−0.3 dB per 25 °C levels) for all modes. The temperature impact on P_{sat} level for all modes is limited and the saturated output power always remains over 16 dBm. The same behavior is noticeable over peak efficiency. PAE_{max} levels are decreasing when the temperature is improved. The maximum efficiency decrease is 3.3% and 4.6% in high-linearity and high-gain modes, respectively. Even at 125 °C, the PAE_{max} value is over 20%. We can hence conclude that temperature has a slight influence over output power and efficiency.

The power gain response is different than the two previous parameters. In fact, at $Vb_{CS} = 0$ V, the tendency is a gain improvement with temperature. From $Vb_{CS} = 0.4$ V, the trend is reversing and the gain is decreased with the temperature. The gain dynamic range between high-linearity and high-gain mode is then reduced with the temperature. However, fine grain gain control is available at all temperatures of operation and performed gain levels are still high.

Finally, P_{DC} levels depending on temperature and body-biasing have been measured. It is remarkable that the P_{DC} level is minimum at ambient temperature.

The highest the temperature and the Vb_{CS} are, the highest the P_{DC} levels are. In high-linearity mode, the result is a P_{DC} increase of 8.2 mW while a 9.4 mW maximum augmentation is achieved in high-gain mode.

It is noticeable that at 125 °C, several changes compared to the general behavior tendencies are observed. This can be explained by the measurements conditions. Indeed, the temperature induced chip dilatation made the RF and DC probe contact less efficient.

The measured performances modifications induced by the temperature improvement can be explained by its physical impact over the chip. Two distinct effects are occurring.

When the temperature is improved, the internal transistor parameters are modified. This affects the V_T that is decreasing with the temperature. This induced V_T reduction is leading to a higher drain current level for a fixed V_{DS}, in the same way that it can be done with body-biasing at ambient temperature. Therefore, it induces the higher power consumption observed at higher temperature. This V_T variation affects both stages and does not have the same impact over performances for all Vb_{CS} values. In fact, for $Vb_{CS} = 0$ V, the V_T variation induces higher gain as the operating class is drifting from class AB to class A. For $Vb_{CS} = 1.65$ V, the PA is already in class A. Therefore, the V_T shift results in a drift from class A to over class A operation, in a mode where the power consumption is higher while gain level is reduced.

Concerning P_{sat} variations, they are linked to the PA routing and that is why the same behavior in function of temperature is observed for all Vb_{CS} values. Indeed, the copper resistivity is improving with the temperature. Therefore, the drain access resistance (balun parasitic resistance) of the power stage is improved. This reduces the total V_{DS} value of this stage for fixed current conditions. Therefore, the saturated output power is reduced.

These large-signal measurements illustrate that a 100 °C operation temperature augmentation leads to output power and efficiency levels reduction, while power consumption is improved, for all modes. The gain dynamic between the two extreme modes is also reduced. Nevertheless, these variations are limited and are not critical. Furthermore, the level of performances reached at 125 °C is still high. Another noticeable point is the fact that the power amplifier is not degraded with temperature. It has been observed during the measurements process that after 125 °C measurements, if the temperature is then set to 25 °C, no drift over performances is observed compared to classical 25 °C operation for both extremes modes, demonstrating the design and technology robustness over temperature variations during measurements.

8.3.6.2 Small-Signal Measurements from 25 to 125 °C

Small-signal analysis for several temperature conditions has also been carried out in order to determine the induced impact over S-parameters.

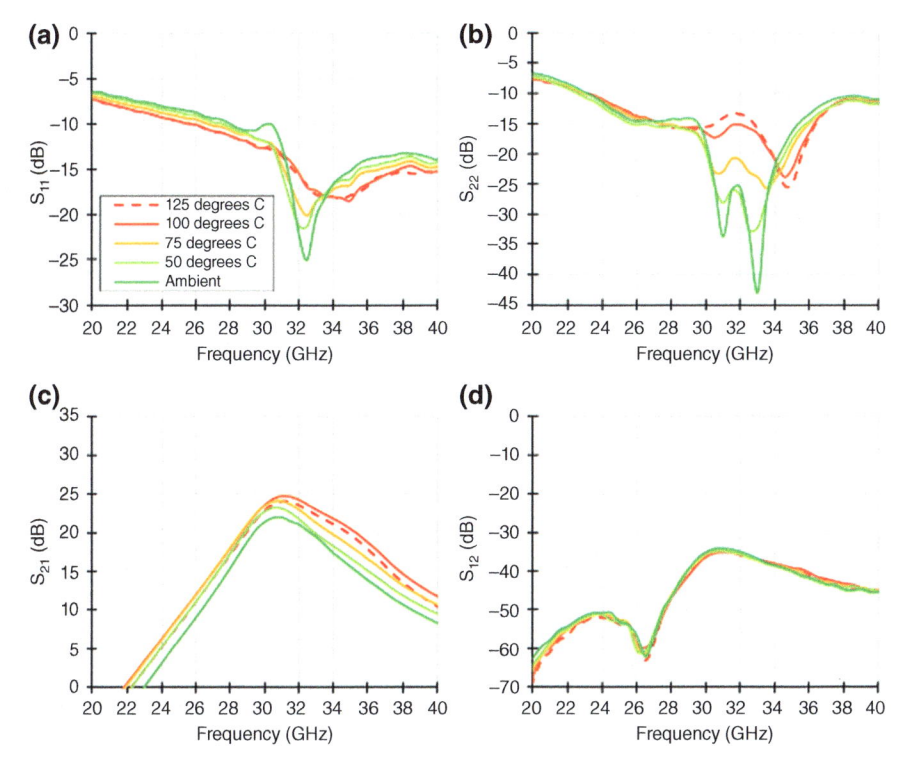

8 Millimeter-Wave Power Amplifiers for 5G Applications in 28 nm FD-SOI... 211

Fig. 8.41 Small-signal measurements from 25 to 125 °C in high-linearity mode: reflection coefficients S11 (**a**) and S22 (**b**), transmission coefficients S21 (**c**) and S12 (**d**)

S-parameters curves in high-linearity mode and high-gain mode are exhibited in Figs. 8.41 and 8.42, respectively. Each curve corresponding to a different temperature value from 25 to 125 °C. In all temperature conditions, input and output stay well matched over the frequency range of interest in all modes. Both S11 and S22 temperature behaviors can be compared to S11 and S22 variations caused by body-bias tuning. It confirms the temperature induced V_T variation behavior as explained in the previous section.

The S21 curves confirm this behavior and the values obtained in the previous section. As expected, higher power gain level and slightly improved $BW_{-3\,dB}$ are obtained with higher temperature levels for $Vb_{CS} = 0$ V, while the inverse tendency occurs for $Vb_{CS} = 1.65$ V. S12 curve illustrates that the isolation is not impacted by temperature variations and remains higher than 35 dB under all temperature conditions. This is conferred by the balanced topology that provides a robust high reverse isolation.

Large-signal and small-signal temperature performances analysis have been performed. They show that the temperature impact over intrinsic transistor parameters and chip routing causes only slight output power, efficiency, and gain tuning range degradation. Satisfying matching, isolation and stability conditions are achieved for

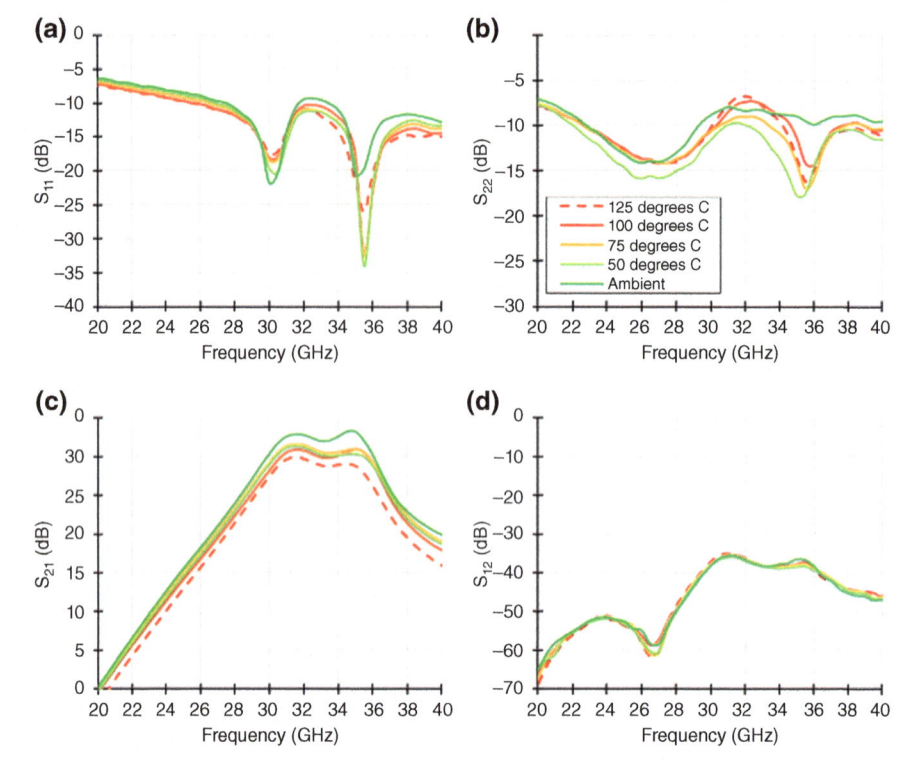

Fig. 8.42 Small-signal measurements from 25 to 125 °C in high-gain mode: reflection coefficients S11 (**a**) and S22 (**b**), transmission coefficients S21 (**c**) and S12 (**d**)

a 100 °C temperature span. The fine grain gain control is still achievable over all temperature conditions. We can hence conclude that the proposed power amplifier design presents robustness to industrial temperature range with good performance levels achieved at 125 °C.

8.3.7 On-Wafer Variability Statistical Study

During the power amplifier design, efforts have been made in order to reduce the impact of process induced on-wafer variability. In order to estimate these variations regarding individual circuit's performances, a series of large-signal measurements have been conducted. For this purpose, 13 on-wafer occurrences of the same circuit at different locations on one wafer have been measured under the same conditions for both high-gain and high-linearity modes. The same conditions as in the previous analysis have been used at ambient temperature. The values of peak efficiency, saturated output power, and power gain have been measured for each circuit. Therefore, average and standard deviation values have been calculated for each parameter in both operating modes. Results of this study are reported in

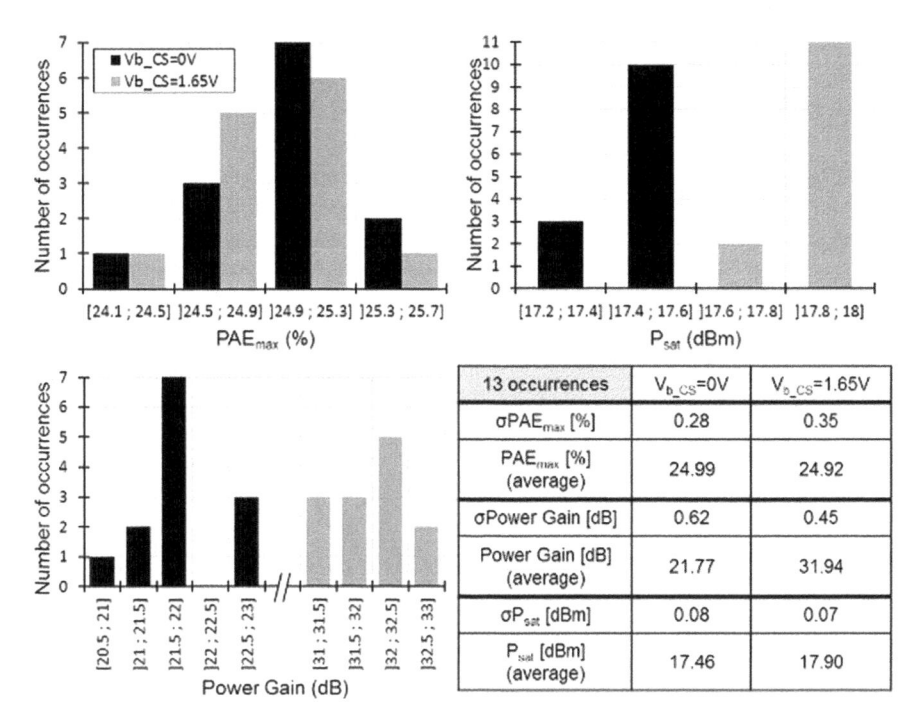

8 Millimeter-Wave Power Amplifiers for 5G Applications in 28 nm FD-SOI... 213

Fig. 8.43 Statistical PAE$_{max}$, power gain, and P$_{sat}$ measurements over 13 on-wafer occurrences at 31 GHz. Identical bias and supply conditions

Fig. 8.43, featuring PAE$_{max}$, P$_{sat}$, and power gain values range versus corresponding number of occurrences leading to these values. Corresponding average and standard deviation values are also calculated and reported.

First, let us focus on high-linearity mode. A PAE$_{max}$ average value of 24.99% with a very low corresponding standard deviation of 0.28% is reported. Concerning power gain, an average value of 21.77 dB and a standard deviation of 0.62 dB have been measured. In addition, a very slight standard deviation is achieved over P$_{sat}$ with 0.08 dBm for an average value of 17.46 dBm.

For the high-gain mode, a PAE$_{max}$ average value of 24.92% and a low standard deviation of 0.35% are measured. Regarding the power gain, average power gain level of 31.94 dB with a standard deviation of 0.45 dB is calculated. Finally, an average P$_{sat}$ value of 17.9 dBm with a very slight standard deviation of 0.07 dBm is reported.

The calculated standard deviation from measurements over 13 different on-wafer locations is very low for P$_{sat}$ and is under 0.1 dBm in both high-linearity and high-gain extreme modes. Similarly, a very low peak efficiency standard deviation under 0.5% for both modes is observed. The highest reported standard deviation is for power gain value and is under 1 dB in all Vb_{CS} conditions from 0 to 1.65 V.

These relatively low variations between measured performances over several chips are achieved thanks to the choices made early in the design process in order to reduce the process variability risks. Indeed, as generally the process variations are occurring over active devices arrays, the choice of dummies transistors integration around the useful transistors reduces this risk as they are more likely to be impacted because of their placement at the edge of the active devices array. The choice of a non-minimal gate length of 60 nm and the sub-division of large total width transistors in elementary transistor cells reduce the chance of a process induced gate length drift in each transistor. Furthermore, the wide RF path routing reduces the impact of a small metal layers width difference that can happen during manufacturing. It has been observed during the measurement process that the gain value variation can come to on-wafer mmW probe placement on RF pads that can slightly differ from one measurement to another, the power gain is the most sensitive parameter observed for this effect. Nevertheless, all standard deviations obtained are low, even for the power gain. These results reflect the robustness of the design and 28 nm FD-SOI technology to process spread and consolidate the choices made during the design to reduce local process variations, like the dummies transistors insertion aside each useful transistor stage.

8.4 Comparison and Discussion Regarding mmW PA State of the Art

This section now compares the measured performances of the power amplifier proposed in this work with the state of the art. We mentioned the future wireless networks requirements regarding power amplifiers. For this purpose, it has been discussed that high efficiency, reduced power consumption, and good linearity are needed while frequency and/or mode configurability is interesting for SoC implementation. Furthermore, good linear performances concerning power levels and gain are necessary. We first provide a state-of-the-art overview.

Wang and Xiao [18] presents a 10–25 GHz dual band power amplifier exhibiting a single-ended 2-stages cascode in 180 nm CMOS technology. Low performances are reached in terms of output power, efficiency, and gain levels. High-power consumption is caused by the higher supply voltage of the 2-stages cascode topology. It illustrates the difficulty to achieve satisfying performances over several bands and the need for advanced technology nodes for efficient and performant power amplifier implementation.

Two 29 GHz power amplifiers are presented in [7] and exhibit high output power and efficiency levels. However, the high supply voltage requested by the single-ended stacked multi-gate topology that permits to achieve these performances is leading to prohibitive levels of power consumption which is not desirable in an energy saving context for future wireless networks.

A solution for 30 GHz highly efficient and low power consumption power amplifier targeting 5G is presented in [16]. However, the use of common-source topology ensuring the low power consumption limits the achievable output power and gain levels.

The limit of common-source topology for high output power is also illustrated in [14]. In fact, two differential power amplifiers are proposed in this reference. One is exhibiting a common-source topology, while the other is featuring cascode configuration. The use of transistor stacking allows to double the supply voltage level and enhance output power, gain, and efficiency levels compared to the common-source. The single-stage topology limits the number of passive devices for matching and differential-to-single conversion and so the induced impact over efficiency. Both circuits are featuring a second harmonic control that is enabling efficient and linear operations. However, low gain values are achieved in both PAs, due to the single-stage topology.

Furthermore, none of these power amplifiers is featuring configurability while it has been demonstrated in [10] that high performances and power gain control are achievable at 60 GHz. In fact, the realized power combining allows high output power level, while the segmented biasing of the common-source transistors is leading to high-linearity operations. This circuit is implemented in 28 nm FD-SOI and uses the back-gate for both segmented biasing and gain configurability.

In the state of the art,[2] there was a lack of power amplifiers featuring efficiency and good performances regarding gain, output power, power consumption, and configurability for frequency bands around 30 GHz to leverage future wireless networks implementation. It is hence in this context that the presented work addresses these remaining challenges targeting SoC implementation. Furthermore, as we developed in previous sections, an implementation in phased array systems is subject to environment variations that can lead to performances degradation. None of the power amplifiers available in the state of the art presents a solution to overcome this issue.

The reference [4] addresses the multi-band challenge. It is featuring a transformer-based Doherty topology implemented in 130 nm SiGe technology. This multi-band operation is leveraged by the use of wideband passive devices and varactor loaded transmission lines at the input of both main and auxiliary paths to dynamically select the desired band of operation and covers several potential 5G bands. However, while good output power levels are achieved, the limited efficiency and gain levels illustrate that high performances are still difficult to obtain over all bands.

Another wideband power amplifier has been proposed in [1]. In this reference, the continuous class F operation enabled by a tuned network and the limitation to a single-stage topology allows high efficiency performances over a wide frequency range of operation. However, this topology is leading to low output power and gain levels, illustrating that efficiency and high performances are hard to obtain simultaneously with wideband operations.

As stated previously, configurability in mode is an interesting alternative solution for SoC implementation. The reference [17] presents an evolution of the power amplifier referenced in [16] and shares the same 2-stage differential common-source topology, implemented in 40 nm CMOS technology. Digital variable gain control with 9 dB dynamic range has been implemented and makes this reference the only

---

[2]January 2018.

gain power amplifier with reconfigurable gain, with our work, of the 30 GHz state of the art. Excepted this ability, the power amplifier presented in [17] presents the same advantages and drawbacks than [16]: high efficiency, low power consumption but limited output power.

The performance trade-off between output power, gain, efficiency, and power consumption is also illustrated by [5]. The Doherty configuration allows high power levels thanks to power combining. Good power gain and linearity levels are measured while a limited peak efficiency is achieved. Furthermore, this topology is leading to high-power consumption, even if the supply voltage is limited to 1 V.

While these references try to address the state-of-the-art challenges concerning efficiency and performances enhancement and/or configurability, none of these addresses the issue of environment variation for phased array implementation. This is the topic of the work presented in [13]. This reference exhibits a balanced topology in 28 nm FD-SOI CMOS technology, providing robustness to input and output impedance variations. The robustness is explored in order to move the biasing limits forward the actual security margins in case of VSWR to enhance global power amplifier performances. While good levels of output power and power consumption are achieved, a limited gain is measured. Furthermore, the efficiency levels are limited by the losses encountered in the passive devices on the RF path.

The reference [3] presents a power amplifier implemented in 14 nm FinFET technology node. This amplifier has been designed for higher 5G frequency bands around 70 GHz. Two modes corresponding to different operating conditions are presented. The "normal mode" provides low power gain at 66 GHz, while a "high-gain mode" provides a higher but still limited power gain value. Furthermore, despite the use of design techniques such as capacitive neutralization, low efficiency, linearity and output power levels are achieved in both modes. These performances do not meet the 5G requirements and illustrate that further research is needed in this technology node to implement a power amplifier targeting SoC implementation for future wireless networks.

Now that we have defined the state of the art, it is necessary to compare the measured performances. For this purpose, for each circuit of the state of the art, several parameters including power gain, P_{sat}, $P_{1\,dB}$, PAE_{max}, PAE@6 dB P_{sat} Back-off, $BW_{-3\,dB}$, and P_{DC} are reported or estimated from the available data in the literature when stated. The number of stages, the technology, and their area footprint are also reported. Furthermore, to compare circuits one to another, figures of merit FOM are commonly used in the literature. For power amplifiers comparisons, the ITRS FOM, for International Technology Roadmap for Semiconductors, is generally used. This figure of merit can be calculated using Eq. (8.15) where power gain is expressed in linear form and P_{sat} is expressed in Watt. All these values for selected circuits are provided in Fig. 8.44 [15][3].

---

[3]Reference [15] has been added after the writing and is not discussed in the following sections.

8 Millimeter-Wave Power Amplifiers for 5G Applications in 28 nm FD-SOI...

	Our Work		[15]$ ISSCC'18	[4] ISSCC '17	[17] ISSCC '17	[5] RFIC '17	[1] RFIC '17	[7] JSSC '16	[16] ISSCC '16	[14] IMS '16	[18] EL '15	[13] LASCAS '17	[10] ISSCC '15	[3] RFIC '17
Technology	28nm UTBB FD-SOI CMOS		130nm SiGe	130nm SiGe	40nm CMOS	28nm CMOS Bulk	65nm CMOS	45nm SOI CMOS	28nm CMOS Bulk	28nm CMOS	180nm CMOS	28nm FD-SOI CMOS	28nm FD-SOI CMOS	14nm FinFET CMOS
Number of Stages	2		1	2	3	2	1	1	2	1	2	2	3	3
Operating Mode	High Lin.	High Gain	-	28GHz Band	Max. Gain	-	-	-	-	-	-	-	High Lin.	High Gain	"High Gain"
Supply Voltage [V]	0.7"/1.98"	0.7"/1.9 8"	4	1.5	1.1	1	1.1	5.2	1.15	1.1	3	2	1	1	1
Frequency [GHz]	31	31	28	28	27	32	29	29	30	28	25	28	60	61	71
Power Gain [dB]	21.9	32.6	14*	18.2	22.4	22	10	13	16.3	10	6.8	17.5	15.4	35	16.7
P_{sat} [dBm]	17.3	17.9	23	16.8	15.1	19.8	14.8	24.8	15.3	14.8	10	18.4	18.8	18.9	7.4
P_{1dB} [dBm]	15.3	11.6	20*	15.2	13.7	16	13.2	21	14.3	14	9	15.4	18.2	15	2
PAE_{max} [%]	24.7	25.5	41.4	20.3	33.7	21	46.4	26	36.6	36.5	4.8	12.4	21	17.7	8.9
PAE_{1dB} [%]	21	10	40*	19.5	27.5*	12*	22*	18*	35.8	35.2	4*	9.5*	21	9	-
$PAE_{6dB\,Psat\,Backoff}$ [%]	11.5	10.4	34.7	14*	15*	8*	-	11*	-	24*	2*	6*	8*	6*	4.5
BW_{-3dB} [GHz]	4.5 (14%)	6 (18%)	-	16.4 (52%)	7 (24%)	6 (18%)	8 (26%)	11 (38%)	4 (13%)	-	4* (16%)	4* (14%)	10 (17%)	7 (12%)	7.4 (10%)
P_{DC} [mW]	76.1	140.2	50*	71*	30.3	250*	57*	448*	20.1	-	150	154	74	331	-
Active Area [mm²]	0.508		0.56**	1.76**	0.23	0.59	0.12	0.24	0.16	0.28	0.864**	0.66	1.988	0.162	0.26
ITRS FOM [W.GHz²]	1,932	26,925	1,626	503	1,381	3,255	117	1,436	476	746	1.5	405	161,671		115

$Stage 1 (S1) supply, "Stage 2 (S2) supply.   *Estimated from reported figures. **With Pads. $Reference added after the writing. Not discussed in the following paragraphs

Fig. 8.44 mmW PA state-of-the-art comparison table

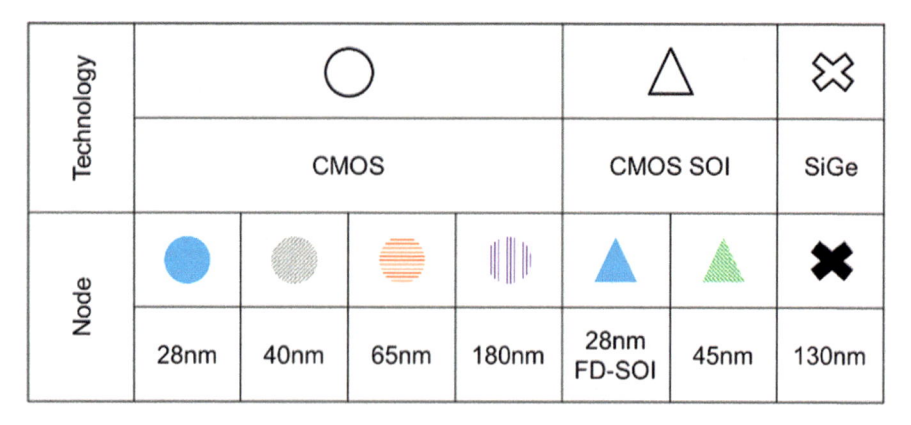

Technology	○				△		✳
	CMOS				CMOS SOI		SiGe
Node	●	●	●	●	▲	▲	✖
	28nm	40nm	65nm	180nm	28nm FD-SOI	45nm	130nm

Fig. 8.45 Technology classification used in Figs. 8.46, 8.47, 8.48, and 8.49

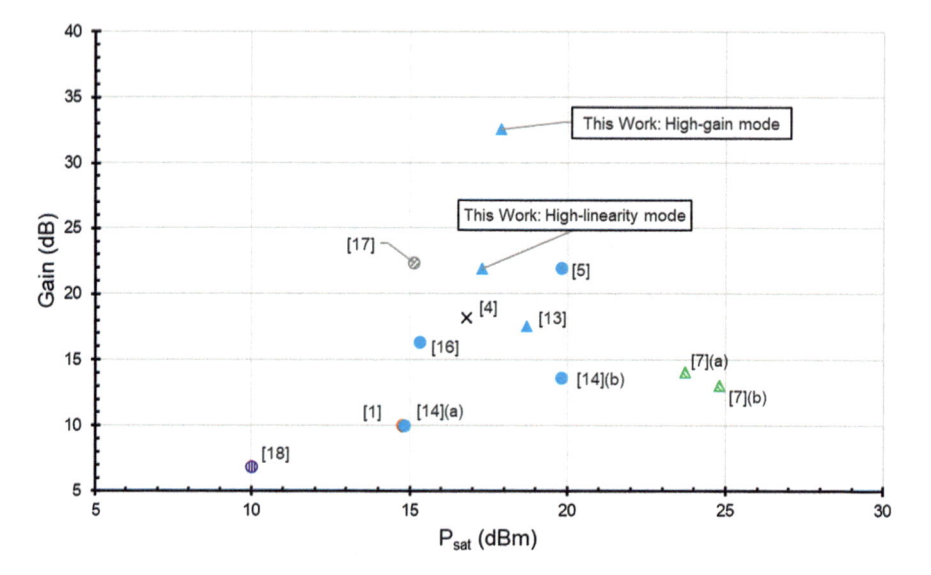

Fig. 8.46 State-of-the-art comparison: power gain level versus P_{sat}

$$\text{ITRS FOM} = \text{Gain[lin]} \times P_{sat}[W] \times \text{freq[GHz]}^2 \times \text{PAE}_{max}[\%] \quad (8.15)$$

In order to provide a more focused comparison with the state of the art, we propose in the Figs. 8.46, 8.47, 8.48 and 8.49 the reported power gain, PAE_{max}, P_{DC}, and ITRS FOM, respectively, versus P_{sat} of each circuit referenced around 30 GHz in Fig. 8.44. The references are sorted by technology type and nodes following the classification detailed in Fig. 8.45.

Several remarks can be done about performance levels, targeted application, and the state of the art. This is discussed in the following paragraphs.

8 Millimeter-Wave Power Amplifiers for 5G Applications in 28 nm FD-SOI... 219

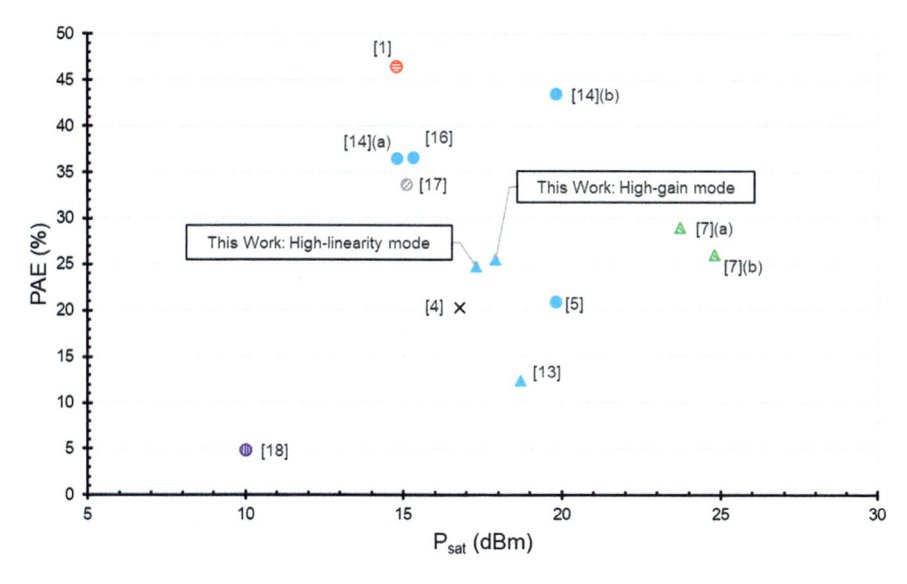

Fig. 8.47 State-of-the-art comparison: PAE versus P_{sat}

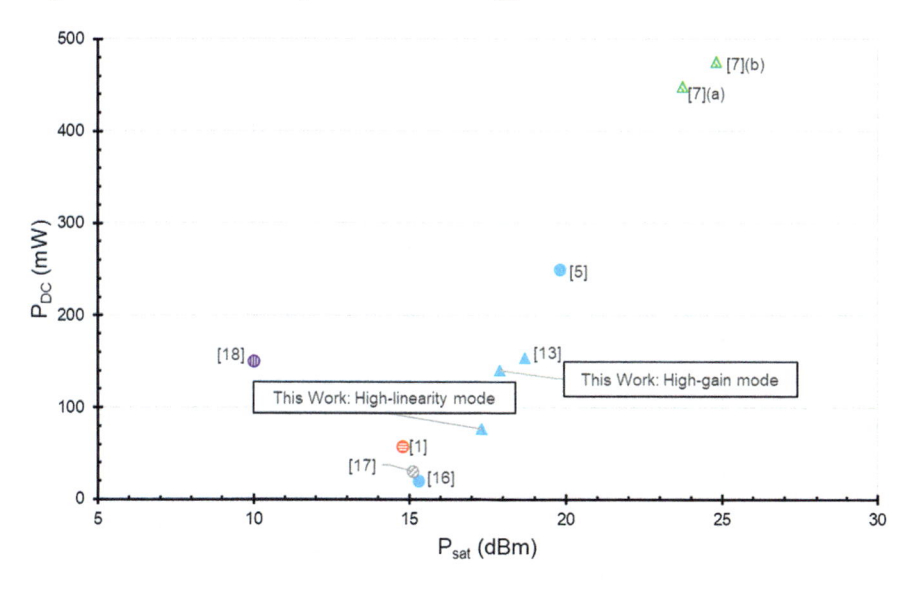

Fig. 8.48 State-of-the-art comparison: P_{DC} versus P_{sat}

We exposed previously the necessity of highly efficient, low consumption, linear power amplifiers to fulfill 5G applications requirements. For beamforming phased array, as multiple amplification paths will be implemented at the same base-station for FD-MIMO, the output power will be distributed. Therefore, high P_{sat} performance for each amplification path is not mandatory, while high gain is expected.

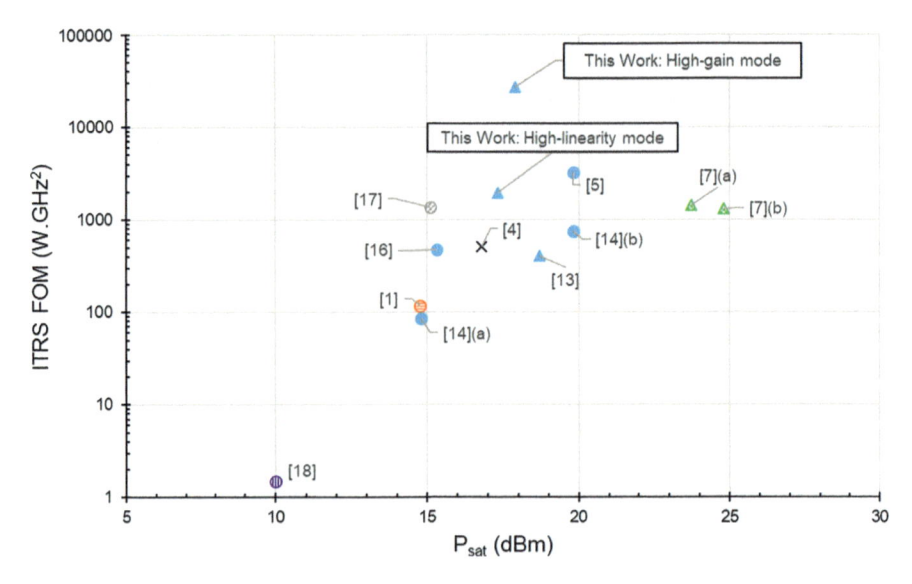

Fig. 8.49 State-of-the-art comparison: ITRS FOM versus P_{sat}

The performances achieved by our designed power amplifier meet these needs. Indeed, we present a sufficient P_{sat} around 17.5 dBm due to the limited achieved power consumption in all operating modes with a maximum of 140 mW. High gain is achieved in high-linearity mode, while the high-gain mode shows the highest reported gain value of the state of the art around 30 GHz, 10 dB higher than [17] that is the second highest reported gain value in this frequency range. Concerning linearity, good performances regarding $P_{1\,dB}$ are achieved in high-linearity mode, with 2 dBm between 1 dB compression point and saturation.

Compared to previous design [10] implemented in 28 nm FD-SOI, the main performance improvement is higher efficiency levels in all modes and the fact that no efficiency degradation is occurring. Furthermore, the balanced topology confers additional robustness to external conditions variations. Compared to [13] featuring balanced topology in the same technology, the efficiency is enhanced while additional body-biasing based gain configurability is demonstrated. Higher gain is also achieved in all modes.

In the state of the art, several trends concerning performances can be identified. Most PAs achieving high output power levels present a limited gain [7](a), [7](b), [14](b). They also suffer from high-power consumption in order to achieve this high output power even if efficiency levels are good. The circuits with a high PAE generally suffer from low P_{sat} levels, linked with the low power consumption that they achieve [1], [14](a), [17]. It is also noticeable that the performances reported for [17] are obtained in the highest gain mode. Therefore, this reference achieves the same level of gain in its highest configuration as our design in our lowest gain configuration. This is illustrating that, a trade-off generally exists between achieved performances for the different applications targeted.

In the proposed power amplifier, this trade-off is relaxed as good levels of PAE and output power are reported, simultaneously with a limited power consumption, while high-gain levels are measured. The very high gain measured in high-gain mode allows to achieve the highest ITRS FOM reported around 30 GHz.

Concerning technologies, it is noticeable that CMOS SOI implemented power amplifiers permit to achieve the highest levels of gain and saturated output power, while the circuits in CMOS technology present the highest peak PAE values. The only circuit implemented in SiGe shows "middle-class" performances while no power consumption is reported. Finally, it is clearly highlighted with [18] that 180 nm CMOS technology node cannot offer the level of performances necessary for this kind of applications.

References

1. S.N. Ali, P. Agarwal, S. Mirabbasi, D. Heo, A 42–46.4% PAE continuous class-F power amplifier with Cgd neutralization at 26–34 GHz in 65 nm CMOS for 5G applications, in *2017 IEEE Radio Frequency Integrated Circuits Symposium (RFIC)*, Honolulu (2017), pp. 212–215
2. G. Bree, Classic designs for lumped element and transmission line 90-degree couplers, in *September 2007 High Frequency Electronics* [Online]. Available: https://www.highfrequencyelectronics.com/Sep07/HFE0907_Tutorial.pdf
3. S. Callender, S. Pellerano, C. Hull, A 73 GHz PA for 5G phased arrays in 14 nm FinFET CMOS, in *2017 Radio Frequency Integrated Circuits Symposium (RFIC)*, Honolulu (2017), pp. 402–405
4. S. Hu, F. Wang, H. Wang, A 28 GHz/37 GHz/39 GHz multiband linear Doherty power amplifier for 5G massive MIMO applications, in *2017 IEEE International Solid-State Circuits Conference (ISSCC)*, San Francisco (2017), pp. 32–34
5. P. Indirayanti, P. Reynaert, A 32 GHz 20 dBm-PSAT transformer-based Doherty power amplifier for multi-Gb/s 5G applications in 28 nm bulk CMOS, in *2017 IEEE Radio Frequency Integrated Circuits Symposium (RFIC)*, Honolulu (2017), pp. 45–48
6. International Communication Union, WRC-2015 Final Acts [Online]. Available: https://www.itu.int/itu_mt_main/catalog/productDetail.jsf?area=R-ACT-WRC.12-2015&sort=TD&wec-appid=EBOOKSHOP_B2B&page=C5E446D490AE478D934F1B09C8B1941C&itemKey=001E4F34E9B11ED5BDC89AF911367041&show=12&view=row&wec-locale=en_US
7. J.A. Jayamon, J.F. Buckwalter, P.M. Asbeck, Multigate-cell stacked FET design for millimeter-wave CMOS power amplifiers. IEEE J. Solid State Circuits **51**(9), 2027–2039 (2016)
8. V. Knopik, B. Moret, E. Kervé, Integrated scalable and tunable RF CMOS SOI quadrature hybrid coupler, in *2017 12th European Microwave Integrated Circuits Conference (EuMIC)*, Nuremberg (2017), pp. 159–162
9. A. Larie, Conception d'amplificateurs de puissance hautement linéaires à 60 GHz en technologies CMOS nanométriques, Ph.D. Dissertation, University of Bordeaux (2014)
10. A. Larie, E. Kervé, B. Martineau, L. Vogt, D. Belot, A 60 GHz 28 nm UTBB FD-SOI CMOS reconfigurable power amplifier with 21% PAE, 18.2 dBm P1dB and 74 mW PDC, in *2015 IEEE International Solid-State Circuits Conference (ISSCC) Digest of Technical Papers*, San Francisco (2015), pp. 1–3
11. T. Lehmann, F. Hettstedt, R. Knochel, Reconfigurable PA networks using switchable directional couplers as RF switch, in *2007 European Microwave Conference*, Munich (2007), pp. 1054–1057

12. B. Moret, Amplificateur de puissance autonome pour applications OFDM et beamforming de la 5G aux fréquences millimétriques en technologie CMOS avancée, Ph.D. Dissertation, University of Bordeaux (2017)
13. B. Moret, V. Knopik, E. Kerhervé, A 28 GHz self-contained power amplifier for 5G applications in 28 nm FD-SOI CMOS, in *2017 IEEE 8th Latin American Symposium on Circuits & Systems (LASCAS)*, Bariloche (2017), pp. 1–4
14. B. Park, D. Jeong, J. Kim, Y. Cho, K. Moon, B. Kim, Highly linear CMOS power amplifier for mm-wave applications, in *2016 IEEE MTT-S International Microwave Symposium (IMS)*, San Francisco (2016), pp. 1–3
15. B. Rabet, J. Buckwalter, A high-efficiency 28 GHz outphasing PA with 23 dBm output power using a triaxial balun combiner, in *2018 IEEE International Solid – State Circuits Conference – (ISSCC)*, San Francisco (2018), pp. 174–176
16. S. Shakib, H.C. Park, J. Dunworth, V. Aparin, K. Entesari, A 28 GHz efficient linear power amplifier for 5G phased arrays in 28 nm bulk CMOS, in *2016 IEEE International Solid-State Circuits Conference (ISSCC)*, San Francisco (2016), pp. 352–353
17. S. Shakib, M. Elkholy, J. Dunworth, V. Aparin, K. Entesari, A wideband 28 GHz power amplifier supporting 8×100 MHz carrier aggregation for 5G in 40 nm CMOS, in *2017 IEEE International Solid-State Circuits Conference (ISSCC)*, San Francisco (2017), pp. 32–34
18. S. Wang, C.Y. Xiao, Concurrent 10.5/25 GHz CMOS power amplifier with harmonics and inter-modulation products suppression. IEEE Electron. Lett. **51**(14), 1058–1059 (2015)

Chapter 9
An 802.15.4 IR-UWB Transmitter SoC with Adaptive-FBB-Based Channel Selection and Programmable Pulse Shape

David Bol and Guerric de Streel

9.1 Introduction

Smart sensors for the Internet-of-Things (IoT) embed four main functions: sensing of the physical world, data processing, wireless communications, and power management as represented in Fig. 9.1. To avoid the economical cost and environmental footprint of battery replacement [1], IoT smart sensors should operate on energy-harvesting sources such as micro PV cells or thermoelectric generators. This put a serious constraint on the average system power consumption: the sensing, processing, and communication functions should fit in a 1–100 μW power budget, depending on the harvester type and size, as well as on the conditions of the environment. Among these functions, sensing can be performed in a pretty low-power way with energies down to 8 pJ/sample for audio sensing [2], 17 pJ/pixel sample for image sensing [3], and 2 pJ/sample for ECG biomedical sensing [4]. Wireless communications are much less energy efficient for several reasons including protocol overhead and physical limitations of the air as a communication channel [5]. In IoT smart sensors, communications are typically asymmetric with more data to be transmitted than data to be received. The transmit operation is thus critical with respect to energy efficiency. State-of-the-art short-range WPAN radio transceivers consume down to 3.7nJ per raw bit transmitted over the air for Bluetooth Low-Energy (BLE) [6], which leads to 10nJ/bit at the application level when considering the protocol overhead [5]. With these solutions, it is possible to meet the energy-harvesting power budget for simple applications with low data

---

D. Bol (✉)
ICTEAM Institute, Université catholique de Louvain, Louvain-la-Neuve, Belgium
e-mail: david.bol@uclouvain.be

G. de Streel
Imec, Heverlee, Belgium
e-mail: guerric.destreel@imec.be

© Springer Nature Switzerland AG 2020
S. Clerc et al. (eds.), *The Fourth Terminal*, Integrated Circuits and Systems,
https://doi.org/10.1007/978-3-030-39496-7_9

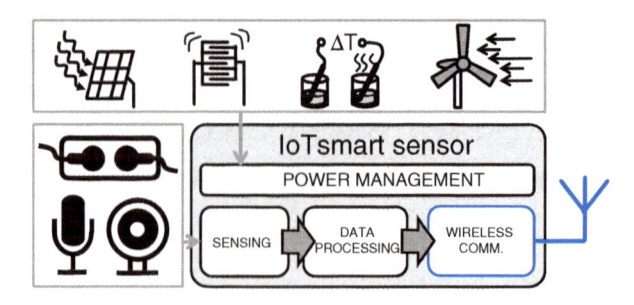

Fig. 9.1 General architecture of an IoT smart sensor

rates below 10 kbps such as motion-based room occupancy monitoring [7] but the transmitter power remains prohibitive in applications with medium continuous data rates in the 0.1–10 Mbps range. As an alternative, local data processing directly in the IoT smart sensor can be used to limit the wireless communications data rates [8] by extracting the relevant information from the sensed data. Ultra-low-power (ULP) design is thus required for local data processing and this is enabled by FD-SOI technology for microcontrollers [9], memories [10], and application-specific processors [11]. However, local data processing comes at the expense of flexibility, computational power, and interaction between smart sensors in short-range networks. Some applications thus cannot fully rely on local data processing and still require medium data rates in the range of 0.1–10 Mbps.

In this chapter, we study how the unique wide-range back-bias control in FD-SOI can be used to build ULP radios by focusing on their transmitter (TX) side. We apply this idea to the design of a TX SoC for impulse-radio ultra-wideband (IR-UWB) communication codenamed SleepTalker. IR-UWB communication is an efficient physical-layer (PHY) solution for high-data rate, short-range, and low-power communications due to the duty-cycled nature of the output signal as well as the potential for low-complexity and low-power TX architectures [12]. These characteristics can lead to strong power savings for asymmetric communications in smart sensor applications and have thus been the driving force behind the development of the IEEE 802.15.4 UWB PHY standard [13]. This standard has two distinct bands: 3–5 GHz and 6–10 GHz, and combines binary burst pulse position modulation (BPPM) with binary phase-shift keying (BPSK) modulation. It provides data rates from 0.11 to 27.24 Mbps. The highest data rate encodes each symbol in just one 2 ns pulse, which results in the lowest energy consumed per bit bit but also the shortest communication range. For the lowest data rates, multiple 2 ns pulses are concatenated in bursts. Each burst encodes one symbol to increase range at the cost of the energy consumed per bit. The burst position within the BPPM time slot is controlled by a pseudo-random sequence. This burst pulse position scrambling improves the spectral smoothness of the output power by limiting spectral lines to meet spectral regulations.

SleepTalker TX SoC can operate in the three channels of the 3–5 GHz 802.15.4 UWB low band with carrier frequencies of 3.5/4.0/4.5 GHz, respectively.

SleepTalker is capable of both BPPM and BPSK modulation and covers seven data rates from 0.11 to 27.24 Mbps, compliant with the IEEE 802.15.4 UWB standard.

In IR-UWB communications, the challenge is to optimize the power consumption of the three common TX functions (frequency synthesis, modulation, and power amplification) while complying with the FCC regulation mask and enabling the wide data rate range of 802.15.4 UWB standard. In SleepTalker, we use a digital TX architecture to exploit the back biasing capability of FD-SOI for implementing these functions at ULP as follows.

- *Feature 1:* a body-biased-controlled ring oscillator (BBCO) is used as a low-jitter local oscillator (LO) with instantaneous startup. This enables the use of a PLL-free digital TX architecture with aggressive LO duty cycling within the BPPM time slot to reduce jitter accumulation and power consumption.
- *Feature 2:* tuning of the BBCO to oscillate at the carrier frequency (CF) of the selected 802.15.4 UWB channel is automatically performed on-chip with a programmable forward back-bias (FBB) generator.
- *Feature 3:* programmable pulse shaping is implemented by a digital power amplifier (PA) for meeting spectral compliance with the FCC regulation with back-bias control of the TX output power.
- *Feature 4:* continuous-time NMOS/PMOS current matching is performed by an adaptive FBB loop with a back-gate-driven amplifier for voltage range compatibility.
- *Feature 5:* ultra-low-voltage (ULV) operation down to 0.55 V of the high-frequency TX up to 4.5 GHz is enabled by 28 nm FD-SOI process with wide-range FBB up to 1.8 V for NMOS (BBN) and −1.8 V for PMOS (BBP), as explained in Chap. 3.

This chapter is organized as follows. The general digital TX architecture operated at 0.55 V is presented in Sect. 9.2. The TX design is explained in Sects. 9.3–9.5, which focus on the innovations enabled by the back biasing technique: duty-cycled frequency synthesis based on the BBCO (*Features 1* and *2*), pulse-shaping digital PA with back-bias controlled output power (*Feature 3*), and FBB generators with current-matching loop (*Feature 4*). The capability of FD-SOI to operate high-speed logic at ULV (*Feature 5*) is not discussed in this chapter, see Chaps. 2 and 3 for more details on this. The architecture of the full SoC including the back-bias drivers and the NMOS/PMOS current-matching loop *Feature 4* is presented in Sect. 9.6 with measurement results in Sect. 9.7. Let us mention that this chapter is largely inspired from [14, 15] where SleepTalker was first presented.

9.2 Architecture of the Digital TX

Figure 9.2 shows 802.15.4 UWB PHY symbols with BPPM and BPSK modulation, which highlights the highly duty-cycled nature of the 802.15.4 UWB output signal. Indeed, the duty cycle is only 0.78 and 3.12% with data rates using a pulse repetition

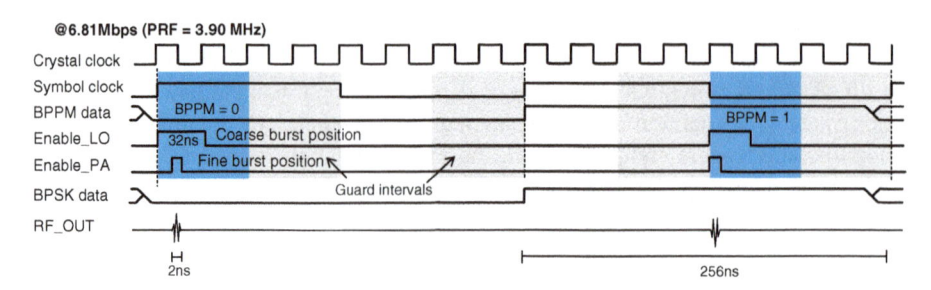

Fig. 9.2 Principle of the aggressive hierarchical duty cycling within the BPPM time slot. The LO is active for one cycle of the 31.25 MHz crystal clock (coarse burst position) and the PA is active only during the pulse burst emission (fine burst position)

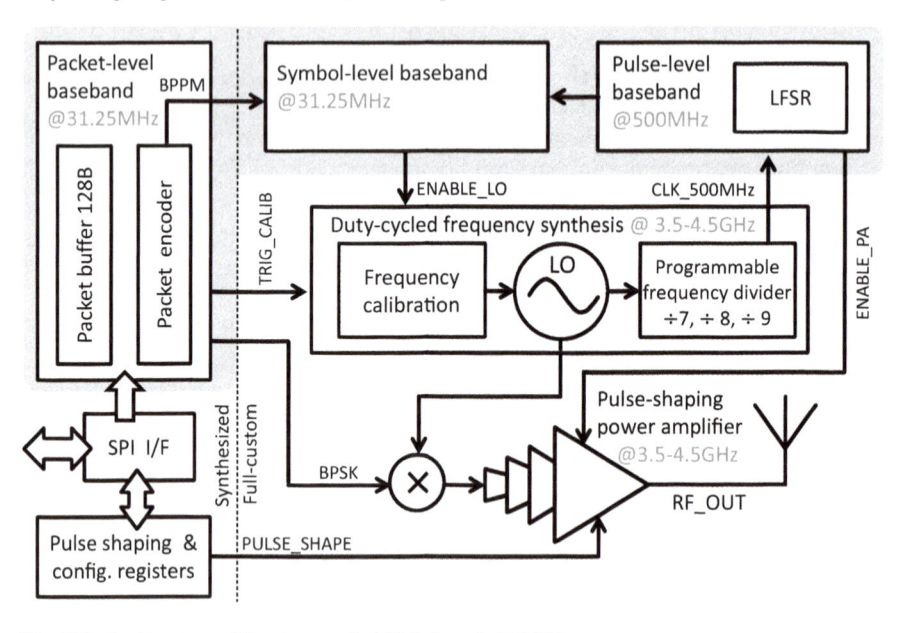

Fig. 9.3 Architecture of the duty-cycled PLL-free digital TX

frequency (PRF) of 3.9 and 15.6 MHz, respectively. The architecture of the proposed digital TX shown in Fig. 9.3 implements this 802.15.4 UWB PHY with aggressive hierarchical duty cycling to leverage this characteristic. The LO generates the 3.5–4.5 GHz carrier frequency depending on the selected channel with both positive and negative polarities. The pulse-shaping power amplifier (PA) performs the BPSK modulation by selecting the pulse burst polarity. BPPM modulation is performed by enabling the rest of the TX only during the used BPPM time slot and thus disabling it in the guard intervals and the unused time slot. Within this BPPM time slot, the coarse burst position is computed with a 32 ns time resolution corresponding to the crystal-clock cycles, with the output of the 802.15.4 linear feedback shift register (LFSR) implementing the pulse burst position scrambling. This enables the LO

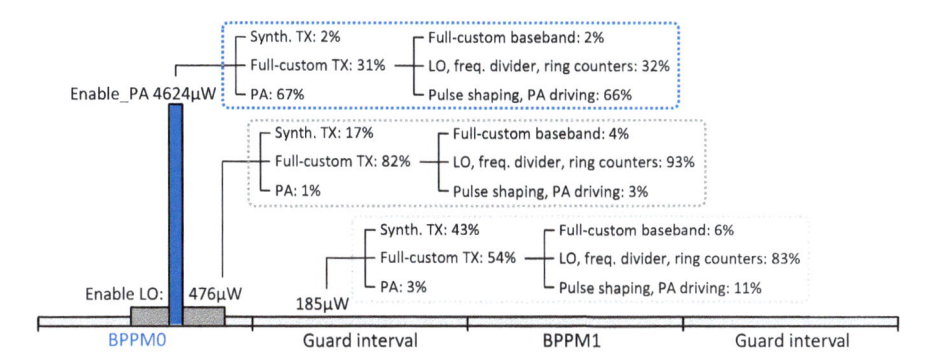

9 An 802.15.4 IR-UWB Transmitter SoC with Adaptive-FBB-Based Channel... 227

Fig. 9.4 Simulated power breakdown between the different parts of the TX depending on the duty cycling mode during the symbol duration

(Enable_LO signal) only during this coarse burst position for durations between 16 and 64 ns depending on the selected data rate. The fine pulse burst position has a 2 ns time resolution and is performed by the pulse-level baseband processor, which is clocked at 500 MHz by the LO clock divided. This enables the PA (Enable_PA signal) for durations between 2 and 64 ns depending on the selected data rates only when the pulses need to be emitted. Such an aggressive duty cycling of the PA and the LO directly results in strong power savings as the subsequent TX logic is clocked either directly by the LO output or by its 500 MHz divided clock.

The LO frequency is calibrated periodically as triggered by the TRIG_CALIB signal as will be detailed in Sect. 9.3. The PA performs pulse shaping (see Sect. 9.4) with the pulse shape programmable through configuration registers. A 128-byte packet buffer is embedded with forward error-correction encoding (Reed–Solomon) in the packet encoder. An SPI interface is used for sending data and programming configuration registers including the pulse shape, the channel, and the data rate. The 31.25 MHz packet-level baseband processor is synthesized from standard cells along with the SPI interface and configuration registers. The rest of the TX results from a full-custom transistor-level design.

Figure 9.4 illustrates the interest of duty cycling for reducing the power consumption of the TX during a symbol. The baseline power consumption when the LO and the PA are disabled is 185 µW. This rises to 476 µW when enabling the LO and thus the pulse-level baseband processor. When the PA is enabled, the power is 4.6 mW, i.e., 25× the baseline power.

9.3 Duty-Cycled Frequency Synthesis

As the 802.15.4 UWB bursts are BPSK modulated, there is a strong requirement on the frequency accuracy and stability for correct coherent demodulation. Following the analysis from [16], we can define an upper bound for the total phase noise at the end of the burst in order to ensure correct demodulation of the BPSK-encoded data. By taking some margin into account, we set the maximum phase noise allowable to $\pi/4$ which corresponds to a maximum accumulated jitter of 27.7 ps after 32 ns at 4.5 GHz.

The aggressive duty cycling of the LO prevents the use of a PLL because of its prohibitive settling time and forces the choice of an architecture based on a free-running LO. LC-based oscillators can offer low jitter at the cost of high power, large area, and large settling time. To avoid these drawbacks, previous low-power TX relied on ring oscillators (ROs) duty-cycled to a quarter of the symbol period. However, in the case of low data rate, the quarter symbol period is quite long and leads to high jitter accumulation due to the degraded phase noise of ROs compared to LC oscillators. Figure 9.5 shows that a current-starved differential RO at nominal 1 V supply voltage suffers from a 10 ns integrated jitter for a run time corresponding to the longest symbol duration (0.11 Mbps rate). As shown in Fig. 9.5a, the aggressive within-BPPM LO duty cycling from Sect. 9.2 significantly limits the jitter accumulation. Nevertheless, it is still prohibitive for BPSK data encoding. Therefore, to get rid of the current-starving delay control of the delay elements, we further improve jitter by first using a back-bias control of their delay. To do so, independent and symmetric FBB voltages are applied to NMOS (BBN_LO voltage) and PMOS (BBP_LO voltage) transistors. The LO is thus a BBCO. Second, we use a differential multipath RO based on feedforward loops, similar to the one proposed in [17]. The 7-stage architecture illustrated in Fig. 9.5 improves both the switching speed as well as the noise performance thanks to its fast transitions. The proposed BBCO is disabled by slightly modifying two of the seven delay elements in the chain to implement NOR and NAND functions in order to gate the oscillation in both the main and feedforward paths. They are sized to feature the same delay as

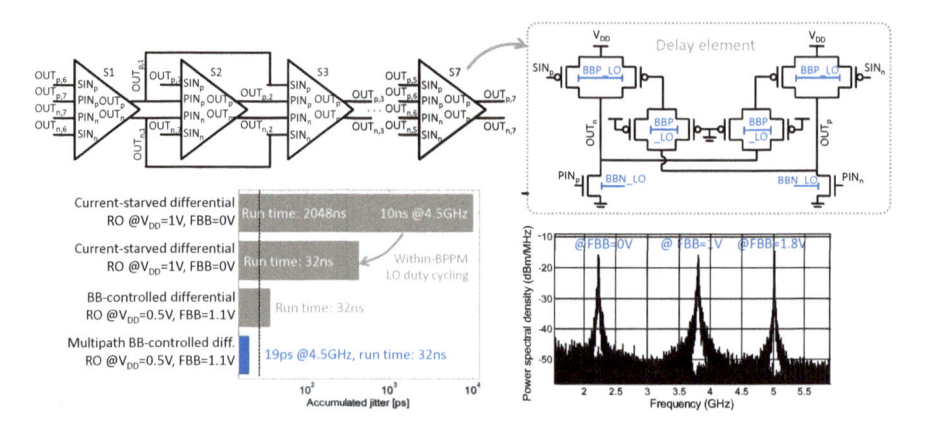

Fig. 9.5 Local oscillator: architecture of the proposed multipath back-bias-controlled oscillator (BBCO) with schematic of the delay element. Its frequency is controlled through the FBB voltages applied to the NMOS (BBN) and the PMOS (BBP). Simulated accumulated jitter and power spectral density

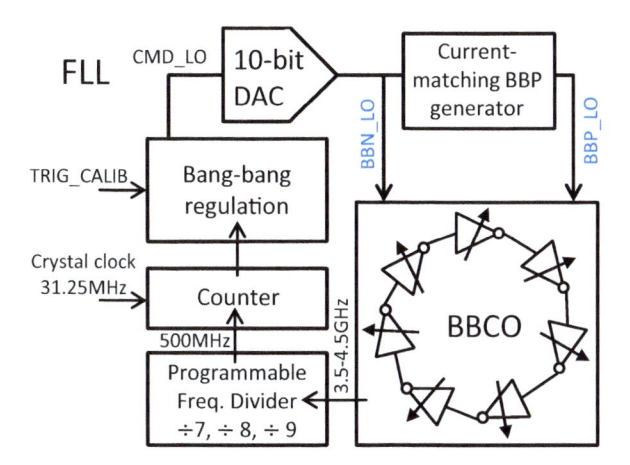

Fig. 9.6 Block diagram of the frequency-locked loop (FLL) performing LO frequency calibration through adaptive FBB

the regular delay elements. Figure 9.5 also shows the wide BBCO frequency range from 2.2 to 5 GHz by sweeping the FBB between 0 V (BBN = BBP = 0 V) and 1.8 V (BBN = −BBP = 1.8 V). This is of course a unique feature of FD-SOI technology thanks to the good control of the transistor V_t through FBB.

The free-running BBCO LO is integrated inside a duty-cycled frequency-locked loop (FLL) represented in Fig. 9.6 to periodically calibrate its frequency. The BBCO output is divided by the programmable frequency divider to generate the pulse-level 500 MHz clock. A control logic compares the divided LO output to a reference crystal clock at 31.25 MHz and adapts the BB voltage command CMD_CF in a bang-bang fashion to lock the LO frequency on the desired 3.5/4.0/4.5 GHz CF of the selected channel. A 10-bit DAC generates the NMOS BB voltage (BBN_LO) and a current-matching loop adapts the PMOS BB voltage (BBP_LO) to keep the NMOS/PMOS ON current ratio constant as described in Sect. 9.5. The CF calibration is performed once before packet transmission.

9.4 Pulse-Shaping Digital Power Amplifier

FCC regulations impose to limit emissions below −75 dBm/MHz of power spectral density (PSD) in the 960–1610 MHz band, which makes the design of digital PA difficult without high-order off-chip filtering. Indeed, as studied in [12], capacitive coupling of the digital PA output alone does not provide a sufficient roll-off to meet the FCC regulation limit when transmitting in the 3–5 GHz UWB bands. In SleepTalker, we designed a fully digital pulse-shaping filter integrated within the PA to improve the roll-off. Figure 9.7 illustrates the spectral impact of shaping a rectangular pulse to a Gaussian pulse with the same duration. For a Gaussian pulse

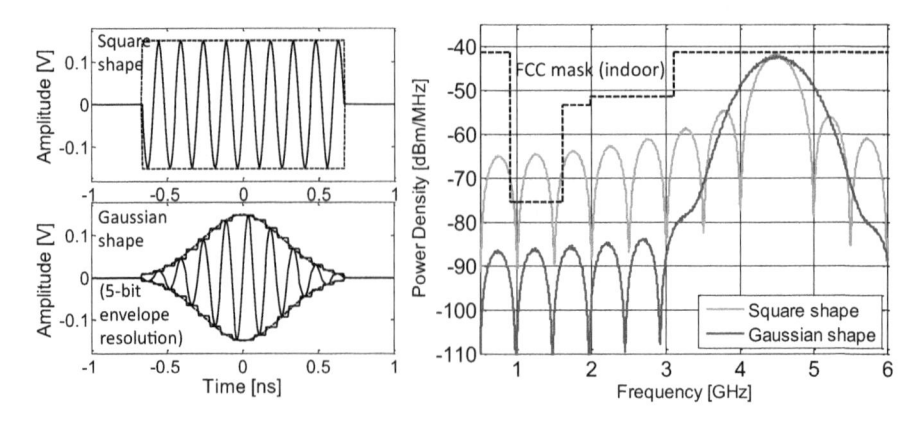

Fig. 9.7 Left: continuous-time square and Gaussian pulse shapes (envelope quantized at 5-bit resolution). Right: impact of the pulse shape on the output PSD. For the same maximum power inside the band, the square-shaped pulse features significant PSD in the prohibited 960–1610 MHz band. The Gaussian-shaped pulse results in much lower out-of-band emissions

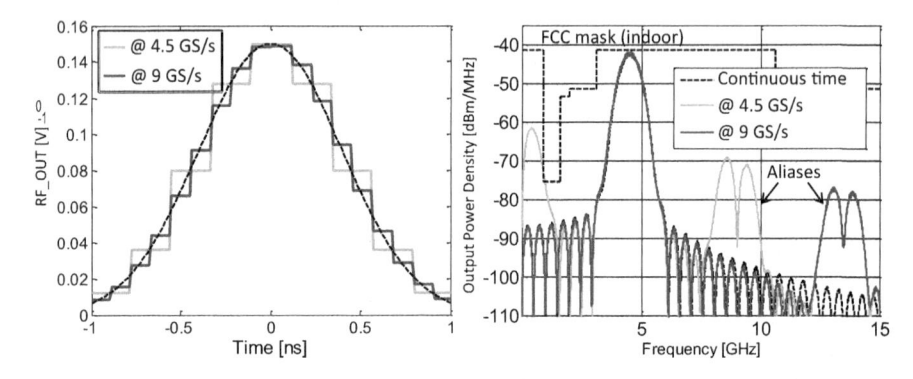

Fig. 9.8 Impact of the pulse shape temporal resolution. Left: Gaussian pulse shape sampled at 4.5 (@ f_{LO}) or 9 (@ $2 \times f_{LO}$) GS/s. Right: spectral folding caused by the sampling frequency of the shape. A sampling at f_{LO} leads to a folding peak at DC and spectral content in the 0.96–1.61 GHz band, which violates the FCC regulation mask. Sampling at $2 \times f_{LO}$ eliminates the folding around DC

shape with its envelope quantized at 5-bit resolution, the spectral content in the prohibited 960–1610 MHz band can be reduced by more than 20 dB. To implement the shaping functionality, the digital power amplifier is split in independent parallel stages to act as a 32-stage current DAC, leading to the 5-bit envelope resolution.

Figure 9.8 illustrates the impact of the pulse shape temporal resolution on the output PSD. If the sampling is performed at the LO central frequency (222 ps temporal resolution at 4.5 GHz), spectral folding causes significant power close to the prohibited 960–1610 MHz band. To avoid this problem, the pulse-shaping digital PA is operated at $2\times$ the LO frequency, which pushes the spectral folding away from the restricted bands.

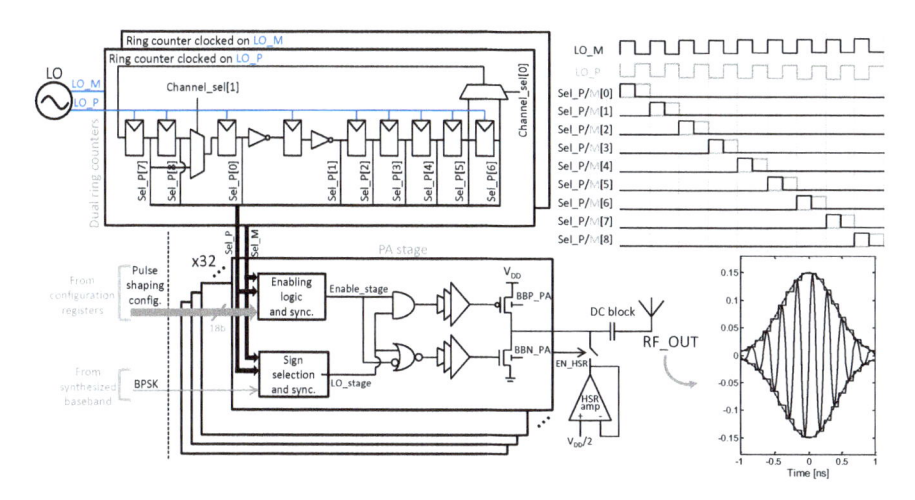

Fig. 9.9 Architecture of the programmable pulse-shaping digital PA. The pulse-shaping functionality is implemented by a 5-bit 9 Gs/s current DAC controlled by the ring counters outputs. The PA output is precharged at $V_{DD}/2$ by a high-slew-rate (HSR) amplifier and a DC blocking MiM capacitor filters the DC value. Right: stage-select signal sequence generated by the ring counters and simulated RF output waveforms

The complete pulse-shaping digital PA architecture is shown in Fig. 9.9. The dual ring counters are clocked on the LO positive and negative edges, respectively, and generate 7 to 9 enable signals with duty cycles varying from 1/7 to 1/9 depending on the selected channel from 3.5 to 4.5 GHz. The symbol-level baseband generates the BPSK-encoded and scrambled data controlling the polarity of the pulses. The 5-bit 14–18-sample programmable envelope of the pulse shape is stored in the pulse-shaping configuration registers. They are converted to thermometer code to control the 32 parallel PA stages. The ring counter output signal is used to control the cyclic selection of the 14–18 samples of the pulse shape and to control the enable of each of the 32 stages. The BPSK-encoded data are used to select the LO phase to use, which modulates the pulse polarity. The PA output stage is a tri-state inverter sized so that the 32 parallel stages with mid-range back biasing are able to drive the antenna with a peak-to-peak voltage of 350 mV at 4.5 GHz. The tri-state output stages offer the advantage of being fully turned off to save power between pulse bursts. Programmable FBB in the last stage of the PA is used to tune the RF output power and to compensate for PVT variations by controlling the current of the PA output stage.

For high PA efficiency, its output swing needs to be close to rail-to-rail with a DC component at $V_{DD}/2$. However, the delay mismatch of the signals driving the output-stage NMOS and PMOS transistors can lead to asymmetric pulses leaving a DC voltage at the end of the pulse different from $V_{DD}/2$. To alleviate this effect, the PA output is precharged to $V_{DD}/2$ between the pulse bursts. The precharge amplifier needs to drive the output node back to $V_{DD}/2$ in less than 16 ns, which is the duration of 1 guard interval in the fastest data rate, which leads to a slew rate of

17 V/μs. For this purpose, we designed a high-slew-rate (HSR) amplifier based on the super-class-AB OTA topology from [18] (Fig. 9.10). It features adaptive biasing of the input differential pair depending on the differential-input large signal to automatically boost the bias current and achieve high slew rate while maintaining a low quiescent current. It also features a local common-mode feedback first described in [19] implemented with two 10 kΩ matched resistors in order to boost the current efficiency. The HSR is supplied at 1.2 V to reach the stringent slew-rate requirement. Two common-source stages level shift the inputs from the 0.55 V domain for voltage compliance and an enabling signal EN_HSR disconnects the HSR output to the output of the PA stages during the pulse bursts as shown in Fig. 9.9. Figure 9.10 illustrates the adaptive biasing as well as Monte-Carlo post-layout simulations of the tri-state outputs converging after the end of a pulse.

9.5 FBB Generation and Current-Matching Loop

The generation of the back-bias voltages for NMOS transistors (BBN_LO and BBN_PA) is implemented by two capacitive DACs depicted in Fig. 9.11 over a [0;1.65 V] voltage range. The BBN voltage of the full-custom TX is generated by a 10-bit DAC to achieve fine tuning of the central frequency to the selected 802.15.4 UWB channel with 5 MHz frequency resolution. The output buffer is a 2-stage Miller-compensated OTA with cascoding to limit the impact of supply noise on the LO frequency accuracy. As the output of capacitive DACs drifts due to the switch leakage, a periodic refresh of the DAC is mandatory to avoid frequency drift during transmission. To do so, the packet-level baseband logic refreshes the DAC before each 802.15.4 packet, composed of maximum 83 preamble symbols and 1209 data symbols. At the lowest 0.11 Mbps data rate, the packet duration exceeds 10ms, which imposes heavy constraints on the switch leakage. To relax these design constraints, for this specific data rate, the DAC refresh is performed during each guard interval. The BBN voltage applied to the PA output stage does not require high accuracy as it is used only to compensate the impact of PVT variations on the output power by increasing or decreasing the peak-to-peak amplitude of the pulses. To generate this BBN, a 5-bit capacitive DAC is used with a similar architecture, output buffer, and refresh pattern.

To generate the back-bias voltages for the PMOS transistors (BBP_LO and BBP_PA), we designed continuous-time current-matching loops shown in Fig. 9.11 and similar to [3]. Replicas of an inverter or a PA output stage are used to determine the correct BBP that needs to be applied to balance the current between NMOS and PMOS transistors. An output buffer supplied by -1.8 V is used to drive the BBP back gate capacitance. The amplifier input range is 0 and 0.55 V, while the output can be driven down to -1.8 V leading to a voltage compatibility challenge even when using 1.8 V I/O transistors. To circumvent this, the Miller-compensated 2-stage OTA output buffer uses a bulk-driven NMOS input pair. This allows voltage range compatibility between the OTA output buffer and the replicas without the use

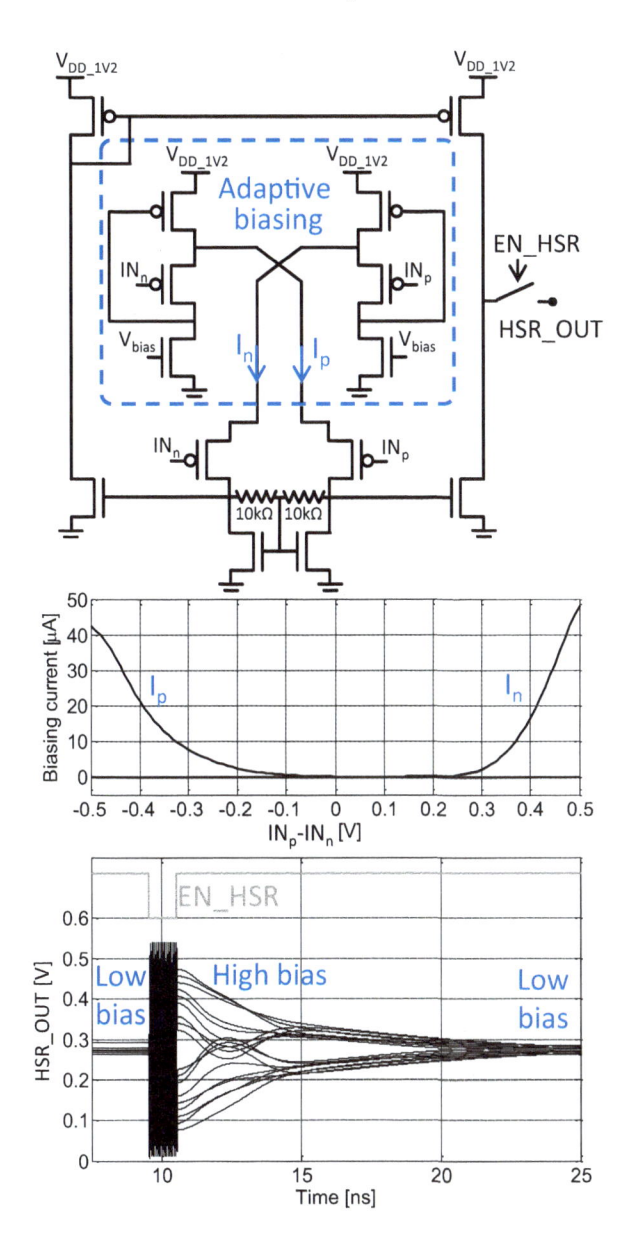

Fig. 9.10 Schematic of the HSR amplifier along with post-layout simulation results of the adaptive biasing and output node, which is quickly driven to $V_{DD}/2$ after the pulse over Monte-Carlo runs

of explicit level shifting, by implementing the output buffer with thick-oxide 1.8 V I/O transistors and the 0.55 V replica with thin-oxide core devices. Let us mention that the DACs are also implemented with thick-oxide 1.8 V I/O transistors.

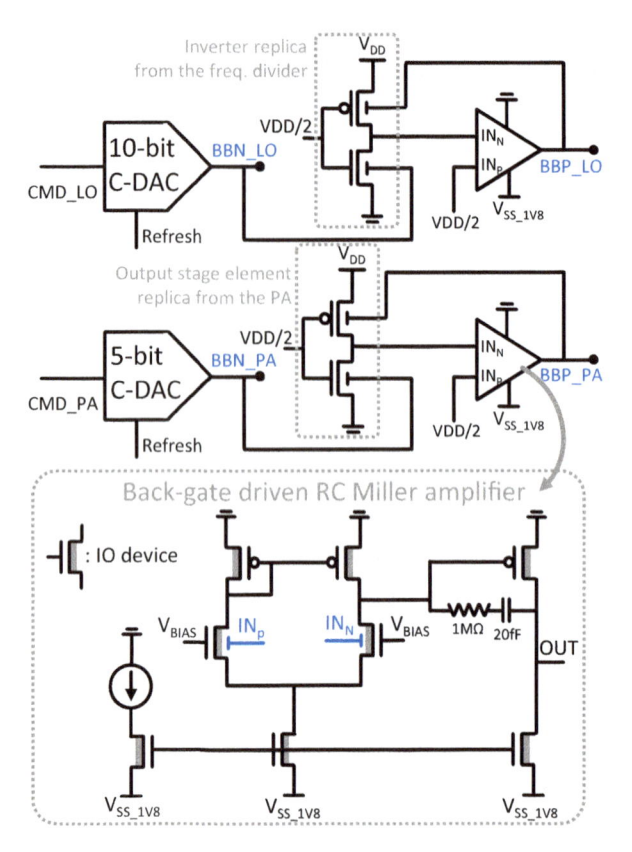

Fig. 9.11 FBB generator architecture. BBN is generated by DACs to tune the LO frequency for channel selection and the PA output power. The corresponding BBP voltages are generated from the BBNs through a current-matching structure with a back-gate-driven amplifier

9.6 SoC Integration

The SleepTalker SoC shown in Fig. 9.12 requires a single 31.25 MHz reference clock and a single 1.2 V external supply. The TX from Sect. 9.2 is supplied at 0.55 V by an on-chip switched-capacitor voltage regulator (SCVR) with high-density MiM caps. For speed concern at 0.55 V, only low-V_t (LVT) devices are used to implement both the full-custom and the synthesized TX blocks. The synthesized TX has zero back bias (ZBB, i.e., with BBN and BBP tied to ground) and uses standard cells with upsized gate length as the clock frequency is moderate (31.25 MHz). As the full-custom 0.55 V TX operates at higher frequencies (500 MHz and 3.5–4.5 GHz), minimum gate length is used with the FBB voltages for NMOS and PMOS transistors that are used to control the BBCO (BBN_LO and BBP_LO). Charge pumps generate the ± 1.8 V supplies for the FBB drivers. An always-on sleep controller supplied at 1.2 V is used to manage sleep mode by disabling the SCVR,

9 An 802.15.4 IR-UWB Transmitter SoC with Adaptive-FBB-Based Channel...   235

Fig. 9.12 SleepTalker SoC architecture

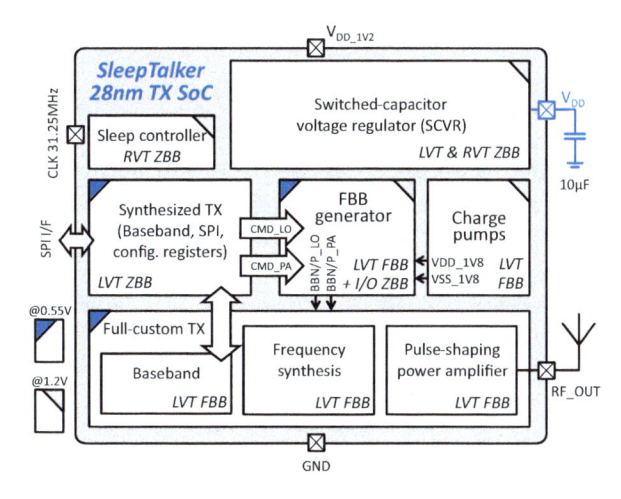

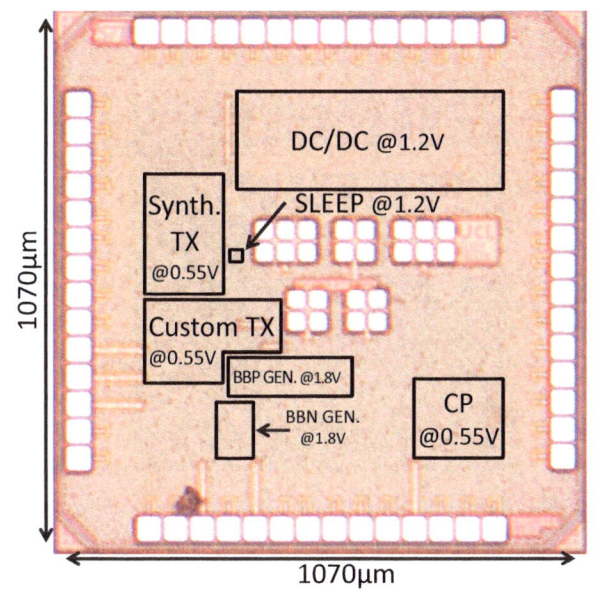

Fig. 9.13 SleepTalker die microphotograph. The SoC core area is 0.93 mm^2 and the TX only occupies 0.095 mm^2

thereby power gating the 0.55 V power domain. To reduce its leakage power, the sleep controller is implemented with regular-V_T devices and zero back bias. The transition from sleep mode to active is completed in less than 1ms and can be done between 2 packet transmissions.

The *SleepTalker* testchip was designed and manufactured in a 10-metal 28 nm FD-SOI CMOS process with high-density MIM capacitance option. The 0.93 mm^2 die in Fig. 9.13 has a total active area of 0.55 and 0.095 mm^2 for the TX.

9.7 Measurement Results

The characteristics of the FBB generation through the BBN DACs and the BBP current-matching loop are represented in Fig. 9.14. The achievable measured BBN and BBP ranges are, respectively, [0;1.6 V] and [0;−1.7 V]. In these measurements, the current-matching loop automatically applies a stronger FBB to the PMOS transistors than to the NMOS transistors compared to the NMOS FBB. This indicates a strong imbalance between NMOS and PMOS in the near-threshold domain, which was already reported in [20]. Figure 9.14 further illustrates this NMOS/PMOS imbalance by showing the BBN/BBP gap for 8 measured dies originated from a single wafer. Without the current-matching adaptive FBB, this NMOS/PMOS imbalance would lead to strong robustness degradation. Nevertheless, the BBP saturation at −1.7 V suggests that the imbalance is not fully compensated when BBN reaches [0.6;1 V] (depending on the die). For next-generation ULP SoCs in 28 nm FD-SOI, we recommend designing an asymmetric FBB generator with a

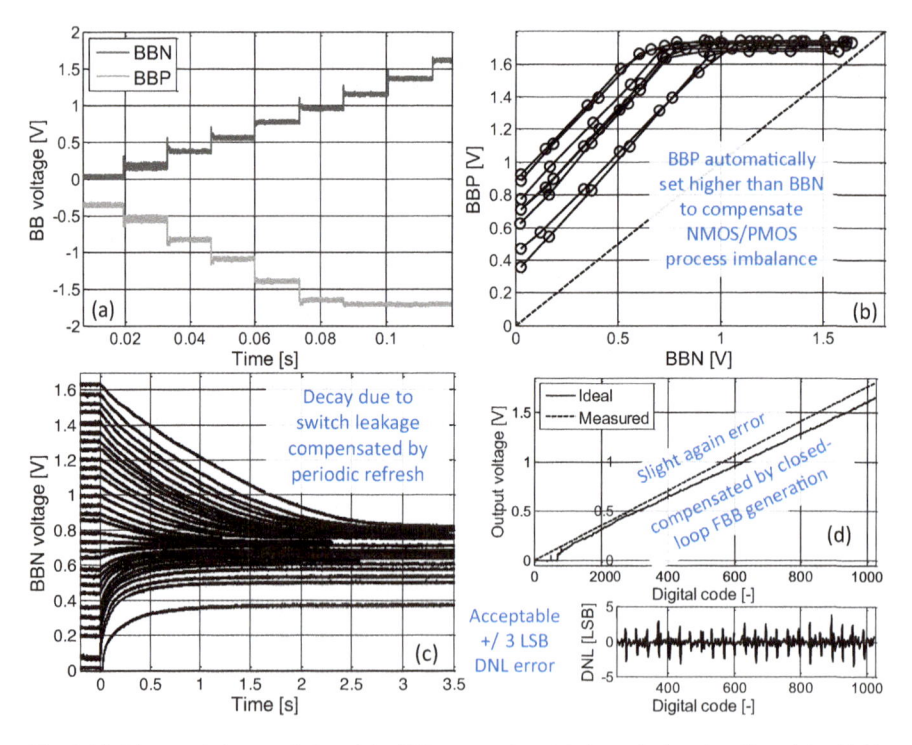

Fig. 9.14 Measured characteristics of the FBB generation: (**a**) dynamic functionality of the 10-bit BBN DAC with the BBP current-matching loop, (**b**) static BBN/BBP transfer function from the current-matching loop showing the die-to-die variations mismatch for 8 dies, (**c**) decay of the 10-bit BBN DAC for various digital input codes and (**d**) DNL performance of the 10-bit BBN DAC

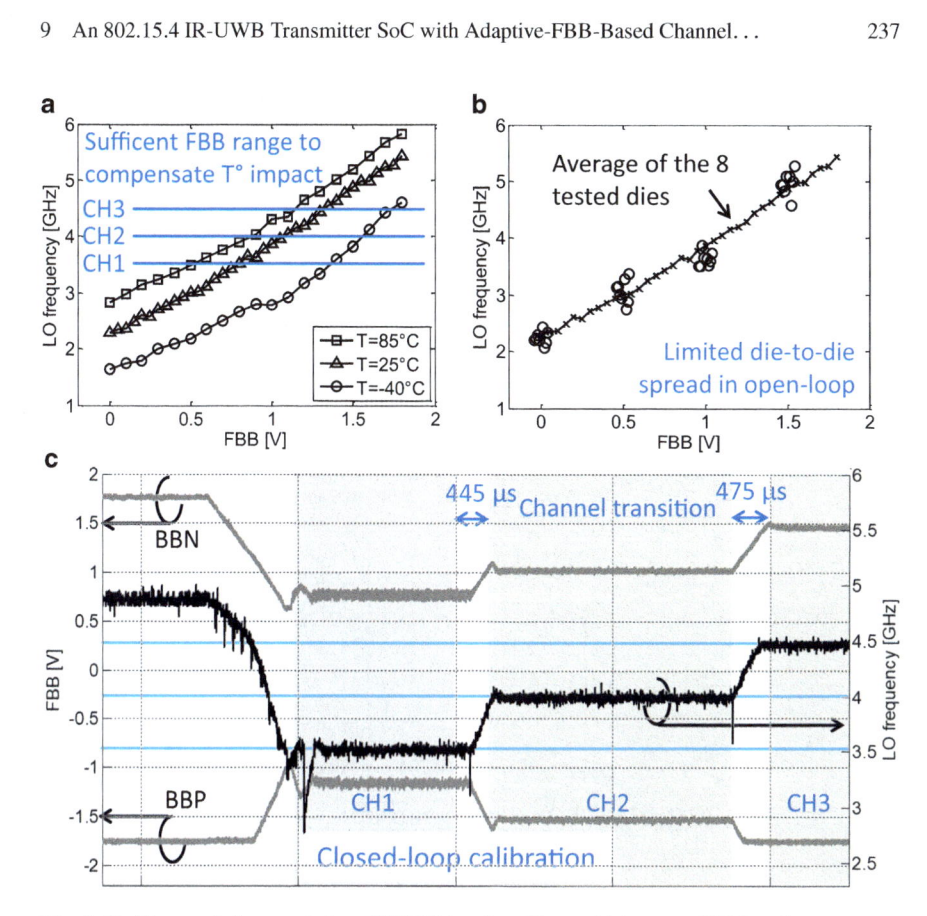

Fig. 9.15 Measured characteristics of BBCO local oscillator: (**a**) impact of FBB and temperature on the LO frequency of a typical die, (**b**) die-to-die variations of LO frequency for 8 tested dies, (**c**) closed-loop calibration of the LO frequency for a typical die

wider range for the PMOS transistors than for the NMOS transistors to fix this imbalance at ULV, see Sect. 3.2.5, and practically implemented in SleepRunner MCU from [9].

We measured the FBB control of the LO carrier frequency through the BBCO. The results are summarized in Fig. 9.15. Despite the aforementioned NMOS/PMOS imbalance, which is not fully compensated by the current-matching BBP loop, the LO frequency range is wide enough to cover the three 802.15.4 UWB channels over the full temperature range. The BBCO gain in typical conditions is in the order of 2 GHz/V. The impact of the FBB is also much stronger than the die-to-die variations. The automatic on-chip LO frequency calibration process performed before a packet transmission is also demonstrated in Fig. 9.15 with less than 500 µs for transition between adjacent channels and a typical lock time around 2 µs for two consecutive packets transmitted in the same channel.

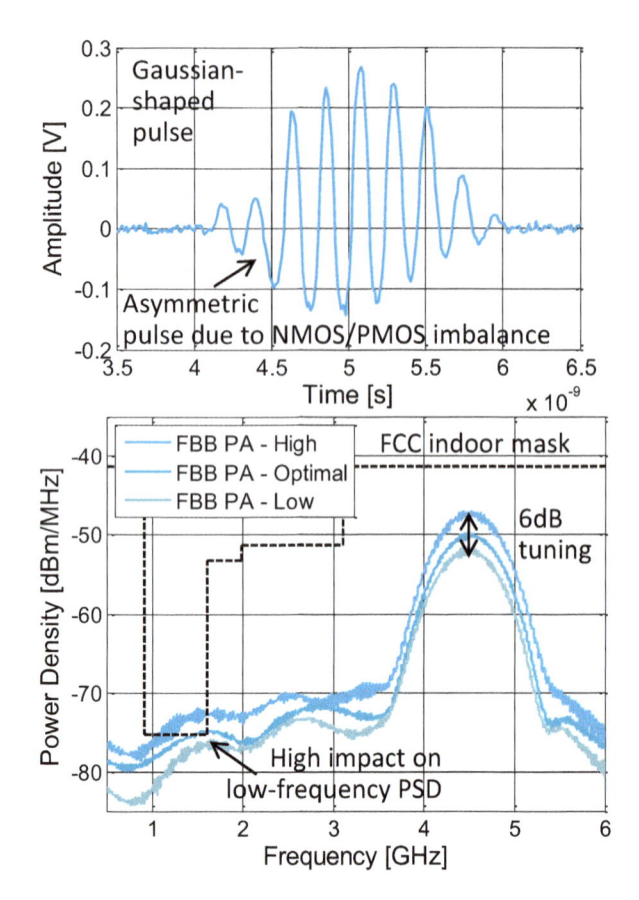

Fig. 9.16 Measured pulse waveform with Gaussian shape and output power spectral density for 27.2 Mbps data rate in channel 3 (4.5 GHz)

The TX functionality was assessed in the seven available data rates and for the three 802.15.4 UWB channels (3.5/4.0/4.5 GHz). We successfully measured pulses at a supply voltage of 0.55 V and validated the pulse-shaping functionality at up to 9 GS/s. The typical transmitted Gaussian-shaped pulse is represented in the time domain in Fig. 9.16 with the associated PSD. The shape is not fully symmetrical, which results in higher PSD in the prohibited 960–1610 MHz band than expected with simulations (Fig. 9.8). This effect is attributed to the strong NMOS/PMOS imbalance, which is not fully compensated by the current-matching BBP loop. As previously explained, this could be fixed with asymmetric FBB (wider BBP range than BBN range) or by optimizing the programmable pulse shape to compensate for this effect. This should be possible thanks to the unique flexibility of the proposed pulse-shaping PA. Figure 9.16 also demonstrates that the output power level can be controlled by the FBB level applied to the PA. The tuning range is limited to 6 dB due to the BBP saturation as soon as BBN reaches 0.6–1 V, depending on the tested die.

9 An 802.15.4 IR-UWB Transmitter SoC with Adaptive-FBB-Based Channel...   239

	Decawave, DW1000, 2014	Ryckaert, JSSC, 2007	Mercier, JSSC, 2009	Liu, RFIC, 2018		This work de Streel, JSSC, 2017		
Technology	90nm CMOS	90nm CMOS	90nm CMOS	22nm FinFET		28nm FDSOI		
TX active area [mm²]	N/A	0.066	0.07	0.085		0.095		
Carrier freq. [GHz]	3.5-6.5	3-10	2.1-5.7	8		3.5-4.5		
Operation mode	Coherent	Coherent	Non-coherent	Non-coherent	Coherent	Coherent		
Datarates [Mbps]	0.11-6.8	16	15.6	0.11		0.11	6.8	27.2
TX active power [mW]	168†	0.65	4.36	0.264	0.3	0.1	0.2	0.38
Average output power [dBm]	N/A	N/A	-16.4	-16.2	-16	>-26‡	-20	-20
TX efficiency	N/A	N/A	0.53%	9.1%	8.3%	>2.5%‡	5%	2.6%
TX energy per bit [nJ]	24.7† (6.8 Mbps)	0.041 (16 Mbps)	0.28 (15.6 Mbps)	2.4 (0.11 Mbps)	2.7 (0.11 Mbps)	0.95 (0.11 Mbps)	0.030 (6.8 Mbps)	0.014 (27.2 Mbps)

† Includes TX and RX power. ‡ The output power is lower at 110 kbps due to lower pulse repetition frequency. It can be increased by configuring the PA FBB to a higher level.

Fig. 9.17 Measured performances compared to the IEEE 802.15.4 IR-UWB TX state of the art

Measured characteristics of the full TX (i.e., all blocks from Fig. 9.3) inside the SleepTalker SoC are summarized in Fig. 9.17 and compared to state-of-the-art IEEE 802.15.4 UWB transmitters. The output power and power consumption are reported for channel 3 (4.5 GHz) for three data rates. The low jitter resulting from the duty cycling of the LO within the BPPM time slot is compatible with a coherent receiver, recovering the bit coded in the pulse phase (BPSK modulation). The output power is -20 dBm for the highest data rates featuring a pulse repetition frequency (PRF) of 15.6 MHz and -26 dBm for the lowest data rates using a PRF of 3.9 MHz. It is limited by the asymmetric pulse shape coming from the strong NMOS/PMOS imbalance and could be overcome with a wider BBP range.

The TX power consumption is 100–380 µW as a function of the data rate. It leads to decent TX efficiency up to 5% despite the limited output power. The resulting energy per bit is lower at high data rates as the power consumption is amortized on more transmitted bits. The data rate to be selected depends on the used RX and of the target range as a lower data rate relies on a higher number of pulses per symbol, which increases the SNR at the RX side and thus mitigates path loss. achieved energy is excellent between 0.95 nJ/bit and 14 pJ/bit, which to the authors' knowledge is better than all the previously reported 802.15.4 UWB TX including the one from [21] in 22 nm FinFET CMOS.

9.8 Conclusions

In a range of IoT applications, the power consumption of smart sensors can be dominated by the wireless data transmission when the continuous data rate exceeds 100 kbps. This prevents energy-harvesting operation. In this chapter, we have shown through the design of an IR-UWB transmitter SoC that the exploitation of the unique wide-range back biasing capability of FD-SOI can also be used to significantly save power for RF circuits. The way we used the back biasing capability is the control of the carrier frequency and the output power with a body-biased-controlled oscillator and a digital PA, respectively. The BBCO allows to get rid of current-starving technique to control its frequency, which improves jitter at low power. Moreover, the BBCO features instantaneous startup, which can be used to aggressively duty cycle the frequency synthesis circuit of RF transceivers to further save power. The measured energy per bit reaches record values for 802.15.4 UWB communication, values between 950 and 14 pJ/bit depending on the selected data rate.

Acknowledgements The authors of this chapter would like to acknowledge all the designers of SleepTalker at UCLouvain: François Stas, Thibault Gurné, Charlotte Frenkel, François Durant and Thomas Haine. They also would like to thank Denis Flandre (UCLouvain) for advising with respect to the sizing of analog primitives in the FBB generation, as well as Andreia Cathelin, Philippe Cathelin, and Frédéric Hasbani for multiple chip reviews. Finally, they would like to thank STMicroelectronics for chip fabrication.

References

1. D. Bol, G. de Streel, D. Flandre, Can we connect trillions of IoT sensors in a sustainable way? A technology/circuit perspective, in *IEEE SOI-3D-Subthreshold Microelectronics Technology Unified Conference (S3S)* (2015), pp. 1–3
2. S. Jeong et al., Always-on 12-nW acoustic sensing and object recognition microsystem for unattended ground sensor nodes. IEEE J. Solid-State Circuits **35**(1), 261–274 (2018)
3. N. Couniot, G. de Streel, F. Botman, A. Kuti Lusala, D. Flandre, D. Bol, A 65 nm 0.5 V DPS CMOS image sensor with 17 pJ/frame.pixel and 42 dB dynamic range for ultra-low-power SoCs. IEEE J. Solid-State Circuits **50**(12), 2419–2430 (2015)
4. P. Harpe, H. Gao, R. van Dommele, E. Cantatore, A. van Roermund, A 0.20 mm^2 3 nW signal acquisition IC for miniature sensor nodes in 65 nm CMOS. IEEE J. Solid-State Circuits **51**(1), 240–248 (2016)
5. D. Bol, "Ultra-low-power wireless communications for IoT smart sensors", invited talk in the "Sensors and Energy Harvesting" tutorial of IEEE European Solid-State Circuits Conference (2018)
6. Y.-H. Liu, A 3.7mW-RX 4.4mW-TX fully integrated Bluetooth low-energy/IEEE802.15.4/proprietary SoC with an ADPLL-based fast frequency offset compensation in 40nm CMOS, in *Proceedings of IEEE International Solid-State Circuit Conference* (2017), pp. 236–237
7. R. Dekimpe, P. Xu, M. Schramme, D. Flandre, D. Bol, A battery-less BLE IoT motion detector supplied by 2.45-GHz wireless power transfer, in *Proceedings of IEEE Power and Timing Modelling Symposium* (2018), 8p.

8. D. Bol, "Ultra-Low-Power SoCs for Local Sensor Data Processing", invited talk in the forum "Intelligent Energy-Efficient Systems at the Edge of IoT" of IEEE International Solid-State Circuits Conference (2018)
9. D. Bol, M. Schramme, L. Moreau, T. Haine, P. Xu, C. Frenkel, R. Dekimpe, F. Stas, D. Flandre, A 40-to-80 MHz sub-4 μW/MHz ULV cortex-M0 MCU SoC in 28 nm FDSOI with dual-loop adaptive back-bias generator for 20 μs wake-up from deep fully retentive sleep mode, in *Proceedings of IEEE International Solid-State Circuits Conference* (2019), 2p.
10. T. Haine, D. Flandre, D. Bol, An 8-T ULV SRAM macro in 28nm FDSOI with 7.4 pW/bit retention power and back-biased-scalable speed/energy trade-off, in *Proceedings of IEEE SOI-3D-Subthreshold Microelectronics Technology Unified Conference (S3S)* (2018), 2p.
11. B. Moons et al., ENVISION: a 0.26-to-10TOPS/W subword-parallel dynamic-voltage-accuracy-frequency-scalable convolutional neural network processor in 28nm FDSOI, in *Proceedings of IEEE International Solid-State Circuit Conference* (2017), pp. 246–247
12. P. Mercier, D. Daly, A. Chandrakasan, An energy-efficient all-digital UWB transmitter employing dual capacitively-coupled pulse-shaping drivers. IEEE J. Solid State Circuits **44**(6), 1679–1688 (2009)
13. IEEE Standard for Local and metropolitan area networks, Part 15.4: Low-Rate Wireless Personal Area Networks (LR-WPANs), IEEE 802.15.4-2011 standard (2011)
14. G. de Streel, F. Stas, T. Gurné, F. Durant, C. Frenkel, A. Cathelin, D.Bol, SleepTalker: a ULV 802.15.4a IR-UWB transmitter SoC in 28-nm FDSOI achieving 14 pJ/b at 27 Mb/s with channel selection based on adaptive FBB and digitally programmable pulse shaping. IEEE J. Solid-State Circuits **52**, 1163–1177 (2017)
15. G. de Streel, F. Stas, T. Gurne, F. Durant, C. Frenkel, D. Bol, SleepTalker: a 28 nm FDSOI ULV 802.15.4a IR-UWB transmitter SoC achieving 14 pJ/bit at 27 Mb/s with adaptive-FBB-based channel selection and programmable pulse shape, in *IEEE Symposium VLSI Circuits (VLSI)* (2016), 2p.
16. J. Ryckaert, G. Van der Plas, V. De Heyn, C. Desset, B. Van Poucke, J. Craninckx, A 0.65-to-1.4 nJ/burst 3-to-10 GHz UWB all-digital TX in 90 nm CMOS for IEEE 802.15.4a. IEEE J. Solid-State Circuits **42**(12), 2860–2869 (2007)
17. Y. Eken, J. Uyemura, A 5.9-GHz voltage-controlled ring oscillator in 0.18-μm CMOS. IEEE J. Solid-State Circuits **39**(1), 230–233 (2004)
18. A. Lopez-Martin, S. Baswa, J. Ramirez-Angulo, R. Carvajal, Low-voltage super class AB CMOS OTA cells with very high slew rate and power efficiency. IEEE J. Solid-State Circuits **40**(5), 1068–1077 (2005)
19. J. Ramirez-Angulo, M. Holmes, Simple technique using local CMFB to enhance slew rate and bandwidth of one-stage CMOS op-amps. Electron. Lett. **38**(23), 1409–1411 (2002)
20. G. de Streel, D. Bol, Impact of back gate biasing schemes on energy and robustness of ULV logic in 28 nm UTBB FDSOI technology, in *IEEE International Symposium on Low Power Electronics and Design (ISLPED)* (2013), pp. 255–260
21. R. Liu et al., A 264-μW 802.15.4a-compliant IR-UWB transmitter in 22 nm FinFET for wireless sensor network application, in *Proceedings of IEEE RFIC Conference* (2018), pp. 164–167

Chapter 10
Body-Bias Calibration Based Temperature Sensor

Martin Cochet, Ben Keller, Sylvain Clerc, Fady Abouzeid, Andreia Cathelin, Jean-Luc Autran, Philippe Roche, and Borivoje Nikolić

10.1 Introduction

The temperature of a circuit directly affects the leakage and speed characteristics of the devices, resulting in an impact on the digital logic power and timing. Just like process, voltage, and aging, those temperature changes can be compensated through the adaptive body-biasing techniques introduced in Chap. 3. However, before such actuation, a temperature sensor is needed first.

Such a sensor needs to have a small area and power consumption, as well as a wide range of voltage operation, so that it can be integrated within a complex SoC with minimum overhead. Moreover, it must keep a good accuracy even across

---

M. Cochet (✉)
STMicroelectronics, Crolles, France

Department of Electrical Engineering and Computer Sciences, University of California, Berkeley, CA, USA

Aix-Marseille University & CNRS, IM2NP (UMR 7334), Marseille, France

IBM Research, Yorktown Heights, NY, USA
e-mail: martin.cochet@ibm.com

B. Keller
NVIDIA Research, Santa Clara, CA, USA

S. Clerc · F. Abouzeid · A. Cathelin · P. Roche
STMicroelectronics, Crolles, France

J.-L. Autran
Aix-Marseille University & CNRS, IM2NP (UMR 7334), Marseille, France

B. Nikolić
Department of Electrical Engineering and Computer Sciences, University of California, Berkeley, CA, USA

© Springer Nature Switzerland AG 2020
S. Clerc et al. (eds.), *The Fourth Terminal*, Integrated Circuits and Systems,
https://doi.org/10.1007/978-3-030-39496-7_10

process variations. An approach consists in using multiple points of calibration [1, 14, 15, 20]. On the other hand, compensating process variation through a single-point calibration reduces testing time and cost.

In FD-SOI body-biasing can be used to compensate a wide range of threshold shift caused by process variation. Hence we propose a temperature sensor including a single-point calibration and bias compensation. It is composed of an oscillator supplied by a single low dropout regulator (LDO) which provides at the same time supply noise rejection and body-bias calibration with minimal power and area overhead. Last, on-chip digital logic provides the frequency-to-temperature readout, including linearization of the frequency-temperature characteristic of the sensor.

This chapter first section includes an overview of temperature sensors roles, specifications, and conventional implementations for on-chip monitoring, followed in Sect. 10.2 by a description of the proposed sensor principle, which relies on FD-SOI specific body-bias tuning for calibration. Section 10.3 covers the circuit level details of the sensor implementation and Sect. 10.4 presents measured results in a full SoC chip.

10.1.1 Temperature Sensor Requirements for Modern SoCs

Most modern digital SoCs now include temperature sensors as part of the dynamic voltage and frequency scaling control implemented by their power management unit. Hence, a compact sensor design that can be easily embedded within the SoC is important.

10.1.1.1 Need for Temperature Monitoring of Digital SoCs

Temperature monitoring is needed to account both for external environment changes and internal self-heating. Those sources of temperature change have different temporal and spatial variations. The former has a time constant in the order of the minute to hour and is uniform across the chip, while the latter can occur within a millisecond and is centered around high-power hotspots.

Those temperature changes directly impact digital timing and power, as well as analog SoC building blocks. The logic delay to temperature dependency is dependent on supply voltage, small near the zero temperature coefficient (ZTC see Appendix B.1.3, about 0.85 V in FD-SOI 28 nm LVT process), but becomes significant in near-threshold operation (about $3\times$ delay change between 0 and 100 °C at 0.4 V). The effect of temperature on power is consistent across voltages, resulting in an order of magnitude increase in leakage between 25 and 100 °C. Last, reliability can be impacted as an increase in temperature results in an exponential increase in electromigration [17].

10.1.1.2 Integrated Temperature Sensor Requirements

Hence, multiple temperature sensors must be scattered across the SoC to capture possible localized hotspots. Such a sensor can be designed in two parts: multiple probes, or front-ends which generate the signal of interest (e.g., a current or frequency proportional to the temperature), and a centralized processing element which provides the temperature readout to the power management unit. Different requirements and metrics can be used to evaluate the sensor's performance.

Area Smaller probes reduce floorplanning overheads and routing congestion, which enables temperature sensing closest to the possible logic hotspots. The smallest probe as of 2018 occupies only 30 μm^2 in a 65 nm process [10].[1]

Power Consumption which applies to both the probe and processing element. It can be normalized with respect to the sampling rate in the form of the joules per sample (J/Sa).

Voltage Range The integrated sensors should be able to track the core voltage scaling to avoid the use of a cumbersome dedicated supply. Ranges of 0.85–1.1 V for [1] and 0.5–1 V for [11] have been reported. This operating range is also related to the sensor accuracy. A post-calibration change in the supply voltage could impact the sensor output, leading to an erroneous change in temperature estimate. The power supply rejection of the sensor is expressed as the supply sensitivity of temperature, in °C/V.

Accuracy and Repeatability The quality of the sensor output depends of the accuracy i.e. the difference between the average of the sensor readings and the actual temperature, as well as the repeatability, i.e., the rms of the distribution of consecutive measurements. As the sensor performance is process dependent, the accuracy should be reported for multiple dies, ideally from different wafers.

Resolution The sensor resolution T_r is defined as the minimum change of temperature it can detect. For digital outputs, this is limited by the temperature value of the least significant bit (LSB). However, if the sensor's repeatability is larger than the LSB resolution, then the effective resolution is equal to the repeatability.[2]

Calibration and References Sensors can require a one or two points per die calibration to compensate for individual mismatch in the circuit's response. Batch level high order curvature correction can be applied. Two-point calibration increases testing costs and time as the die needs to be brought to two different temperatures for testing. Last, some sensors require fixed references for proper calibration, such as a stable clock, voltage, or current. This may not be an overhead if the SoC already

---

[1] Further metric performances of the sensors are compiled in Table 10.3 at the end of this chapter.
[2] Quantitatively, $T_r = \max((T_{\max} - T_{\min})/2^N, T_{rms})$, where $[T_{\min}; T_{\max}]$ is the sensor's range, N the number of output bits, and T_{rms} the measurement repeatability.

needs such reference; however, the sensitivity to a change in the reference must be estimated (e.g., temperature error caused by a given amount of jitter).

10.1.2 Current Methods for Temperature Sensors Design and Process Compensation

Several different sensor architectures have been proposed in the last decades, offering different trade-offs between accuracy, integrability, power, and calibration process.

10.1.2.1 Analog Bandgap

The first class of temperature sensors historically dates back to the BJT era. It takes its inspiration from the 1964 invention of the bandgap voltage reference [8] and is perfected in 1974 as the Brokaw three terminal bandgap reference [3].

The general principle relies on the linear dependency with temperature of the base–emitter voltage V_{BE} of a BJT for a given collector current. Most circuit topologies use a differential approach in reading out V_{BE} to minimize process and voltage sensitivity. One can refer to [4] for a detailed description of operation.

This approach is known for its accuracy even without multiple point calibration; however, BJTs do not benefit from advanced process nodes and require a higher voltage to operate than their CMOS counterparts [7].

10.1.2.2 Resistor-Based

Another analog approach consists in taking advantage of the temperature coefficient of resistivity (TCR) of resistors which is mostly unaffected by process. By measuring the generated voltage across such a thermistor with a fixed current, a proportional to absolute temperature (PTAT) voltage can be generated.

Different approaches can be used to accurately readout that voltage, but the trade-off for their high accuracy is a large sensor and especially front-end size, (e.g., 44,000 [14] to 720,000 μm^2 [13]).

10.1.2.3 Thermal Diffusivity

An alternative method is based on the thermal properties of temperature diffusion within the silicon substrate. As the substrate is lightly doped, its thermal diffusivity D is almost independent of process variation and has a known temperature dependence ($D \propto T^{1.8}$)[21].

The thermal diffusivity (TD) sensor includes a heater supplied by a periodic current I_{IN} at a frequency f and a thermopile, with a known spacing L. The thermopile converts back the temperature change into a voltage V_{OUT}. A sensing circuit then measures the phase difference Φ between I_{IN} and V_{OUT} caused by the delay of propagation through the substrate distance L. Φ provides an almost linear temperature readout $\Phi = L\sqrt{f/D} \propto T^{0.9}$ [18].

This method offers good accuracy at single-point calibration [16], but requires careful layout and multi-physics simulation for validation across technology nodes.

10.1.2.4 Digital Differential CMOS

Several approaches have been proposed to take full advantage of digital CMOS technology area and power scaling. Such digital sensors include a temperature to frequency (or temperature to time) probe in a form of an inverter delay line or ring oscillator (RO), which can be converted to a digital readout with a simple digital counter.

To reduce process and supply errors, a differential topology is used, where two oscillators have similar supply and process frequency dependency, but different temperature–frequency curves. Different designs rely on supply difference [15], change in gates W/L [1], or differently biased current starved inverters [9]. The optimal approach is very technology dependent.

Their small area and digital nature usually comes at the cost of significant mismatch, which requires a two-point calibration.

10.1.2.5 Single-Ended Digital Temperature Sensor (SED-THS)

To circumvent the need for two oscillators design, a single-ended sensor approach (SED-THS) has been proposed [19]. It assumes that the process P and temperature T effects on logic delay D are uncorrelated, i.e., $D(T, P) = f(T) \cdot g(P)$, so that a single-point normalization at a known T_0 provides a process-independent readout:

$$D(T, P)/D(T_0, P) = f(T) \quad (10.1)$$

This approach however does not provide supply noise rejection and had only be validated in an older 0.18 μm technology. In advanced nodes, the process changes could also affect the temperature-delay response of the oscillator, for example, in case of V_{th} shift.

10.2 Body-Bias Compensated Oscillator Principle

The rest of this chapter presents the proposed temperature sensor, which leverages FD-SOI body-biasing capabilities to overcome the shortcomings of the SED-THS in terms of voltage and process susceptibility.

10.2.1 Uncompensated SED-THS

The first option to limit the voltage-dependency of the oscillator delay is to protect it through a voltage regulator, such as a LDO. It provides a rejection of supply noise as well as a possible extension of the supply range (which can be treated as DC voltage noise).

10.2.1.1 Basic Principle

Figure 10.1 presents an example of such a design. A ring oscillator with a given delay function $D(T, P)$ (Eq. (10.1)) and supplied by a LDO regulator comprises the sensor probe, and a simple digital counter and logic provides the processing element.

Assuming independent process and temperature delay, the output can be generated simply by inverting Eq. (10.1):

$$T = T_0 + \frac{\partial T}{\partial D} \frac{D - D_0}{D_0} \qquad (10.2)$$

where T_0 is the calibration point and $D_0 = D(T_0, P)$.

10.2.1.2 Simulated Performance

This approach has been simulated in 28 nm FD-SOI process using standard cell inverters and an ideal LDO to test the independence of process and temperature. The voltage was chosen to be 0.75 V, enough below the ZTC to provide a ±5%

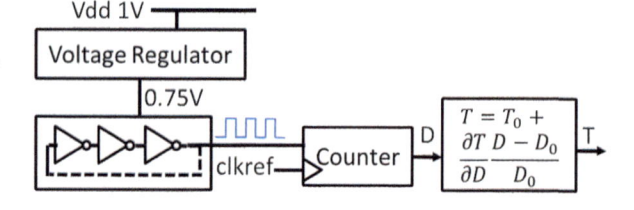

Fig. 10.1 Concept block diagram of a single-ended oscillator-based temperature sensor

delay change across the temperature range. As seen in Fig. 10.3a, the absolute delay changes by up to 25% across corners. However, when normalized to the calibration point $T_0 = 25\,°C$, the delays slopes are much more similar.

Yet, the small change in normalized delay is enough to result in errors up to $5.1\,°C/-3.4\,°C$ across corners. Moreover, the sensor is still very sensitive to supply noise. A change in only $\pm 1\,mV$ at the supply also results in a $\pm 3\,°C$ error. This would put very high constraints for the LDO's power supply rejection ratio (PSRR) requirements.

10.2.2 Bias-Compensated SED-THS

In order to reduce the process and voltage sensitivity relative to the temperature sensitivity, a lower operating supply, further from the ZTC can be chosen. However, this increases process variation, as near-threshold inverters are more sensitive to V_{th} changes. As seen previously, body-biasing can be used to compensate such voltage variations.

10.2.2.1 Use of Body-Bias Compensation

A direct approach would be to include a body-bias generator in the sensing probe to generate $Vdds$ and $gnds$ levels to compensate the PMOS and NMOS process changes. However, this approach has several complications.

First a full body-bias generator, including negative voltages for LVT-FBB requires complex charge pumps and regulators, as seen in Chaps. 11 and 15. Such a circuit could not fit in the tight area and power budget of a sensor front-end.

Moreover, this approach would require independent monitoring of the PMOS and NMOS process corner to provide differentiated $Vdds$ and $gnds$.

10.2.2.2 Supply-Bias Merged Oscillator

Instead a simpler approach has been proposed [5, 6], which uses an NMOS-only oscillator, and ties the $gnds$ net to the LDO-generated supply vdd, as illustrated in Fig. 10.2.

This approach does away with the need to generate a negative $Vdds$ supply and the LDO provides the body-bias $gnds$ generation, without the need of a separate body-bias generator. Last, there is no need for separate PMOS and NMOS process monitor, as the oscillator itself can be used for calibration.

The calibration is set at a single point T_0, where the oscillator period is measured and compared to the TT CAD expected value D_0. Rather than a simple normalization as in the previous SED-THS, the $Vdd = Gnds$ is tuned to match the target D_0. In practice, this means that for a slow-N corner (i.e., higher V_{th}) the

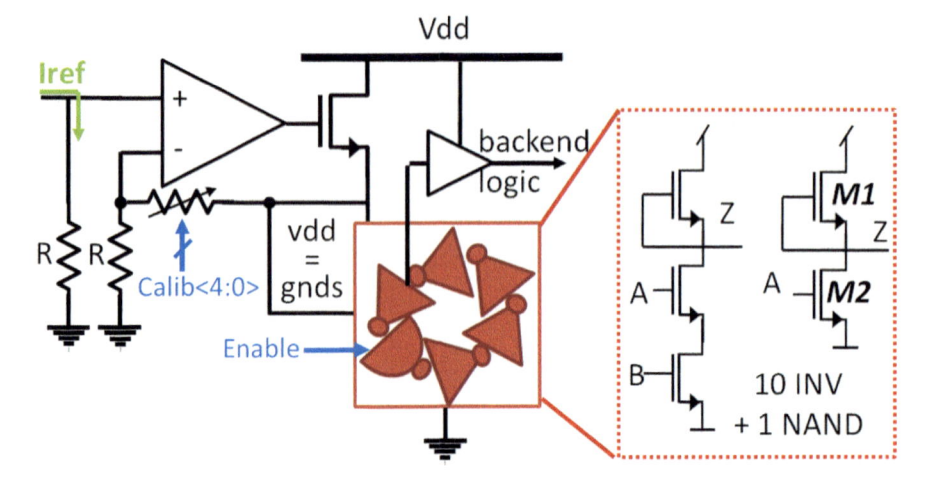

Fig. 10.2 Components of the single-ended body-bias compensated oscillator for temperature sensing. ©2018 IEEE reprinted with permission

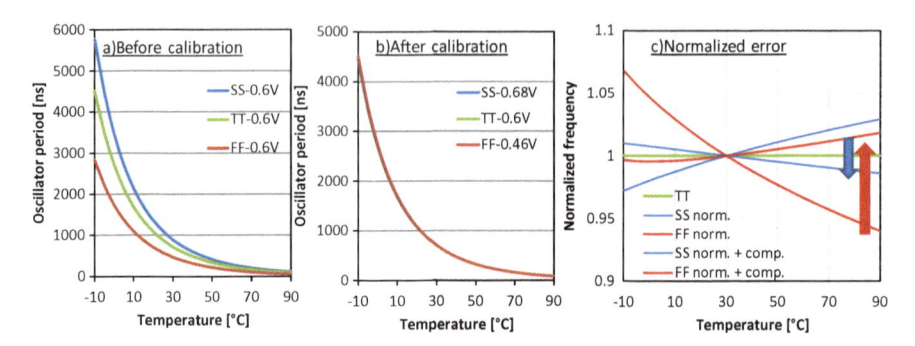

Fig. 10.3 Simulated period of the probe across process corners and temperatures before (**a**) and after (**b**) calibration, and resulting normalized frequency change (**c**) with respect to the TT corner. ©2018 IEEE reprinted with permission

measured delay pre-calibration D will be larger than D_0. To reach the target, the calibration will apply a larger $Gnds$ to reach the target D_0, hence the V_{th} will be decreased towards its nominal value.

As the oscillator is designed with NOMS-only devices, the pull-up transistor M1 which dominates the delay is always in sub-threshold operation $V_{GS} = 0\,\text{V}$, resulting in a high temperature sensitivity (about 44x change in delay between -10 and 85 °C), which limits the relative impact of voltage and process effects, even for an above threshold 0.6 V nominal supply.

As seen in Fig. 10.3, the temperature delay of the oscillator is very non-linear. This makes a direct conversion of the measured frequency to temperature more complicated. However, thanks to the high digital integration in advanced nodes such

10.2.2.3 Oscillator Performance

First the oscillator itself is simulated. As seen in Fig. 10.3c, using a simple normalization (constant Vdd and $gnds$ across corners) would lead to a $-6.8\%/+6.0\%$ change in frequency, which translates to a $-0.9\,°C/+1.6\,°C$ temperature error. The adaptive supply and bias approach reduces the mismatch to $-1.8\%/+1.4\%$, i.e., $-0.6\,°C/1.4\,°C$ temperature error.

10.3 Circuit Implementation

In addition to the probe, the circuit requires bias generations, LDO design, and digital backend processing. Figure 10.4 presents a top level block diagram of the system. A current mirror splits a single 48 µA current reference into 1 µA references, distributed to each of the 11 probes. Then the probes generated oscillating output are sent to a centralized digital processing logic, which converts the oscillation frequency to a digital representation and serially operates the time-to-temperature conversion.

10.3.1 Probe Detailed Implementation

Due to the NOMS-only implementation, the pull-up transistor M1 (Fig. 10.2) is very weak, resulting in an oscillation below full swing. To provide full CMOS levels,

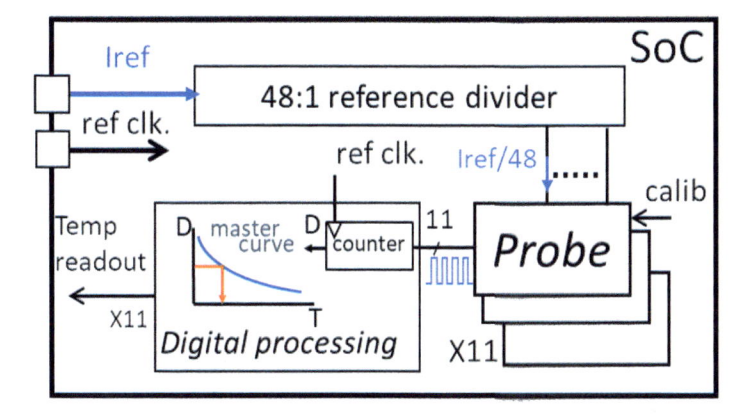

Fig. 10.4 Block diagram of the full temperature sensor system

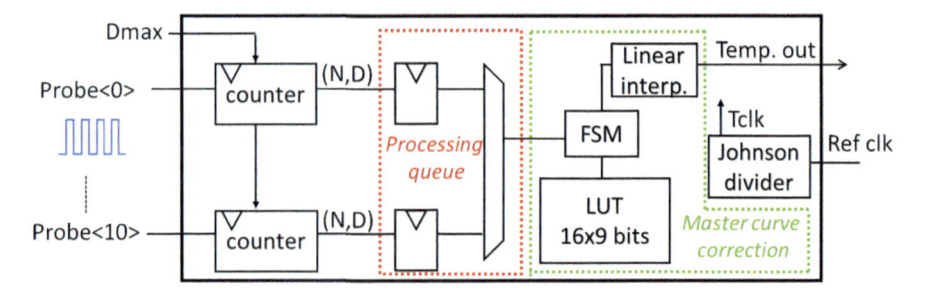

Fig. 10.5 Block diagram of the digital processing backend. ©2018 IEEE reprinted with permission

the signal needs to be amplified, filtered, and level-shifted. Note the signal is also inverted, so that the period is measured between two falling edges, which are much better defined than the low slew-rate rising edges.

The LDO is designed to offer a good PSRR, with minimum power and area overhead, as the load current to the ring oscillator is extremely small (under 500 nA in nominal PVT). First, the pass-gate is an NMOS device biased to the highest supply $gnds = Vdd$ to reduce resistance for a given area. The feedback comparator is designed as a simple 6-transistor complementary self-biased differential amplifier [2] rather than a full operation amplifier, whose driving stage would not be needed as the load (the pass transistor's gate) has a very high impedance.

The resistors are polysilicon devices and their value is set as a trade-off between area and biasing power consumption.

10.3.2 Digital Processing

The digital logic is needed both to provide a time-to-digital conversion of the probes' oscillator output and to correct for the high non-linearity of the probe's response. Its block diagram is illustrated in Fig. 10.5.

The general principle of the first stage consists in measuring the number of periods of the oscillator T_{RO} in a fixed number N_s of periods of a fixed clock T_{clk}, which results in a sampling period $T_{spl} = N_s \cdot T_{clk}$. A first approach consists in simply counting the number N of full periods T_{RO} in a T_{spl} period, as illustrated in Fig. 10.6. This results in a quantization error of T_{RO}/T_{spl} which becomes large at low temperatures where T_{RO} increases. A second approach consists in counting the number of T_{clk} periods in a single T_{RO} period D. The quantization error, T_{clk}/T_{RO}, increases when the higher temperature results in a smaller T_{RO}. Hence a third method is proposed: the number D of T_{clk} periods in all the N full T_{RO} periods which fit in the sampling window T_{spl} is generated. The (D,N) output pairs allow

Fig. 10.6 Compared methods of measuring the oscillator period. ©2018 IEEE reprinted with permission

Measurement	N only	D only	N and D
Waveforms	(N=1, N=2)		
	T_{spl}	T_{clk}, D	T_{spl}, D
Quantization error	T_{RO}/T_{spl}	T_{clk}/T_{RO}	T_{clk}/T_{spl}
−10°C	99.5%	0.2%	0.2%
25°C	18.8%	1.1%	0.2%
85°C	2.2%	9.2%	0.2%

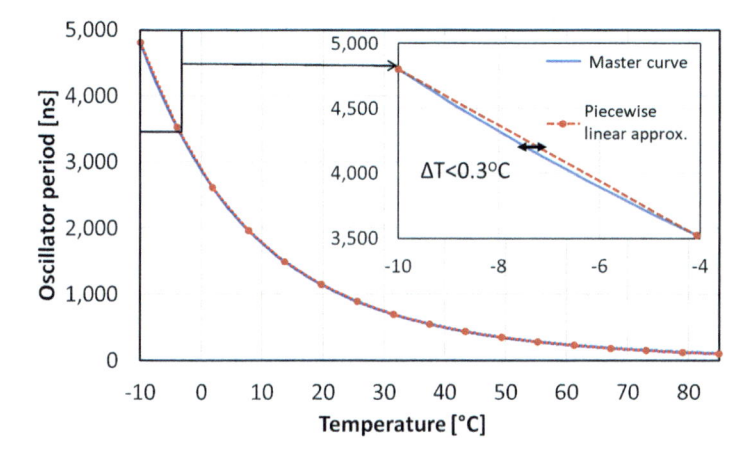

Fig. 10.7 16-point piecewise-linear approximation of the master curve

the reconstruction of the $T_{RO} = D \cdot T_{clk}/N$ with a quantization error of T_{clk}/T_{spl}, constant across T_{RO} values. The counting of the (D,N) pair is done in parallel for all 11 probes, then is processed serially by the next stage.

The second part of the circuit generates the temperature readout on 10 bits. An analytical model of the delay-temperature curve could be made and processed by a general purpose circuit, as in [15]. On the other extreme, a full look-up table containing the 11-bit input 10-bit output values of all the points could be made. However both options result in excessive area and power overhead. An intermediate approach was used instead: the 4 MSB of the output are generated from a 16-entry look-up table, while the 6 LSB are computed from a linear interpolation of the two closest LUT values.

As seen in Fig. 10.7, this piece-wise approximation of the curve results in a simple implementation, a fast 5-cycle decoding, and keeps the accuracy within 0.3 °C.

10.3.3 Noise Analysis Methodology

Last, the different sources of error of the sensor can be identified and simulated. The DC components (global and local process, DC voltage changes) affect the accuracy, while supply noise, device noise, and references error affect the measurement repeatability. The detail of the simulated contributions is presented in Table 10.2.

Process simulations across corners of the full system show a contribution of global and local (rms) process variation of only 0.55 °C and 0.24 °C, respectively, thanks to the proposed body-bias compensation scheme.

The impact of the DC supply noise (i.e., the voltage operating range) is mitigated by the LDO. As the current mirrors providing the reference share the same supply, their impact also has to be accounted for. Table 10.1 shows the system's combined PSRR across frequencies. The DC rejection is above 50 dB at DC, but reduces with frequency because of capacitive coupling. Across corners it reaches a worst case of 30.9 dB at the sensor's sampling frequency F_{spl}. As the oscillator integrates the supply noise over one sampling period $1/F_{spl}$, frequencies above F_{spl} get filtered (Table 10.2).

The device noise can be estimated through periodic steady state periodic noise (PSS-PNoise) simulations. The simulated jitter was used as a target specification to size the oscillator's transistors.

Last, the sensor relies on two references, I_{ref} and T_{clk} whose sensitivity must be estimated. The strong superlinear temperature dependency of the oscillation period limits the sensitivity to an error in the measured frequency to 0.21 °C/% at 25 °C (worst case of 0.36 °C/% at 85 °C). This relaxes the jitter and frequency drift constraints on the reference clock. The reference sensitivity is 0.32 °C/%, or 0.66 °C/μA, referred to the pre-divided 48 μA reference.[3]

Table 10.1 Simulated PSRR of the full system across corners

Corner		RCtyp TT	RCmin SS	RCmax FF
Calibration code		"01111"	"11001"	"00110"
VRO (mV)		634	750	530
PSRR at F_{spl} (dB)	−10 °C	30.9	31.3	31.6
	25 °C	33.5	34	34.1
	85 °C	40.5	41.2	41.5

Table 10.2 Summary of the sources of error, mitigation strategies, and estimated performance

	Process variation		Supply noise			Device noise
	Global	Local	DC	$<F_{spl}$	$>F_{spl}$	
Mitigation	Calibration		LDO		Averaging	Design (sizing)
Affects	Accuracy		Repeatability			
Error (°C)	0.55	0.24	0.23	0.11	0.1	0.07

---

[3]This value can be derived in the following steps: an error in the I_{ref} directly translates to an equal relative error in the generated supply, i.e., $dI/I = dV/V$; then the voltage–frequency sensitivity

10.4 Manufactured Chip

To validate the design, the sensor system has been fully designed, laid out, and fabricated in 28 nm FD-SOI process.

10.4.1 Sensor Layout

The sensor was laid out to maximize its integration and minimize the overhead resulting from the spacing of differently biased wells. Figure 10.8 presents its layout as well as cross sections illustrating the NWells (NW) placement. Note that the Deep NWell (DNW) under the differential amplifier and logic is not needed for functionality, but reduces substrate noise on the logic PMOS. A less conservative option would be to remove the DNW. This would allow to reduce the spacing from 2.5 to 1 µm, reducing the probe area by 10%, but resulting in potentially more noise. Last, a compensation capacitor was placed at the gate of the LDO pass transistor to improve its stability. It is designed as a MOM finger capacitor on higher metal levels on top of the probe to limit area overhead, resulting in a total probe area of 225 µm².

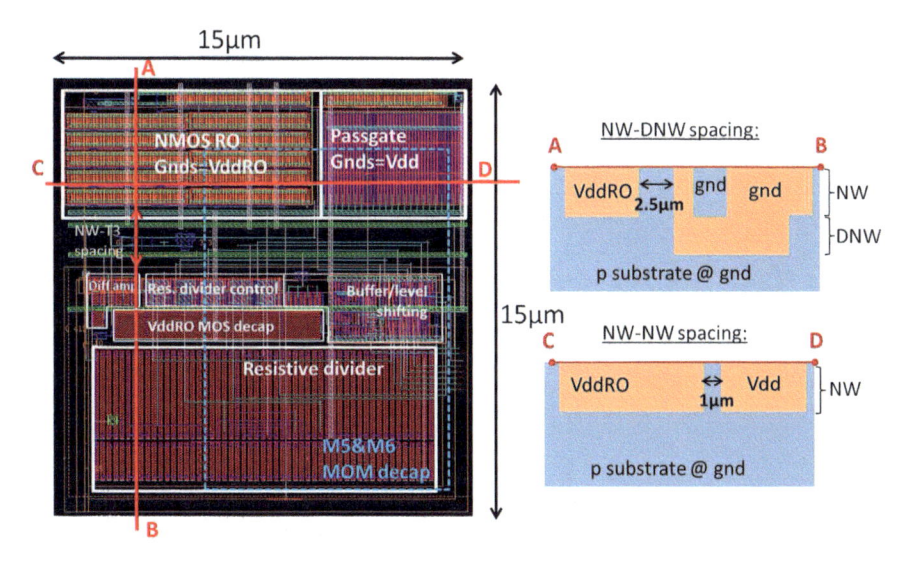

Fig. 10.8 Left: layout of the sensor probe. Right: cross sections illustrating the wells spacing

can be estimated from simulations to 1.49 ($dF/F = 1.49\,dV/V$, typical PVT corner); last, the frequency–temperature sensitivity is the metric previously seen of 0.21 °C/%, ($dC = 21.3\,dF/F$).

10.4.2 Validation of Calibration

As described in Sect. 10.2.2, the main assumption of the body-bias compensation is that the slope of the delay-temperature response will change across corners, hence a simple normalization is not enough to provide accurate outputs. Rather, increasing the level of $Vdd = gnds$ in the sensor (i.e., forward biasing) decreases the V_{th} and decreases the sensor's relative delay-temperature response.

To validate this principle, the oscillator's frequency of 77 sensors across 11 dies is first measured at nominal temperature (30 °C) with default calibration. Then the relative slope of the frequency response $F(75\,°C)/F(0\,°C)$ is measured.

As seen in Fig. 10.9, there is a significant correlation r on each die and among the whole population between the two parameters. That is, slow sensors (i.e., with a low $F(30\,°C)$) have a higher relative delay-temperature dependency $F(75\,°C)/F(0\,°C)$, and fast sensors have a lower relative delay-temperature response.

With our calibration method, the slow sensors are sped up by forward biasing, which in turn reduces their relative delay-temperature response. This results in a calibration of the sensor both at the point of calibration (30 °C) and across temperatures.

10.4.2.1 Measured Accuracy

The accuracy of the sensor is then measured across 21 probes on $[-5\,°C, 85\,°C]$ temperature range.[4] As illustrated in Fig. 10.10, the accuracy is of $-2.5\,°C/1.2\,°C$,

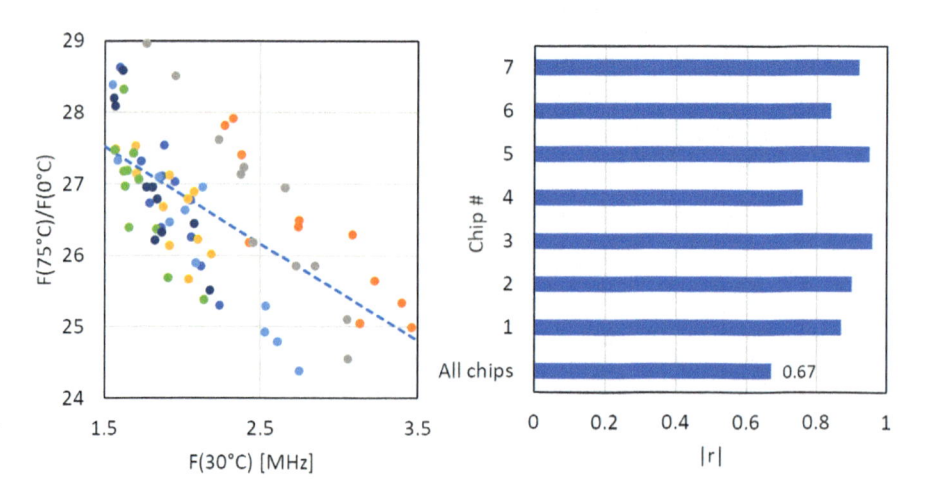

Fig. 10.9 Measured correlation r between the probes frequency at room temperature, and temperature sensitivity. ©2018 IEEE reprinted with permission

---

[4]The proposed architecture's operating range is limited by the exponential delay-temperature trend: for wider ranges, the change of frequency between T_{min} and T_{max} makes the design impractical.

or $-3.3\,°C/1.9\,°C$ for a worst case of plus or minus three standard deviations around the mean (3σ). A systematic correction can be implemented to subtract the average measured error across the chips for each temperature, for example, by fine tuning the LUT values. Applying this correction reduces the error to $-1.4\,°C/1.3\,°C$ min/max and $-2.2\,°C/2.6\,°C$ (3σ).

The sensor repeatability can be traded off with conversion rate (and energy per sample) by averaging out the result of N samples. This reduces the rms error by $\sqrt{N}$ while sampling time and energy per sample increase by a factor N. An averaging value $N = 5$ was found optimal to provide a rms resolution of $0.76\,°C$ and an energy of 2.0 nJ/Sa at 27.8 kSa/s.

Last, the supply rejection of the probe can be measured by sweeping its supply over a wide operating range, as seen in Fig. 10.10. The LDO guarantees an error within $+0.58\,°C/-0.52\,°C$ across 0.62–1.2 V (i.e., a $1.9\,°C/V$ sensitivity) which makes this sensor a good candidate for wide voltage range SoCs.

10.4.2.2 Example of SoC Integration

As explained in the introduction, the target application of our sensor is to provide a low overhead, medium accuracy for full SoC monitoring. To illustrate this the 11 probes are integrated into a full dual core RISC-V SoC, with 4 sensors per core and 3 sensors in the periphery.

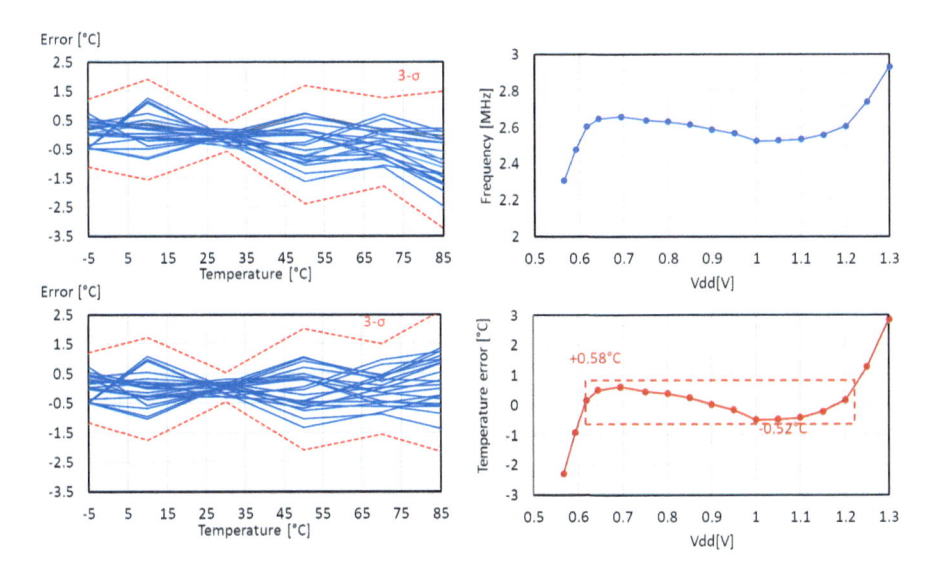

Fig. 10.10 Measured accuracy of the body-bias compensated temperature sensor. Left: measured accuracy of the sensors before and after systematic correction. Right: measured frequency and corresponding error across voltage range. ©2018 IEEE reprinted with permission

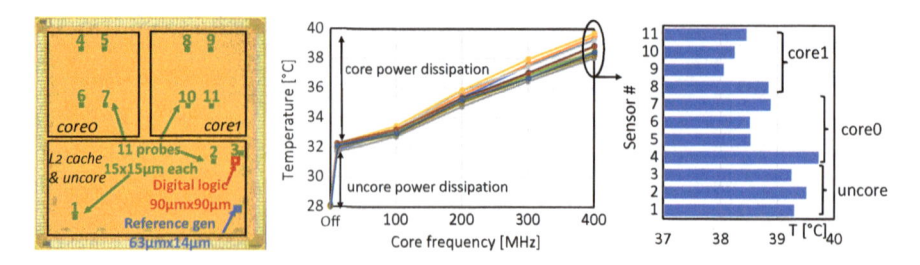

Fig. 10.11 Left: integration of the probes in a full SoC. Middle measured temperature for different activities. Right: temperature gradient at maximum activity across sensors. ©2018 IEEE reprinted with permission

Figure 10.11 presents the corresponding layout and measurement of self-heating. When the core0 runs a high-power computation benchmark, both a global activity-dependent heating of up to 10–12 °C is detected, as well as a local gradient between cores.

10.4.2.3 State-of-the-Art Summary

The performance of the proposed body-bias compensated as well as state-of-the art, as of 2018, temperature sensors is presented in Table 10.3, as well as in the Fig. 10.12 showing accuracy-area and power-resolution scatter plots.

The proposed approach results in an area of $11,482\,\mu m^2$, for the full system, i.e., $1044\,\mu m^2$ when normalized per probe, while keeping an accuracy on par with SoC temperature monitoring requirements and other CMOS sensors performance.

10.5 Conclusion

This chapter has illustrated an application of FD-SOI body-biasing versatility for local process compensation of temperature sensing oscillators. The use of an NOMS-only oscillator design enables a simple compensation scheme where the single polarity well bias can be generated by the same circuit as the regulated supply.

This approach proves more accurate than relying on calibration by normalization only, and has less testing overhead than a two-point calibration scheme. A custom piecewise-linear approximation of the temperature-delay function is implemented with low digital area overhead. It results in an extremely compact ($1.0\,k\mu m^2$/probe) sensor able to accommodate a wide operating range (0.62–1.2 V) and satisfying power (2.0 nJ/Sa) and accuracy.

Table 10.3 Comparison of the proposed architecture with state-of-the-art temperature sensors

	This work	[19]	[1]	[20]	[9]	[15]	[16]	[14]	[7]
Method	Single CMOS	Single CMOS	Diff CMOS	Single CMOS	Diff. CMOS	Diff CMOS	TD	Res	BJT
Technology	28 nm FD-SOI	130 nm	65 nm	180 nm	65 nm	40 nm	40 nm	65 nm	28 nm
Calibration	1 pt (+2pt on 1 die)	1 pt	2 pt	2 pt	1 pt	2 pt + batch	0pt/1 pt	2 pt	0pt/1 pt
Supply	0.62–1.2 V	1.2 V	0.85–1.1 V	1.2 V	1 V	1, 0.5 V	0.9–1.2 V	0.6–1.2 V	1.1–2 V
Supply sensitivity [°C/V]	1.9	NA	34	0.13	NA	NA	2.8	0.28	<0.33
External references	LDO Iref, Clock ref.	Clock ref.	None	Clock ref.	LDO ref., Clock ref.	None	None	Clock ref.	None
Temp range	−5 to 85 °C	0 to 100 °C	0 to 100 °C	−20 to 100 °C	0 to 110 °C	−40 to 100 °C	−40 to 125 °C	−45 to 85 °C	−20 to 130 °C
Systematic correction	No / Yes	No	Yes	Yes (quadratic)	No	No	No	Off-chip linearization	No
Accuracy [°C]	−2.5/+1.2 / −1.4/+1.3	−1.8/+2.3	−0.9/+0.9	−0.22/+0.19	−1.5/+1.5	−0.95/+0.97	±1.4/±0.75	−1.6/+1	±1.8/±0.8
Resolution [°C rms][a]	0.76	NA	0.3	0.07	0.94	0.12	0.36[d]	0.12	0.59
Conversion rate	27.8 kS/s[b]	5 kS/s	45 kS/s	0.125 kS/s	496 kS/s	50 kS/s	1 kS/s	100 kS/s	31.2 kS/s
Power	56 µW (0.9 V)[b]	1200 µW	154 µW[c]	0.075 µW[c]	500 µW	241 µW	2500 µW	47.2 µW	28.8 µW[e]
Energy	2.0 nJ/S[b]	240 nJ/S	3.4 nJ/S[c]	0.6 nJ/S[c]	1.0 nJ/S	4 nJ/S	2500 nJ/S	0.47 nJ/S	0.9 nJ/S[e]
Area [µm²]	1.0 k[b]	120 k	8.2 k	8.9 k[c]	8.2 k	58 k	1.65 k	44 k	3.8 k

[a] The resolution corresponds to the worst of the quantization step and of the rms repeatability
[b] Normalized per probe. The aggregated power, conversion rate, and area over the 11 probes are 616 µW, 305.8 kS/s, and 11.482 µm², respectively
[c] Does not include linear correction. Excluding digital processing logic, this work uses 0.1 nJ/S per probe and 307 µm² per probe
[d] Another sensor geometry achieves 0.24 [°C rms] resolution with ±2.3/±1.05 °C (0/1 pt trim) accuracy
[e] Power/energy at 1.8 V

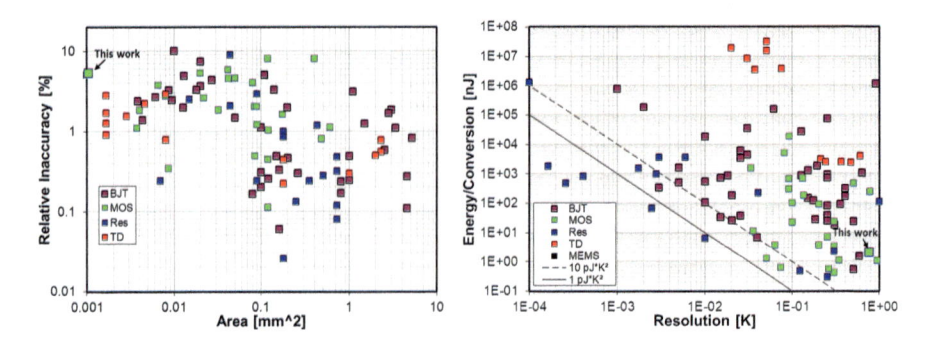

Fig. 10.12 Comparison of published sensors performance, adapted from [12]

References

1. T. Anand, K.A.A. Makinwa, P.K. Hanumolu, A VCO based highly digital temperature sensor with 0.034 C/mV supply sensitivity. IEEE J. Solid State Circuits **51**(11), 2651–2663 (2016)
2. M. Bazes, Two novel fully complementary self-biased CMOS differential amplifiers. IEEE J. Solid State Circuits **26**(2), 165–168 (1991)
3. A. Brokaw, A simple three-terminal IC bandgap reference. IEEE J. Solid State Circuits **9**(6), 388–393 (1974)
4. P. Brokaw, How to make a bandgap voltage reference in one easy lesson. Integrated Device Technology, Technical Report 2011 [Online]. Available: https://www.idt.com/document/whp/how-make-bandgap-voltage-reference-one-easy-lesson-paul-brokaw
5. M. Cochet, Temperature Sensor in Energy Efficiency Optimization in 28 nm FD-SOI: Circuit Design for Adaptive Clocking and Power-Temperature Aware Digital SoCs (Doctoral Dissertation) (2016), pp. 105–130
6. M. Cochet et al., A 225 μm^2 probe single-point calibration digital temperature sensor using body-bias adjustment in 28 nm FD-SOI CMOS. IEEE Solid-State Circuits Lett. **1**(1), 14–17 (2018)
7. M. Eberlein, I. Yahav, A 28 nm CMOS ultra-compact thermal sensor in current-mode technique, in *2016 IEEE Symposium on VLSI Circuits (VLSI-Circuits)*, Honolulu (2016), pp. 1–2
8. D. Hilbiber, A new semiconductor voltage standard, in *IEEE International Solid-State Circuits Conference*. Digest of Technical Papers. 1964, vol. VII (1964), pp. 32–33
9. S. Hwang, J. Koo, K. Kim, H. Lee, C. Kim, A 0.008mm^2 500 /spl mu/W 469 kS/s frequency-to-digital converter based CMOS temperature sensor with process variation compensation. IEEE Trans. Circuits Syst. Regul. Pap. **60**(9), 2241–2248 (2013)
10. S. Kim, M. Seok, A sub-50 μm^2, voltage-scalable, digital-standard-cell-compatible thermal sensor frontend for on-chip thermal monitoring. J. Low Power Electron. Appl. **8**(2), 16 (2018)
11. A. Mahfuzul Islam, J. Shiomi, T. Ishihara, H. Onodera, Wide-supply-range all-digital leakage variation sensor for on-chip process and temperature monitoring. IEEE J. Solid State Circuits **50**(11), 2475–2490 (2015)
12. K.A.A. Makinwa, Smart Temperature Sensor Survey, [Online]. Available: http://ei.ewi.tudelft.nl/docs/TSensor_survey.xls. Accessed 2 April 2018
13. S. Pan, Y. Luo, S.H. Shalmany, K.A.A. Makinwa, 9.1 A resistor-based temperature sensor with a 0.13pJ·K2resolution FOM, in *2017 IEEE International Solid-State Circuits Conference (ISSCC)*, San Francisco (2017), pp. 158–159
14. H. Park, J. Kim, A 0.8-V resistor-based temperature sensor in 65-nm CMOS with supply sensitivity of 0.28 °C/V, IEEE J. Solid State Circuits **53**(3), 906–912 (2018)

15. M. Saligane, M. Khayatzadeh, Y. Zhang, S. Jeong, D. Blaauw, D. Sylvester, All-digital SoC thermal sensor using on-chip high order temperature curvature correction, in *2015 IEEE Custom Integrated Circuits Conference (CICC)*, San Jose (2015), pp. 1–4
16. U. Sönmez, F. Sebastiano, K.A.A. Makinwa, 11.4 1650 μm^2 thermal-diffusivity sensors with inaccuracies down to ±0.75 °C in 40nm CMOS, in *2016 IEEE International Solid-State Circuits Conference (ISSCC)*, San Francisco (2016), pp. 206–207
17. C.-C. Teng, Y.-K. Cheng, E. Rosenbaum, S.-M. Kang, iTEM: a temperature-dependent electromigration reliability diagnosis tool. IEEE Trans. Comput. Aided Des. Integr. Circuits Syst. **16**(8), 882–893 (1997)
18. C. van Vroonhoven, K. Makinwa, A CMOS temperature-to-digital converter with an inaccuracy of +/−0.5 c (3/spl sigma) from −55 to 125c, in *IEEE International Solid-State Circuits Conference, 2008*. Digest of Technical Papers (2008), pp. 576–637
19. K. Woo, S. Meninger, T. Xanthopoulos, E. Crain, D. Ha, D. Ham, Dual-DLL based CMOS all-digital temperature sensor for microprocessor thermal monitoring, in *IEEE International Solid-State Circuits Conference – Digest of Technical Papers, 2009* (2009), pp. 68–69, 69a
20. K. Yang et al., 9.2 A 0.6nJ −0.22/ + 0.19°C inaccuracy temperature sensor using exponential subthreshold oscillation dependence, in *2017 IEEE International Solid-State Circuits Conference (ISSCC)*, San Francisco (2017), pp. 160–161
21. C. Zhang, K. Makinwa, The effect of substrate doping on the behaviour of a CMOS electrothermal frequency-locked-loop, in *TRANSDUCERS 2007 – International Solid-State Sensors, Actuators and Microsystems Conference, 2007*. (2007), pp. 2283–2286

Chapter 11
System Integration of RISC-V Processors with FD-SOI

Ben Keller, Borivoje Nikolić, Brian Zimmer, Martin Cochet, Yunsup Lee, Jaehwa Kwak, Alberto Puggelli, Milovan Blagojević, Ruzica Jevtić, Pi-Feng Chiu, Stevo Bailey, Palmer Dabbelt, Colin Schmidt, Hanh-Phuc Le, Po-Hung Chen, Nicholas Sutardja, Rimas Avizienis, Andrew Waterman, James Dunn, Brian Richards, Philippe Flatresse, Andrei Vladimirescu, Andreia Cathelin, Elad Alon, and Krste Asanović

11.1 SoC Design in FD-SOI

FD-SOI technology presents unique opportunities for system integration to achieve resilient and energy-efficient operation of digital systems. The use of the free, open RISC-V instruction set architecture (ISA) for processor design allows custom experimentation and tight integration of power management features. By integrating voltage regulators, clock generators, resilient SRAMs, and body-bias generators into a single die, the system can respond rapidly to variation and implement closed-loop feedback algorithms for energy savings at high performance. The Raven-3

B. Keller (✉) · B. Zimmer
NVIDIA Research, Santa Clara, CA, USA
e-mail: benk@nvidia.com

B. Nikolić · C. Schmidt · J. Dunn · B. Richards · E. Alon · K. Asanović
Department of Electrical Engineering and Computer Sciences, University of California, Berkeley, CA, USA
e-mail: bora@eecs.berkeley.edu

M. Cochet
STMicroelectronics, Crolles, France

Department of Electrical Engineering and Computer Sciences, University of California, Berkeley, CA, USA

Aix-Marseille University & CNRS, IM2NP (UMR 7334), Marseille, France

IBM Research, Yorktown Heights, NY, USA

Y. Lee · A. Waterman
SiFive, San Mateo, CA, USA

J. Kwak
LG Electronics, Santa Clara, CA, USA

A. Puggelli
Apple Inc., Cupertino, CA, USA

© Springer Nature Switzerland AG 2020
S. Clerc et al. (eds.), *The Fourth Terminal*, Integrated Circuits and Systems, https://doi.org/10.1007/978-3-030-39496-7_11

and Raven-4 testchips were developed to explore these concepts and prove the practicality of efficient system integration in FD-SOI [1].

11.1.1 Raven-3

The Raven-3 testchip features energy-efficient integrated voltage regulation, adaptive clocking, resilient SRAMs, and a fully featured application processor [2]. A block diagram of the system is shown in Fig. 11.1.

The application core in Raven-3 is Rocket Chip, a 64-bit in-order energy-efficient processor. The Hwacha vector accelerator can accelerate common computational kernels, achieving energy-efficient dataflow via its systolic operation and tight coupling with the Rocket core. The SRAMs in Rocket's instruction and data caches, as well as the Hwacha register file, are implemented as custom-designed, highly resilient macros that can operate at low voltages. Together, these components comprise the core digital logic of the system.

The core voltage is supplied by an integrated switched-capacitor (SC) DC-DC converter. 1.0 and 1.8 V supplies are downconverted to rippling output voltages

---

M. Blagojević
Infineon Technologies, Neubiberg, Germany

R. Jevtić
University CEU-San Pablo, Madrid, Spain

P.-F. Chiu
Western Digital, San Jose, CA, USA

S. Bailey
Mojo Vision, Saratoga, CA, USA

P. Dabbelt
Google, Mountain View, CA, USA

H.-P. Le
Department of Electrical and Computer Engineering, University of California, San Diego, La Jolla, CA, USA

P.-H. Chen
National Chiao Tung University, Hsinchu, Taiwan

N. Sutardja
Danger Devices, Berkeley, CA, USA

R. Avizienis
Meyer Sound, Berkeley, CA, USA

P. Flatresse
Dolphin Design, Isére, France

A. Vladimirescu
ISEP, Paris, France

A. Cathelin
STMicroelectronics, Crolles, France

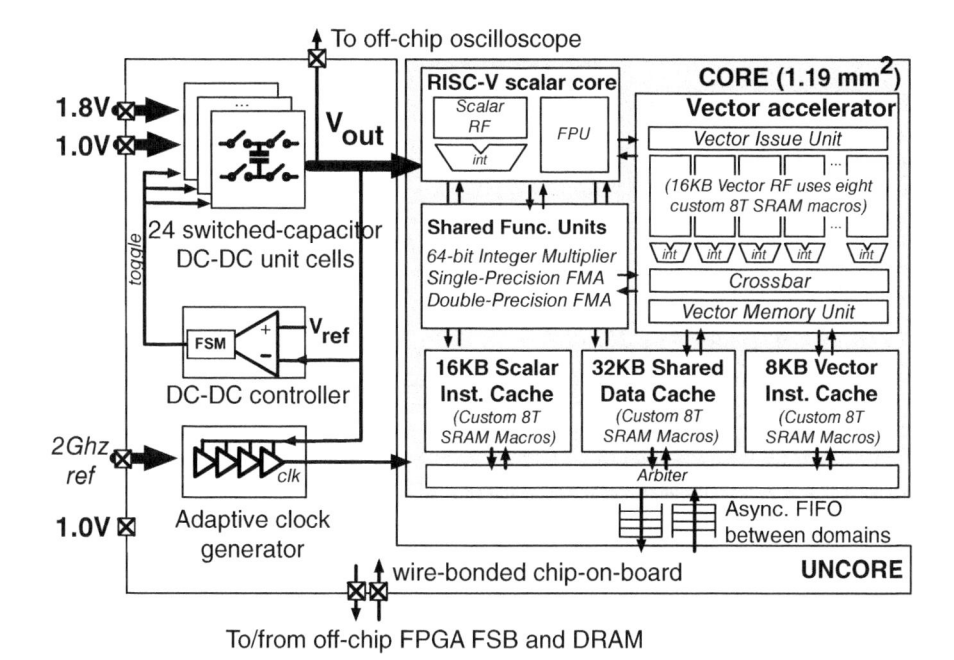

Fig. 11.1 A block diagram of the Raven-3 testchip [2] (© 2016 IEEE, reprinted with permission)

averaging 0.9, 0.67, and 0.5 V depending on the reconfigurable converter topology. The 1.0 V supply can also be passed directly to the core in a bypass mode. The converter is subdivided into 24 90 × 90 μm unit cells that are located near the core voltage area.

The core clock is generated by an adaptive clock generator that selects clock edges from a 16-phase DLL output as determined by the delay through a replica critical path supplied by the core voltage. The replica path is made up only of inverters, with the depth of the path programmable via selectable muxes. The generated clock is supplied to all register and SRAM sinks in the core voltage domain.

The IP blocks and their control registers, IO cells, and interface logic comprise the uncore voltage domain, which operates at a fixed 1 V and fixed frequency. Because the uncore and core may be asynchronous depending on the operating mode, bisynchronous FIFOs guard all communication between them (see Fig. 11.2). Level shifters are inserted on signals crossing the domains to ensure correct operation as the core voltage varies.

A multivoltage and multiclock design flow was used to construct the processor. Figure 11.3 shows the processor floorplan, with the larger core voltage domain in red separated from the smaller uncore voltage domain to the right of the chip. The custom SRAMs were manually placed within the core voltage domain. The SC unit cells surround the core to minimize voltage drop. Two layers of thick upper-layer metal were dedicated to a power grid, where the core voltage and ground each utilize

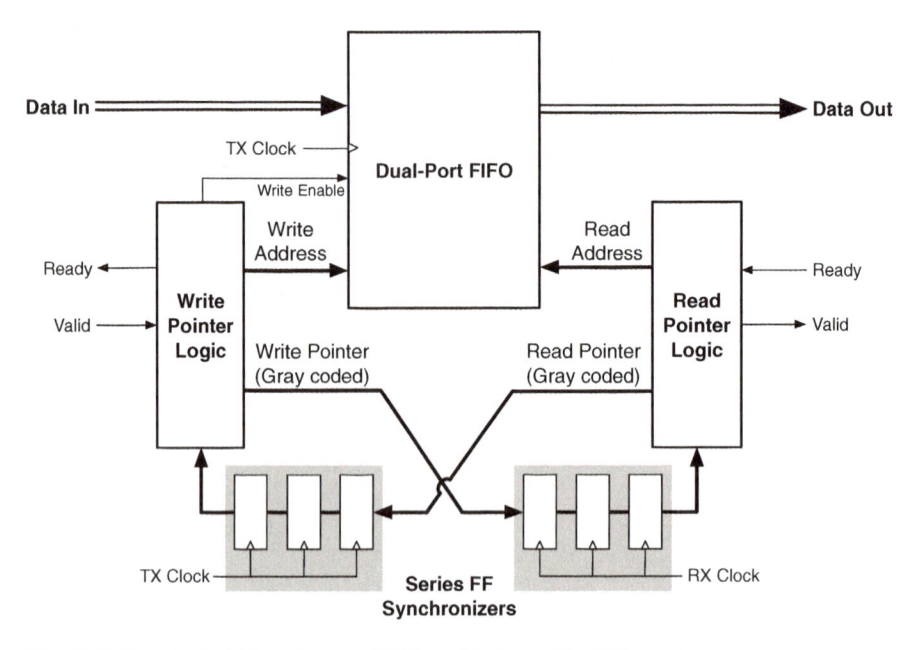

Fig. 11.2 The standard bisynchronous FIFO used in Raven-3 and Raven-4

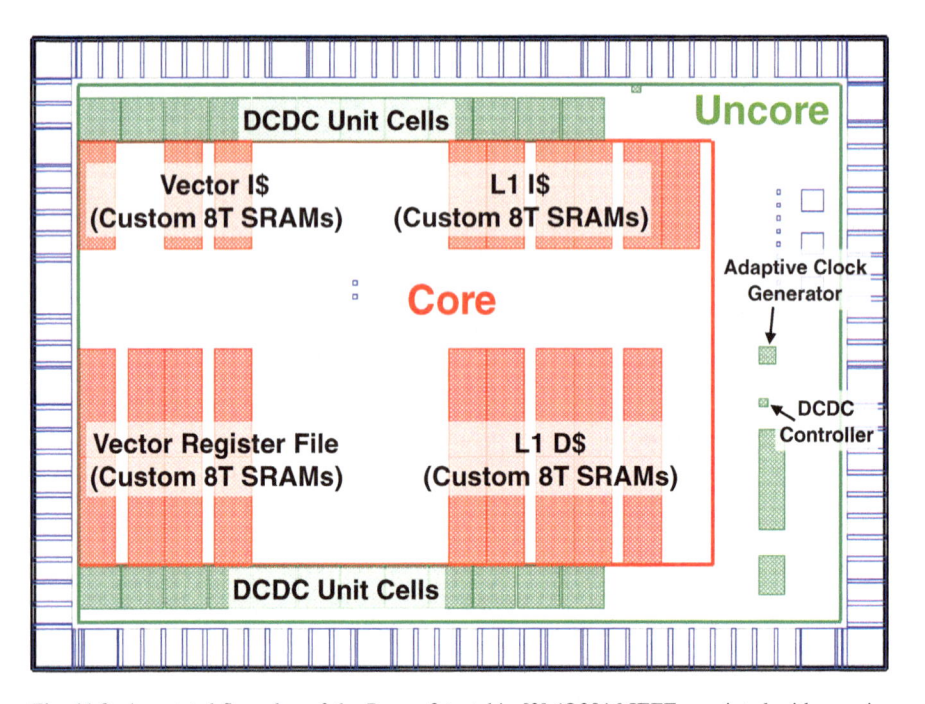

Fig. 11.3 Annotated floorplan of the Raven-3 testchip [2] (©2016 IEEE, reprinted with permission)

11 System Integration of RISC-V Processors with FD-SOI 267

Fig. 11.4 Annotated die micrograph of the Raven-3 testchip [2]

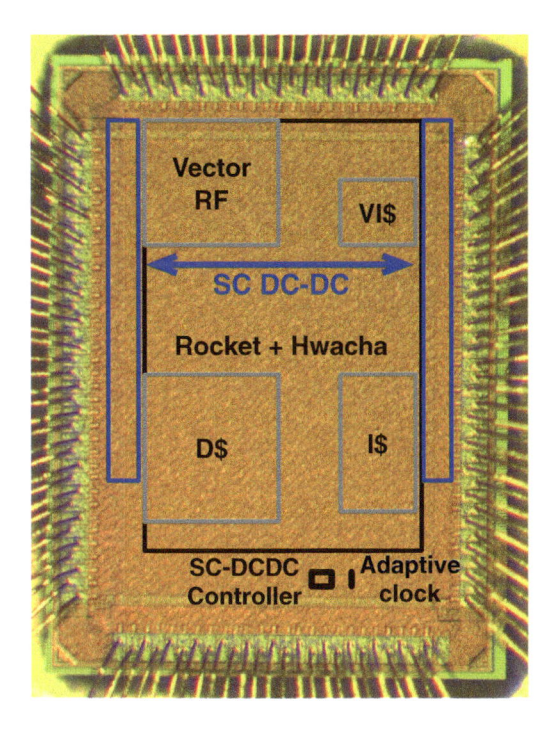

25% of the chip area in each layer. Outside the core, these core voltage rails are not necessary, so the input voltages to the converters use the majority of the power routing resources to connect power coming from the pad frame to the converters. An annotated die photo of the chip, which was fabricated in 28 nm ultra-thin body and BOX fully depleted silicon-on-insulator (UTTB FD-SOI) technology [3], is shown in Fig. 11.4.

11.1.2 Raven-4

The Raven-4 testchip builds upon the Raven-3 design with design changes and additional features [4]. The integrated voltage regulation and adaptive clocking systems are improved and supplemented with an integrated power management unit allowing fast adaptive voltage scaling (AVS) algorithms to be demonstrated on die. An integrated body-bias generator, power monitor, and measurement circuits enable more extensive power management and system characterization. A block diagram of the testchip is shown in Fig. 11.5.

The core voltage domain contains an updated version of the Rocket core implemented in Raven-3. The Hwacha vector accelerator is improved with the inclusion of additional math resources, doubling the peak rate of computation. Additionally, separate long-latency functional units are dedicated to the coprocessor

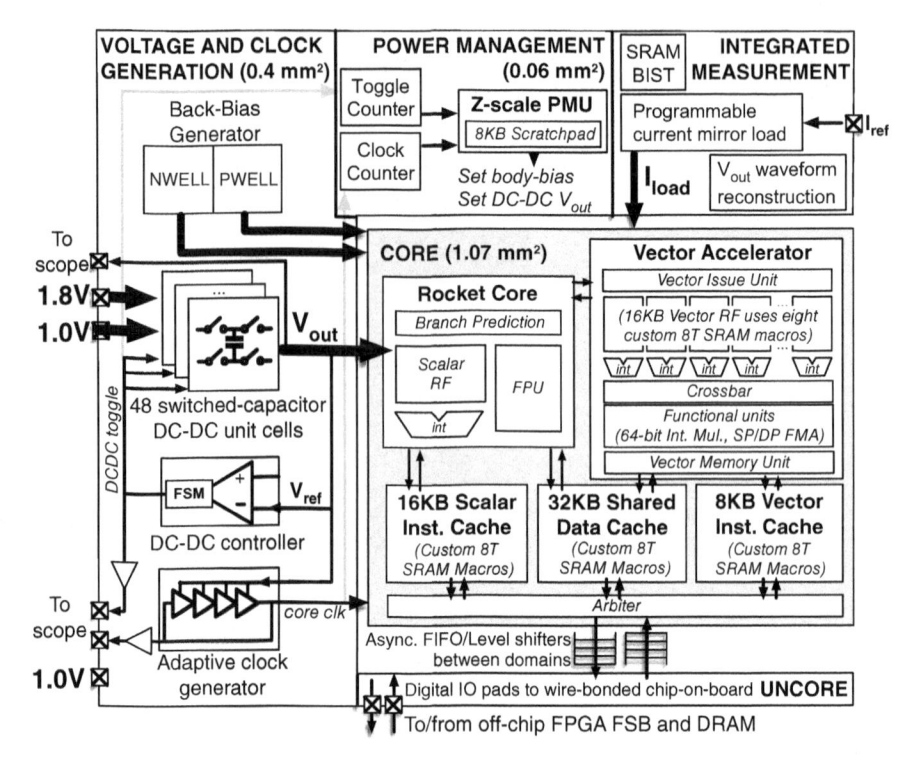

Fig. 11.5 A block diagram of the Raven-4 testchip [5] (©2016 IEEE, reprinted with permission)

rather than being shared with the Rocket core. The core SRAMs are implemented using the same custom 8T macros to enable low-voltage operation. Level shifters and bisynchronous FIFOs guard the crossings to the uncore clock domain, which contains the PMU, control registers, IP, and IO logic.

The integrated voltage regulator and clocking systems that supply the core are also improved from Raven-3. The flying capacitance of the SC regulators is doubled to 48 $90 \times 90\,\mu m$ unit cells, enabling improvements in conversion efficiency. Decoupling capacitance was added to improve the integrity of the 1.0 and 1.8 V inputs to the SC unit cells and help offset power delivery issues caused by the wirebond packaging of the chip. An improved, free-running clock generator is implemented alongside the version employed in Raven-3 for direct measurement comparison. Unlike the Raven-3 design, the replica timing circuit is comprised of multiple types of standard cells, each with an independently selectable depth.

A second, smaller processor serves as the power management unit (PMU) for the design. The PMU processor is based on Z-scale, a tiny, three-stage RISC-V implementation that is nonetheless fully programmable via the RISC-V software toolchain. The application core and the PMU can communicate directly via interprocessor interrupts, and each has a register mapped directly into the control and status register (CSR) space of the other system, allowing arbitrary data to be

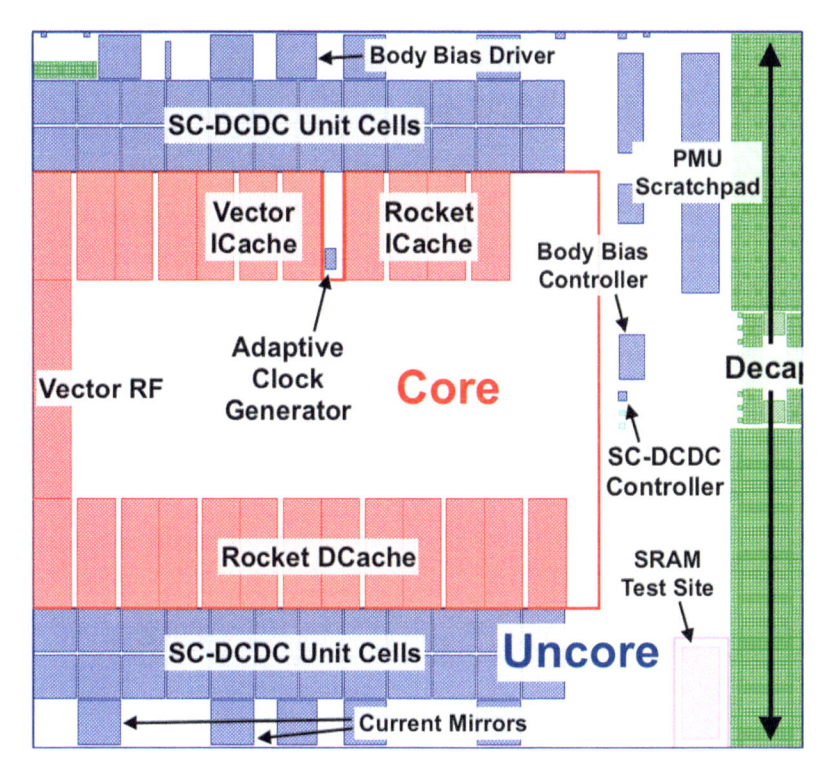

Fig. 11.6 Annotated floorplan of the Raven-4 testchip [4] (©2017 IEEE, reprinted with permission)

communicated between the two cores. The processor maps the control registers for the SC converters, adaptive clock generator, and other IP into its CSR space, which enables programs to directly manipulate the voltage and frequency of the chip. In addition, the core clock and the SC toggle clock are read by counters, and the counter values are synchronized into the uncore domain. Successive reads to this second counter enable a rapid estimate of core power consumption for active power management.

Additional circuits extend the power management and measurement capabilities of the SoC. The threshold voltage of the logic in the core voltage domain can be manipulated by an integrated body-bias generator. Fine tuning resolution and fast response allow adaptive body-bias to be incorporated into power management techniques. The core clock and SC toggle clock are also connected to output drivers for direct observation.

Figure 11.6 shows the floorplan of the SoC. The design is partitioned into two voltage areas, with the core voltage area supplied by the SC converters placed centrally. Numerous additional voltages and clocks are defined to supply both the core and the various analog and mixed-signal blocks that make up the SoC. To reduce core insertion delay, the clock generator itself was placed near the center of

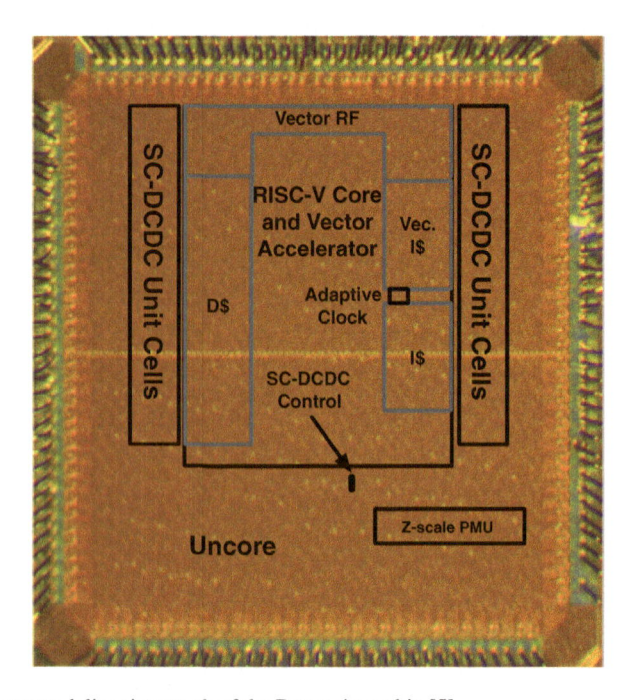

Fig. 11.7 Annotated die micrograph of the Raven-4 testchip [5]

the core area, and a "peninsula" of the uncore voltage domain was extended to allow the routing of control signals from the block to the top level of the design hierarchy. The location of the core clock multiplexer, which allowed the selection of different core clock sources for test, was specified explicitly and placed near the center of the core area. These improvements combined to reduce core insertion delay by several hundred picoseconds. An annotated die photo of the testchip, which was fabricated in 28 nm UTBB FD-SOI, is shown in Fig. 11.7.

11.2 RISC-V Processors

The RISC-V instruction set was developed as a free and open architecture that defines simple base instructions and optional extensions [6]. Without licensing requirements, the ISA encourages experimentation and allows for the flexibility needed to design tightly integrated FD-SOI systems. Raven-3 and Raven-4 implement the Rocket scalar processor and the Hwacha vector accelerator. In addition, Raven-4 implements a small Z-scale processor for power management.

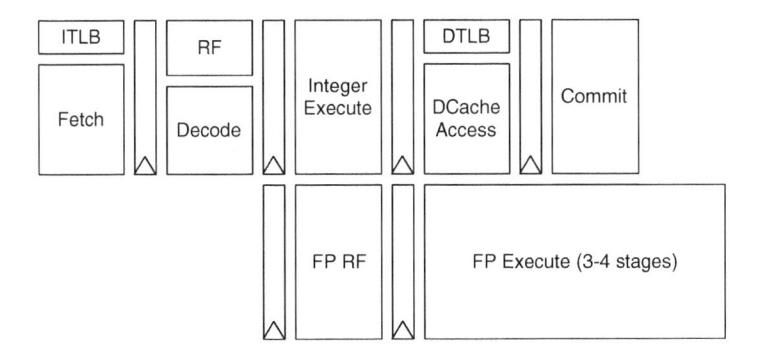

Fig. 11.8 A simplified pipeline diagram of the Rocket processor [7]

11.2.1 Rocket Chip

Rocket Chip is a freely available in-order RISC-V implementation. The processor, an instance of the Rocket Chip Generator [7], is a 64-bit five-stage single-issue pipeline (see Fig. 11.8). The pipeline is carefully designed to minimize the impact of long clock-to-output delays of SRAM macros. For example, the pipeline resolves branches in the memory stage to shorten the critical path through the bypass path, but relies on extensive branch prediction (a 64-entry branch target buffer, a 256-entry two-level branch history table, and a two-entry return address stack) to mitigate the increased branch resolution penalty. The blocking 16 KiB instruction cache is private to the Rocket core, while the nonblocking 32 KiB data cache is shared between the scalar core and a vector coprocessor designed to accelerate data-parallel workloads. The Rocket core has a memory-management unit that supports page-based virtual memory. Both caches are virtually indexed and physically tagged, and have separate TLBs that are accessed in parallel with cache accesses. The core has an IEEE 754-2008-compliant floating-point unit that executes single- and double-precision floating-point operations, including fused multiply add (FMA) operations, with hardware support for subnormal numbers.

11.2.2 Hwacha Vector Processor

The Hwacha vector accelerator, shown in Fig. 11.9, is a decoupled single-lane vector unit tightly coupled with the Rocket core that implements a custom RISC-V vector extension [8]. Hwacha executes vector operations temporally (split across subsequent cycles) rather than spatially (split across parallel datapaths), and has a vector length register that simplifies vector code generation and keeps the binary code compatible across different vector microarchitectures with different numbers of execution resources. The Rocket scalar core sends vector memory instructions and

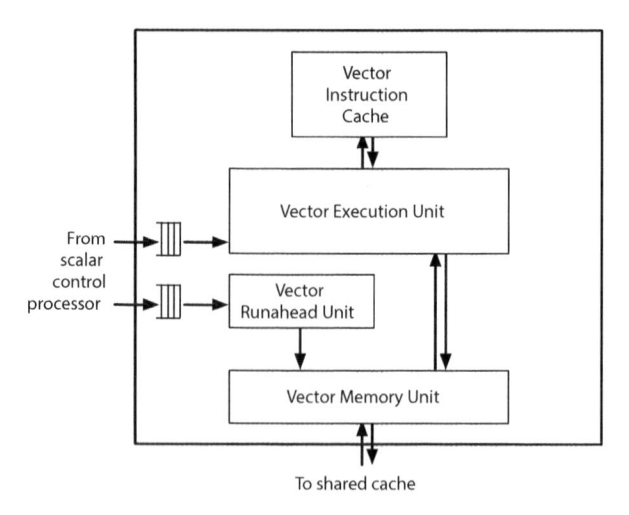

Fig. 11.9 A block diagram of the Hwacha vector accelerator [2]

vector fetch instructions to the vector accelerator. A vector fetch instruction initiates execution of a block of vector arithmetic instructions. The vector execution unit (VXU) fetches instructions from the private 8 KiB vector instruction cache (VI$), decodes instructions, clears hazards, and sequences vector instruction execution by sending multiple micro-ops down the vector lane. The vector lane consists of a banked vector register file built out of two-ported SRAM macros, operand registers, per-bank integer ALUs, and long-latency functional units. Multiple operands per cycle are read from the banked register file by exploiting the regular access pattern with operand registers used as temporary space [8]. The vector memory unit (VMU) supports unit-strided, constant-strided, and gather/scatter vector memory operations to the shared L1 data cache. Vector memory instructions are also sent to the vector runahead unit (VRU) by the scalar core. The VRU prefetches data blocks from memory and places them in the L1 data cache ahead of time to increase performance of vector memory operations executed by the VXU. The resulting vector accelerator is more similar to traditional Cray-style vector pipelines [9] than SIMD units such as those that execute ARM's NEON or Intel's SSE/AVX instruction sets, and delivers high performance and energy efficiency while remaining area efficient.

11.2.3 Z-Scale

Z-scale is a 32-bit 3-stage single-issue in-order RISC-V processor (see Fig. 11.10) that is suitable as a power management unit (PMU) due to its simplicity and small footprint. The core forgoes caches in favor of an 8 KiB 128-bit-wide scratchpad memory. The PMU supports the RV32IM instruction extensions and is fully programmable via the RISC-V software toolchain. It is designed to reside in a

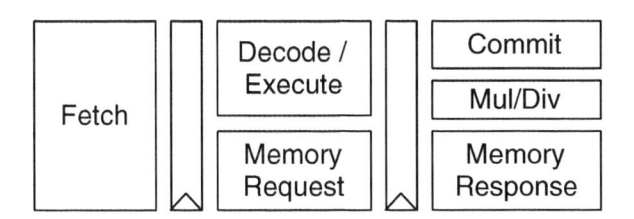

11 System Integration of RISC-V Processors with FD-SOI 273

Fig. 11.10 The Z-scale processor pipeline [4]

Table 11.1 Comparison of the two Raven-4 processors [4]

Feature	Rocket (excluding Hwacha)	Z-scale
Instruction set	RISC-V RV64G	RISC-V RV32IM
Pipeline	Five-stage single-issue in-order	Three-stage single-issue in-order
Memories	32 KB DCache, 16 KB ICache	8 KB unified scratchpad
Standard cell count	196K	18K
Area (Cells + Memories)	0.461 mm^2	0.053 mm^2

fixed-voltage "always-on" domain for use in controlling the operating states of the remaining domains in the system.

The three-stage design minimizes gate count while enabling sufficient performance to enable fine-grained power management. The first stage of the pipeline fetches an instruction out of a 128-bit line buffer that reduces read-port contention on the single-ported scratchpad by storing four consecutive RISC-V instructions (the value of the program counter is calculated in the previous cycle). The second stage of the pipeline decodes a RISC-V instruction, reads the register file, and executes an ALU instruction. Branches are resolved in the second stage, so the instruction in the fetch stage is flushed when a branch is taken. Writeback is isolated into the third stage of the pipeline, reducing the fanout delay on the write port of the register file. The result in the third stage is bypassed to the second stage, eliminating the need for stalls in some cases. The memory stage and the multiplication/division pipeline stages are also in the third pipeline stage, although only one of these three subsystems will be active at a time, as their use triggers a stall in the second stage until the result of the instruction is written back to the register file. The multiply and divide units minimize hardware resources such that only one bit of the operation is completed each cycle, and so 32 cycles are required to compute any multiplication or division result. Loads and stores directly access the scratchpad; since the scratchpad is 128 bits wide, the pipeline must properly swizzle the load data and store data. An extra pipeline register was added in the arbiter between instruction and data memory requests to eliminate a long critical path and speed the achievable cycle time of the design. The entire design (excluding the scratchpad memories) uses just 18K gates, making the implementation overhead small relative to large application processors, accelerators, and caches. A comparison between the Rocket and Z-scale implementations in Raven-4 is shown in Table 11.1.

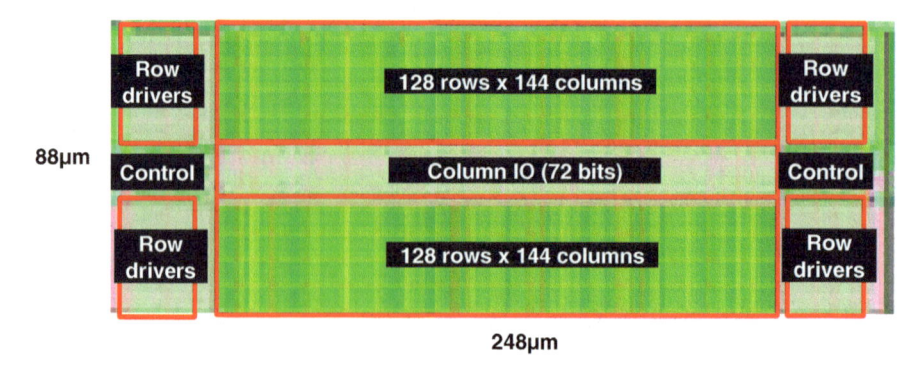

Fig. 11.11 Layout of the custom 8T SRAM macro [4] (©2017 IEEE, reprinted with permission)

11.3 Energy-Efficient SRAMs

Operation at low voltages is critical to achieving maximum energy efficiency, but on-chip SRAM typically limits the minimum operating voltage of the entire system as the small transistors in SRAM bitcells are especially vulnerable to process variation. Raven-3 and Raven-4 implement custom 2 KB 8T-based SRAM macros as shown in Fig. 11.11. It is logically organized as 512 entries of 72 bits (64 bits + 8 possible ECC bits) and physically organized as two arrays of 128 rows by 144 columns with two-to-one physical interleaving. Low-voltage operation is enabled by the 8T bitcell, where each transistor is larger than the equivalent high-density 6T bitcell, and by the FD-SOI process which reduces the threshold voltage variation [10, 11]. While the arrays also implemented a negative bitline write assist, the assist was not necessary to achieve minimum voltage operation.

11.4 DC-DC Converters

The implementation of AVS to improve system energy efficiency requires the delivery of multiple different voltages to different areas of the chip, and the particular voltages used must be dynamically reconfigurable. Typically, one or two input voltages will be downconverted by one or more voltage regulators to some number of lower voltages for use by the different voltage areas of the SoC. By building regulators into the same die as the SoC they are supplying, system costs are reduced, transition times are shortened due to the smaller passives, and many voltages can be generated from just a few external supplies, ameliorating any issues with IO count. However, integrated regulators must achieve high conversion efficiency while minimizing area overhead.

Switched-capacitor voltage regulators are a favorable choice for integrated voltage regulation because high-quality integrated capacitors are much easier to

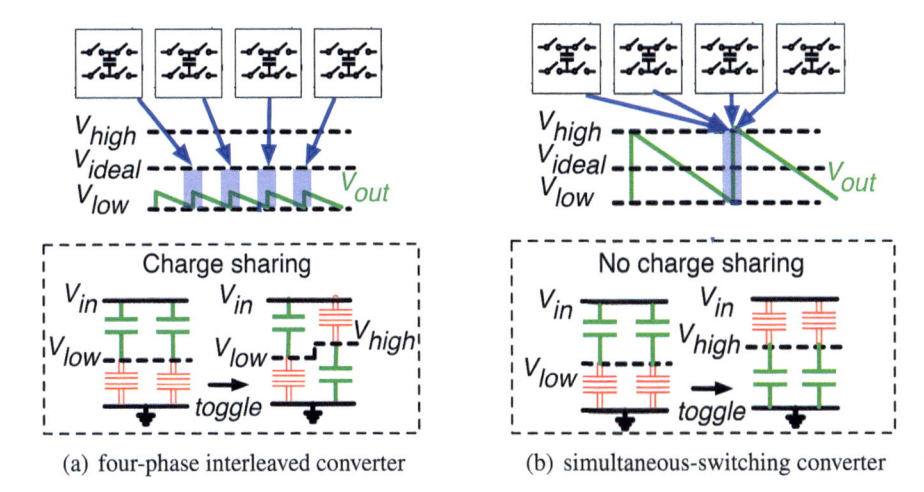

Fig. 11.12 Interleaved and simultaneous-switching switched-capacitor DC-DC converters [2]. In the four-phase interleaved converter shown in (**a**), each unit cell switches out of phase with the others, resulting in a relatively flat output voltage but incurring charge-sharing losses. In the simultaneous-switching converter shown in (**b**), all unit cells switch at once, eliminating charge-sharing losses but resulting in a large output voltage ripple

implement than inductors, while switching regulator efficiencies exceed those of linear regulators. However, switched-capacitor regulators typically achieve conversion efficiencies that are not high enough to justify their overhead. Designing a switched-capacitor regulator with reasonably high efficiency is a critical component of a realistic FG-AVS system.

Because each switching event in a switched-capacitor circuit is perturbing its output voltage with a discontinuous addition of charge, the output of switched-capacitor regulators tends to ripple. Traditional switched-capacitor designs employ a technique known as interleaving to suppress this voltage ripple and produce a relatively flat output supply that is well-suited for synchronous digital logic [12–17]. By dividing the total flying capacitance in the system into many smaller unit cells, and switching each of these unit cells out of phase with the others, the relative amount of charge delivered onto the supply with each switching event is small, and the output voltage ripple is minimized as shown in Fig. 11.12a. High interleaving phase counts of 16, 32, or even greater can be used to suppress the ripple and produce a steady output voltage [18–20]. However, this interleaved approach suffers from charge-sharing losses as each unit cell shares charge across the flying capacitance of the others when it switches. These intrinsic charge-sharing losses comprise up to 40% of the overall energy losses in the voltage conversion [21].

The conversion losses of switched-capacitor voltage regulators are composed of four parts, three of which depend on the switching frequency of the design [19]. The intrinsic charge-sharing loss P_{cfly} of switched-capacitor designs is inversely proportional to the switching frequency of the design, because a slower switching frequency means more charge being transferred for each switching event. The bottom-plate conduction loss P_{bottom} caused by the parasitic capacitance of the flying capacitor to ground and the parasitic gate capacitance P_{gate} of the switches are both directly proportional to the switching frequency. The conduction loss P_{cond} through the switches does not depend on switching frequency. Switched-capacitor designs typically operate at the optimal switching frequency to minimize the total losses in the system, operating fast enough to reduce charge-sharing losses but not so fast that the parasitic losses dominate.

An alternative approach called simultaneous switching is shown in Fig. 11.12b. Instead of interleaving the unit cells, the entire flying capacitance of the regulator is switched at once, eliminating all charge sharing between different unit cells present in the interleaved approach [2]. The elimination of this loss term also means that the only losses remaining in the system are either directly proportional to switching frequency or agnostic to it, so the switching frequency of the system can be reduced, further shrinking remaining losses. Figure 11.13 shows analytical results comparing the losses from an interleaved system with those of a simultaneous-switching regulator driving an ideal resistive load. The simultaneous-switching switched-capacitor DC-DC (SS-SC) converter is able to achieve total losses of less than 10% by slowing its switching frequency. In real systems, the load is not purely resistive, but instead has some capacitive component, so charge-sharing losses are not eliminated entirely. Nonetheless, the energy savings over the interleaved approach remain substantial, and the optimal switching frequency is lower in the SS-SC design.

Simultaneous-switching designs reduce the loss components associated with switched-capacitor regulation, but they generate an output voltage that ripples substantially around an average. This rippling supply is not well-suited to a traditional synchronous digital system operating at a fixed clock frequency, because the digital clock must operate at a slower frequency corresponding to the lowest voltage of the ripple in order to guarantee safe operation. This would result in wasted energy as the rippling voltage exceeds its minimum, as shown in Fig. 11.14a, and these over-voltage losses would exceed the savings from the elimination of charge sharing. Systems employing SS-SC regulators therefore require the clock supplied to the digital load to adapt to the changing voltage in order to achieve reasonable efficiencies [21]. As shown in Fig. 11.14b, by changing the clock on a cycle-by-cycle basis to match the instantaneous operating conditions of the load, the over-voltage losses can be eliminated. Raven-3 and Raven-4 each implement this combination of simultaneous-switching conversion and adaptive clocking.

Traditional switched-capacitor regulators achieve peak efficiencies when supplying output voltages at discrete levels determined by their topology. While SS-SC regulators do not generate a fixed output voltage, their efficiency still peaks at a particular average output level. As a wide range of voltages are desirable for FG-AVS, a reconfigurable switching network with a topology similar to that of [19] was implemented. The design can operate in four different modes that generate voltages from fixed 1 and 1.8 V supplies. The 1.8 V 1/2 mode downconverts the 1.8 V input

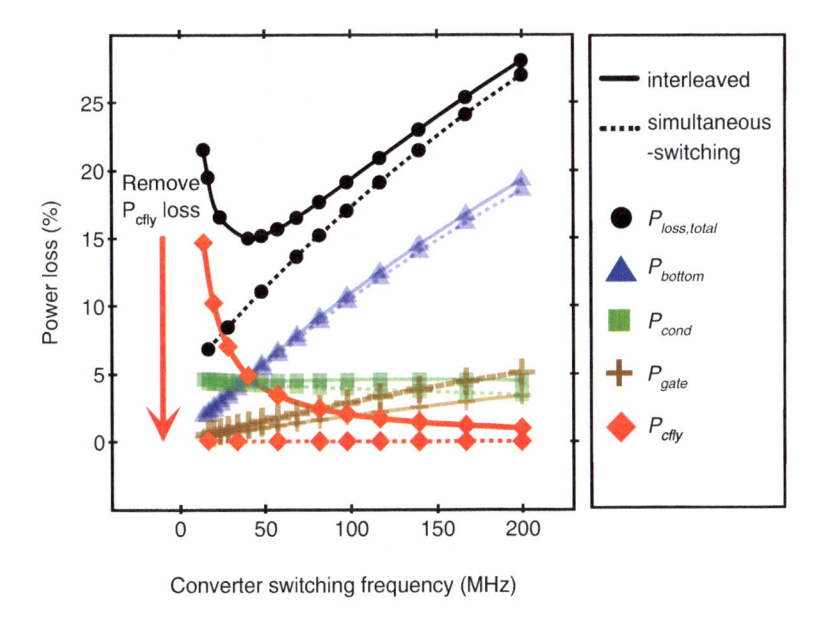

11 System Integration of RISC-V Processors with FD-SOI 277

Fig. 11.13 Analytical results showing the efficiency improvement of the simultaneous-switching switched-capacitor voltage regulator [2] (©2016 IEEE, reprinted with permission)

in a 2:1 ratio, resulting in an average output voltage around 900 mV. The 1 V 2/3 mode and 1 V 1/2 mode each downconvert the 1.0 V input, resulting in average output voltages of roughly 667 and 500 mV, respectively. A bypass mode connects the 1 V input directly to the output rail. The design re-uses the flying capacitance and switches, reconfiguring the switching pattern between each topology to avoid area overhead (see Fig. 11.15).

The phases and configurations of the switches in the SS-SC regulator are set by a finite-state-machine controller. This logic is responsible for configuring and reconfiguring the topology of the regulator, as well as switching between the two operating phases to pump charge onto the output of the converter. The regulator topology is determined by setting a control register that, when changed, triggers a reconfiguration between voltage modes. To generate the toggle clock that is distributed to the switches in the regulator, a comparator circuit acting on a 2 GHz clock detects when the regulated output voltage drops below a fixed external reference. When the comparator triggers, control logic produces the next toggle clock edge, switching the SS-SC toggle clock phase and causing more charge to be supplied, boosting the voltage. As a different reference voltage is needed for each mode, different comparators must be implemented to function in the appropriate voltage ranges (see Fig. 11.16). In the event that one switching event does not sufficiently increase the generated voltage to bring it above the reference voltage,

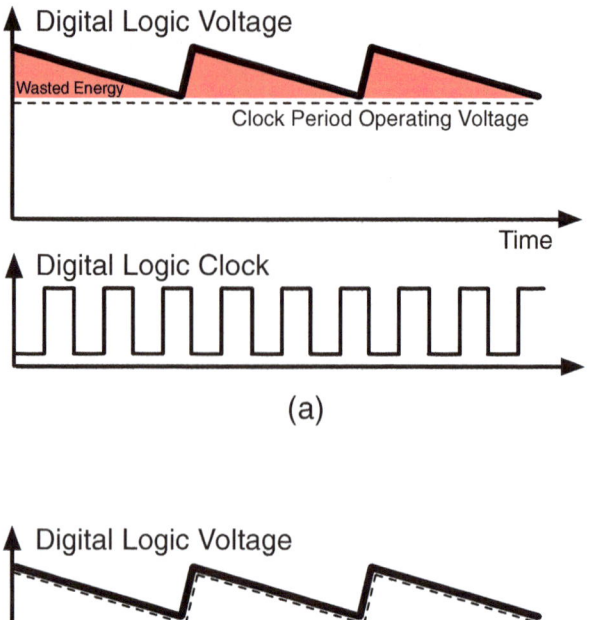

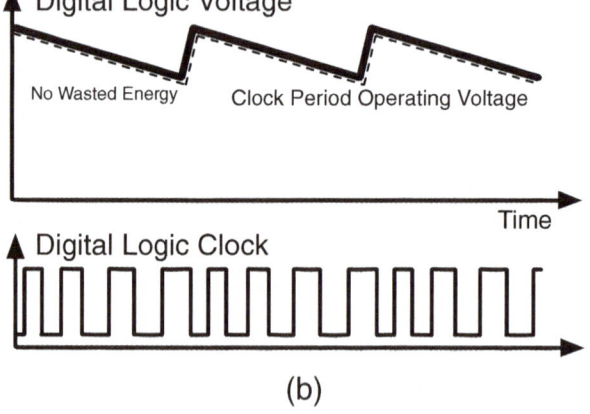

Fig. 11.14 Sample voltage and clock waveforms of SS-SC systems. In (**a**), the clock does not adapt to the voltage ripple, so it must be margined for the lowest supply voltage, resulting in wasted energy at higher voltages. In (**b**), the clock adapts to the voltage ripple, allowing the circuit to speed up when the voltage is higher than the minimum

additional logic triggers further switching events after a delay, ensuring that the voltage will eventually be boosted back up to nominal levels.

11.5 Body-Bias Generation

The ultra-thin body and box (UTBB) FD-SOI technology features a high sensitivity (>70 mV/V) of transistor threshold voltages to body-bias. Forward well bias further

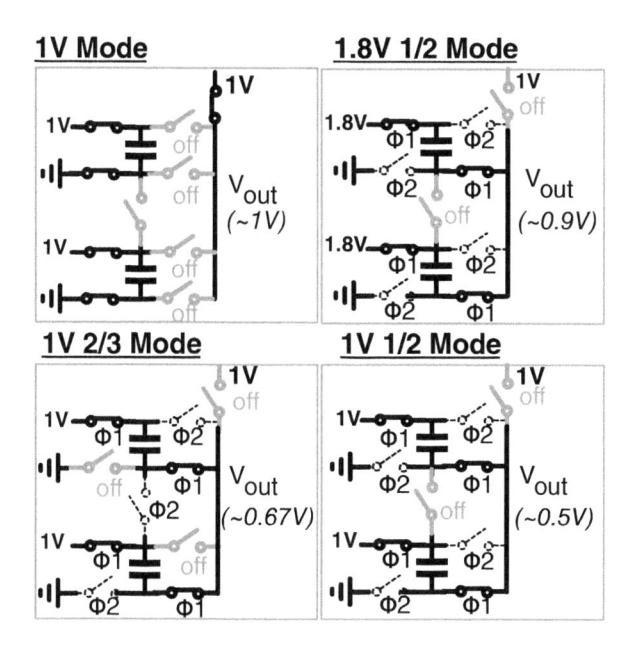

11 System Integration of RISC-V Processors with FD-SOI 279

Fig. 11.15 The four topologies of the reconfigurable SS-SC converter [2] (©2016 IEEE, reprinted with permission)

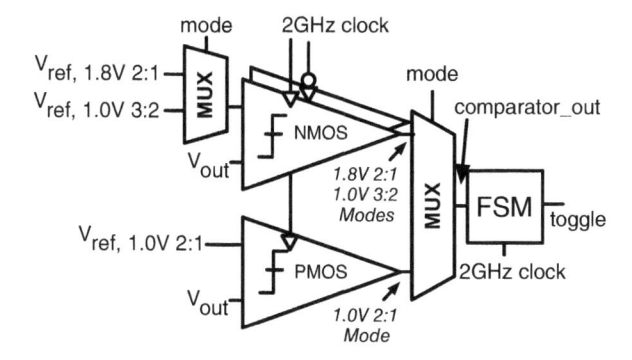

Fig. 11.16 A diagram of the comparators used in the SS-SC controller [2] (©2016 IEEE, reprinted with permission). Different comparators are needed to accommodate the voltage ranges of the references for the three different switching modes

lowers thresholds and offers the opportunity for trading off active for leakage power to improve overall energy efficiency. Optimal body-bias can be used for run-time compensation of process, aging, temperature, and supply voltage variations, and to improve energy efficiency.

In contrast to adaptive voltage scaling (AVS), enabling adaptive body-bias (ABB) requires only two additional power grid lines. The load for body-biasing is almost purely capacitive with low static current. Optimal body-bias can be used for run-time compensation of process, aging, temperature and supply voltage variations, or to improve energy efficiency [22], but a compact, low-power, fast body-bias generator (BBG) is required. Raven-4 implements a fully integrated switched-capacitor solution with a fine voltage step and sub-100 ns response time that enables fast, fine-granularity threshold control [23].

The block diagram of the body-bias generator is shown in Fig. 11.17, illustrating its two main modules, the controller and the driver unit. The controller contains a body voltage sensor and control logic. Multiple drivers are placed across the chip to provide rapid response. The controller has two main operating modes: transition or ON-mode and keep-the-value or STEADY-mode. In the ON-mode, a body-bias sensor is employed. This is a direct voltage sample-and-compare circuit clocked at 1–2 GHz, enabling sub-ns closed-loop control over the driver units. A p-well sampler employs a −1:0.8 switched-capacitor structure with 20% gain-error to compensate for the lower PMOS body factor. The outputs of the analog upper- and lower-bound rail-to-rail comparators generate mutually exclusive charge or discharge commands.

Once the voltages settle inside the bounds, the control unit switches to low-power STEADY-mode and a 0.5 MHz clock. The analog reference voltages are generated with 5-bit two-path poly-resistor DACs in the slow path and a combination of MOS and poly resistors in the fast path. This enables 2 ns settling of the new upper-bound/lower-bound values and only 400 nA of DAC static consumption in STEADY-mode. The size of upper-bound/lower-bound window is programmable with a default value of 58 mV. When new digital target values are received, or when voltages leak out of the window, the fast clock unit is engaged within a cycle and the control unit switches into ON-mode to set new body-bias voltages.

The driver units receive four 1-bit digital commands from the control unit to toggle the charging and discharging of the transistor wells. The n-well drivers are PMOS and NMOS power switches that provide 0–1.8 V range and high slew rates. Figure 11.18 shows the negative voltage generator needed for the p-well, which is based on a switched-capacitor (SC) charge pump. The driver implements a 1:1 SC charge pump with a negative bootstrap circuit for the negative gate drive signal G_{bot} [24]. The flying capacitor consists of a MOS/MOM stack and occupies the lowest five metal layers to achieve 8 fF/mm density at 1.8 V. The driver unit design enables straightforward distribution across large body-bias domains for a fast and uniform charging profile.

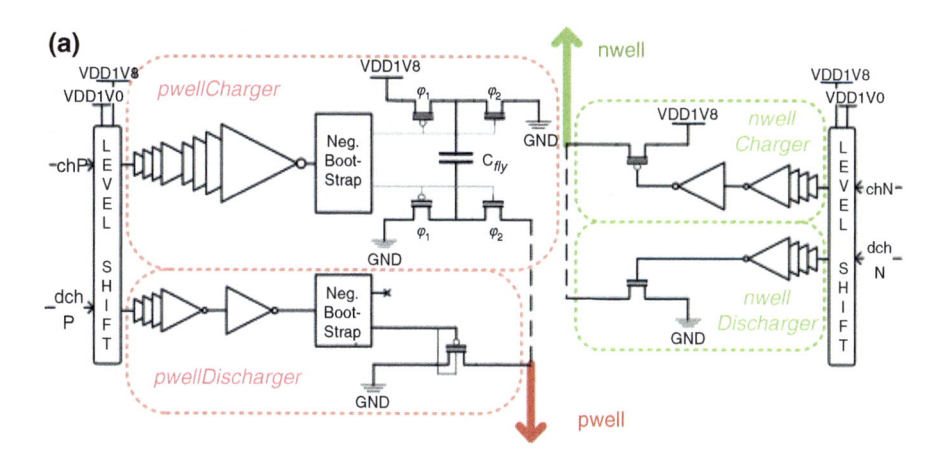

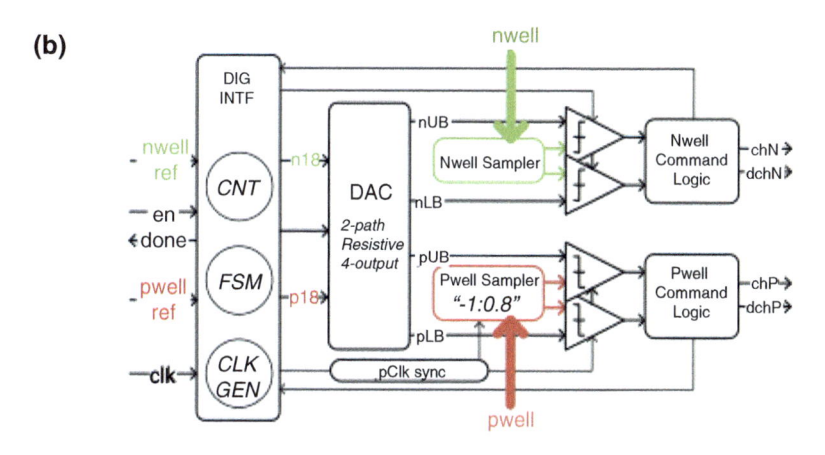

11 System Integration of RISC-V Processors with FD-SOI 281

Fig. 11.17 Block diagrams of the controller (**a**) and driver (**b**) of the integrated body-bias generator [23] (©2016 IEEE, reprinted with permission)

Fig. 11.18 The negative voltage generator used to generate p-well voltages [23] (©2016 IEEE, reprinted with permission)

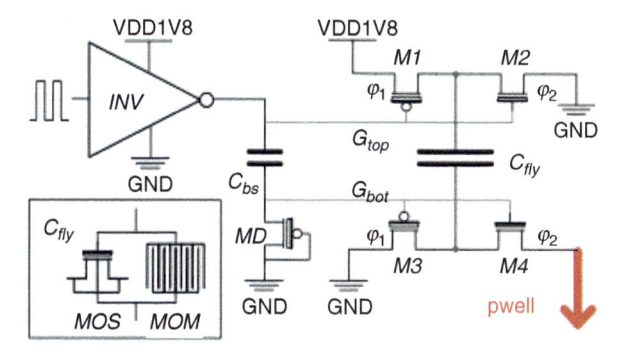

11.6 Clock Generation

Because SS-SC converters generate a rippling supply voltage, only a clock that adjusts to this ripple on a cycle-by-cycle basis to supply the digital logic can avoid wasted energy in these systems. Accordingly, two local adaptive clock generators were implemented that can be paired with the SS-SC converter for FG-AVS systems.

11.6.1 DLL-Based Adaptive Clocking

A block diagram of the clock generator implemented in Raven-3 is shown in Fig. 11.19 [25]. The clock generator has a DLL, a tunable replica circuit (TRC), and a controller. An external global clock supplies the DLL, which generates the equally spaced reference clock phases. By using the reference signals from the DLL and the estimated clock delay from TRC, the control logic generates the clock output signal by selecting one of the DLL outputs. A watchdog circuit assures that metastability does not interrupt the clock.

The DLL contains a voltage-controlled delay line (VCDL), a phase detector, and a charge pump with loop filter. The VCDL converts a single-ended global 2 GHz clock signal into eight differential signals of the same frequency. The

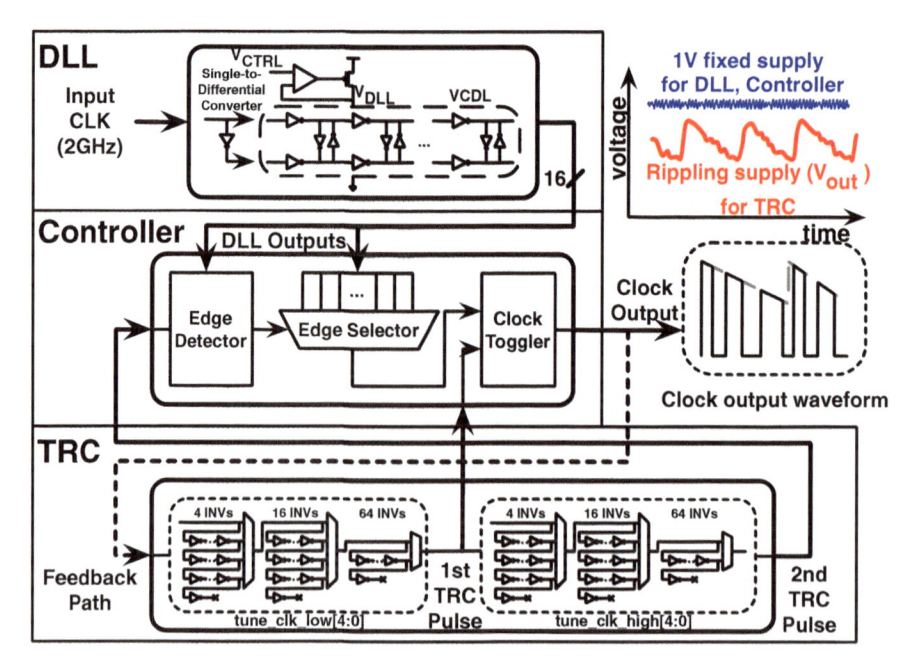

Fig. 11.19 A block diagram of the DLL-based adaptive clock generator [2] (©2016 IEEE, reprinted with permission)

single-ended input is converted to a differential pair using an inverter followed by cross-coupled differential inverter chains. The cross-coupled differential delay unit is composed of two inverters for the main path, and two small inverters that connect the outputs of the main inverters to minimize the phase mismatch between the output nodes.

The tunable replica circuit shown is placed in the processor core that is adaptively clocked. The TRC is designed as a configurable inverter chain, and has two units that are independently controlled to set the duty cycle of the output clock. Each unit consists of 124 identical inverter cells with a 5-bit control signal. The configuration resolution is 4 FO1 delays, where FO1 is the fanout-of-one inverter delay. Four inverter cells compose the resolution step, and the control signal determines how many resolution steps are used in the delay path. The first unit sets the high phase of the clock, which starts at the rising edge and generates the falling edge, while the second unit is for the low phase. With this feature, the period and the duty cycle can be independently adjusted. The TRC inverters are connected to the processor supply in order to track the critical path under voltage variations, while other logic remains in an isolated 1 V domain. Therefore, 4 FO1 resolution is guaranteed at any supply voltage level, while the total delay is self-adjusted as a function of the supply. TRC resolution is related to the phase resolution of the DLL. TRC resolution finer than 4 FO1 would eventually be quantized by the DLL with the resolution of 31.25 ps, and will be ignored after passing the quantizer. In the particular implementation, the 4 FO1 is about 25 ps at 1 V.

At the top level, the output clock signal is connected to the TRC input. Two TRC units delay the rising edge of the input successively by the estimated critical path of the processor through appropriate digital calibration of the number of delay stages, and convert the delayed edges to pulses of short duration for the controller by using the pulse generators (PG). The level shifters are located between the TRC and the PG, because the TRC units are on the processor voltage domain while the PG and the controller operate at 1 V. The TRC goes into sleep mode to reduce the energy consumption when not in use by de-asserting the enable signal. Since the TRC delay directly determines the next clock period without any additional logic, the circuit achieves a response time of a single cycle.

The controller is composed of an edge detector, an edge selector, and a clock flip-flop. The DLL references and the TRC pulses are the inputs to this block. The clock output signal is generated from the flip-flop followed by a large clock buffer. The flip-flop data input is connected to VDD, its asynchronous reset input is connected to the first TRC output pulse, and its clock input is supplied by the edge selector.

Since the DLL references and the TRC output pulse are asynchronous, the flip-flops in the edge detector may occasionally fail to sample the inputs, resulting in metastability. To avoid this, a watchdog block is implemented. This logic continuously monitors the status of the clock output signal, and generates an extra pulse for the clock flip-flop input when the clock output signal remains zero for longer than two clock delays of the previous period. The clock output (CLKOUT) is synchronized to the global 2 GHz clock (CLKREF) which is also used in the DLL. The sampled clock output is connected to the gate logic which generates the pulse, PU, that is asserted for a cycle when the sampled clock output goes high. The free-running counter is reset when PU is asserted, and sends the value to the reference. If the clock output is metastable, the counter value exceeds the previous counter value stored in the reference without the PU assertion. Although two clock signals are asynchronous, it is possible to estimate the clock skew because the CLKOUT is quantized by the DLL outputs which originated from CLKREF. In the design phase, the skew is adjusted to minimize metastability.

The clock generator is designed to conform to standard cell design rules to enable tight integration with the core logic. By using same design style, and thus by embedding the block directly in the processor, the core and the clock generator can share operating conditions such as voltage and temperature. The signal paths for the DLL references are manually routed, and the wire lengths are carefully matched to minimize any possible discrepancy in the delay intervals. The signal paths in the controller are routed in the same manner. The watchdog is synthesized and automatically placed and routed.

11.6.2 Free-Running Adaptive Clocking

A block diagram of the clock generator circuit implemented in Raven-4 is shown in Fig. 11.20 [4]. The adaptive clock generator is free-running and does not require any external reference. The circuit generates clock edges via a D flip-flop, which is set or reset with pulse generators to toggle between one and zero. After a rising clock edge is generated, the edge propagates through the first delay unit and then triggers a reset pulse on the flip-flop, causing the output to fall. This edge then continues through the second delay unit and triggers a clock pulse on the flip-flop, resulting in a rising edge at the output. This design allows the duty cycle of the generated clock to be adjusted.

The adaptive clock generator is supplied by the same rippling output voltage that is used to power the digital logic in the system. The delay units therefore respond to voltage changes similarly to the digital logic; the propagation delays that trigger each clock edge face the same environment as the digital logic paths in the design. Each delay unit is comprised of four delay banks, and each bank consists of a chain of a particular cell in addition to multiplexers and control logic that allow the number of cells in the delay chain to be selected. The first bank uses a custom buffer cell design to balance rise and fall times. Each of the remaining banks consists of a standard cell that was commonly observed in the synthesized critical paths of the digital logic. Each bank uses a different combination of pMOS/nMOS ratio and gate length, as these characteristics affect the voltage-frequency relationship of the cells. By selecting different mux settings, the delay paths can be tuned until they

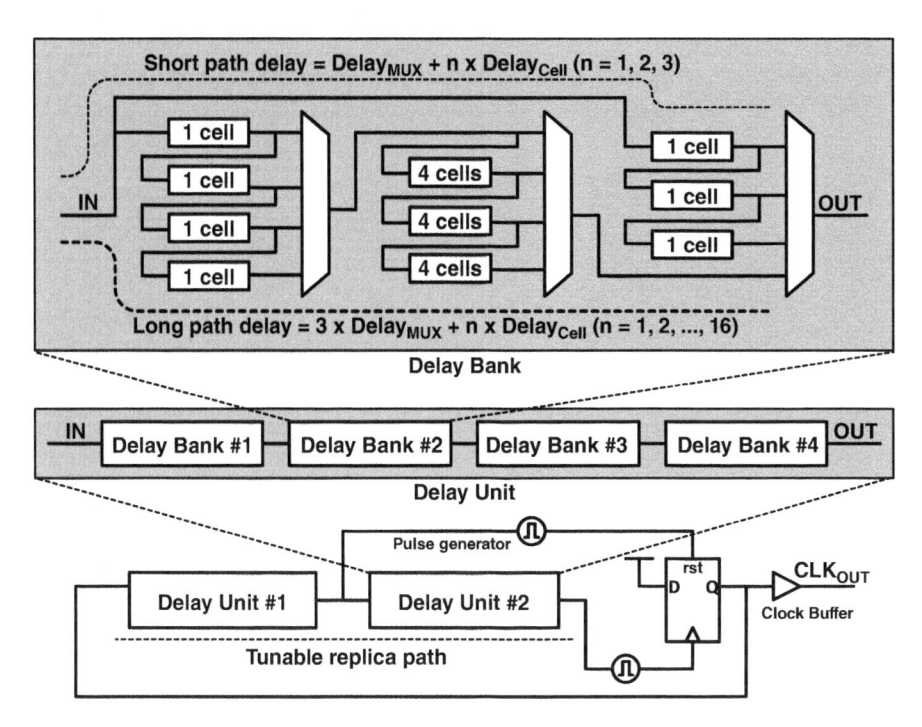

Fig. 11.20 A block diagram of the free-running adaptive clock generator [4, 5] (©2016 IEEE, reprinted with permission)

match the delay characteristics of the critical paths in the digital logic supplied by the generated clock, ensuring that the clock edges are generated at the appropriate frequency for the instantaneous voltage produced by the regulator.

11.7 System Performance

The Raven-3 and Raven-4 testchips were implemented in 28 nm FD-SOI technology. Measurement results demonstrate the effectiveness of the various systems implemented in the two testchips, showing the utility of FD-SOI technology in designing energy-efficient processor systems. Table 11.2 shows a summary of the chip and key measurement results of the Raven-3 testchip, and Table 11.3 summarizes key features and measurement results of the Raven-4 testchip.

Table 11.2 A summary of key results and details from the Raven-3 testchip [2]

Technology	28 nm FD-SOI
Die area	1305 × 1818 μm (2.37 mm^2)
Core area	880 × 1350 μm (1.19 mm^2)
Converter area	24 × 90 μm × 90 μm (0.19 mm^2)
Voltage	0.45–1 V (1 V FBB)
Frequency	93–961 MHz (1 V FBB)
Power	8–173 mW (1 V FBB)
SC density	11.0 fF/μm^2
SC power density	0.35 W/mm^2 @ 88% efficiency

Table 11.3 A summary of key results and details from the Raven-4 testchip [4]

Technology	28 nm FD-SOI
Die area	1665 × 1818 μm (3.03 mm^2)
Core area	895 × 1193 μm (1.07 mm^2)
Standard cells	568K
Converter area	48 × 90 μm × 90 μm (0.39 mm^2)
Core voltage	0.48–1 V (bypass mode)
Core power	1.2–231 mW (bypass mode)
Core frequency	20–797 MHz (bypass mode)
Peak energy efficiency	41.8 GFLOPS/W (1/2 1 V mode)
Conversion efficiency	82–89%
AVS transition time	<1 μs
Peak AVS energy savings	39.8%

11.7.1 Raven-3 Measurement Results

Figure 11.21 shows oscilloscope traces of the rippling core voltage domain for all four SS-SC regulator configurations. As noted in Sect. 11.4, simultaneous-switching converters do not completely eliminate charge sharing because the load itself has a capacitive component. This results in observed average output voltages that are lower than the ideal ratios. Figure 11.22 shows the impact of adjusting the lower-bound reference voltage V_{ref} on conversion efficiency. Changing the reference voltage changes both the average output voltage and the switching rate of the converter, which affects the conversion efficiency. Each switching mode has a particular V_{ref} that achieves maximum efficiency at a given load. Different processor loads correspond to different optimal V_{ref} because the processor load affects the switching frequency of the regulators.

For all possible converter topologies with adaptive clocking, the processor successfully boots Linux and runs user applications, demonstrating that complex digital logic operates reliably with an intentionally rippling supply voltage. Figure 11.23 shows the rapid (<20 ns) transitions between different voltage modes. The application core can continue to operate through these mode transitions, demonstrating the utility of integrated regulators for FG-AVS.

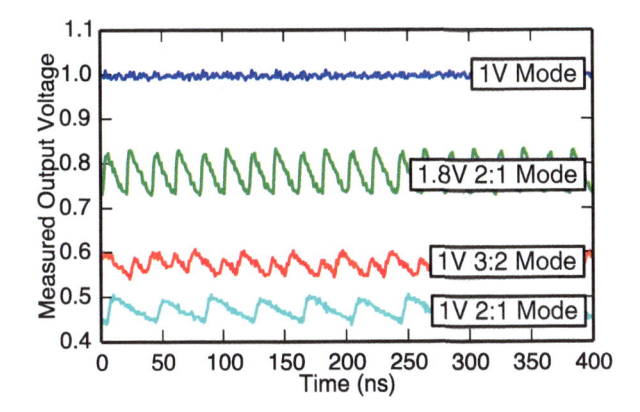

11 System Integration of RISC-V Processors with FD-SOI   287

Fig. 11.21 Oscilloscope traces of the voltages generated in the four SS-SC operating modes [2] (©2016 IEEE, reprinted with permission)

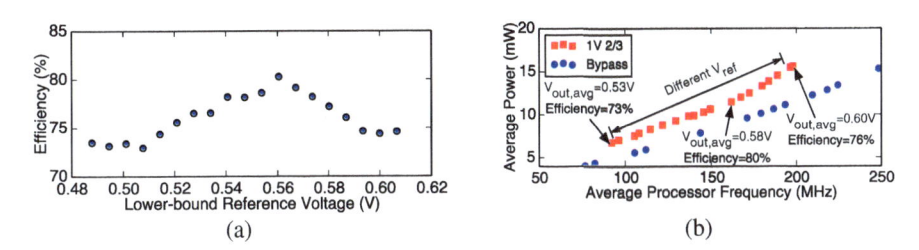

Fig. 11.22 Figures showing the effect of different lower-bound reference voltages on system operation [2] (©2016 IEEE, reprinted with permission). (**a**) Shows the system conversion efficiency as the voltage is swept, and (**b**) shows the effect of different V_{ref} on average output voltage and frequency

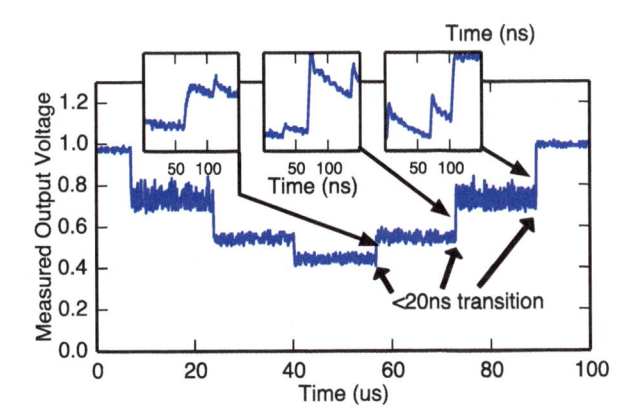

Fig. 11.23 An oscilloscope trace showing the generated core voltage as the SS-SC converter rapidly cycles through each of the four operating modes [2] (©2016 IEEE, reprinted with permission)

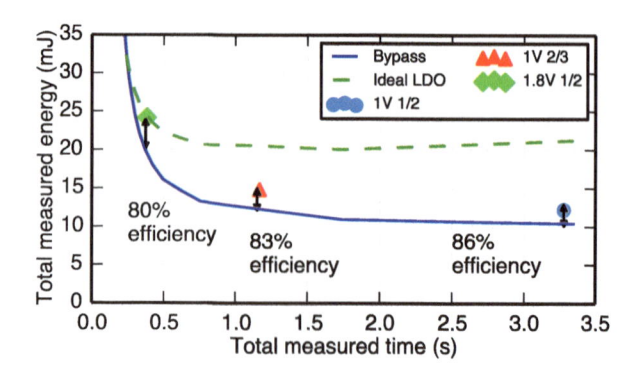

Fig. 11.24 A plot demonstrating the system conversion efficiency of the three SS-SC switching modes [2] (©2016 IEEE, reprinted with permission)

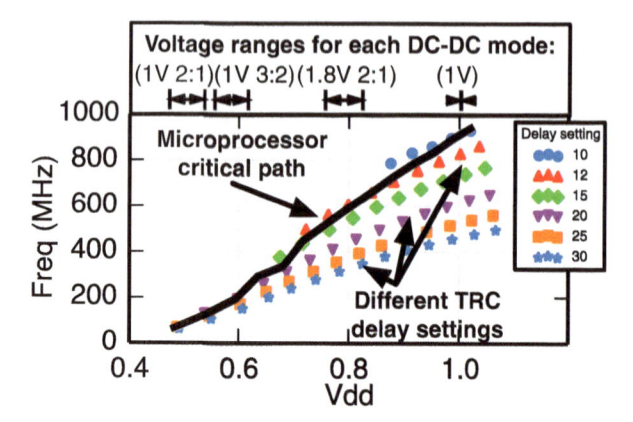

Fig. 11.25 Measurements of the adaptive clock generator frequency over different voltages and replica path delay settings [2] (©2016 IEEE, reprinted with permission)

System conversion efficiency measurements include all sources of overhead, including non-idealities in the adaptive clock. As shown in Fig. 11.24, the measured voltage conversion efficiency ranges from 80 to 86% for the different output voltage modes.

Figure 11.25 shows measured average frequency for different delay settings for the tunable replica circuit across a range of operating voltages (supplied using the bypass mode). Annotations above the plot indicate the approximate voltage ranges seen in each SS-SC voltage mode. Because the inverter-based replica path delay characteristics do not match the critical paths of the processor, a single delay setting poorly tracks the processor critical path over the entire voltage range. At higher voltages, a shorter replica path delay best tracks the processor critical path, while at lower voltages, a longer replica path delay best tracks the processor critical path. However, manual calibration of specific delay settings for each SS-SC voltage mode allows reasonably accurate tracking within the relatively small ripple of that mode.

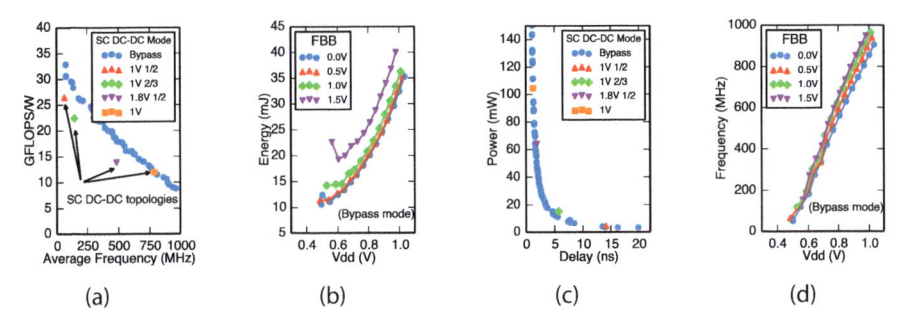

11 System Integration of RISC-V Processors with FD-SOI

Fig. 11.26 Plots showing the measured performance (**a**), energy (**b**), power (**c**), and frequency (**d**) of the processor core in bypass mode and with the switching regulators enabled [2] (©2016 IEEE, reprinted with permission). (**a**) and (**c**) show the effect of different SC-DCDC mode settings, while (**b**) and (**d**) show the effect of forward body-bias

Figure 11.26 shows various energy-delay curves for the application processor. Energy efficiency is measured by measuring total core energy consumption while executing a fixed-length double-precision floating-point matrix multiplication kernel on the vector accelerator. Figure 11.26a, c shows measurements both under bypass mode and when accounting for the losses of the switching regulation modes. By using the on-chip converter to generate the lowest output voltage, the system achieves a peak efficiency of 26.2 GFLOPS/W. The FD-SOI technology of the testchip enables up to 1.8 V of FBB to be safely applied, which trades off improved performance for increased leakage [26]. Figure 11.26b and d shows the impact of FBB on frequency and energy. In each of the plots, each point represents the best achievable operating frequency at a given voltage. The integrated system is able to achieve high energy efficiencies under the proposed voltage regulation and clocking scheme, demonstrating the potential utility of FG-AVS.

11.7.2 Raven-4 Measurement Results

Figure 11.27 shows measured waveforms of the core voltage and core clock during a transition between voltage modes. The core clock was measured via the output driver, which was not present on Raven-3. The core clock adapts automatically to the voltage ripples within each mode and to the large voltage transition between modes. Table 11.4 shows the peak system conversion efficiency of each of the three switching SS-SC modes. The adaptive clock is tuned at each voltage setting by sweeping the settings of its replica delay path and choosing the fastest setting that still results in correct core functionality. The adaptive clocking system provides a large improvement in system conversion efficiency because the core is able to operate at a higher average frequency as the supply voltage ripples, reducing the amount of energy required to complete the same amount of work. When supplied by

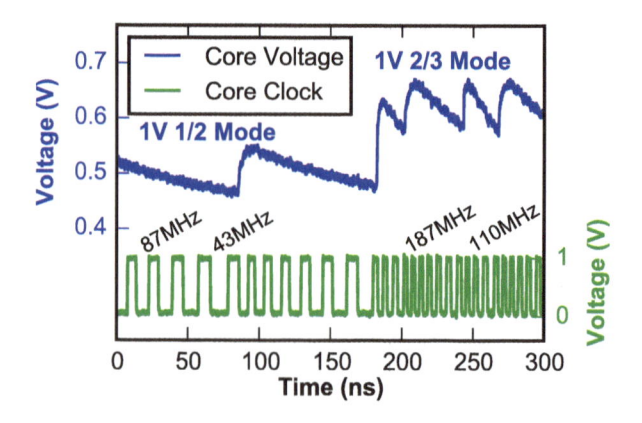

Fig. 11.27 Oscilloscope traces of the core voltage and clock during an SS-SC mode transition [4, 5] (©2016 IEEE, reprinted with permission)

Table 11.4 Measured system conversion efficiencies achieved by the Raven-4 system [4]

DC-DC mode	Efficiency (adaptive clock)	Efficiency (fixed clock)
1/2 1.8 V	88.7%	76.6%
2/3 1.0 V	85.0%	75.4%
1/2 1.0 V	81.6%	61.9%

Fig. 11.28 A shmoo chart showing processor performance while executing a matrix-multiply benchmark under a wide range of operating modes [5] (©2016 IEEE, reprinted with permission). The number in each box represents the energy efficiency of the application core as measured in double-precision GFLOPS/W

the SS-SC converter, the processor achieves a peak energy efficiency of 41.8 double-precision GFLOPS/W running an FMA microbenchmark on the vector coprocessor in 1/2 1 V mode. The processor is able to boot Linux and run user programs while powered by the rippling supply voltage and adaptive clock.

Figure 11.28 shows the processor functionality across a wide range of voltages and frequencies. The SS-SC converter is placed into bypass mode for characterization, allowing the measurement of processor performance under fixed voltage

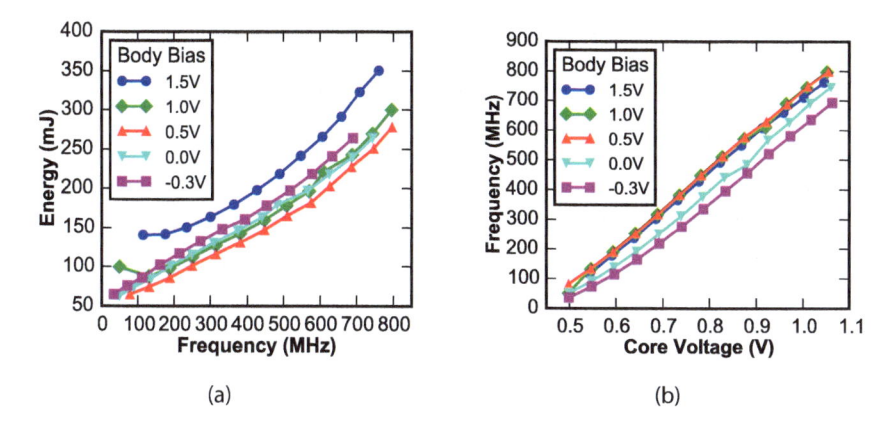

Fig. 11.29 Plots showing the effects of FBB on core energy (**a**) and frequency (**b**) [5] (©2016 IEEE, reprinted with permission)

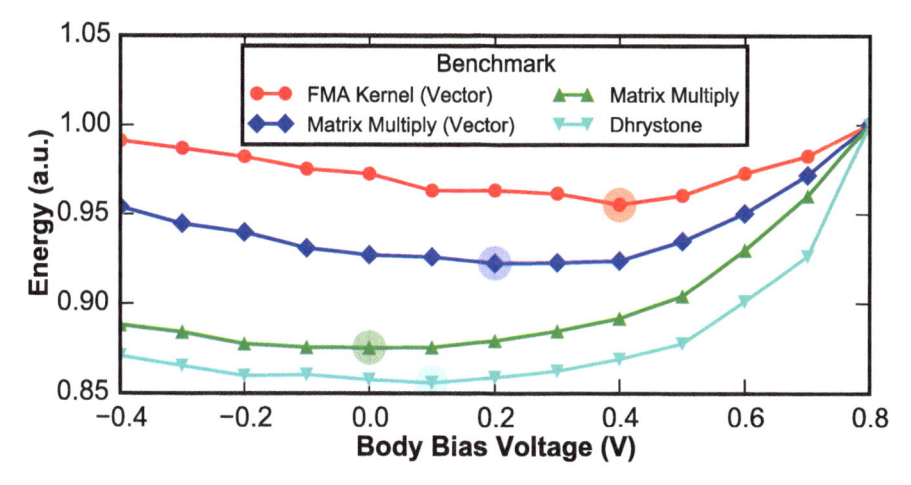

Fig. 11.30 The effect of FBB on different benchmarks with a supply voltage of 0.6 V (bypass mode) [5] (©2016 IEEE, reprinted with permission). The total energy consumed by each benchmark has been normalized so the relative effects of body-bias can be compared. The minimum-energy point is highlighted for each benchmark

and frequency. The best energy efficiency in bypass mode of 54.0 double-precision GFLOPS/W is achieved at 500 mV and 40 MHz. Figure 11.29 shows the best frequency achievable at each operating point and the total energy consumed by a fixed-duration matrix-multiply benchmark at that operating point. The application of FBB increases performance but results in higher leakage power. The FBB voltage that achieves minimum energy depends on the proportion of switching power to leakage power and is therefore benchmark-dependent. Figure 11.30 shows that the energy of more computationally intensive benchmarks is minimized at higher body-bias settings.

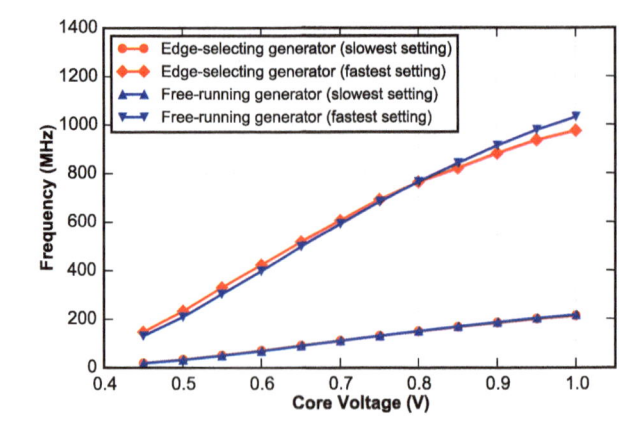

Fig. 11.31 Measured comparison of the two clock generators implemented in the Raven-4 testchip [4] (©2017 IEEE, reprinted with permission)

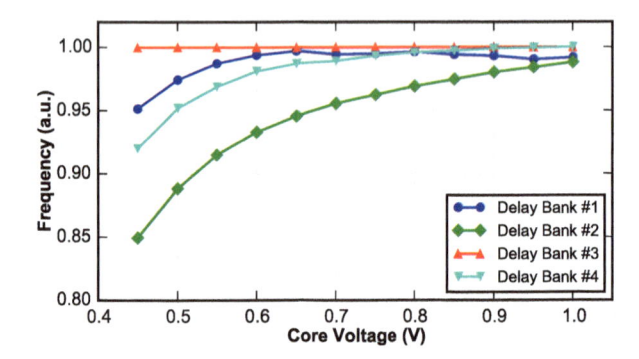

Fig. 11.32 Measurement results showing the relative difference in voltage-dependent frequency behavior of the four delay banks in the adaptive clock generator [4] (©2017 IEEE, reprinted with permission)

Figure 11.31 compares the generated frequency of the free-running adaptive clock generator with the design from Raven-3. Both designs reliably generate a core clock that can supply the core area while a workload is executed on the application core. The two designs track voltage similarly at the slowest delay settings, but the tracking varies at the fastest setting because the free-running design oscillates entirely in the variable-voltage domain, while part of the timing loop in the edge-selecting design is level-shifted to the fixed 1 V supply. The free-running generator achieves the same functionality as the adaptive clocking scheme from Raven-3 while simplifying the design and reducing area and power overhead.

Figure 11.32 compares the voltage-dependent delay characteristics of the four delay banks, normalized to the delay of the custom buffer cells in Bank 3. The result for each bank was measured by recording the frequency of the generated clock after selecting the maximum delay through that bank and the minimum delay through the remaining banks. The cells with small pMOS/nMOS ratios and larger gate lengths

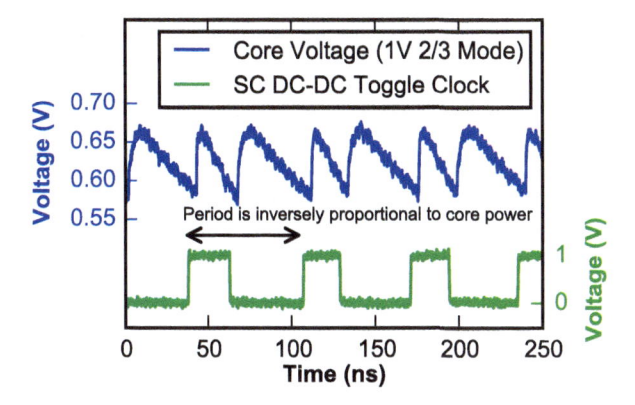

Fig. 11.33 Oscilloscope traces showing the rippling core supply voltage and the SS-SC toggle clock [5] (©2016 IEEE, reprinted with permission)

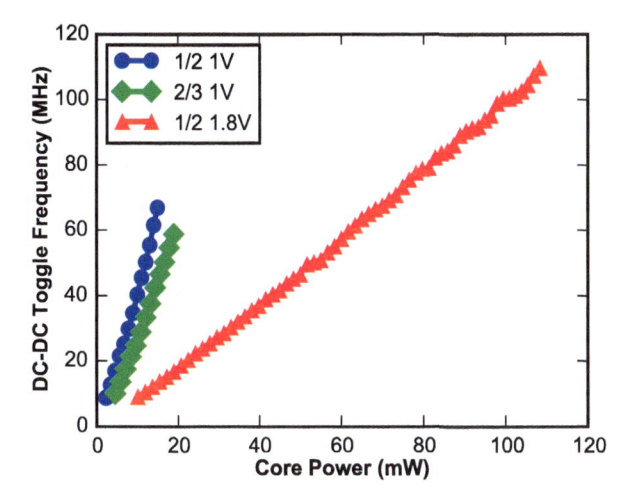

Fig. 11.34 Measurement results showing the correlation between SS-SC toggle frequency and measured core power [4] (©2017 IEEE, reprinted with permission)

have larger delays at lower voltages. The wide variation in voltage-dependent delays between the different delay banks (up to 18% at 0.45 V) validates the need for multiple different standard cells to achieve accurate critical path tracking.

Figure 11.33 shows the measured core voltage and SS-SC toggle clock, and Fig. 11.34 compares the core power measured by the bench equipment with the SS-SC toggle frequency measured by the integrated counter. The correlation is monotonic and approximately linear for each of the three conversion modes, confirming the practicality of using this toggle frequency to estimate core power and system load.

The programmable PMU allows the implementation of a wide variety of power-management algorithms to improve energy efficiency. Several different experiments demonstrate the flexibility of the system in implementing common energy-saving techniques. In one experiment, dithering between two voltages to achieve an arbitrary target core frequency is implemented. To implement this algorithm, the PMU first calibrates the system by using the core clock counter to measure the average operating frequency at each voltage mode. Then a target frequency is provided to the PMU, which polls the core clock counter and dithers the voltage setting to achieve the target frequency in aggregate. The results of the experiment are shown in Fig. 11.35. Without dithering, the processor would need to operate only in the higher mode to guarantee that the performance target is met, which would consume up to 40% more energy than the dithered approach.

The choice of hopping frequency presents a tradeoff between increased fidelity to the target effective frequency and the more frequent occurrence of transition overheads, which can increase energy consumption. In the testchip SS-SC implementation, the energy cost associated with transitions between voltage modes is small because the processor continues to operate as the clock frequency adjusts during the mode transition. Accordingly, a hopping frequency of approximately 6 μs was chosen. This frequency is still quite fast, but is slow enough that transition energy costs can be neglected.

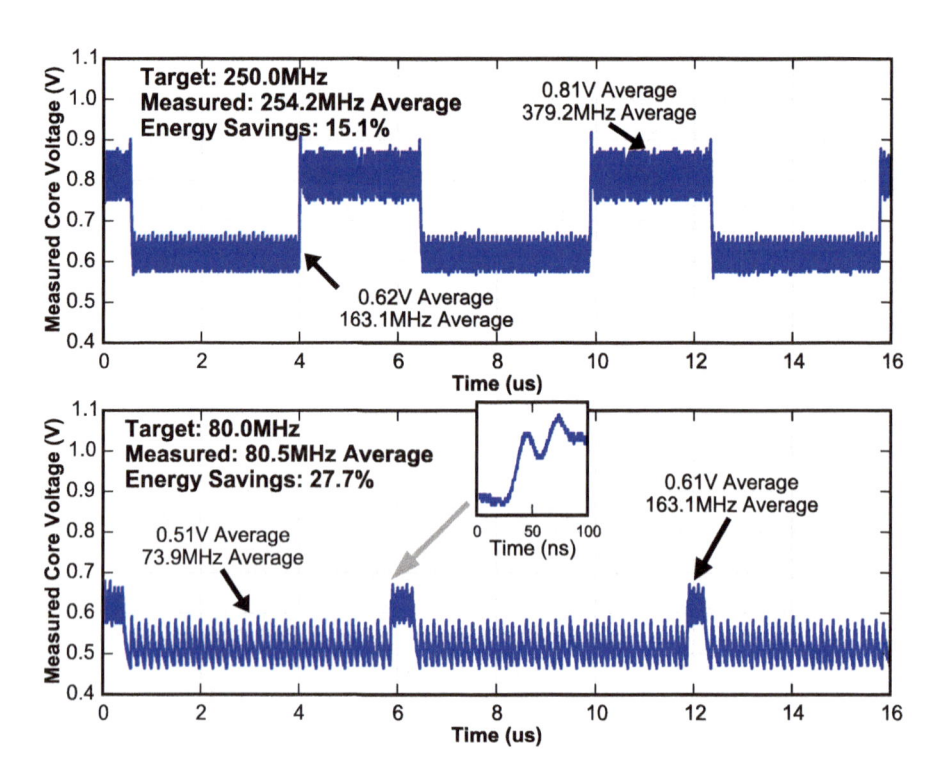

Fig. 11.35 Oscilloscope traces showing the core voltage as the frequency hopping algorithm is applied with two different frequency targets [5] (©2016 IEEE, reprinted with permission)

Voltage dithering can suffer reduced conversion efficiency compared to continuous regulation. This effect is explored in Fig. 11.36. Each measured point in the figures represents the completion of a fixed-length matrix-multiply benchmark. The dithering algorithm enables a wide operating range for the core, bounded only by the lowest and highest voltage settings of the SS-SC converter. Figure 11.36a shows measurement results of system conversion efficiency, similar to the measured results of Raven-3 as shown in Fig. 11.24. Instead of just three conversion points, however, the interpolated time-energy points measured from dithering between each of the neighboring voltages at a fixed ratio are also shown. Figure 11.36b quantifies the effects of dithering on system conversion efficiency. Two different dithering programs were run on the PMU. The first program simply switches the voltage mode setting of the SS-SC converters, without changing any other system settings; the delay settings of the replica paths in the adaptive clock generator were tuned to the best setting that could function across both operating modes. The results are shown by the red points in the figure. Because the best setting of the replica paths changes according to operating mode, the conversion efficiencies of this approach are less than optimal for part of the dithering range. The second program switches both the voltage mode and the delay settings of the replica paths according to a pre-characterization of the best adaptive clock setting associated with each voltage mode. This program is able to speed the generated clock at the higher voltage settings, leading to higher conversion efficiencies. In all, the efficiencies of the second program range from 70 to 100%, depending on the voltage mode and dithering ratio. Conversion efficiencies while dithering are not as high as the efficiencies of the fixed operating modes in Table 11.4 because the external voltage reference used by the comparator cannot be tuned for a particular SS-SC mode. This implies that in some cases it is more efficient to operate at a single mode than to dither, even if the performance target is somewhat exceeded by the average operating frequency at the fixed mode.

A second experiment demonstrates power envelope tracking, a common requirement for systems that must operate within a user-specified power budget. Figure 11.37 shows the results of a power-management algorithm executed on the PMU that maximizes core performance within a user-specified power budget. The power-management program polls an externally writeable control register that stores the absolute power limit for the program. The PMU core then monitors the SS-SC toggle counter to continuously estimate core power using the quadratic model described in [27] with pre-characterized coefficients. If the estimated core power is above the specified limit, core frequency is decreased, and if it is below the limit, core frequency is increased. In this way, the best possible performance is automatically obtained while the user-specified power budget is respected.

The PMU can also use the integrated counters to coordinate fine-grained adaptive voltage scaling (AVS) on-chip without any explicit guidance from the programs running on the processor. In the algorithm used in this experiment, core power is used as a marker of program phase. When core power is higher, the core is likely executing a compute-intensive program region, and a high voltage is best suited to a race-to-halt strategy. When the core power is lower, the core is likely waiting for off-

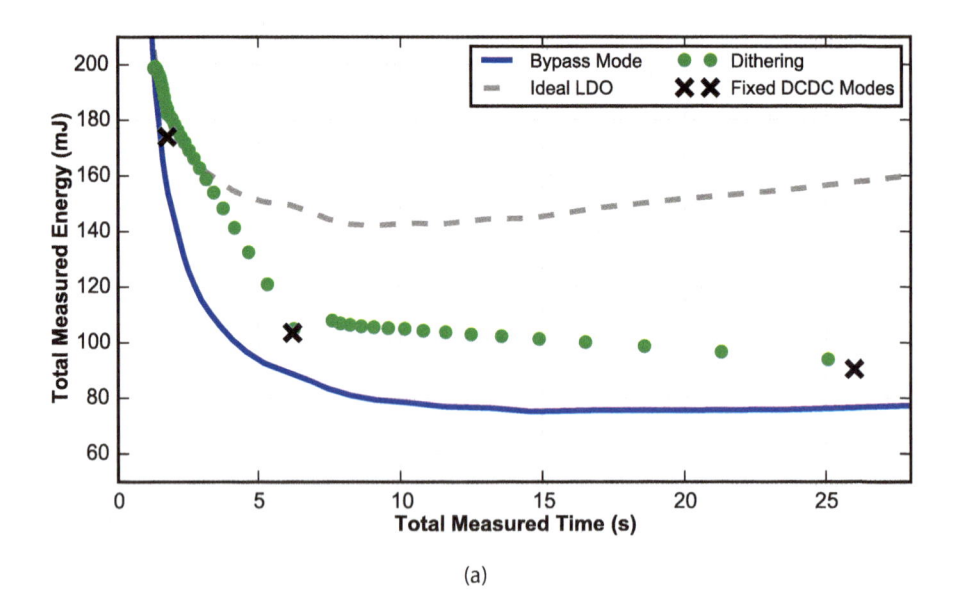

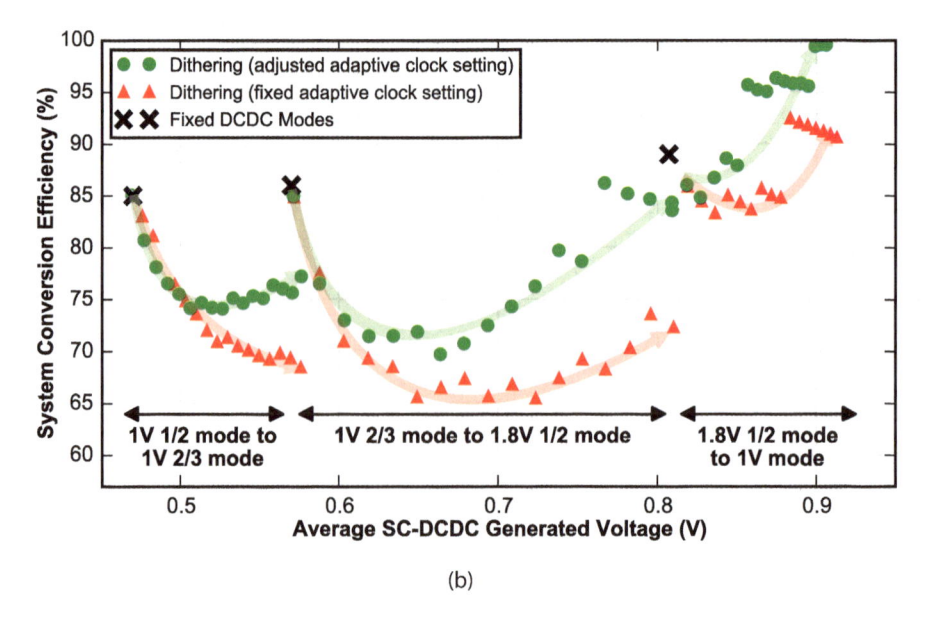

Fig. 11.36 Plots showing the effect of voltage dithering on system conversion efficiency [4] (©2017 IEEE, reprinted with permission). (**a**) Compares the energy cost of dithering to the bypass mode baseline (in blue), which represents 100% efficient regulation. The dithering operating points linearly interpolate completion time and energy between the fixed SS-SC modes. (**b**) Shows the measured system conversion efficiencies under voltage dithering. The results in green show the benefit of re-tuning the replica timing circuit in the adaptive clock generator after each SS-SC mode transition

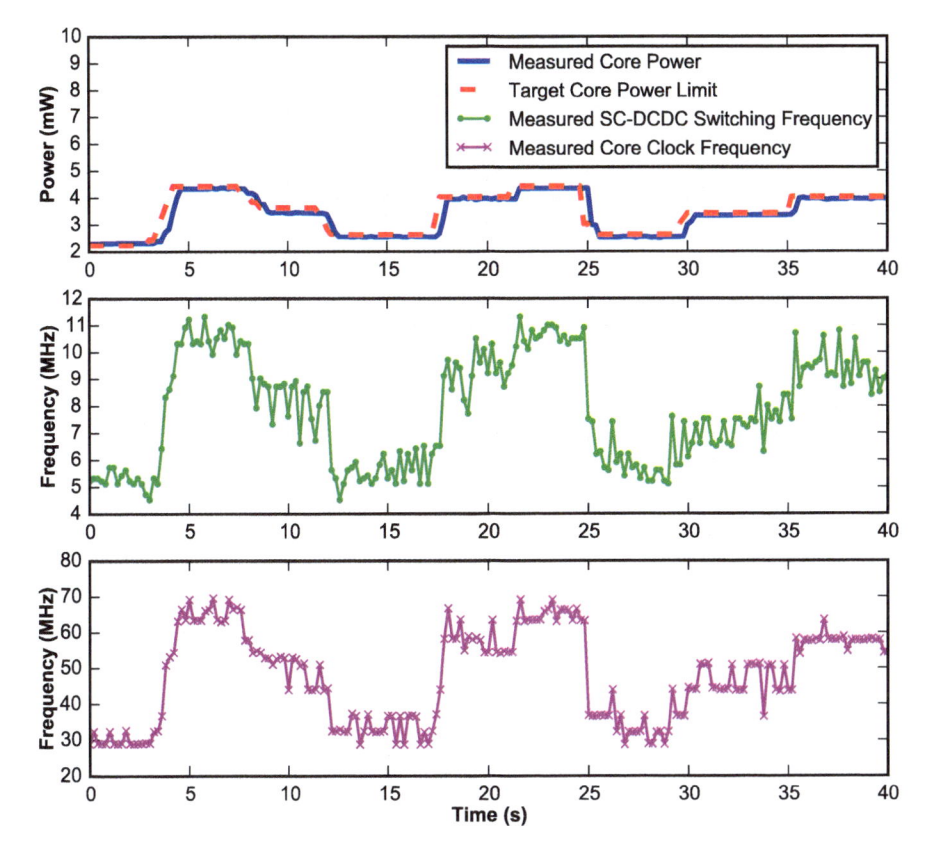

11 System Integration of RISC-V Processors with FD-SOI

Fig. 11.37 Measurement results showing a power envelope tracking program executing on the PMU [4] (©2017 IEEE, reprinted with permission). The frequency of the core is adjusted in response to measured changes in core power so that the core operates as quickly as possible without exceeding a time-varying, user-specified power budget

chip communication in a memory-bound program region, and energy can be saved with minimal performance impact by reducing the voltage. In this experiment, the application core runs a synthetic benchmark that alternates between the compute-intensive and idle phases at a timescale of tens of microseconds. Figure 11.38 shows the core voltage measured during the execution of the benchmark and the AVS power-management algorithm. The algorithm switches the core voltage between the 1.8 V 1/2 mode and the 1 V 2/3 mode, actuated by core power estimates determined by continuously polling the SS-SC toggle counter. When the core voltage is high and the toggle rate drops below a threshold, this corresponds to an idle program period, so the PMU reduces the core voltage to save energy. When the core voltage is low and the toggle rate exceeds a threshold, the workload has increased and the PMU increases the core voltage. The system is able to detect changes in workload in less than 1 μs and adjust the core voltage in response. Without integrated voltage

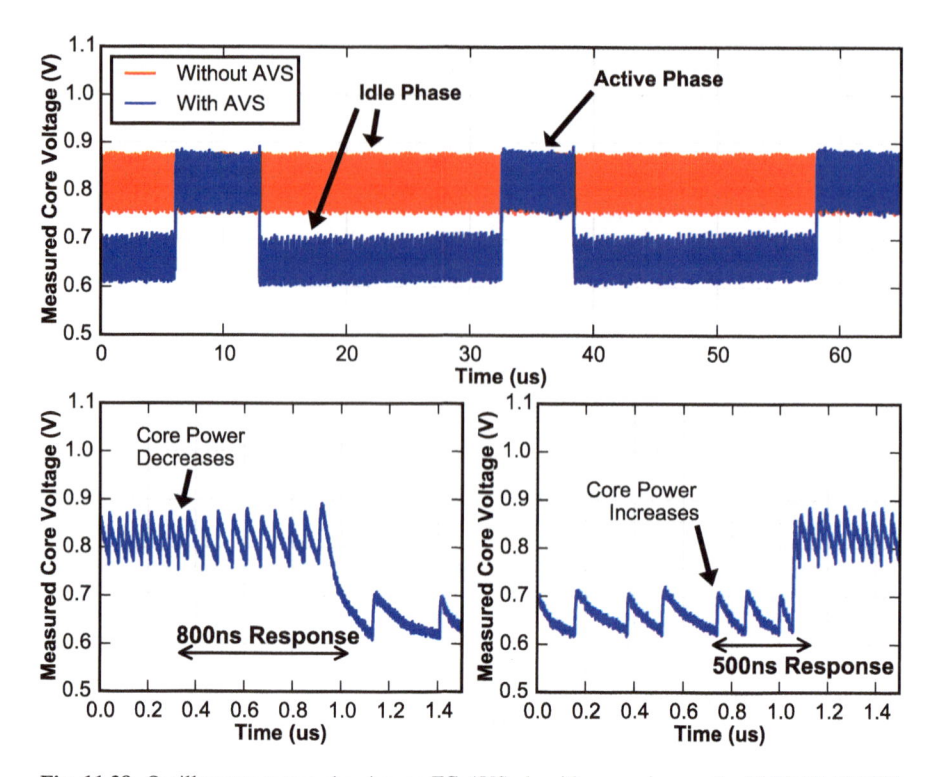

Fig. 11.38 Oscilloscope traces showing an FG-AVS algorithm running on the PMU [4] (©2017 IEEE, reprinted with permission)

regulators and power management, the system would not be able to respond within the timescales of the workload variation. The results of the power-management algorithm are therefore compared against continuous operation in the higher voltage mode, which would otherwise be required to meet the same performance target. The power-management algorithm reduces the energy consumed by 39.8%, and the fast response incurs negligible (<0.2%) performance penalty compared with this baseline because the fast response minimizes the time spent in the lower voltage mode during a compute-bound region. This experiment demonstrates the efficacy of fine-grained AVS at improving energy efficiency with fast workload tracking.

The integrated body-bias generator achieves up to 1.48 V of positive body-bias in the n-well and up to 1.2 V of negative body-bias in the p-well. The n-well voltage achieves 58 mV resolution and the p-well voltage achieves 72 mV resolution, corresponding in each case to roughly 5 mV of threshold voltage step. Figure 11.39 shows the dynamics of charging and discharging of the wells. The n-well reaches high slew rates of −80 and +65 mV/ns during charging and discharging. The p-well voltage ramps down from 0 to −1.3 V in 160 ns with one driver unit and in 90 ns with two units, switching at 1 GHz in both cases. The p-well discharges from 1.4

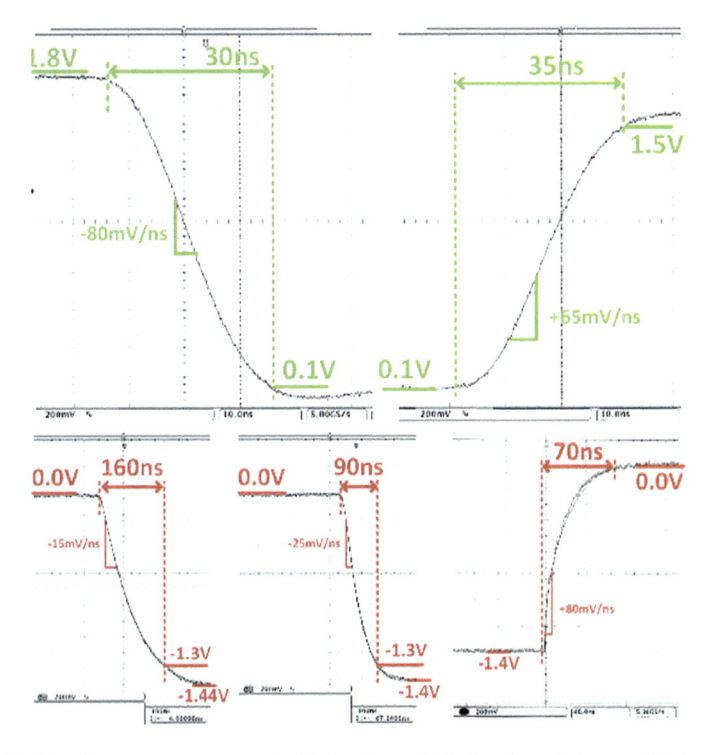

11 System Integration of RISC-V Processors with FD-SOI

Fig. 11.39 Transient measurements of well charging and discharging of the p-well (**a**) and n-well (**b**) voltages [23] (©2016 IEEE, reprinted with permission)

Fig. 11.40 Maintaining a constant body-bias with occasional recharge to compensate for leakage [23] (©2016 IEEE, reprinted with permission)

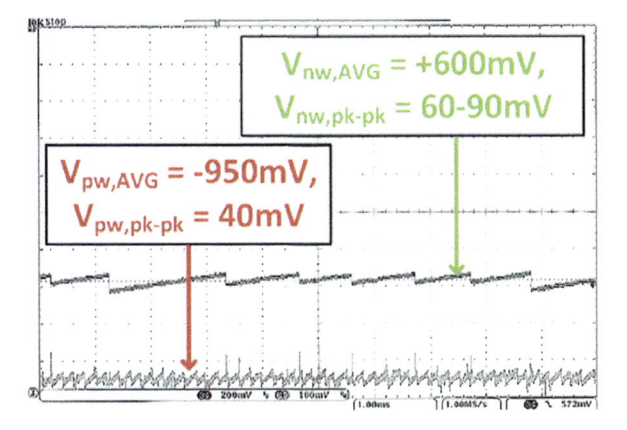

$V_{nw,AVG} = +600mV$, $V_{nw,pk-pk} = 60\text{-}90mV$

$V_{pw,AVG} = -950mV$, $V_{pw,pk-pk} = 40mV$

Fig. 11.41 Adaptive body-bias compensation to supply voltage drift [23] (©2016 IEEE, reprinted with permission). A constant frequency is maintained by adjusting body-bias as supply voltage changes

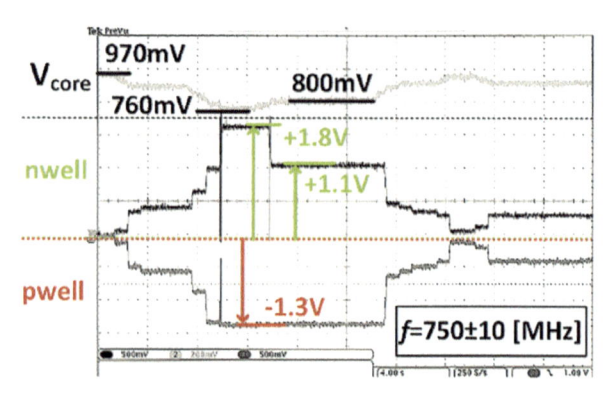

to 0 V in 70 ns with only one unit operating. During the ON-mode the BBG drives currents in the range of 40–200 mA, while in STEADY-mode it sources only 3.3 and 1 μA from the 1.8 and 1 V supplies, respectively. High ON-mode currents are averaged over long periods of STEADY-mode and contribute less than 2 μA in total average current. Figure 11.40 illustrates the maintenance of target n-well and p-well voltages with short recharge phases when the well voltages drift due to leakage. In this case, ON-mode recharge of 5 ns duration occurs every 1 and 0.2 ms for the n-well and p-well, respectively. Figure 11.41 demonstrates the variation-compensation and energy-efficiency optimization capability of the BBG. The frequency counters are used to extract the switching frequency of the integrated clock generator. This frequency is influenced both by the core supply voltage and the body-bias effect. A feedback loop adjusts the body-bias voltage so that a constant frequency is maintained even as supply voltage changes. The design is able to maintain the target frequency within 1% while dynamically changing the digital supply voltage in the range of 760–970 mV. Note that the p-well and n-well voltages can be asymmetric to achieve robust compensation.

Acknowledgements Fabrication of the Raven-3 and Raven-4 testchips was donated by STMicroelectronics. The research presented in this chapter was supported by the Berkeley Wireless Research Center, the Berkeley ASPIRE Lab, DARPA PERFECT Award Number HR0011-12-2-0016, Intel ARO, AMD, SRC/TxACE, Marie Curie FP7, the NSF GRFP, and the NVIDIA Fellowship.

References

1. B. Keller, Energy-efficient system design through adaptive voltage scaling. Ph.D. dissertation, Department of Electrical Engineering and Computer Sciences, University of California, Berkeley, December 2017
2. B. Zimmer, Y. Lee, A. Puggelli, J. Kwak, R. Jevtić, B. Keller, S. Bailey, M. Blagojević, P.F. Chiu, H.P. Le, P.H. Chen, N. Sutardja, R. Avizienis, A. Waterman, B. Richards, P. Flatresse, E. Alon, K. Asanović, B. Nikolić, A RISC-V vector processor with simultaneous-switching switched-capacitor DC-DC converters in 28 nm FDSOI. IEEE J. Solid State Circuits **51**(4),

930–942 (2016)
3. P. Flatresse, B. Giraud, J. Noel, B. Pelloux-Prayer, F. Giner, D. Arora, F. Arnaud, N. Planes, J. Le Coz, O. Thomas, S. Engels, G. Cesana, R. Wilson, P. Urard, Ultra-wide body-bias range LDPC decoder in 28nm UTBB FDSOI technology, in *IEEE International Solid-State Circuits Conference Digest of Technical Papers, February* (2013), pp. 424–425
4. B. Keller, M. Cochet, B. Zimmer, J. Kwak, A. Puggelli, Y. Lee, M. Blagojević, S. Bailey, P.F. Chiu, P. Dabbelt, C. Schmidt, E. Alon, K. Asanović, B. Nikolić, A RISC-V processor SoC with integrated power management at submicrosecond timescales in 28 nm FD-SOI. IEEE J. Solid State Circuits **52**(7), 1863–1875 (2017)
5. B. Keller, M. Cochet, B. Zimmer, Y. Lee, M. Blagojević, J. Kwak, A. Puggelli, S. Bailey, P.F. Chiu, P. Dabbelt, C. Schmidt, E. Alon, K. Asanović, B. Nikolić, Sub-microsecond adaptive voltage scaling in a 28 nm FD-SOI processor SoC, in *Proceedings of the European Solid-State Circuits Conference, September* (2016), pp. 269–272
6. The RISC-V ISA [Online]. Available http://riscv.org
7. K. Asanović, R. Avizienis, J. Bachrach, S. Beamer, D. Biancolin, C. Celio, H. Cook, D. Dabbelt, J. Hauser, A. Izraelevitz, S. Karandikar, B. Keller, D. Kim, J. Koenig, Y. Lee, E. Love, M. Maas, A. Magyar, H. Mao, M. Moreto, A. Ou, D.A. Patterson, B. Richards, C. Schmidt, S. Twigg, H. Vo, A. Waterman, The rocket chip generator. Department of Electrical Engineering and Computer Sciences, University of California, Berkeley. Technical Report UCB/EECS-2016-17 (2016)
8. Y. Lee, A. Waterman, R. Avizienis, H. Cook, C. Sun, V. Stojanovic, K. Asanović, A 45nm 1.3GHz 16.7 double-precision GFLOPS/W RISC-V processor with vector accelerators, in *Proceedings of the European Solid-State Circuits Conference, September* (2014), pp. 199–202
9. R.M. Russell, The CRAY-1 computer system. Commun. ACM **21**(1), 63–72 (1978)
10. N. Planes, O. Weber, V. Barral, S. Haendler, D. Noblet, D. Croain, M. Bocat, P. Sassoulas, X. Federspiel, A. Cros, A. Bajolet, E. Richard, B. Dumont, P. Perreau, D. Petit, D. Golanski, C. Fenouillet-Beranger, N. Guillot, M. Rafik, V. Huard, S. Puget, X. Montagner, M.A. Jaud, O. Rozeau, O. Saxod, F. Wacquant, F. Monsieur, D. Barge, L. Pinzelli, M. Mellier, F. Boeuf, F. Arnaud, M. Haond, 28nm FDSOI technology platform for high-speed low-voltage digital applications, in *Proceedings of the IEEE Symposium on VLSI Technology, June* (2012), pp. 133–134
11. B. Zimmer, O. Thomas, S.O. Toh, T. Vincent, K. Asanović, B. Nikolić, Joint impact of random variations and RTN on dynamic writeability in 28nm bulk and FDSOI SRAM, in *Proceedings of the IEEE European Solid-State Device Research Conference, September* (2014), pp. 98–101
12. S. Clerc, M. Saligane, F. Abouzeid, M. Cochet, J.-M. Daveau, C. Bottoni, D. Bol, J. De-Vos, D. Zamora, B. Coeffic, D. Soussan, D. Croain, M. Naceur, P. Schamberger, P. Roche, D. Sylvester, A 0.33V/-40C process/temperature closed-loop compensation SoC embedding all-digital clock multiplier and DC-DC converter exploiting FDSOI 28nm back-gate biasing, in *IEEE International Solid-State Circuits Conference Digest of Technical Papers, February* (2015), pp. 150–151
13. R. Jain, S. Kim, V. Vaidya, J. Tschanz, K. Ravichandran, V. De, Conductance modulation techniques in switched-capacitor DC-DC converter for maximum-efficiency tracking and ripple mitigation in 22nm tri-gate CMOS, in *Proceedings of the IEEE Custom Integrated Circuits Conference, September* (2014), pp. 1–4
14. J. Jiang, Y. Lu, W.H. Ki, U. Seng-Pan, R.P. Martins, A dual-symmetrical-output switched-capacitor converter with dynamic power cells and minimized cross regulation for application processors in 28nm CMOS, in *IEEE International Solid-State Circuits Conference Digest of Technical Papers, February* (2017), pp. 344–345
15. S. Kim, Y.-C. Shih, K. Mazumdar, R. Jain, J. Ryan, C. Tokunaga, C. Augustine, J. Kulkarni, K. Ravichandran, J. Tschanz, M. Khellah, V. De, Enabling wide autonomous DVFS in a 22nm graphics execution core using a digitally controlled hybrid LDO/switched-capacitor VR with fast droop mitigation, in *IEEE International Solid-State Circuits Conference Digest of Technical Papers, February* (2015), pp. 154–155

16. M.K. Song, L. Chen, J. Sankman, S. Terry, D. Ma, A 20V 8.4W 20MHz four-phase GaN DC-DC converter with fully on-chip dual-SR bootstrapped GaN FET driver achieving 4ns constant propagation delay and 1ns switching rise time, in *IEEE International Solid-State Circuits Conference Digest of Technical Papers, February* (2015), pp. 302–304
17. C.K. Teh, A. Suzuki, A 2-output step-up/step-down switched-capacitor DC-DC converter with 95.8% peak efficiency and 0.85-to-3.6V input voltage range, in *IEEE International Solid-State Circuits Conference Digest of Technical Papers, January* (2016), pp. 222–223
18. T.M. Andersen, F. Krismer, J.W. Kolar, T. Toifl, C. Menolfi, L. Kull, T. Morf, M. Kossel, M. Brundli, P. Buchmann, P.A. Francese, A sub-ns response on-chip switched-capacitor DC-DC voltage regulator delivering 3.7W/mm2 at 90% efficiency using deep-trench capacitors in 32nm SOI CMOS, in *IEEE International Solid-State Circuits Conference Digest of Technical Papers, February* (2014), pp. 90–91
19. H.-P. Le, S. Sanders, E. Alon, Design techniques for fully integrated switched-capacitor DC-DC converters. IEEE J. Solid State Circuits **46**(9), 2120–2131 (2011)
20. G.V. Piqué, A 41-phase switched-capacitor power converter with 3.8mV output ripple and 81% efficiency in baseline 90nm CMOS, in *IEEE International Solid-State Circuits Conference Digest of Technical Papers, February* (2012), pp. 98–100
21. R. Jevtić, H.P. Le, M. Blagojević, S. Bailey, K. Asanović, E. Alon, B. Nikolić, Per-core DVFS with switched-capacitor converters for energy efficiency in manycore processors. IEEE Trans. Very Large Scale Integr. VLSI Syst. **23**(4), 723–730 (2015)
22. J.W. Tschanz, J.T. Kao, S.G. Narendra, R. Nair, D.A. Antoniadis, A.P. Chandrakasan, V. De, Adaptive body bias for reducing impacts of die-to-die and within-die parameter variations on microprocessor frequency and leakage. IEEE J. Solid State Circuits **37**(11), 1396–1402 (2002)
23. M. Blagojević, M. Cochet, B. Keller, P. Flatresse, A. Vladimirescu, B. Nikolić, A fast, flexible, positive and negative adaptive body-bias generator in 28nm FDSOI, in *Proceedings of the Symposium on VLSI Circuits, June* (2016), pp. 61–62
24. Y. Tsukikawa, T. Kajimoto, Y. Okasaka, H. Miyamoto, H. Ozaki, An efficient back-bias generator with hybrid pumping circuit for 1.5 v drams, in *Proceedings of the Symposium on VLSI Circuits, May* (1993), pp. 85–86
25. J. Kwak, B. Nikolić, A self-adjustable clock generator with wide dynamic range in 28 nm FDSOI. IEEE J. Solid State Circuits **51**(10), 2368–2379 (2016)
26. D. Jacquet, F. Hasbani, P. Flatresse, R. Wilson, F. Arnaud, G. Cesana, T.D. Gilio, C. Lecocq, T. Roy, A. Chhabra, C. Grover, O. Minez, J. Uginet, G. Durieu, C. Adobati, D. Casalotto, F. Nyer, P. Menut, A. Cathelin, I. Vongsavady, P. Magarshack, A 3 GHz dual core processor ARM Cortex-A9 in 28 nm UTBB FD-SOI CMOS with ultra-wide voltage range and energy efficiency optimization. IEEE J. Solid State Circuits **49**(4), 812–826 (2014)
27. M. Cochet, A. Puggelli, B. Keller, B. Zimmer, M. Blagojević, S. Clerc, P. Roche, J.L. Autran, B. Nikoli, On-chip supply power measurement and waveform reconstruction in a 28nm FD-SOI processor SoC, in *Proceedings of the IEEE Asian Solid-State Circuits Conference, November* (2016), pp. 125–128

Part III
Body-Bias Deployment in Mixed-Signal and Digital SoCs

This part will describe Body-Bias deployment: design solution synthesis, voltage generation infrastructure and design integration.

Chapter 12
Timing-Based Closed Loop Compensation

Ricardo Gomez Gomez and Sylvain Clerc

12.1 Closed-Loop Timing Monitoring

12.1.1 Introduction

In an open-loop BB-compensation system, a precharacterized association of PVTA information and the required body-bias voltage is stored and accessed at runtime to regulate the substrate voltage. In a closed-loop timing compensation system, the monitored circuit's propagation delay is incorporated in a feedback loop, and the difference between the monitored value and the target propagation delay (given by the circuit targeted frequency) is minimized through the tuning of the body-bias. This approach benefits from the fact that an association of process and environmental data and the circuit's speed is a complex and sometimes an inaccurate exercise, especially in scaled nodes and low-voltage designs.

Figure 12.1a shows the basic building blocks of a closed-loop control system. The parameter to be controlled is monitored by a sensor and compared to the reference value, often called setpoint. The system then minimizes the difference between these two values by generating a control signal which drives the actuator subsystem, finally inducing a difference in the controlled system's variable. Figure 12.1b identifies the building blocks of Fig. 12.1a in the context of variation compensation with body-bias, hence making the controlled variable to be the product's maximum

---

R. G. Gomez (✉) · S. Clerc
STMicroelectronics, Crolles, France
e-mail: ricardo.gomezgomez@st.com; sylvain.clerc@st.com

© Springer Nature Switzerland AG 2020
S. Clerc et al. (eds.), *The Fourth Terminal*, Integrated Circuits and Systems,
https://doi.org/10.1007/978-3-030-39496-7_12

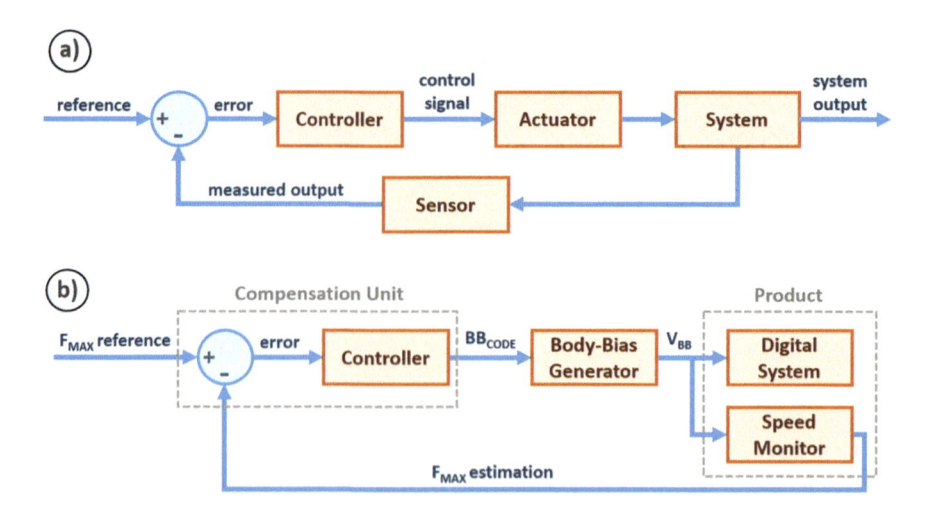

Fig. 12.1 Building blocks of a generic closed-loop system (**a**) and a closed-loop delay monitoring system based on body-bias (**b**)

frequency, F_{MAX}, and the control action to be performed through the body-bias adjustment, V_{BB}.[1]

Ideally, designers would like the controlled variable to be the actual F_{MAX} of the product, so that the product speed would directly be aligned to the targeted specification. However, this would require the live measurement of the product F_{MAX}, which is nearly unfeasible without interrupting functionality. This impediment is avoided through the integration of speed monitors, that is, sensors whose speed is correlated with the product F_{MAX}. This sensor data can be used to estimate the product speed at any given PVTA state. By doing so, the compensation system can indirectly control the circuit's speed without impacting the system availability.

This makes a closed-loop control system based on delay monitoring and body-bias to be composed by:

- One or more speed monitor(s) that estimate the product's F_{MAX}.
- A compensation unit which:

    – Interfaces the on-chip speed monitors and the body-bias generator.
    – Implements the control function (discussed on Sect. 12.3) to generate the required body-bias value based on the difference the reference ($F_{MAX,REF}$) and the estimated F_{MAX}.

- A body-bias generator to supply the compensated circuit with the requested substrate voltage.

---

[1] For the sake of simplicity, the body-bias voltage of the PMOS devices ($V_{BB,P}$) and the NMOS devices ($V_{BB,N}$) are jointly referred to as V_{BB}.

12.2 Speed Monitors

The speed monitor's characteristics have a significant impact on those of the closed-loop compensation system. This makes the choice of speed monitor a fundamental basis for the implementation of the complete system. In this section, we review the most commonly used monitors that are used to estimate the circuit's F_{MAX}. While a generic monitor ranking that covers all possible use cases cannot be concluded, we will finish this section by a comparative evaluation covering key criteria such as matching accuracy, power/performance/area overhead, etc.

12.2.1 Ring Oscillator Based Monitors

One of the most widely used solution to monitor the circuit performance is ring oscillator (RO) based monitors. In such structures, a set of digital logic cells acting as delay stages are connected in a ring fashion (Fig. 12.2 top) to achieve a continuous oscillation.[2] The oscillating frequency $F_{RO,osc}$ of a ring oscillator approximately becomes

$$F_{RO,osc} = \frac{1}{\left(2 \sum_{k=0}^{k=N} \tau_{d,k}\right)} \quad (12.1)$$

being N the number of elements and $\tau_{d,k}$ the propagation delay due of the k element.

The ring oscillator is integrated in the design under test (DUT), sharing the same supply V_{DD}, body-biasing, and TA environmental conditions. This makes the delay $\tau_{d,k}$ of each delay element to be subject to similar PVTA variations than the ones influencing the DUT. The delay variation correspondence between the ring oscillator and the DUT makes the oscillating frequency $F_{RO,osc}$ to be correlated with the product's maximum speed, making the RO become an F_{MAX} estimator.

The speed tracking capability of the RO depends on how much the delay elements in the RO's datapath match the logic gates limiting the DUT's maximum frequency. Hence, by making the RO's delay elements correspond to the gates on the DUT's critical paths, designers can trade-off monitor re-usability with better F_{MAX} matching. This is detailed in Fig. 12.2, where a range of RO's datapath composition strategies are shown:

- In **1**, the ring oscillator's datapath is built by using a single type of logic gate (e.g., an inverter, NAND, NOR, etc.). This path presents the highest re-usability as it is totally design independent at the cost of a lower matching with the DUT's

---

[2]Under the assumption that the ring's combinational path is inverting, that is, $DATA_{RO,out} = \overline{DATA_{RO,in}}$.

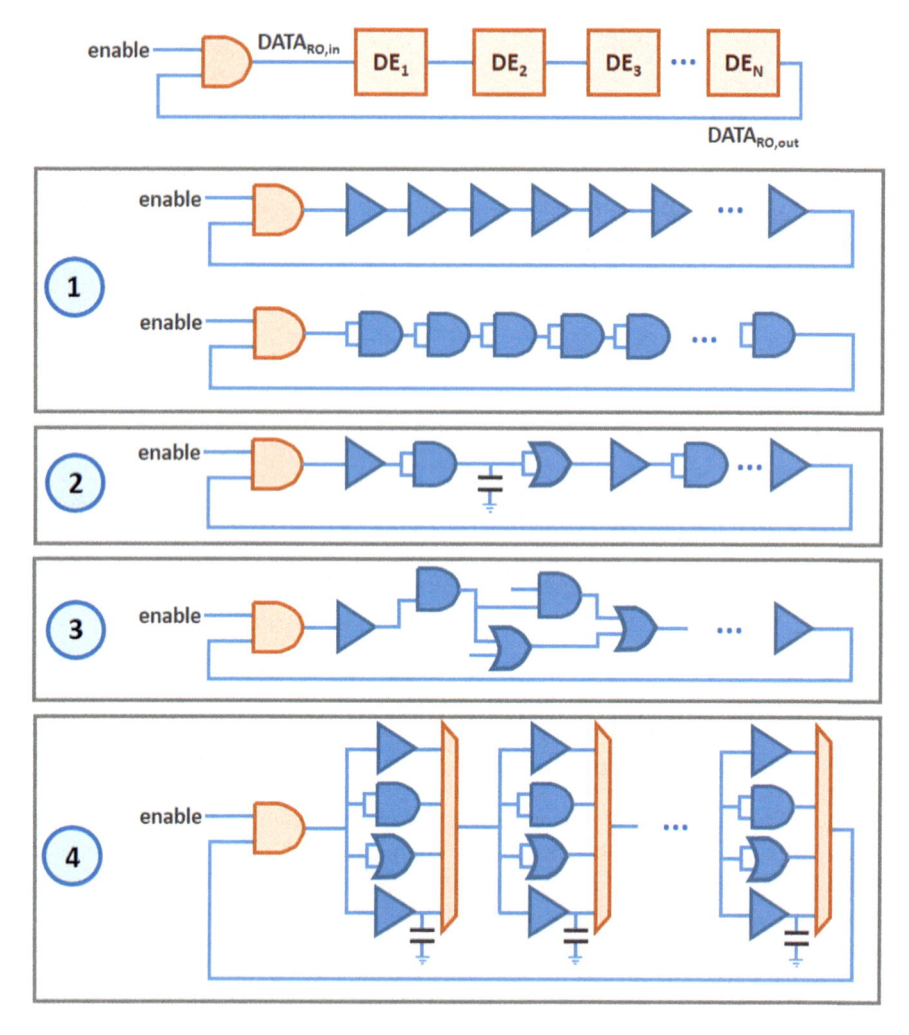

Fig. 12.2 Generic ring oscillator based monitor (top), various RO's datapath compositions (middle), and programmable-RO based monitor (bottom)

F_{MAX}. Without modifying the RO composition, monitor-DUT's matching can be improved by combining the data coming from several rings [1] at the cost of engineering complexity and higher area overhead.
- In circuit **2**, a path composed by a pre-defined set of INV, NAND, NOR, and interconnect delay elements is implemented, aiming a more complete path composition w.r.t. the previous circuit.
- In circuit **3**, the critical path of a implemented design is extracted and used to build the RO's datapath. This path can be replicated from the same application-type circuit (e.g., an ALU coming a generic CPU) to improve the matching of similar circuits while allowing some degree of re-usability, or be extracted

from the DUT during static timing analysis (STA), achieving the best possible matching, but implying that the monitor needs to be re-implemented every time there is a significant variation on the critical path's composition.
- In circuit **4**, a RO-based monitor includes different types of logic gates that can be combined based on programmable selection logic, which drives the multiplexers on the RO's datapath. This arrangement allows the monitor to be calibrated based on post-silicon engineering data to better improve the product's F_{MAX}, achieving both a good matching and high re-usability at the cost of higher design complexity and area overhead.

12.2.2 Tunable Replica Circuits

In ring oscillator based monitors, the contribution of flip-flops to the critical path's maximum frequency is not incorporated, as only the combinational logic is replicated on the monitor's datapath. As a result, deeply pipelined designs with a significant proportion of FF's clock-to-output and setup-timing constraints on the maximum frequency are not properly estimated through ring oscillator monitors.

In tunable replica circuits (TRCs), both the FF at the beginning (launching FF) and at the end (capturing FF) of the path are incorporated on the monitor's datapath, making the TRC emulate the DUT's critical path that limits the achievable frequency.

The composition of a TRC, depicted in Fig. 12.3, can be detailed as follows:

- A launching register that triggers the data transition at the FF's clock active edge. Additional logic can be added to generate a selectable rise/fall data transition.
- A datapath composed by a programmable amount of logic gates to track the DUT's datapath delay variations. Based on whether the combinational datapath combines more than one type of digital gate or not, TRCs can be classified as having multi-cell type path composition [2] and single-cell type path composition [3].
- A set of delay cells in the tunable delay section to adapt the datapath's length and extend the monitor's dynamic range. Tuning delay cells are commonly buffers or

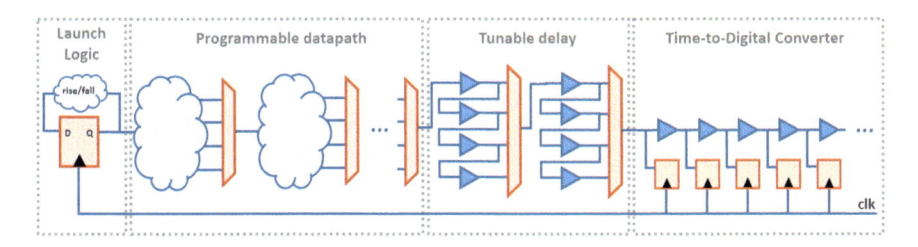

Fig. 12.3 Building blocks of a tunable replica circuit

inverter pairs due to its low area impact, balanced rise and fall delays, and high tuning granularity.
- A time to digital converter (TDC) to provide a thermometer code representing the position of the data transition at the arrival of the next active clock edge. This reading is then used to evaluate the propagation speed across the logic path within a clock period. Depending on the use-case, the TDC depth can range from few bits as in [2], where the output is used as **early/ok/late** warning signals, to a higher depth [3–5], commonly used to achieve a higher dynamic range or a better resolution.

12.2.2.1 Multi-Cell Type Path Composition

A multi-cell type path composition TRC is depicted in Fig. 12.4. In this case, the datapath is built as a cascade of delay stages composed by a selectable number of the same logic gates, being different among stages. The aim of this type of composition is to emulate, as close as it is possible, the actual cell composition of the critical path. The main advantage of this strategy with respect to the single-cell type composition is the possibility to adapt the datapath's length with the same type of cells used to track the DUT's variations.

12.2.2.2 Single-Cell Type Path Composition

Single-cell type path composition simplifies the TRC calibration process by integrating several pre-defined paths with a single type of gate cell (Fig. 12.5), drastically reducing the amount of programmable combinations that can be used. By doing so, the dominant delay component limiting the DUT's F_{MAX} (NMOS-series with NAND, PMOS-series with NOR, RC-dominant with interconnect, etc.) can be tracked at a lower engineering effort.

12.2.3 Endpoint Monitors

While RO and TRC monitors relate the target circuit's F_{MAX} to the propagation delay across logic that intentionally matches the one existing at the DUT's, endpoint

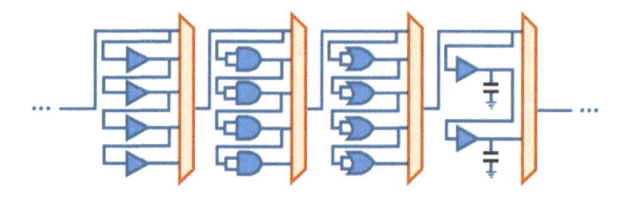

Fig. 12.4 TRC's datapath consisting of a multiple type of cell

Fig. 12.5 TRC's datapath consisting of a single type of cell

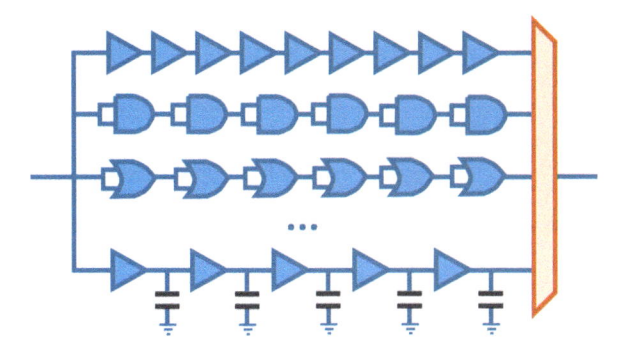

monitoring techniques do so by probing the DUT's endpoints and generating a warning signal whenever the detected timing slack falls below a given limit. It can be implemented by two methods:

- **Custom Canary Register**: The baseline FF is substituted by a custom FF, which integrates the capability to generate a warning signal whenever a transition on the data input occurs within a certain observation window [6–8]. The main advantage of this solution is the lower area impact with respect to the register replication method. However, this strategy requires the custom design of the canary-FF, which can be a significant design effort considering the number of FFs existing in any standard cell's offer. Furthermore, in case a different timing slack margin is required, the whole canary-FF offer needs to be re-designed.
- **Register Replication**: In this implementation, each monitored FF is replaced by two FFs registering the same data, being the data input of one of them (the canary-FF) delayed by a fixed amount of time. Then, an XOR operation between the outputs of both FFs generates a warning signal if the canary-FF has registered different data than the non-delayed FF, evidencing that the timing slack at the monitored endpoint is below the implemented limit. The main advantage of this solution w.r.t. the previous one is that it does not require the design of a custom FF, allowing the users to utilize the FFs existing on the standard cell's offer. On the contrary, the replication of the FF plus the added delay element has a higher area impact than the custom canary register solution.

The warning signals generated at the endpoints are then aggregated by means of an or-tree to generate a unified warning signal. In some cases, the location and number of violating endpoints are used at system level to drive the compensation controller.

12.2.4 Critical Path Replica

Figure 12.6 shows the critical path replica (CPR) sensing principle. The DUT's monitored path is replicated and instrumented by generating a data transition on the

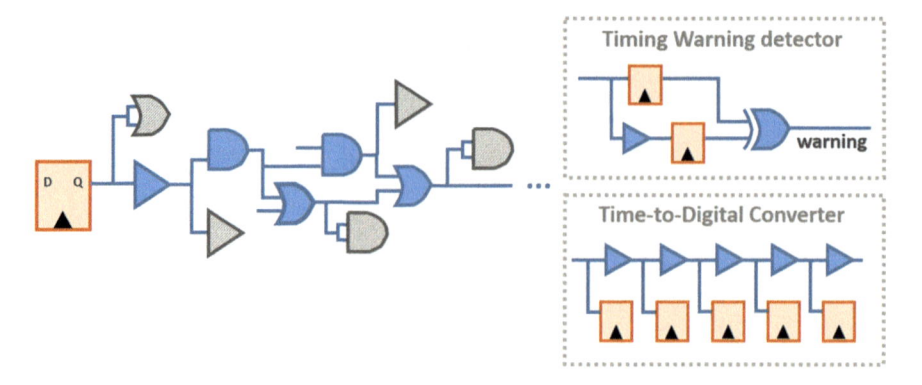

Fig. 12.6 Critical path replica sensor with either timing warning detection or time-to-digital conversion

launching FF. The replicated cells' inputs are tied to the values giving the worst propagation delays as extracted from STA analysis. The effect of the gate's fanout on the critical path is incorporated through the addition of dummy cells, which are depicted in grey color in Fig. 12.6. The propagation delay through the CPR's datapath can be sensed with either a simple timing warning detector composed by a capturing FF and a canary-FF, or a TDC, which enables a multi-valued speed read. To incorporate local variation effects, CPRs can be instantiated close from the monitored paths [9], or dispersed across the chip [10], achieving a robust sampling of the chip-wide critical path variations.

12.2.5 Monitor Calibration

Any mismatch between the timing monitor and the DUT implies an error on the F_{MAX} estimation, which needs to be covered by a timing margin to prevent insufficient compensation, which would otherwise lead to a functional failure. However, the timing margins accounting for monitor-to-circuit miscorrelations reduce the achievable gains of this compensation approach. For this reason, it is important to incorporate a calibration step to increase the monitor accuracy and reduce the miscorrelation margin of the closed-loop system.

In Fig. 12.7 three different types of monitor calibration are illustrated, starting from a CAD-only perspective where the monitor components and settings are fixed before fabrication, to a die-per-die adjustment during high volume manufacturing (HVM).

In Fig. 12.7a, canary-FF monitoring is used as example of the CAD-only perspective. Relying on STA-extracted paths ranking and/or spice simulations, canary-FFs are placed and their detection windows' size is fixed without the capability to tune them after fabrication.

12 Timing-Based Closed Loop Compensation   313

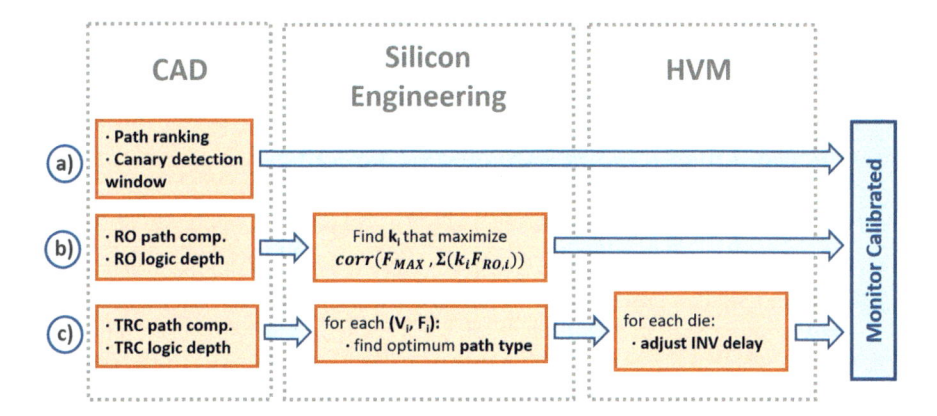

Fig. 12.7 Timing monitors' calibration

In Fig. 12.7b, the multi-RO monitoring approach (as described in Sect. 12.2.1) exemplifies the silicon engineering monitor calibration. At design time, CAD-data is used to set the RO's logic depth and composition according to the desired specifications. Then, during silicon engineering, a representative set of dies of all process corners is used to obtain the optimal pondering factors k_i of the ROs oscillation frequencies $F_{RO,i}$ that maximizes the correlation with the product's F_{MAX} throughout all dies and corners.

Finally, Fig. 12.7c illustrates an example where a per-die tuning is added on top of the silicon engineering calibration and CAD. As in the previous case, CAD tools and data are used to set the TRC path composition and logic depth (please refer to Sect. 12.2.2.2). Then, during silicon engineering, one path type per voltage (V_i) and frequency (F_i) pair is found. To do so, the tracking of each type of path across the temperature range is evaluated for all sample chips, and the path with the smallest tracking error is chosen for that particular (V_i, F_i) pair. Finally, during high volume manufacturing phase, the adjustable INV delay that obtains a TDC mid-point read under test conditions is chosen die-per-die and used during the die's lifetime.

12.2.5.1 Timing Margin Elaboration

To ensure a reliable closed-loop operation under non-idealities (e.g., limited feedback loop response time), a delay guard band needs to be added on top of the estimated F_{MAX} by the speed monitor. The optimal sizing of this delay guard band is key to avoid an excessive margin degrading the DUT's performance while ensuring a reliable operation. The elaboration of the timing margin to be added can be described as follows:

First, a timing margin needs to be added to account for the mismatch between the monitor and the DUT. Indeed, while the application of post-silicon tuning techniques can effectively improve the quality of the F_{MAX} estimation, a perfect monitor-to-DUT correlation is not possible. The miscorrelation can be quantified as the frequency error ($F_{MAX,diff}$) that results from the difference between the $F_{MAX,monitor}$ and the product's F_{MAX}. In order to ensure a reliable control operation, the timing margin added on top of the monitor read needs to be, at least, the largest $F_{MAX,diff}$ across all product's PVTAB corners (Eq. (12.2)).

$$F_{margin,mismatch} \geq \max_{\{P,V,T,A,B\}} (F_{MAX,diff}) \quad (12.2)$$

Second, a timing margin needs to account for the non-instantaneous response of the compensation system to a drift in the DUT's F_{MAX}. The margin to be added depends on whether the compensation system gates the clock once a drift has been detected or not. In the case where the clock is gated while changing the bias, the systems' time to react (T_{react}, Eq. (12.3)) is set by time it takes for the monitor to have a valid timing reading ($T_{monitor}$) and for the clock gating to be effective (T_{clock_gate}). In the case where the clock is not gated (Eq. (12.4)), T_{react} corresponds to the time it takes to counter-measure the monitored speed drift. This time is then set by $T_{monitor}$, $T_{controller}$ (the time for the controller to process the control action and generate the BB_{code} request), and T_{bbgen} (the body-bias generator's timing response).

$$T_{react} = T_{monitor} + T_{clock\_gate} \quad (12.3)$$

$$T_{react} = T_{monitor} + T_{controller} + T_{bbgen} \quad (12.4)$$

$$T_{margin,react} \geq T_{react} * \max_{\{V,T,A\}} \left(\frac{\partial D_{prop}}{\partial t}\right) \quad (12.5)$$

In order to ensure a reliable operation, designers need to add a timing margin $T_{margin,react}$ to account for the non-zero T_{react} of the compensation system. This timing margin needs to be, at least, equal to the timing slack difference caused by the fastest delay (D_{prop}) degradation taken place during T_{react} (Eq. (12.5)). This is better explained with Fig. 12.8. $D_{PROP,ERROR}$ stands for the propagation delay that causes a timing error at the product's frequency and $D_{PROP,WARN}$ is the propagation delay limit after which a warning flag is generated by the monitor. In the figure, the delay of the critical path of the system (in orange) is degraded by the fastest variation source $\max(\frac{\partial D_{prop}}{\partial t})$ until the system reacts. The resulting slack degradation is then the minimum timing margin on top of $D_{PROP,ERROR}$ so that, in the worst-case, the system can compensate the delay degradation before a timing error is produced.

Finally, the total margin to be added on top of the speed monitor read is

$$T_{margin} = T_{margin,mismatch} + T_{margin,react} \quad (12.6)$$

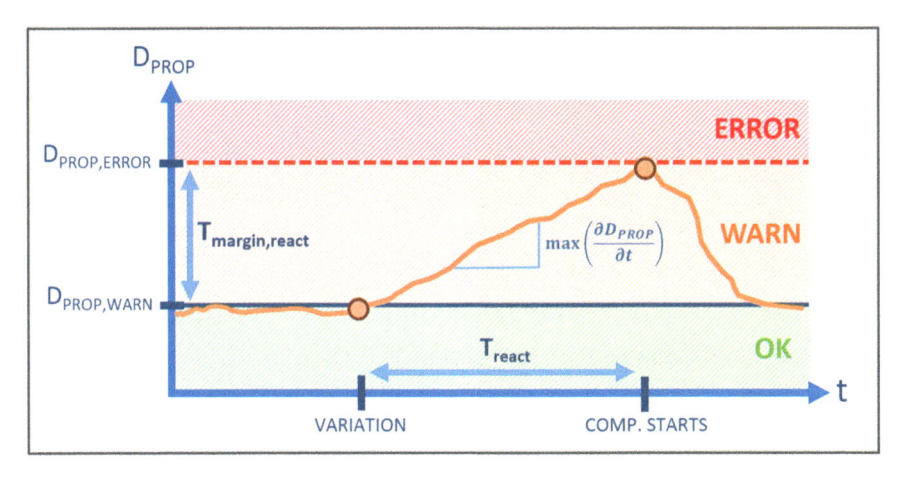

Fig. 12.8 Relation between loop system's T_{react} and the monitor's timing margin $T_{margin,react}$

12.2.6 Monitor Evaluation

The optimal monitoring choice strongly depends on the targeted context and DUT characteristics, making the *one speed monitor for all circuits* approach unrealistic. In this section, we assist the monitors' trade-off discussion by outlining some qualitative aspects of the monitors presented so far, aiding designers in their monitoring choice. The main criteria used in the monitor comparison are:

- **Accuracy**. As presented in Sect. 12.2.5.1, the F_{MAX} estimation error needs to be added in the form of a fixed margin, degrading the achievable gains of closed-loop timing compensation. Hence, monitors with a better estimation accuracy are favored.
- **Overhead**. The area and power overhead of the monitors offset the gain brought by the flat-margining removal, hence playing a crucial role on the monitoring choice.
- **Acquisition's time**. It refers to the time taken by the monitor to perform the acquisition after a request.
- **Intrusiveness**. The intrusiveness of a monitor is a broad concept that refers to the amount of interference the monitor causes in the DUT design and production cycle. This ranges from implementation's intrusiveness, e.g., whenever a significant implementation change needs to be done to accommodate the monitor, to test intrusiveness, e.g., when a monitoring solution requires the addition or modification of a test methodology during HVM.
- **Re-usability**. It refers to the capability of the monitor to match different DUTs without requiring monitor re-design.
- **Monitor's activation**. Monitoring solutions which require a certain activation condition not controlled by the designer, e.g., related to the monitored endpoint activity, risk to have an unbounded response time which would let a timing error

to occur before a warning has been generated, severely degrading the system's safety level. Hence, monitors whose activation condition can be controlled by the compensation system (e.g., ring oscillators or tunable replica circuits) are favored.

- **Single- vs Multi-Valued Monitor Data**. This criterion refers to the quantity of values generated by the monitor to encode the timing reading. They range from the "warning"/"OK" values generated by canary-based monitors to a multi-valued reading delivered by a TDC or a ring oscillator.

12.2.6.1 Ring Oscillator Based Monitors

The characteristics of the RO-based monitors depend on the way the RO's datapath has been constructed. Single-cell composition RO's are highly re-usable but achieve a low F_{MAX} tracking across PVT corners, while RO's built from a critical path extracted have a better product correlation but are specific for each product. Tunable-composition ROs trade-off design complexity and engineering calibration labor with better re-usability/matching w.r.t. to fixed path composition ROs.

The main limitation of RO-based monitors is the combinational-only composition of the sensing logic. The lack of FF-related delays and timing constraints implies a tracking mismatch with the DUT, which may be significant in deeply pipelined designs with a large portion of the delay given by the register elements.

Furthermore, RO's variation tracking require the accumulation of oscillations during a certain given time, which is usually in the order of hundreds of DUT's clock cycles. While this approach is advantageous during process and aging monitoring, where the acquisition's time-filtering effect can remove the supply noise and temperature fluctuations from the process and aging measurement, it invalidates their utilization for fast and dynamic variations' tracking.

12.2.6.2 Tunable Replica Circuits

The main RO limitations are overcome by Tunable Replica Circuits, which improve the DUT variation's tracking by incorporating the FF's effect on the critical paths. Furthermore, the FF-to-FF construction of TRC monitors enables their incorporation for monitoring fast and dynamic variations, as the ones caused by fast voltage droops. However, TRC's dynamic range is lower when compared to RO monitors: the propagation delay across the TRC datapath needs to correspond to the one at the DUT's critical paths to allow the TDC to capture the position of the edge after one DUT's clock cycle. In spite of the incorporation of a tunable delay section, which increases the monitor's dynamic range, TRC monitoring range is limited when compared to ring oscillators.

12.2.6.3 Endpoint Monitors

The substantially different approach of endpoint monitoring techniques, with the promising on-chip variation tracking capabilities and inherent accuracy of DUT's endpoint slack monitoring has drawn significant attention from the academia. However, the gains brought by this type of techniques are hardly transposable to industrial products at a reasonable cost.

First, monitoring coverage is workload dependent, given by the fact that a timing warning flag can be generated only if there is a data transition arriving to the monitored endpoint. This makes the coverage of the canary registers to be unknown at design time, and only provable for a given workload. As an example, the case of a 4-bit carry-propagate multiplier as depicted in Fig. 12.9 can be considered. HA and FA stand for half- and full-adder, and R_0 to R_7 are the FFs registering the output bits of the multiplier. For the sake of simplicity, the propagation delay from the adder inputs towards the adder outputs (either carry-out or data-out) τ_d is considered to be equal for HA and FA. This makes R_7 the endpoint with the smallest timing slack (a critical path example is highlighted in a discontinuous red arrow). In order to monitor the timing slack of the circuit, a canary-FF can be added in parallel to the R_7 endpoint. However, the safety and coverage of this monitoring cannot be assessed, as the monitor is only able to have a timing slack read when R_7 changes its value, which is dependent on the input terms and thus not known at design time. The same reasoning can be extended to R_6, R_5, etc.

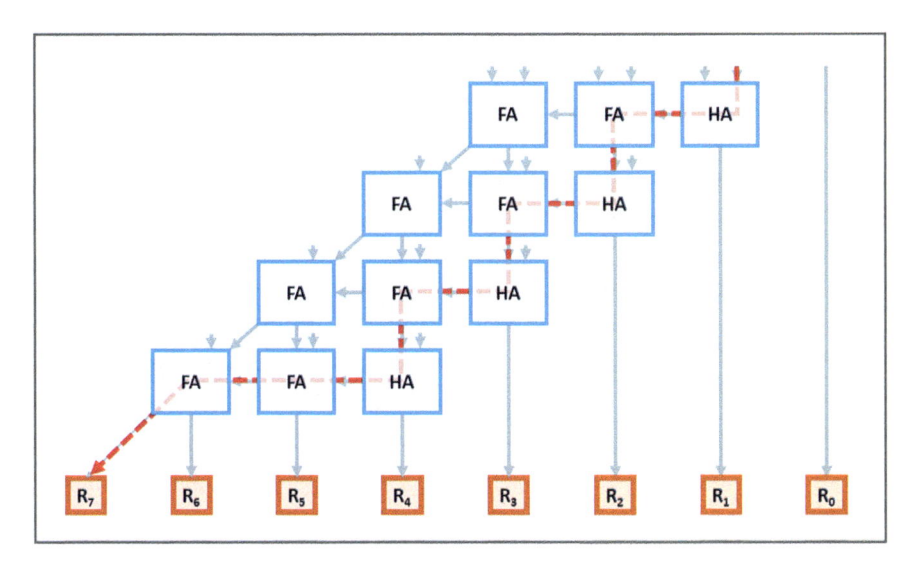

Fig. 12.9 4-bit carry-propagate multiplier circuit. FA and HA stand for Full-Adder and Half-Adder, R_{0-7} are the multiplier's output registers. One of the multiplier's critical path is identified in a red discontinuous line

As the canary-FF monitor activation is not controlled by the designer, but by the actual dataset/workload, the canary-FF monitor coverage cannot be assessed without the need to interrupt functionality to run a workload expressively designed to make sure the monitored endpoints are activated. However, for most applications, the elaboration of functional patterns that activate the desired endpoints is too expensive to justify the implementation of canary-FF monitoring w.r.t. to simpler active-on-demand monitoring approaches. Furthermore, it is seldom the case where the final application software is available at ASIC design time, complicating the approach where STA path-slack database is crossed with functional simulation results to incorporate high-activity endpoints data to the placement algorithm.

Second, the canary-FF monitor placement is driven by the critical path ranking extracted during STA analysis. This makes the placement to be limited to the CAD extracted paths based on a set of global P, V_{DD}, T, A, and V_{BB} extreme corners. However, the critical paths of a typical die at a typical corner are rarely located at the same endpoints w.r.t. the CAD-based critical paths extracted at extreme corners. As the unbiased IPs are not accelerated with BB (unlike the digital logic), critical paths move from pure-logic paths (accelerated with BB) to paths that interface unbiased IPs. This makes circuits containing unbiased IPs (SRAM, LPDDR, etc.) display an alteration of the critical path distribution on corners where body-bias is applied (i.e., $V_{DD,LOW}$, T_{LOW}). Furthermore, the growing contribution of on-chip variation to the paths' timing slacks makes the accurate determination and canary-FF monitoring of the critical paths within all manufactured dies inconceivable at a reasonable cost. Finally, the implementation's leakage recovery phase (where the cells on the non-critical paths are swapped with higher-V_T cells to reduce leakage) makes non-critical paths reduce their positive timing slack and become critical, creating a steep wall-of-slack of a large number of new critical path candidates.

Third, there is also a strong limitation due to the incapability of canary-FF monitors to tune the delay element. The margin window between the timing slack warning and the setup-timing violation is fixed at design time and not tunable at a reasonable cost. This implies:

1. The time to react of the regulation system is fixed at design time prior to the engineering-phase evaluation of the final system. Any deviation from the required value requires re-design, which creates the need to overmargin the delay element to ensure safety will not be compromised, hence degrading the circuit's F_{MAX}.
2. There is no possibility to adjust the timing margin in between different OPPs: the higher variability in near-threshold domain at low voltages necessitates a larger delay element which imposes a significant performance impact at low-variability high performance mode at higher supply voltages.

Finally, while RO monitors and TRCs offer a multi-valued F_{MAX} estimation, either via $F_{osc,RO}$ (ring oscillator) or a multi-bit TDC (tunable replica circuit), the output coming from canary-FF monitors is binary-valued. This difficults the feedback control as there are no early signs of speed drift until it is already close from the setup-timing violation. As an example, the multi-valued TDC output can be used to predict fast voltage droops and activate countermeasures even before the slack gets close to the timing error [3, 4], which is however not possible for canary registers.

12.2.6.4 Critical Path Replicas

CPR-based monitors are characterized by a comparable F_{MAX} tracking w.r.t. canary register monitoring while overcoming many of the challenges endpoint monitoring faces. First, the target path's replication enables the monitor control and activation by the compensation system without any dependency on the workload being run. Furthermore, while CPR composition is also based on a critical path ranking based on STA extraction (and hence is limited to global PVTAB characterizations on extreme corners and critical path reordering), the fact that monitors are implemented independently from the circuit allows designers to integrate the monitors after the target circuit's timing closure without impacting it. The integration of CPR monitors outside the core area, while implying a larger within-die variation in between the functional path and the replicated path, enables designers to integrate more complex delay monitoring techniques such as TDCs and tunable delays. This addition comes without degrading the routability or timing of the monitored path.

The exact path replication in CPRs allow a better path-to-monitor matching w.r.t. TRC monitors, where the monitor's datapath contains multiplexers and selection logic to calibrate the sensor. However, the fact that the CPR cells are fixed from design requires the implementation of large set of CPR instances to cover a wide range of product's paths across all corners, hence ensuring a good critical path coverage. Furthermore, while TRCs can be re-used and calibrated to mimic different F_{MAX} contributors, CPR are fixed and specific for each design, requiring re-implementing the CPR monitor set for every DUT.

12.3 Control Functions

12.3.1 Successive Approximation

In a successive approximation control function, the controller either increases the BB_{code} by one step in case the speed monitor is reading an F_{MAX} below the reference value, or decrease the BB_{code} in case the speed monitor reads a F_{MAX} above the reference value.

The main advantages of this type of control functions are:

- As BB_{code} is incremented/decremented step by step, it ensures no overshoot larger than 1 BB_{code} exists.
- It simplifies the speed sensor design, as it requires only a slower-than-reference / faster-than-reference sensor output.

As a drawback, successively approximating the BB$_{code}$ makes the system incapable to react to quick variations.

12.3.2 Proportional Control

In a proportional control system, the control signal $u(t)$ is proportional to the difference between the desired frequency and the measured frequency as:

$$u(t) = K_p e(t) \tag{12.7}$$

As an advantage over the simple successive approximation function, it enables faster compensation, as the output BB$_{code}$ will depend on the distance between the circuit speed and the target. However, there may be overshoots larger than 1 BB$_{code}$.

12.3.3 Proportional-Integral Control

To remove the residual error existing on single P control systems, an integral term can be added to account for the cumulative value of the error signal $e(t)$, making

$$u(t) = K_p e(t) + K_i \int_0^t e(t) dt \tag{12.8}$$

However, the residual steady-state F$_{MAX}$ error between the target and the actual circuit delay is mainly caused by the body-bias generator step, making the addition of an integral control loop to remove it useless.

12.3.4 Proportional-Derivative Control

The addition of a derivative term aims to reduce the response time of the controller, giving

$$u(t) = K_p e(t) + K_d \frac{de(t)}{dt} \tag{12.9}$$

The main drawback of the proportional derivative (PD) control is, however, the instability and the existence of overshoots. This is especially severe in case the compensation unit modifies the body-bias without stopping the clock, as a large over- or under-shoot may induce a timing violation and cause a system failure.

12.4 Design Example: Process and Temperature Timing-Based Closed Loop Compensation

12.4.1 Architecture

The circuit features an ultra-low voltage SPARC/LEON3 processor able to function at either 0.33 V or 0.45 V, the frequency was, respectively, of 1 MHz and 25 MHz across a temperature range of −40 °C to +40 °C. The CPU's area with Body-Bias compensation was 0.62 mm^2 as a part of a larger demonstration chip[3] of 15 mm^2 (Fig. 12.10).

Fig. 12.10 Ultra low voltage body-bias compensated SPARC microprocessor die photo, ©2015 IEEE. Reprinted with permission

---

[3]This testchip has been used to explore open-loop compensation with body-bias as well. Please refer to Chap. 13 for more details.

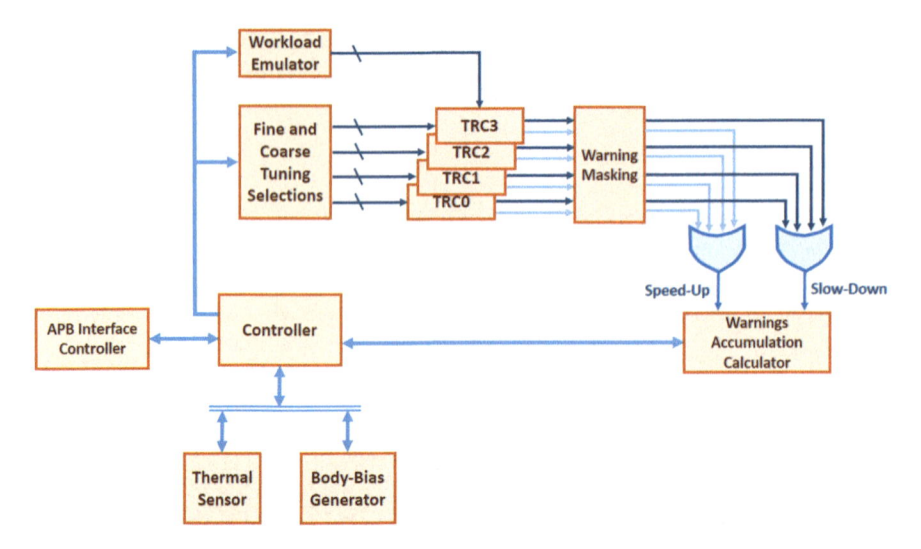

Fig. 12.11 Compensation unit overview

12.4.2 Compensation Unit

The compensation unit (Fig. 12.11) is architected around four instances of timing violation monitors (TRC with multiple-cell type composition) that replicate the critical path of the instrumented design and report to the *Controller* if the system is either too slow, on time, or too fast. The TRC's delay and gate composition can be calibrated through the *Fine and Coarse Tuning Selections* subsystem. At a time, any combination of 1–4 TRCs can be activated by controlling the *Warning Masking* subsystem, which filters out deactivated monitor warnings. The *Workload Emulator* permits exercising different workloads on the TRC monitors by injecting a loop pattern programmable by the *Controller*. The *Warnings Accumulation Calculator* is in charge of accumulating **Speed-Up** and **Slow-Down** warnings coming from the TRC monitors and generate a bias change request in case the amount of **Speed-Up** or **Slow-Down** flags exceed a certain threshold (set by the *Controller*) during a programmable period of time. In case violations are non-monotonic (both **Speed-Up** and **Slow-Down**) during a time window, the violation flag counter is reset. Once the *Warnings Accumulation Calculator* accumulates incoming warnings to generate a bias change request, the *Controller* generates a request to the on-chip *Body-Bias Generator* to either increase or decrease one BB_{CODE} at a time (please refer to Sect. 12.3.1). Temperature-based compensation can also be activated by the *Controller*, which makes use of an integrated *Thermal Sensor* to have an on-chip temperature read.

12 Timing-Based Closed Loop Compensation

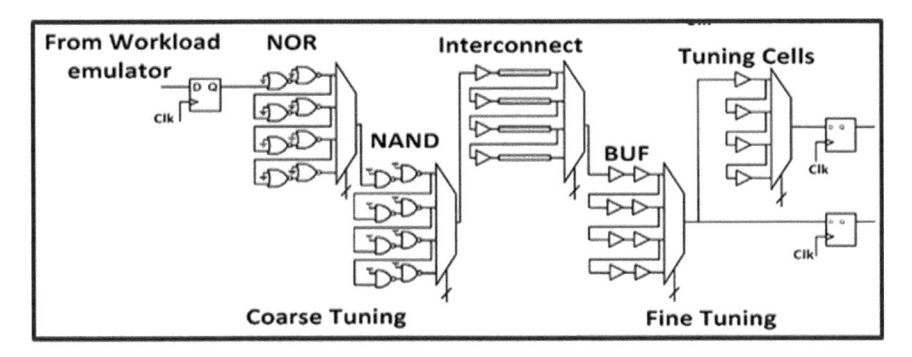

Fig. 12.12 Integrated speed monitor with a critical path composed by a set of NOR, NAND, Interconnect, BUF, and Tuning Cells. ©IEEE 2015 reprinted with permission

12.4.3 Speed Monitor

The speed monitors integrated on this circuit are tunable replica circuit (TRC) with a multi-cell type path composition. Figure 12.12 depicts the integrated speed monitor. A launching FF with the Data pin controlled by the *Workload Emulator* generates a data transition that is propagated through a set of tunable NOR, NAND, Interconnect, and buffer (BUF) gates. The interconnect cell has been composed by a logic buffer plus a twisting metal-1 and metal-2 connection to incorporate a RC-dominated delay component on the monitored path. All inverter gates (i.e., NOR) are placed in pairs to ensure a non-inverting propagation from the launching FF to the capturing FFs. The propagated signal is captured by both capturing FFs with a tunable delay difference set by the *Tuning Cells*, composed by a selectable set of buffers.

The TRC monitor gives a too-slow/ok/too-fast speed read based on the captured value on the TRC FFs. Considering that the timing-correct data at the capturing FFs should be equal to the launching FF's data $data\_in$ (as ensured by the non-inverting tunable path composition), three different situations can take place:

- In case the both the non-delayed and delayed FF registers capture $\overline{data\_in}$, the design is considered to be running too slow, and thus a speed-up signal is generated by the TRC
- In case the both the non-delayed and delayed FF registers capture $data\_in$, the design is considered to be running too fast, and a slow-down flag is generated.
- If the non-delayed FF captures $data\_in$ while the delayed FF captures $\overline{data\_in}$, the design is considered to be running at proper speed and no warning flag is generated by the monitor

12.4.4 Measurements

Figure 12.13 shows the live measurement of the FBB modulation (left) and current (right) when varying the circuit's temperature from +40 °C to −40 °C. The lower circuit speed at low temperature is monitored by the compensation system, which step by step increases the body-bias voltage to recover the speed degradation. The current compensation induced by body-biasing is shown on the right plot in Fig. 12.13. The peak discontinuities correspond to the clock gating taking place before each bias step is applied.

Figure 12.14 shows the frequency and power improvement enabled by the body-bias utilization. There, a large performance improvement from the kHz range to the MHz range was enabled by monitoring the circuit speed and compensating with body-bias. Furthermore, the reduction of body-bias when it is not needed (i.e., at high temperature) makes the BB compensation not have any impact on the power figure.

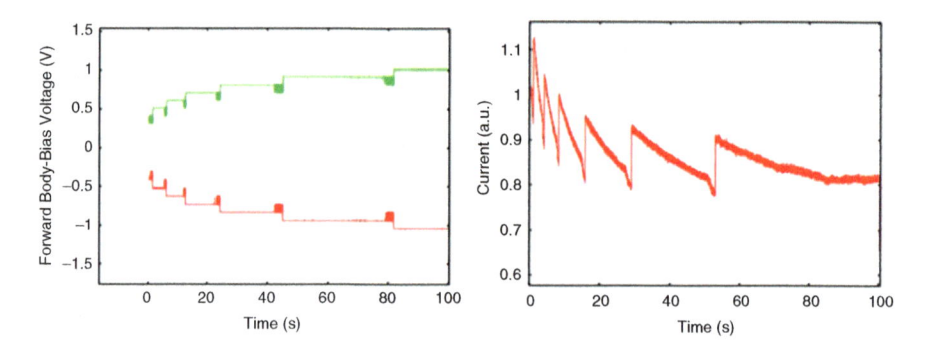

Fig. 12.13 FBB regulation and current evolution over-time when varying the temperature from +40 °C to −40 °C at 0.35 V_{DD} and 1 MHz F_{CLK}. ©IEEE 2015 reprinted with permission

		Baseline	T compensated
F_{MAX}	0.33V @-40C	Few kHz	1MHz
	0.45V @-40C	Few 100kHz	20MHz
P_{LEAK}	0.33V	90μW@20C 0V FBB	90μW@20C 0V FBB
	0.45V	472μW@20C 0V FBB	472μW@20C 0V FBB

Fig. 12.14 F_{MAX} and P_{LEAK} improvement after closed-loop temperature compensation with respect to non-compensated circuit

References

1. J. Kim, K. Choi, Y. Kim, W. Kim, K. Do, J. Choi, Delay monitoring system with multiple generic monitors for wide voltage range operation. IEEE Trans. Very Large Scale Integr. VLSI Syst. **26**(1), 37–49 (2018)
2. S. Clerc, M. Saligane, F. Abouzeid, M. Cochet, J. Daveau, C. Bottoni, D. Bol, J. De-Vos, D. Zamora, B. Coeffic, D. Soussan, D. Croain, M. Naceur, P. Schamberger, P. Roche, D. Sylvester, 8.4 a 0.33v/-40c process/temperature closed-loop compensation SoC embedding all-digital clock multiplier and dc-dc converter exploiting FDSOI 28 nm back-gate biasing, in *2015 IEEE International Solid-State Circuits Conference (ISSCC) Digest of Technical Papers, February* (2015), pp. 1–3
3. M. Cho, S.T. Kim, C. Tokunaga, C. Augustine, J.P. Kulkarni, K. Ravichandran, J.W. Tschanz, M.M. Khellah, V. De, Postsilicon voltage guard-band reduction in a 22 nm graphics execution core using adaptive voltage scaling and dynamic power gating. IEEE J. Solid State Circuits **52**(1), 50–63 (2017)
4. S. Shibahara, C. Takahashi, K. Fukuoka, Y. Kitaji, T. Irita, H. Hara, Y. Shimazaki, J. Matsushima, A 16 nm FinFET heterogeneous nona-core SoC supporting ISO26262 ASIL B standard. IEEE J. Solid State Circuits **52**(1), 77–88 (2017)
5. C. Berry, D. Wolpert, C. Vezyrtzis, R. Rizzolo, S. Carey, Y. Maroz, H. Shi, D. Chidambarrao, C. Jacobi, A. Saporito, T. Strach, A. Buyuktosunoglu, P. Lobo, P. Chuang, P. Owczarczyk, R. Bertran, T. Webel, P.J. Restle, IBM z14: Processor characterization and power management for high-reliability mainframe systems. IEEE J. Solid State Circuits **54**, 1–12 (2018)
6. I. Kwon, S. Kim, D. Fick, M. Kim, Y. Chen, D. Sylvester, Razor-lite: a light-weight register for error detection by observing virtual supply rails. IEEE J. Solid State Circuits **49**(9), 2054–2066 (2014)
7. S. Das, C. Tokunaga, S. Pant, W. Ma, S. Kalaiselvan, K. Lai, D.M. Bull, D.T. Blaauw, RazorII: In situ error detection and correction for PVT and SER tolerance. IEEE J. Solid State Circuits **44**(1), 32–48 (2009)
8. M. Fojtik, D. Fick, Y. Kim, N. Pinckney, D.M. Harris, D. Blaauw, D. Sylvester, Bubble razor: eliminating timing margins in an ARM cortex-M3 processor in 45 nm CMOS using architecturally independent error detection and correction. IEEE J. Solid State Circuits **48**(1), 66–81 (2013)
9. R. Wilson, E. Beigne, P. Flatresse, A. Valentian, F. Abouzeid, T. Benoist, C. Bernard, S. Bernard, O. Billoint, S. Clerc, B. Giraud, A. Grover, J.L. Coz, I.M. Panades, J. Noel, B. Pelloux-Prayer, P. Roche, O. Thomas, Y. Thonnart, D. Turgis, F. Clermidy, P. Magarshack, A 460MHZ at 397mV, 2.6GHZ at 1.3V, 32b VLIW DSP, embedding FMAX tracking, in *2014 IEEE International Solid-State Circuits Conference Digest of Technical Papers (ISSCC), February* (2014), pp. 452–453
10. K. Wilcox, D. Akeson, H.R. Fair, J. Farrell, D. Johnson, G. Krishnan, H. McIntyre, E. McLellan, S. Naffziger, R. Schreiber, S. Sundaram, J. White, 4.8 a 28nm x86 APU optimized for power and area efficiency, in *2015 IEEE International Solid-State Circuits Conference (ISSCC) Digest of Technical Papers, February* (2015), pp. 1–3

Chapter 13
Open Loop Compensation

Sylvain Clerc and Ricardo Gomez Gomez

Open-Loop embedded Body-Bias control is simple.

13.1 Circuits Content

Two circuit designs are reported in this chapter illustrating two method of Body-Bias control: one is ASIC flow based hardware centric and embeds a linear compensation for both voltage and temperature.[1] Its block diagram is displayed in Fig. 13.2.

The other one is software centric: a CPU embeds a thermal sensor and a Body-Bias Generator as peripherals, adjusting the Body-Bias voltage from a C-program look-up table, as illustrated in Fig. 13.1, its die photo is displayed in Fig. 12.10.

The decision between software centric or hardware centric your Body-Bias controller design is application dependent. The factors influencing the choice are listed below:

- Software maintenance with software versatility to be opposed to lower configurability of dedicated hardware
- Post-fabrication tunability need
- CPU bandwidth allocated to Body-Bias control and real time response ability, in the order of milliseconds latency. The hosting CPU should serve the interrupt to change the bias voltage and keep the hosting SoC within its SOF limits (see Sect. 16.3 and previous Chapter's Sect. 12.2.5.1, Fig. 12.8).
- Power performance area arbitration depending on the level of hardware mutualization in your system. For example, in the case of software based regulation, the

---
[1] Bias for voltage and bias for temperature are independent, each related compensation bias can be summed, see Sect. 3.2.

S. Clerc (✉) · R. G. Gomez
STMicroelectronics, Crolles, France
e-mail: sylvain.clerc@st.com

CPU regulating Body-Bias may be also used for other power management unit tasks.

13.2 Mixed ASIC Flow and Software Based Open Loop Body-Bias Controller

The system diagram is displayed in Fig. 13.1. It includes a compensation unit and a power management unit that contains an embedded Body-Bias GENerator, a bias controller interface and a thermal sensor connected together via an APB bus. The bias controller interface is based on a finite state machine and features some configuration and communication registers mapped to the CPU address space. A bias control interrupt is activated every millisecond to read the temperature, and in case a change is detected (as denoted by departure from two temperature bounds), the FSM triggers a bias adjust following an association of bias with temperature which is defined within the C-program of the interrupt handler. The software table associates temperature bounds and bias and the comparison of actual temperature with bounds is done via hardware comparators within FSM.

The design reported here [1] was designed to operate at ultra-low voltage. The temperature compensation enabled to run the CPU at respectively 1 MHz@0.33 V −40 °C and 20 MHz@0.45 V −40 °C. It was showcased during SEMICON 2014, a video of live compensation with frost bomb is available in [2].

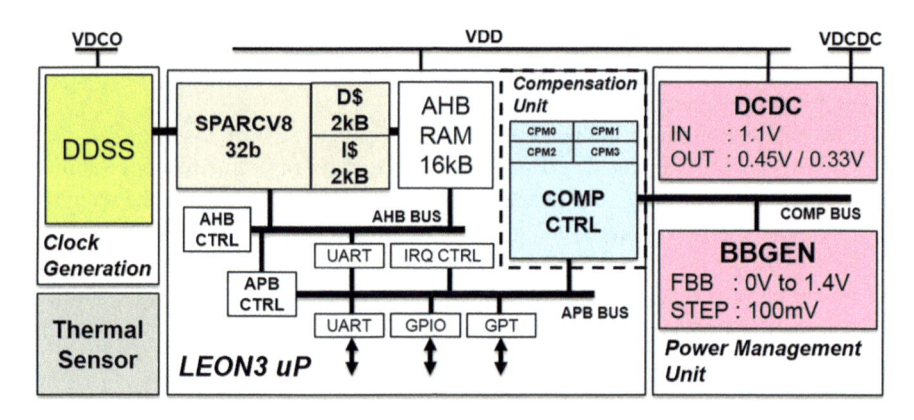

Fig. 13.1 Compensation unit with BBGEN and thermal sensor, the design includes a BBGEN and a thermal sensor, the integrated DC-DC and critical path monitors labelled as CPM in the figure are not used in the case of open loop control, after [1], ©2015 IEEE reprinted with permission

13.3 Full ASIC Flow Based Open Loop Body-Bias Controller

A demonstration circuit embedding an open loop Body-Bias control is displayed in Fig. 13.2. It includes the controller in this section.

Following the assumption that a linear compensation law with both temperature and voltage is acceptable and that these two parameters are independent in the (BB, V, T) space as covered in Sect. 3.2.7, the Body-Bias law will follow the Eq. (3.3), recalled below:

$$BB = BB_P(T) + BB_P(V) + BB_P(a) + BB_{Offset} \quad (13.1)$$

A Body-Bias controller was designed in RTL with the arithmetic operation implemented in fixed-point with 26 bits of precision, 21 for the integer part and −5 for the fractional part, empirically found enough not to saturate computation with −40 °C to 165 °C temperature range and 0.7 V to 1.15 V voltage range encoded by a 12 bits dynamic VTS output.

The design relies on two points of calibration done at test level (see Sect. 16.1).[2] Taking the example of temperature, two points in (BB, T) space are recorded and combined as follows (Fig. 13.3):

$$BB = a \cdot T + b \quad (13.2)$$

Fig. 13.2 An FD-SOI open loop Forward Body-Bias regulation circuit's block diagram. A set of voltage (V) monitors, one temperature (T) sensor and a Body-Bias Generator are integrated in the CPU's domain. These IP's are controlled by a voltage and temperature sensor (VTS) controller and a BBGEN controller, which are accessible via an APB bus from the Body-Bias controller

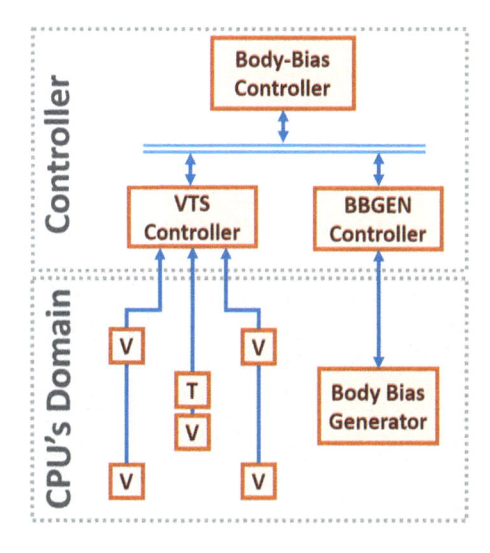

---

[2]Alternatively, to gain test time you could characterize during engineering the Body-Bias slope your system needs covering your production spread and then extrapolate the linear coefficient from single point EWS test.

Fig. 13.3 Illustration of 2-points test read linear Body-Bias law versus temperature

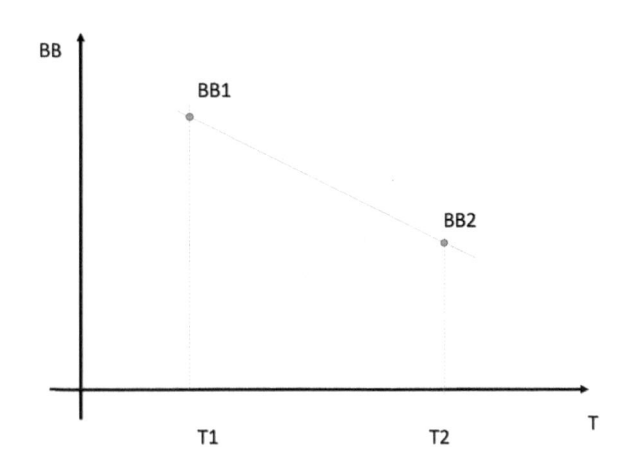

$$a = \frac{BB_2 - BB_1}{T_2 - T_1} \tag{13.3}$$

$$b = \frac{BB_1 \cdot T_2 - BB_2 \cdot T_1}{T_2 - T_1} \tag{13.4}$$

To minimize hardware area, the fixed point division is mutualized and the division is made multi-cycle. The controller FSM computes the bias values for both temperature and voltage and sums them up with an offset which can be used to compensate ageing or N/P imbalance. The power-up sequence is to initialize the bias law linear coefficients and offset, let the BBGEN's charge pumps reach their steady state and then let the machine sets the first bias before the SoC can boot-up.

13.4 Open Loop Controller Solution Design Synthesis

The synthesis of reported Body-Bias controller solutions is displayed in Table 13.1. The cost of the Body-Bias voltage generator, voltage and thermal sensors need to be added in both full ASIC and CPU + ASIC solutions. This makes the total overhead needed to drive 50 mm² of bias island to be 0.38 mm² and 400 μW when bias voltage is at steady state (see Chap. 15 for detailed review of BBGEN design).

Table 13.1 Hardware cost of ASIC and mixed CPU + ASIC Body-Bias controller design

Design	Full ASIC	CPU + ASIC
Gate	22k	5k
Flip-flops	1k	0.5k
Code size	N.A.	1 kB
BBGEN for 50 mm² island	0.3 mm²–100 μW	
Volt + Temp sensor	0.08 mm²–300 μW	

The take-away message is that the overall design cost of a simple linear compensation is in the order of 1k Flip-Flops added to 20k gates in case of full ASIC design or in the order of 1k Flip-Flops and 1k Byte of code in the case of mixed CPU with ASIC dedicated peripherals approach. The only difficulty to integrate Body-Bias control in a SoC is not to fail the STA PVT constellation, this aspect is covered in Chap. 16, this is comparable with AVS compensation situation.

References

1. S. Clerc, M. Saligane, F. Abouzeid, M. Cochet, J.M. Daveau, C. Bottoni, D. Bol, J. DeVos, D. Zamora, B. Coeffic, D. Soussan, D. Croain, M. Naceur, P. Schamberger, P. Roche, D. Sylvester, 8.4 A 0.33V/-40C process/temperature closed-loop compensation SoC embedding all-digital clock multiplier and DC-DC converter exploiting FDSOI 28 nm back-gate biasing, in *2015 IEEE International Solid-State Circuits Conference (ISSCC) Digest of Technical Papers* (2015), pp. 1–3. https://doi.org/10.1109/ISSCC.2015.7062970
2. A. Hars, *SEMICON 2014*. https://www.youtube.com/watch?v=d2roREsk7oI

Chapter 14
Compensation and Regulation Solutions' Synthesis

Ricardo Gomez Gomez

14.1 Body-Bias and Voltage Scaling

14.1.1 Introduction

The principle of applying body-biasing as a post-silicon tuning technique is well known and has been widely applied in bulk technology, either separately [1] or in conjunction with voltage scaling [2]. Nevertheless, the limited body-bias range in bulk CMOS plus the decreasing sensitivity of body-bias modulation with the node scaling trend (going from 280 mV/V in 180 nm to 110 mV/V in 90 nm technology [3]) tipped the scales in favour of voltage scaling, making it the favoured performance tuning approach in bulk technologies.

However, the advances brought by the FD-SOI technology (please refer to Chap. 2) such as improved I_{on} sensitivity to V_{bb} and extended body-bias range have significantly expanded the capabilities of body-bias as a tuning knob, turning the tables in the body-bias vs voltage scaling discussion and re-drawing the region of interest of BB.

14.1.2 Comparison

While both voltage scaling and body-biasing can be used for performance tuning, designers can eventually face the question of which technique to use. Making the right decision implies evaluating which one is capable of tuning the system to reach

---

R. G. Gomez (✉)
STMicroelectronics, Crolles, France
e-mail: ricardo.gomezgomez@st.com

© Springer Nature Switzerland AG 2020
S. Clerc et al. (eds.), *The Fourth Terminal*, Integrated Circuits and Systems,
https://doi.org/10.1007/978-3-030-39496-7_14

the desired goal at the lowest cost possible: it may be required to compensate environmental and process variations without exceeding the power envelope, to reach higher performance at the lowest power and reliability cost (speed boosting), or to reduce energy consumption with the lowest performance loss (low energy mode). The optimal choice that fits all circuits and cases does not exist, but some general trends can be outlined. In this chapter, we examine both tuning techniques to give an insight about which technique may be more appropriate depending on the goal and application context.

14.1.2.1 Variation Compensation

In previous chapters, the effect of process and environmental variations on the circuit specifications has been shown. Then, the capabilities of body-biasing to compensate those variations through closed-loop and open-loop regulation have been demonstrated. However, how the capabilities of body-biasing on variation compensation relate to those of voltage scaling?

The goal of variation compensation is the improvement of the limiting, worst-case ratings of the circuit under variations. Circuit designers care about them because they are later translated into product's specifications and delivered to the customers. In a general case, the worst-case performance, that is, the worst-case die at the worst-case corner sets the performance of the product. This not only applies to performance but in a general manner it comprehends all ratings and specifications of the circuit: as with frequency, the circuit's leakage reported to the customer is the one coming from the leakiest dies at the leakiest corner.[1]

This makes the spread reduction techniques only beneficial under the condition that they have an impact on the reported specifications to the customer. This conflicts with the better-than-worst-case approach where typical- and best-case optimizations lead to an average performance/power improvement, but leaves the worst-case ratings identic, and hence they do not report a gain from the product perspective.

The concept of spread reduction and its effect on the reported product specifications is explained in Fig. 14.1. In (a), a conceptual view of the spread of performance and leakage caused by a generic variation source is depicted. The worst-case product leakage and frequency are identified as $I_{LEAK,A}$ and $F_{MAX,A}$, respectively. Figure (b) presents the same product spread after a better-than-worst-case improvement approach has been performed. There, the typical- and best-case performance optimizations have shifted most of the dies towards the right side, increasing the average performance. However, the performance of the slowest dies is not ameliorated, and hence the product performance limit does not improve.

---

[1] Some exceptions to this worst-case specifications may be found in certain cases such as low power devices for mobile phones. There, the reported consumption usually comes from the typical die and the shipping of some more-consuming dies with a mild battery life degradation is tolerable.

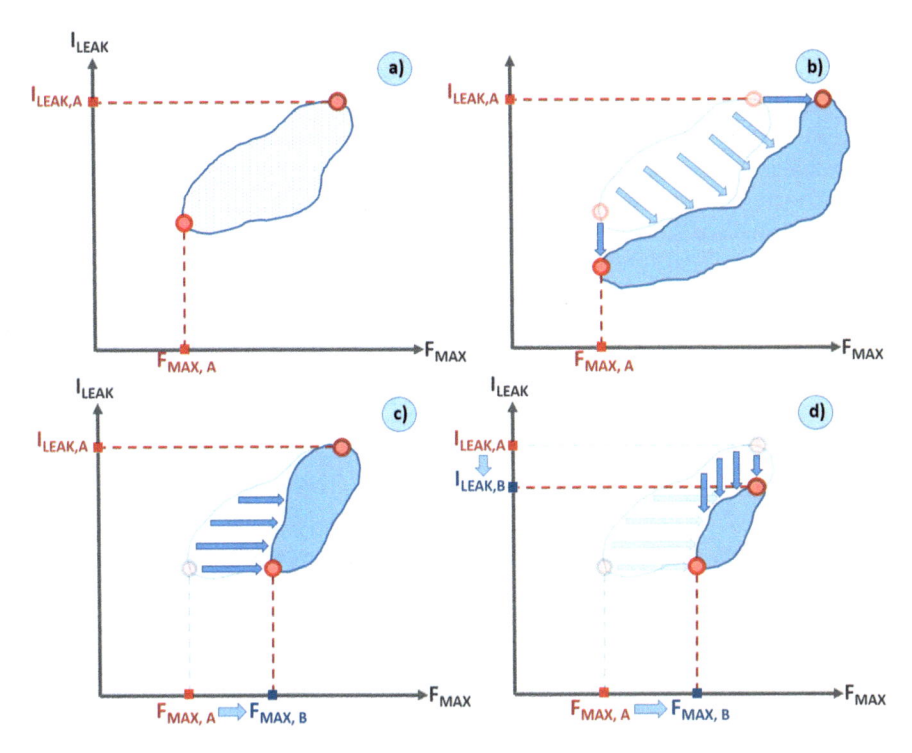

Fig. 14.1 Impact of spread reduction techniques into reported product specifications. (**a**) Die spread in (frequency, leakage) space. (**b**) Better-than-worst-case improvement, all dies but the worst FMAX and the worst leakage ones are improved. (**c**) Improved FMAX worst case. (**d**) Improved FMAX and ILEAK worst cases. ©2019 IEEE. Reprinted, with permission, from R. G. Gomez et al., "Comparative evaluation of Body-Biasing and Voltage Scaling for Low-Power Design on 28nm UTBB FD-SOI Technology," 2019 IEEE/ACM International Symposium on Low Power Electronics and Design (ISLPED)

Similarly, the average leakage has been improved as the product distribution is shifted down, but it has left the worst-case limiting leakage $I_{LEAK,A}$ identical.

Subfigure (c) shows an example where the improvement of the slowest and hence performance limiting dies leads an improvement of the product specifications, going from $F_{MAX,A}$ towards $F_{MAX,B}$. Similarly, in subfigure (d) a compensation approach targeting the worst-case leakage has taken place, improving the product specification leakage from $I_{LEAK,A}$ to $I_{LEAK,B}$. These two cases (c) and (d) depict the goal of variation compensation techniques, that is, the improvement of the worst-case specifications.

14.1.2.2 Process Compensation

To evaluate the costs and benefits of body-biasing and voltage scaling when compensating process variation we have emulated the process spread expected from the manufacturing process of a 28nm FD-SOI technology intended for automotive applications. We have set a 2000-samples Monte Carlo simulation (corresponding to 3-sigma yield at 90% confidence level) including both global and local variations and we have extracted the frequency, dynamic power and leakage at (0.7 V V_{DD}, 0.3 V V_{BB}, 25 °C) of a qFO4 circuit (please refer to Sect. 3.1.2 for a precise description of the test circuit).

This spread is depicted in top section of Fig. 14.2. Here, the process dispersion is represented in a three dimensional plot where the 3D-view of the samples is coloured in blue, and its projection into the axonometric planes has been represented in grey.

Then, we have emulated a per-sample compensation based both on body-biasing (bottom left) and voltage scaling (bottom right). In order to do so, we have set as target the mid-range performance target of the process spread (representative of a TT die) and through an optimization process we have searched the minimum amount of body-bias and supply voltage, respectively, to achieve the desired performance.

From this figure it can be appreciated the higher capability of body-biasing to reduce the spread caused by process variations. While both techniques have been capable of achieving a similar F_{max} spread reduction (with the subsequent improvement of the product's operational frequency), body-biasing has been able to make it without impacting the power specifications. Furthermore, it has reduced the product's leakage set by the worst-case leakage dies (FF) by setting a lower FBB value than the typical sample (0.3 V_{BB}). On the contrary, using supply voltage scaling has caused an increase in the slow samples' dynamic consumption, exceeding the dynamic power worst-case limit and thus degrading the product specifications.

14.1.2.3 Voltage Compensation

The supply voltage reaching the digital logic on the chip may suffer from variations due to the combined effect of several sources such as the regulator tolerance, PCB and package parasitics, and the on-chip's power delivery network (PDN). The voltage fluctuations can range from *fast* voltage droops (such as the ones caused by the instantaneous switching of the logic within a clock cycle) to *slow* changing variations (as it is the case of the voltage tolerance of supply regulators). In conventional designs, to ensure the right performance under such variations, guardbands are used in the form of margins on the voltage stack and/or the clock

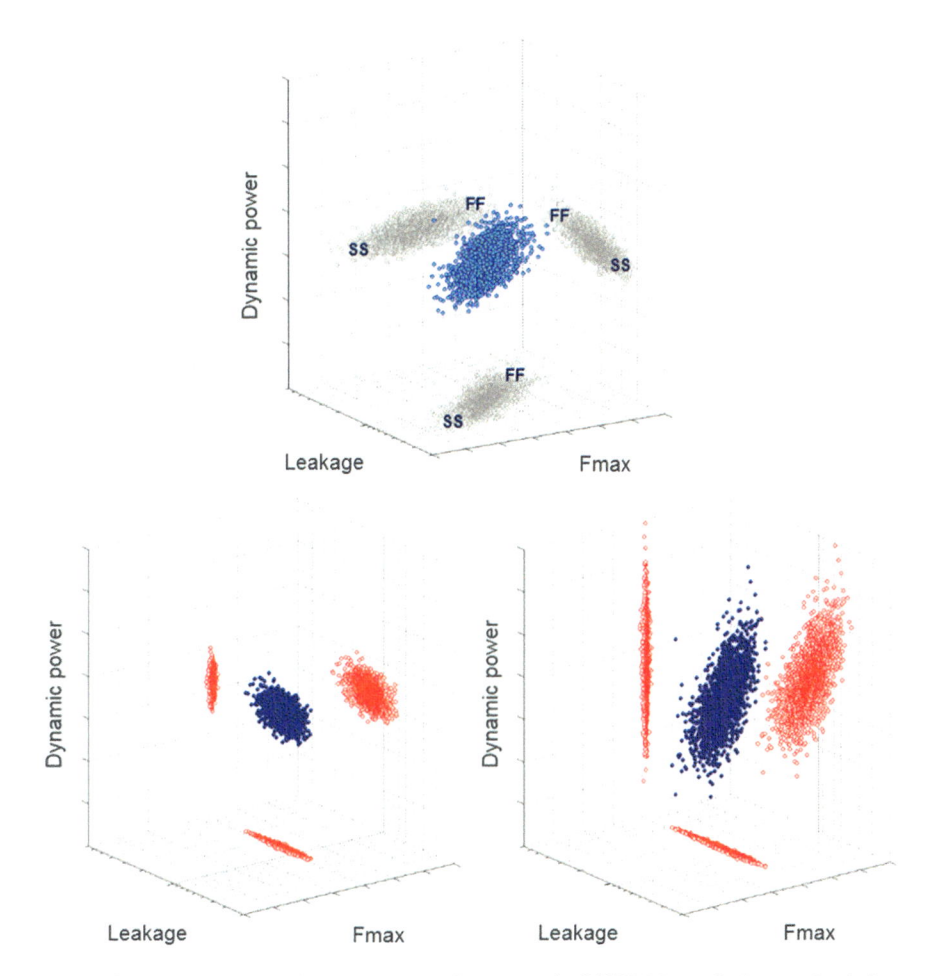

14 Compensation and Regulation Solutions' Synthesis 337

Fig. 14.2 Frequency, dynamic power, and leakage spread of 2000 Monte Carlo samples before compensation (top figure), after body-bias compensation (bottom left), and voltage scaling compensation (bottom right). All three figures have the same axis scale. ©2019 IEEE. Reprinted, with permission, from R. G. Gomez et al., "Comparative evaluation of Body-Biasing and Voltage Scaling for Low-Power Design on 28nm UTBB FD-SOI Technology," 2019 IEEE/ACM International Symposium on Low Power Electronics and Design (ISLPED)

frequency. However, as an alternative to pessimistic guardbands, a regulation loop to compensate *slow* variations on-chip can be embedded.[2]

Figure 14.3a depicts the block level view of a voltage compensation system based on AVS. There, a power management IC supplies a certain V_{DD} on its output terminal. Then, this voltage is delivered through the PCB supply network,

---
[2]Fast voltage droop detection and compensation makes a subject on itself and is not part of our scope here.

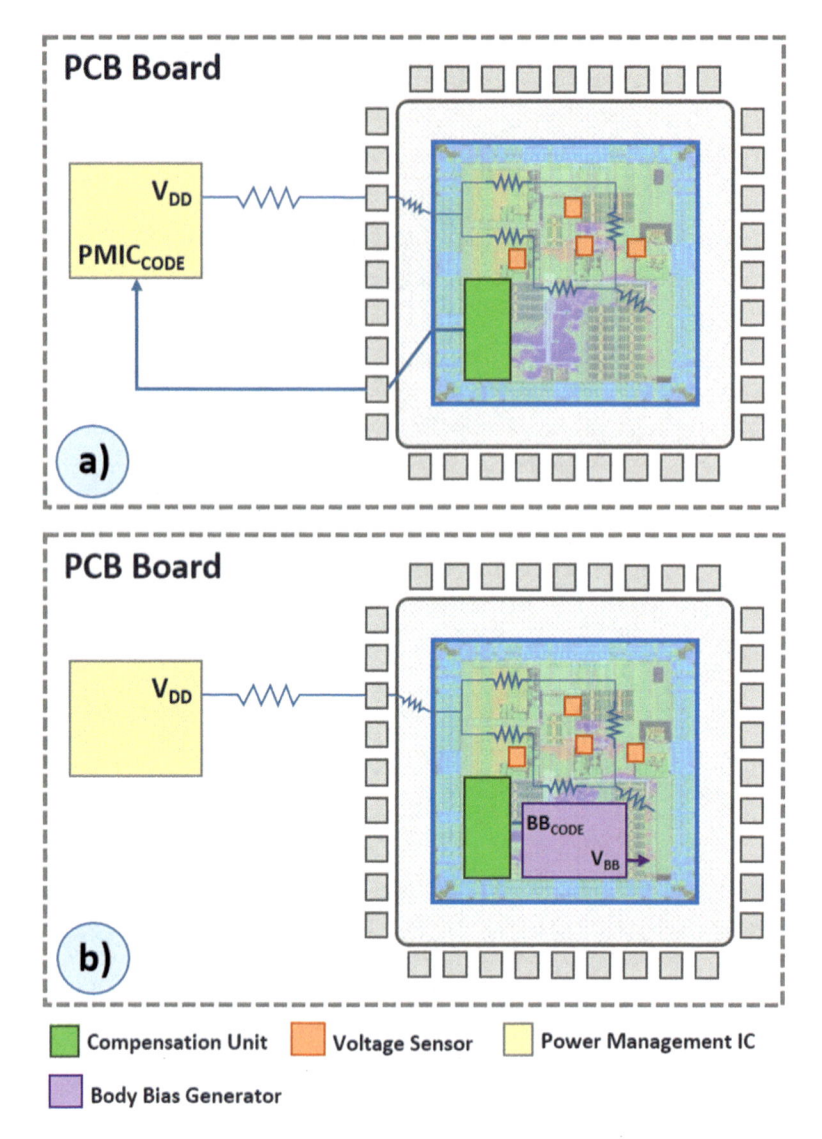

Fig. 14.3 Block diagram of a V compensation system based on (**a**) AVS and (**b**) ABB

the package pins, the IOs, and the PDN of the chip to finally reach the logic. In order to measure the effective supply reaching the logic, a set of voltage sensors have been deployed throughout the logic, relaying the measured voltage to a central compensation unit.

Compensating a drift from the nominal voltage by regulating the supply voltage itself may indeed sound strange. However, in certain cases, static or slow derives and fluctuations on the supply voltage (as it is the case of the LDO tolerance) can

14 Compensation and Regulation Solutions' Synthesis

be compensated by requesting a higher voltage step (in case the measured on-chip voltage is too low) or, in the contrary, a lower voltage step.

Similarly, a voltage compensation system based on body-biasing is depicted in Fig. 14.3b. In this case, the system compensates the slow voltage fluctuations by modulating the body-bias generated by the embedded body-bias generator.

The impact of both techniques on the power envelope will depend on the relative weight of dynamic power and static power on the overall consumption. Requesting an extra V_{DD} code to compensate static drop through the supply distribution will increase the dynamic power of the system, while requesting an extra V_{BB} code will impact the leakage power. It is then up to the designer of the compensation system to choose the right tuning knob in order to reduce the impact on the total power consumption.

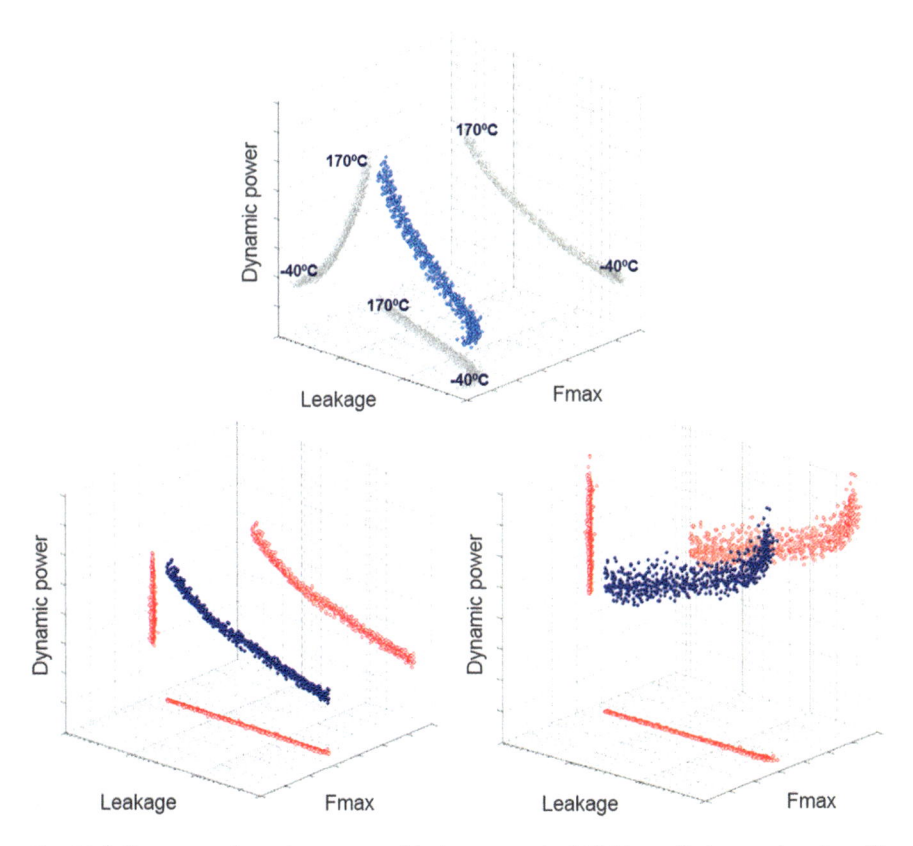

Fig. 14.4 Frequency, dynamic power, and leakage spread of 30 Monte Carlo samples of an SS corner when varying temperature from $-40\,°C$ to $170\,°C$ before compensation (top figure), after body-bias compensation (bottom left), and voltage scaling compensation (bottom right). All three figures have the same axis scale. ©2019 IEEE. Reprinted, with permission, from R. G. Gomez et al., "Comparative evaluation of Body-Biasing and Voltage Scaling for Low-Power Design on 28nm UTBB FD-SOI Technology," 2019 IEEE/ACM International Symposium on Low Power Electronics and Design (ISLPED)

14.1.2.4 Temperature Compensation

To compare the costs and benefits of body-biasing and voltage scaling when compensating temperature variations, we have run a 30-sample Monte Carlo simulation of an SS corner for each 10 °C step from −40 °C to 170 °C at 0.7 V V_{DD} and 0 V V_{BB}. Then, we have extracted the frequency, dynamic power, and leakage as shown in Fig. 14.4.

To emulate body-bias and voltage scaling compensation we have set an optimizer to search the minimum body-bias and supply voltage to achieve the same frequency at the fastest corner, that is, at 170 °C. The resulting frequency, dynamic power, and leakage after reaching the target frequency are depicted in the bottom left plot in the case of body-biasing and in the bottom right for voltage scaling.

While both body-biasing and voltage scaling are able to compensate the temperature effect on the frequency, the former has not impacted the dynamic power nor the leakage. On the contrary, the application of higher supply voltage on the cold (−40 °C) dies has increased their dynamic power consumption over the worst-case dynamic power extracted from the non-compensated dies. As a result, the overall product specifications will worsen as an effect of voltage scaling, making body-biasing the preferred tuning technique to compensate temperature variations.

This benchmark study has been performed using a 28nm FD-SOI technology with a $V_{DD,min}$ below the technologie's temperature inversion point V_{Tinv}, making the slowest corner to be at cold temperature. This makes the application of body-biasing compensation specially interesting, as the leakage increase after the application of forward body-bias at cold temperature does not exceed the worst-case leakage extracted at hot temperature. This is explained in Fig. 14.5, where we show the F_{MAX} (in blue) and the I_{LEAK} (in orange) before (dashed line) and after (solid

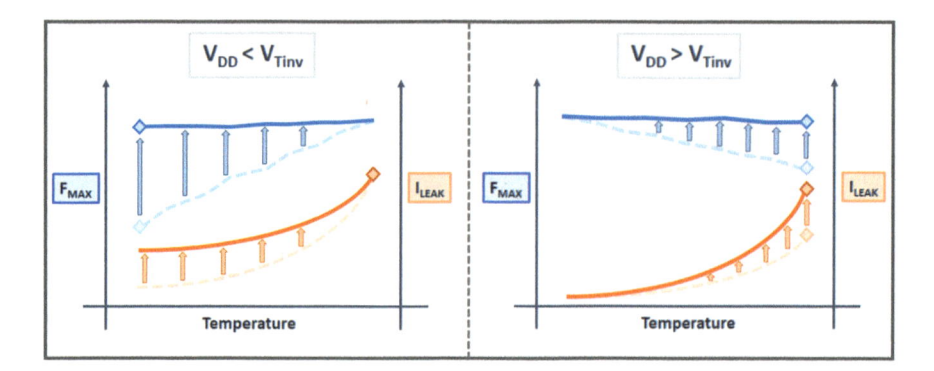

Fig. 14.5 Frequency (in blue) and leakage (in orange) before (dashed line) and after (solid line) body-biasing compensation for supply voltages below and above the temperature inversion point V_{Tinv}. ©2019 IEEE. Reprinted, with permission, from R. G. Gomez et al., "Comparative evaluation of Body-Biasing and Voltage Scaling for Low-Power Design on 28nm UTBB FD-SOI Technology," 2019 IEEE/ACM International Symposium on Low Power Electronics and Design (ISLPED)

line) temperature compensation with body-bias. However, at supply voltages above the temperature inversion point, the slowest corner is located at hot temperature, and then, the application of forward body-bias on the slow corner at hot temperature increases the leakage of the worst-case leakage corner, hence degrading the product specifications.

14.1.2.5 Ageing

We now compare body-biasing and voltage scaling on supressing the ageing effects by measuring their impact on the dynamic and static power of the circuit while maintaining the performance extracted at fresh.

First, we have extracted the delay of the qFO4 at the worst-case fresh delay corner (SS, 0.75 V V_{DD}, BB_{MAX} = 1.00 V V_{BB}, −40 °C). Then, we have extracted the dynamic and static consumption at hot (165 °C) temperature with the minimum body-bias voltage necessary to maintain the previous extracted delay, resulting in this case 540 mV$_{BB}$.

By making use of the aged spice models of the technology we have characterized the NBTI and HCI ageing mechanisms at the (SS, high-V_{DD}, 0 V_{BB}, high-T) stress corner at a fixed activity factor for several timing intervals. Lastly we have extracted the aged-circuit's delay and power at (SS, 0.75 V_{DD}, 1.00 V_{BB}, −40 °C) and (SS, 0.75 V_{DD}, 0.54 V_{BB}, 165 °C), respectively. The extracted results for our baseline circuit are shown in Fig. 14.6.

As expected, there is a circuit delay degradation specially during the first 0.5 stress years, followed by slower degradation beyond this point, almost reaching 5% delay increment after 5 years of ageing stress. The ageing-induced V_{TH} drift has also caused a slight decrease of the dynamic power (due to a reduction of the short-circuit currents) and a considerable reduction of the static power consumption.

The impact of using body-biasing and voltage scaling on the dynamic power of the circuit when compensating ageing is shown in Fig. 14.7. From there, it

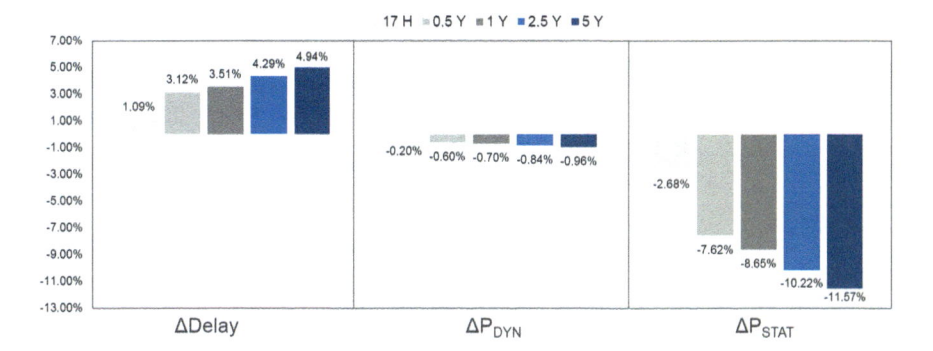

Fig. 14.6 Evolution of the delay, dynamic, and static power consumption of a qFO4 with different ageing profiles

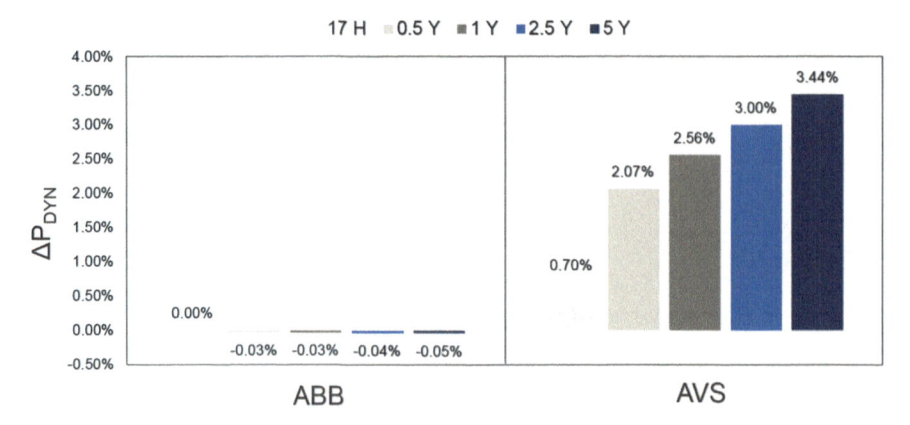

Fig. 14.7 Comparison of the dynamic power consumption after compensating different ageing profiles with body-biasing and voltage scaling

is clear that body-biasing allows the recovery of the lost performance with a negligible impact on the dynamic power (which is not degraded w.r.t. to fresh), while increasing the supply voltage inherently causes a degradation on the dynamic power consumption.

Similarly, Fig. 14.8 shows the impact on the static power consumption of body-biasing and voltage scaling when recovering ageing. In this case, recovering ageing with body-bias has a higher impact on leakage with respect to adaptive voltage scaling. The reason for this effect is the fact that we have applied a symmetrical bias, that is, same voltage on the nwell and pwell. This enables the recovery of the PMOS' V_{TH}, but also lowers the NMOS' V_{TH} as it is forward body-biased, which then increases the leakage currents. In Sect. 14.1.2.7 we discuss the advantages of asymmetrical body-biasing over voltage scaling on targeting asymmetrical variations such as NBTI ageing.

14.1.2.6 Speed Boost

Speed boosting is the technique in which a higher-than-nominal performance is pursued in order to respond to a higher computation demand for a short period of time. As it implies the oversupplying of the circuit, the key point of speed boosting is achieving the maximum performance gain at the lower power and reliability cost.

Voltage scaling is better boosting solution w.r.t. body-biasing. First, there is a stronger performance sensitivity to supply voltage w.r.t. performance sensitivity to body-bias. Furthermore, above the temperature inversion point (located around $0.90\,V_{DD}$ for our example 28nm FD-SOI technology), an excess on body-biasing creates a significant increase in leakage at hot temperature which may degrade the energy efficiency at this corner. In such cases, increasing the supply voltage may be a better choice in terms of energy efficiency.

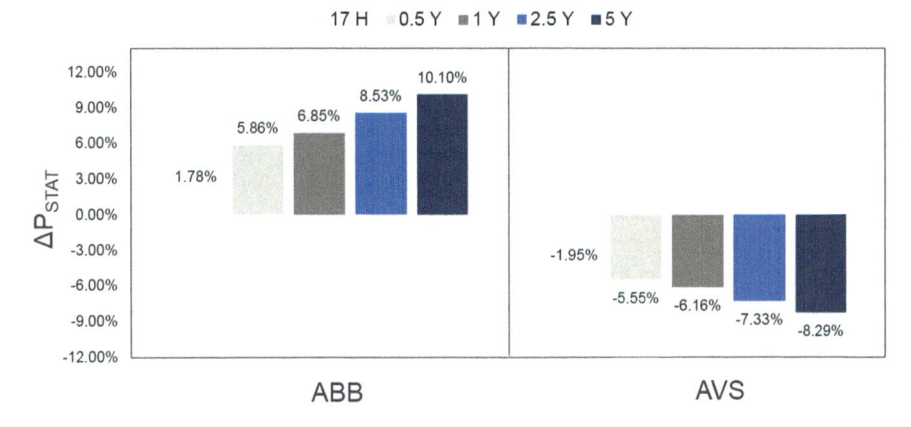

14 Compensation and Regulation Solutions' Synthesis

17 H 0.5 Y 1 Y 2.5 Y 5 Y

Fig. 14.8 Comparison of the static power consumption after compensating different ageing profiles with body-biasing and voltage scaling

However, there are cases where the performance boost with body-biasing is a better choice. An example of this situation are circuits supplied close to the reliability-limiting V_{MAX}. In this mode, an increase of the supply voltage is unacceptable as it surpasses the limit set by the technology's reliability, making body-biasing the only option to further boost performance within reliability-safe limits.

14.1.2.7 Selective N vs P Adjustment

Throughout the previous sections we have considered a symmetric bias, meaning the same body-bias voltage (but with different polarity) applied in both nwell and pwell. However, as it has been introduced, one of the main advantages of body-biasing versus voltage scaling is the capability to selectively tune the P- and N-MOS transistors.

One interesting application of asymmetric bias is the compensation of skewed corners, i.e., SF and FS global corners. There, a FS (Fast-NMOS, Slow-PMOS) corner will suffer an overall performance degradation due to weaker pull-up networks of the digital logic. Raising the supply voltage will mean an increased V_{DS} on the slower PMOS which will recover its strength loss, but will also unnecessarily increase the NMOS' V_{DS}.

On the contrary, the capability to selectively bias either type of transistors allows designers to recover the performance loss due of the skewed corners while reducing the leakage impact that would come with a symmetrical bias.

Asymmetric bias can also reduce the power overhead that comes with the recovery of ageing mechanisms that have a different impact on the NMOS and PMOS devices. One example of such mechanism is the bias temperature instability (BTI), which has a larger impact on PMOS devices (NBTI mechanism) w.r.t. to the one affecting the NMOS devices (PBTI). In previous sections we have shown that these ageing mechanisms can be recovered by both AVS and ABB. However, in both cases, this ageing recovery implied a power overhead due to the unnecessary oversupplying of the NMOS device, whose V_{TH}'s degradation was lower compared to the PMOS degradation. In this scenario, the asymmetric capability of body-bias enables the application of different FBB voltages to the n- and p-well to adapt the performance recovery to the actual device ageing.

CMOS circuits under ionizing radiation's stress, as it is the case in aerospacial applications, are subject to a device degradation which impacts PMOS and NMOS devices differently [4]. In [5], authors demonstrate the capability of asymmetrical body-biasing to individually mitigate the total ionizing dose (TID) effects on NMOS and PMOS transistors fabricated in FD-SOI technology.

14.1.2.8 Multiple Domain Adjustment

The functional, workload, and profile heterogeneity among different units on digital ASICs motivate the implementation of multiple power domains to adapt the supply voltages to the specific requirements of each sub-system. Furthermore, in advanced technology nodes, the existing process, voltage, temperature, and ageing gradients induce different performance and power characteristics across the chip, which can be compensated by implementing a finer on-chip tuning [6].

While both body-bias and voltage scaling can be the base for an on-chip multi-domain adjustment, the costs of the former in terms of regulator integration and inter-domain communication are lower.

First, supplying each domain with a different V_{DD} or V_{BB} requires the independent generation of voltages for each domain, usually done on-chip by using integrated voltage regulators or body-bias generators. The design of a supply regulator with a low area overhead is, however, more challenging than the design of a body-bias generator because of the DC current drained from the supplies, which, in opposition to the capacitive load seen by the V_{BB} generation, makes power conversion efficiency important for supply generators but not for BBGENs. In fact, the BBGEN adaptation to a different (in this case smaller) bias domain size only requires the re-sizing of the drivers to maintain the desired output slew rate.[3] On the contrary, the design of integrated voltage regulators both power efficient and small is more challenging, and is not restricted to the simple re-sizing of the driver as in the BBGEN. Furthermore, while the design of the small-sized BBGEN only impacts the non-critical slew rate, the design of the small integrated voltage regulator must take into account the impact on the conversion efficiency, stability, quiescent current, load regulation range, etc.

---

[3] With the exception of some fast body-bias response required by high-performance processors, the body-bias response is usually not a primarily design target.

Second, interfacing different power domains is also less costly in case of multiple body-bias domains. The reason is that interfacing power domains with different supply voltage requires the addition of level-shifter cells to avoid the impact on the power consumption caused by high-V_{DD} cells driven by low-V_{DD} signals. However, in the case of different body-bias domains, this interfacing can be performed without the need of level-shifter cells with no impact on the power consumption. This eases the integration of different power domains and reduces the cost of a fine-grained adjustment[4] even if well spacing rules need to be taken into account for multi-body-bias domains.

For these reasons we consider body-biasing to be more suitable for a fine-grain domain adjustment, being able to reach higher partitioning granularity at lower cost w.r.t. to multiple-supply domains.

14.1.3 Combination of the Two

Throughout this chapter we have found that depending on the optimization goal and context, either body-bias or voltage scaling is the most suitable compensation solution. In some cases, the application of only one of the previous technique produces satisfactory results. However, for some other cases, it may not be the case. This is especially true for circuits with a variety of operating performance points (OPPs), which operate at a wider range of performance/power characteristics across the supply range. In such cases, we show that the joint application of body-biasing and voltage scaling (whenever it is possible) leads to an optimal compensation solution w.r.t. to a single ABB or AVS approach.

To exemplify this statement we make use of our qFO4 circuit, which is required to perform at two different OPPs: a low power mode running at 0.7 V_{DD} and a high-performance mode at 1.0 V_{DD}, within a specified temperature range from $-40\,°C$ to $165\,°C$.

In Fig. 14.9 we depict the static (left) and dynamic (right) power versus the circuit F_{MAX} of the low power mode (in red) and the high-performance mode (in blue). The slowest (SS) and the fastest and leakiest (FF) corners of the baseline non-compensated circuit are connected with a dotted line. In order to improve the frequency degradation caused by variations, we have compensated the slowest corner with both ABB (square marker) and AVS (diamond marker) to achieve the same F_{MAX} improvement on both cases.

On the low power mode, ABB is the clear winner. As the nominal voltage of 0.7 V_{DD} is well below the temperature inversion point V_{Tinv}, biasing the slowest corner does not impact neither the worst P_{STATIC} point nor the worst $P_{DYNAMIC}$ point, meaning the performance improvement comes without any impact on the power specs (we neglect here the BBGEN and multi-level LDO regulator power

---

[4]It can be assumed that no extra well taps will be needed because they are present in any case.

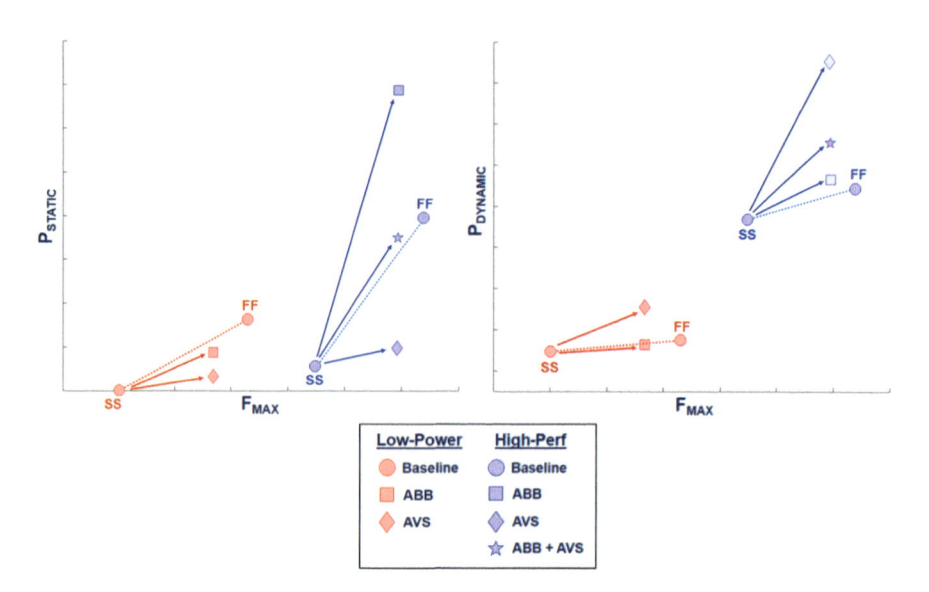

Fig. 14.9 F_{MAX}, P_{STATIC}, $P_{DYNAMIC}$ comparison after ABB, AVS, and ABB+AVS compensation techniques in a qFO4 circuit running at a low-power and high-performance modes. ©2019 IEEE. Reprinted, with permission, from R. G. Gomez et al., "Comparative evaluation of Body-Biasing and Voltage Scaling for Low-Power Design on 28nm UTBB FD-SOI Technology," 2019 IEEE/ACM International Symposium on Low Power Electronics and Design (ISLPED)

overhead). AVS, on the contrary, has an impact on the worst-case dynamic power, worsening the power specs.

On the high-performance mode, the nominal supply of 1.0 V_{DD} is above V_{Tinv}, making the slowest corner to be located at 165 °C. For this reason, as shown in the left side of Fig. 14.9, the acceleration achieved with ABB has a large impact on the leakage consumption. Similarly, the AVS compensation severely impacts the reported dynamic consumption, previously set by the (FF, 165 °C) corner.

However, the AVS-only compensation leaves a considerable leakage margin between the compensated SS corner and the worst-case P_{STATIC} set by FF as shown in Fig. 14.9. This opens the opportunity of increasing the body-bias without impacting the current P_{STATIC} bound, and at the same time reducing the supply voltage to cut down some of the $P_{DYNAMIC}$ overhead. This ABB+AVS combination, depicted in Fig. 14.9, demonstrates the benefits of a combined application of both techniques, enabling the same performance recovery as a ABB-only or AVS-only approach, but without leakage impact and reducing the dynamic overhead of a AVS-only compensation.

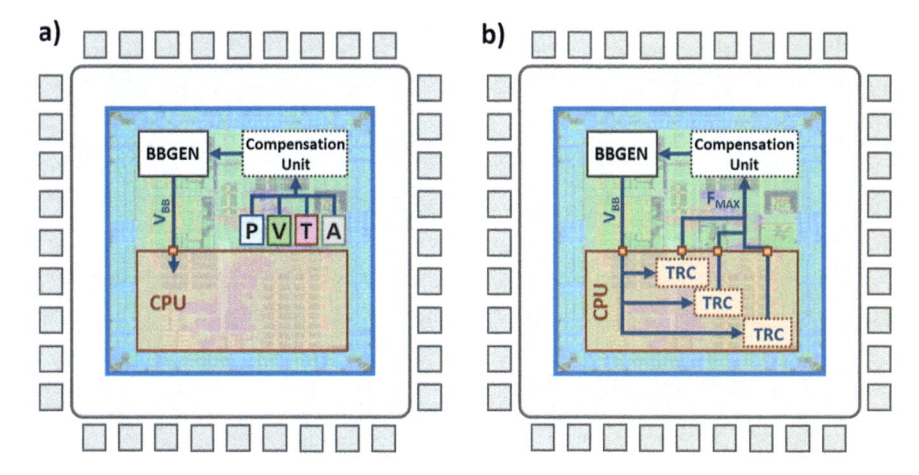

Fig. 14.10 Open-loop compensation system (**a**) and closed-loop compensation system (**b**)

14.2 Body-Bias Control: Closed-Loop vs Open-Loop

Closed-loop and open-loop regulation systems differ on how the control action is performed. In open-loop systems, the control action is independent of the process controlled variable, that is, the circuit's timing. This makes the body-bias levels to be extracted from a pre-characterized association between process and environmental positions and circuit's F_{MAX}. This characterization, usually stored on-chip, is accessed during runtime to fix the body-bias voltage to be applied at the current environmental conditions. In a closed-loop system, the control action is dependent of the process controlled variable. In this approach, the estimated maximum frequency, F_{MAX}, is provided to the loop controller which compensates variations through supply or body-bias modulation.

In Fig. 14.10a, an open-loop compensation system based on body-biasing is depicted. In this strategy, the sensed process, voltage, temperature, and ageing data are the entry values to the compensation unit, which generates the required body-bias voltage to compensate the effect of PVTA variations in the circuit's timing, as described in Chap. 13. In Fig. 14.10b, the closed-loop system counterpart is depicted. In this example, the compensation unit monitors the circuit's F_{MAX} through the use of a tunable replica circuit (please refer to Chap. 12), and uses this information to calculate the required V_{BB} to minimize the difference between the measured and the targeted frequency.

When architecting the compensation system, designers need to balance the costs and benefits of a more complex closed-loop system w.r.t. an open-loop system. Although closed-loop systems have gained a lot of attention from the academia, from a worst-case specification's improvement perspective, open-loop compensation system can perform equally good as the closed-loop in most cases.

The reasons for this statement are:

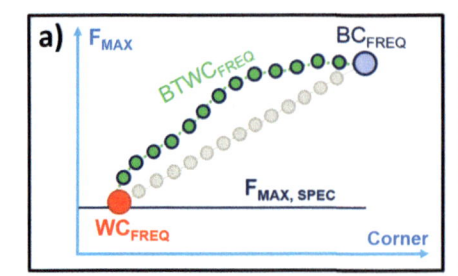

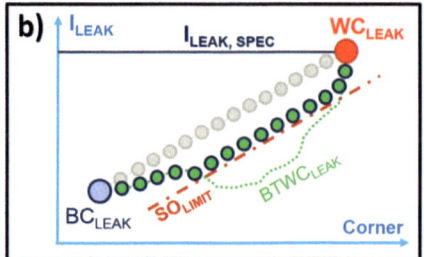

Fig. 14.11 Better-than-worst-case impact on frequency (**a**) and leakage (**b**)

- The improvement brought by closed-loop regulation mainly applies to the better-than-worst-case samples, with a very limited improvement of the gating extreme cases, thus not improving the product specifications.
- All dies must remain inside the signoff space, which is delimited by process, voltage, temperature, body-bias, and ageing. Any excursion from this safe-space is not allowed, and thus the body-bias requested by the closed-loop controller needs to be anyway clamped to the $V_{BB,MIN}$ enforced by the global signoff (P,V,T,B,A) limits.

The first point is better explained in Fig. 14.11a. There, we have represented the F_{MAX} spread of several PVTA corners, sorted from the slowest WC_{FREQ} corner (SS, $V_{DD,LOW}$, $V_{BB,MAX}$, T_{LOW}) to the fastest BC_{FREQ} corner (FF, $V_{DD,HIGH}$, $0\,V_{BB}$, T_{HIGH}). The implementation of a closed-loop control with a speed monitoring sensor matching the circuit's F_{MAX} will be able to, for the majority of samples, remove the margins, reaching a higher F_{MAX} (green coloured $BTWC_{FREQ}$) w.r.t. to a simpler open-loop interpolation (grey coloured samples). However, with respect to the worst-case timing corner, a closed-loop compensation system applying the worst-case budgeted body-biasing will not increase the worst-case performance achieved by the open-loop compensation (given the characterized F_{MAX} of the open-loop system is accurate). However, as the target frequency is set by the worst-case timing point, and remove margins on this worst-case corner, the product frequency will not be improved. In this case, a simpler open-loop control system will achieve the same frequency target without the design and engineering efforts to implement a safe F_{MAX} tracking loop.

The second point is explained in Fig. 14.11b. Again, the product leakage $I_{LEAK,SPEC}$ is set by the worst-case leakage corner, WC_{LEAK}, which makes any improvement on all better-than-worst-case corners irrelevant to the gating specs. However, it may be argued that a closed-loop system matching the circuit F_{MAX} can reduce the power overhead of body-bias compensation by supplying the minimum V_{BB} per die, achieving a better averaged product leakage w.r.t. using an open-loop compensation system. In fact, for those corners and samples where on-chip speed monitors measure a speed value above the $F_{MAX,SPEC}$, the V_{BB} for that particular die/corner could be further reduced to the minimum to decrease leakage. However, as all samples are required to remain within CAD signoff bounds, as soon as the V_{BB} voltage requested by the control system underpasses the signoff limit SO_{LIMIT}, it needs to be clamped to the minimum V_{BB} required by (P,V,T,A) signoff. This requires the implementation of a PVT monitoring scheme (just like an open-loop system) to locate the circuit within the signoff space and protect it from out-of-signoff operation. This reduces the interest of a closed-loop operation, as ultimately the applied V_{BB} on the circuit will be bounded by signoff.

References

1. J.W. Tschanz, J.T. Kao, S.G. Narendra, R. Nair, D.A. Antoniadis, A.P. Chandrakasan, V. De, Adaptive body bias for reducing impacts of die-to-die and within-die parameter variations on microprocessor frequency and leakage. IEEE J. Solid State Circuits **37**(11), 1396–1402 (2002)
2. S.M. Martin, K. Flautner, T. Mudge, D. Blaauw, Combined dynamic voltage scaling and adaptive body biasing for lower power microprocessors under dynamic workloads, in *IEEE/ACM International Conference on Computer Aided Design, 2002. ICCAD 2002, November* (2002), pp. 721–725
3. C.H. Diaz, K.H. Fung, S.M. Cheng, K.L. Cheng, S.W. Wang, H.T. Huang, Y.K. Leung, M.H. Tsai, C.C. Wu, C.C. Lin, M.-C. Chang, D. Tang, Device properties in 90 nm and beyond and implications on circuit design, in *IEEE International Electron Devices Meeting 2003, December* (2003), pp. 2.6.1–2.6.4.
4. D.M. Fleetwood, Evolution of total ionizing dose effects in MOS devices with Moore's law scaling. IEEE Trans. Nucl. Sci. **65**(8), 1465–1481 (2018)
5. M. Gaillardin, M. Martinez, P. Paillet, M. Raine, F. Andrieu, O. Faynot, O. Thomas, Total ionizing dose effects mitigation strategy for nanoscaled FDSOI technologies. IEEE Trans. Nucl. Sci. **61**(6), 3023–3029 (2014)
6. P. Meinerzhagen, C. Tokunaga, A. Malavasi, V. Vaidya, A. Mendon, D. Mathaikutty, J. Kulkarni, C. Augustine, M. Cho, S. Kim, G. Matthew, R. Jain, J. Ryan, C. Peng, S. Paul, S. Vangal, B.P. Esparza, L. Cuellar, M. Woodman, B. Iyer, S. Maiyuran, G. Chinya, C. Zou, Y. Liao, K. Ravichandran, H. Wang, M. Khellah, J. Tschanz, V. De, An energy-efficient graphics processor featuring fine-grain DVFS with integrated voltage regulators, execution-unit turbo, and retentive sleep in 14 nm tri-gate CMOS, in *2018 IEEE International Solid-State Circuits Conference (ISSCC), February* (2018), pp. 38–40

Chapter 15
Body-Bias Voltage Generation

Thierry Di Gilio

15.1 Introduction

When we speak about power management implementation, the question of embedding power management unit or not has to be asked. This is true for body-bias generator (BBGEN) too. Most of the time, the final choice is a trade-off between

1. the cost: silicon area versus on-board components price,
2. the power consumption: efficiency of each solution,
3. way to control and implement the chosen solution.

The choice for point 1 is market dependent. As an example, automotive market is less constraint than mobile market for area availability on the board. Let us consider the case where a project architect decides to use some external components to implement a programmable body-bias voltage generation. For cost reasons, generic components have to be used, and specific place and route have to be achieved on board. Such a cost-efficient solution would work, but it poses other challenges related to complexity and power consumption. This brings us to second point: the power budget.

Even if power consumption is more and more a concern for any chip, mobile market is the one that has the highest constraint when coming to power consumption.

To deepen the topic, let us talk about the differences between external and internal solutions from a control and implementation point of view. Let us first consider external BB solution. We need to look at the connection impedance of both body voltages (negative and positive): as shown in Fig. 15.1 the connection path

---

T. Di Gilio (✉)
STMicroelectronics, Crolles, France
e-mail: thierry.di-gilio@st.com

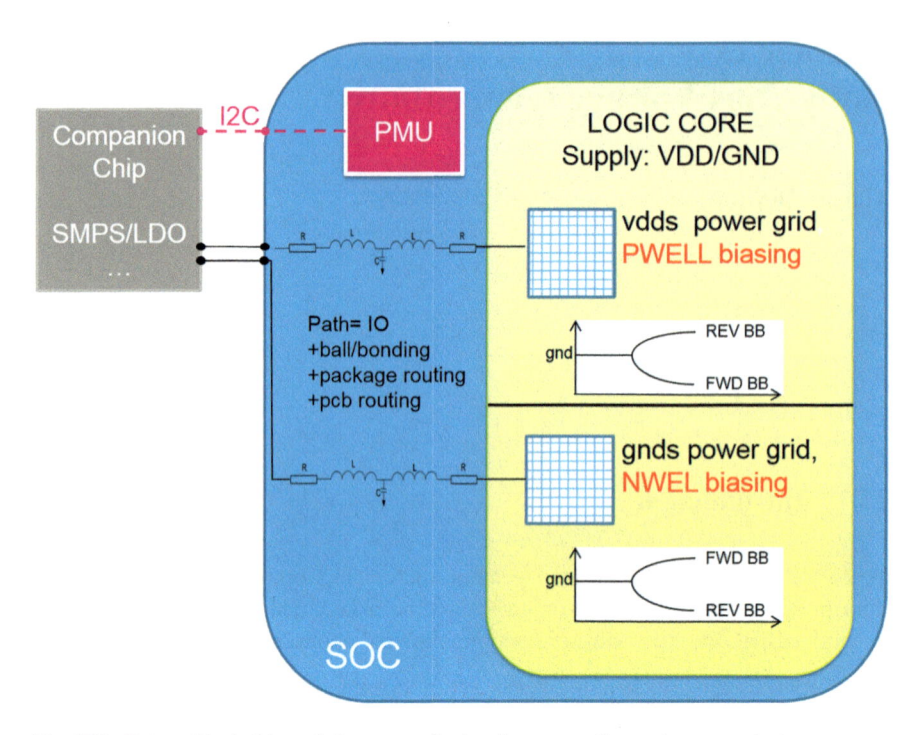

Fig. 15.1 External body-bias solution synoptic showing connection and command wires

includes parasitic resistance, inductance, and capacitance, corresponding to the chip package (bonding, bonding solder, ball soldering, etc.), and PCB routing. This path is not always easy to modelize. This RLC impedance could be an issue for stability and speed. Next we can observe the control bus is serial type (I2C, SPI, etc.). If the number of control bit is high or if the refreshing rate is high, the "slowness" of such a bus could be an issue. Note that in this example, the controller (PMU) is embedded in the SoC. This PMU could be outside the SoC, giving more complexity to the board designer. In the case of an internal body-bias solution, these routing constraints do not exist: control bit buses are parallel and direct, and parasitics are sensibly lower and can be managed (improved) at layout level. This discussion is summarized in Table 15.1.

To be comprehensive on this topic, we should discuss about the purpose of body-biasing (process compensation, speed boost, leakage saving, etc.). Indeed, this can determine the architecture of the BBGEN and thus its power consumption. These considerations will be addressed in Sect. 15.3.

15 Body-Bias Voltage Generation

Table 15.1 Advantage and disadvantage for internal and external body-bias solution

Criteria	Internal BB	External BB
Cost	Additional silicon area	Additional components + additional PCB area + additional I/O SoC pins
Power consumption	Globally lower, but adds to SoC total consumption	Generic component will give higher power consumption
Implementation and control	Full and direct control, low parasitics in connexions, fast reaction	Control generally done via slow serial link, parasitic in body voltage routing

15.2 Load Model Description and Modelling

In Sect. 15.1, we discussed about the differences between internal and external body-bias solution. Whatever the chosen architecture, we have to know precisely the nature of the load that our voltage generator will have to address. This is what will be exposed in this chapter.

15.2.1 Common Load Topology

Three common body-biasing topologies can be described in digital world. The most classic one is for reference: the standard bulk (Fig. 15.2a). Second and third are relative to UTBB well arrangement. One have to distinguish flipped wells, where PMOSFETs are built on P_{well} and NMOSFETs on N_{well} (Fig. 15.2c), from non-flipped wells, where PMOS are on N_{well} and NMOS on P_{well} (Fig. 15.2e). Let us consider an inverter, the simplest gate that can be found in digital libraries. On the Fig. 15.2a–e, we can detail wells arrangement under an inverter for both bulk and UTBB SOI topology (Fig. 15.2a and b). This figure also shows equivalent schematics for all cases (Fig. 15.2b–f). In the bulk case, there is some p+/N_{well} diodes between S/D and N_{well} for PMOS, and some n+/P_{well} diodes between S/D and P_{well}. In some field effect conditions, at the drain side of P and N transistor, one can see a leakage current appearing, named gate induced drain leakage I_{GIDL}. This leakage has to be driven by our BBGEN. In UTBB SOI case, this I_{GIDL} is still present but it is not flowing into wells nodes because of the box isolation. This is a good news for a BBGEN designer. If we now have a look at other components that make a BBGEN load, we find $p-n$ diodes made by N_{well}–P_{well}, Deep N_{well}–P_{well}, Deep N_{well}–Psub, and N_{well}–P_{sub} interfaces. As a function of the topology, of the direction of body-bias effect (reverse/forward), and if we consider that a junction should not be forward biased above 0.3 V (to avoid excessive leakage), we can establish acceptable voltage range for body-bias for each topology. To have a clear view of the situation, Table 15.2 is helpful.

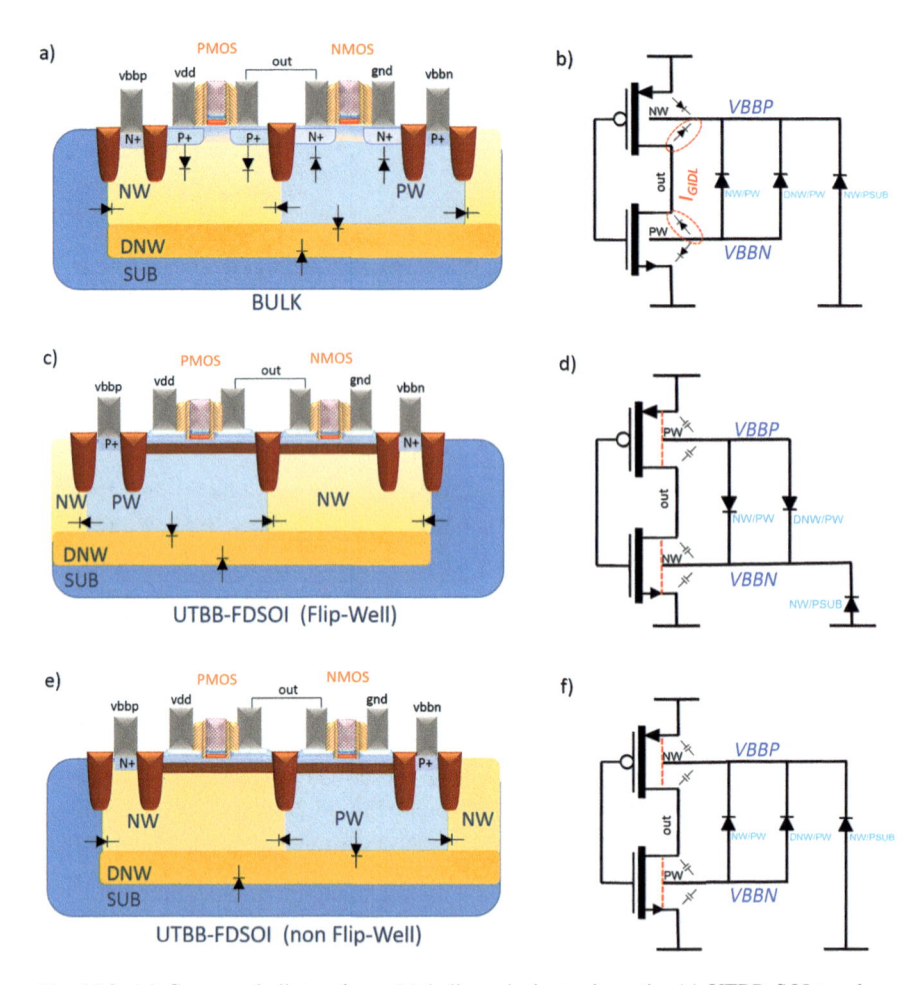

Fig. 15.2 (**a**) Common bulk topology, (**b**) bulk equivalent schematic, (**c**) UTBB SOI topology (flipped well), (**d**) UTBB SOI equivalent schematic (flipped well), (**e**) UTBB SOI topology (non-flipped well), (**f**) UTBB SOI equivalent schematic (non-flipped well)

In the next sections, we will make some useful reminders on diodes currents and capacitances physics, because they constitute the BBGEN's load.

15.2.2 Currents

P_{well}–N_{well} junction current description is well known and describe by Shockley equation (for and ideal diode) [1]:

15 Body-Bias Voltage Generation

Table 15.2 Voltage and $p-n$ biasing mode when applying forward body-bias bulk and UTBB SOI topology

	Bulk	UTBB SOI non-flip well	UTBB SOI flip well
NMOS WELL	P_{well}	P_{well}	N_{well}
PMOS WELL	N_{well}	N_{well}	P_{well}
vbbn range	$-350\,\text{mV} \to 350\,\text{mV}$ $REV \to FWD$	$-350\,\text{mV} \to 350\,\text{mV}$ $REV \to FWD$	$-350\,\text{mV} \to 1.8\,\text{V}$ $REV \to FWD$
vbbp range	$vdd - 350\,\text{mV} \to vdd + 350\,\text{mV}$ $FWD \to REV$	$vdd - 350\,\text{mV} \to vdd + 350\,\text{mV}$ $FWD \to REV$	$-1.8\,\text{V} \to 350\,\text{mV V}$ $FWD \to REV$
Limits	FBB limited to vdd/2	FBB limited to vdd/2	RBB limited to vdd/2

$$I_D = I_S \left(e^{\frac{qV_D}{nkT}} - 1 \right) \quad (15.1)$$

with

- q: the electronic charge,
- V_D: the applied voltage,
- n: the ideality factor (1 for silicon),
- k: the Boltzmann's constant,
- T: the temperature in Kelvin.

kt/q is also named thermal voltage V_{th}. At room temperature (300 K), $V_{th} = 25.9\,\text{mV}$. Figure 15.3 shows the common behaviour of current flowing through a junction as a function of the voltage across this junction, and this is the BBGEN static load current.

15.2.3 Capacitances

There are two kinds of capacitance in a $p-n$ junction. One is named diffusion capacitance C_{Diff}, because it is created by charges flowing by diffusion mechanism through the junction, and second one is named transition capacitance C_T and is related to depletion region width. Diffusion capacitance occurs in a forward biased $p-n$ junction diode. Transition capacitance exists in both reverse and forward biased diode, but it is negligible versus the diffusion capacitance in forward.

$$C_T = \frac{\varepsilon A}{W} \quad (15.2)$$

where

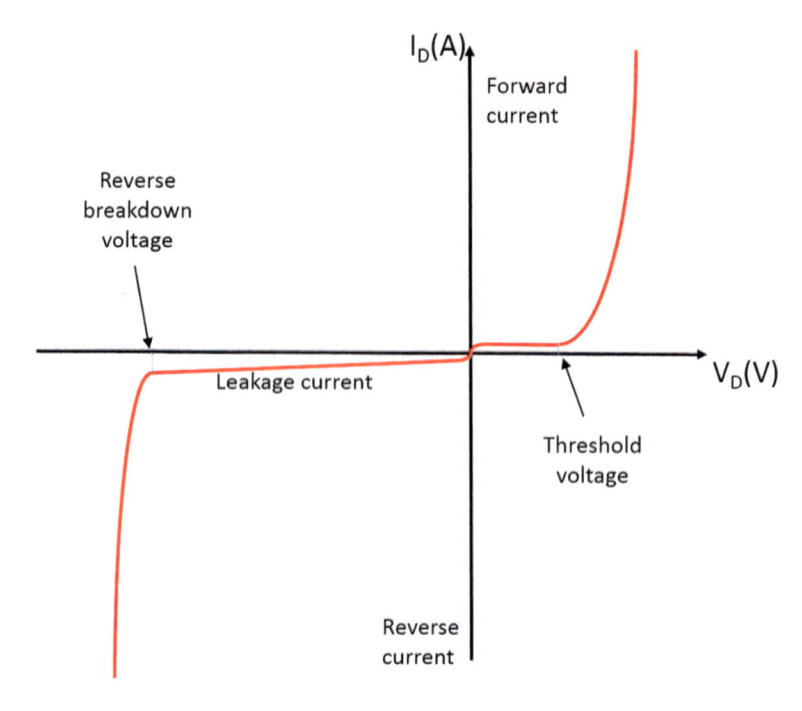

Fig. 15.3 I(V) curve characteristic of a $p - n$ junction

ε = Permittivity of the semiconductor
A = Area of plates or p-type and n-type regions
W = Width of depletion region
W is a function of doping of p and n sides (N_A and N_D, respectively):

$$W = \sqrt{\frac{2\varepsilon}{q}\left(\frac{N_A + N_D}{N_A N_D}\right) V_{bi}} \qquad (15.3)$$

where V_{bi} is the potential barrier that is a direct consequence of the emergence of positively and negatively charged areas under the effect of the carrier dissemination.

Let us now consider the diffusion capacitance. Its expression can be obtained from:

$$C_D = \frac{dQ}{dV} \qquad (15.4)$$

where dQ is the change in number of minority carrier stored outside the depletion region when a change in voltage across the diode, dV is applied. If τ is the mean lifetime of charge carriers, then a flow of charge Q yields a diode current I and is given by $I = Q/\tau$. In case of a forward current, using (15.1), C_D can be derived:

15 Body-Bias Voltage Generation

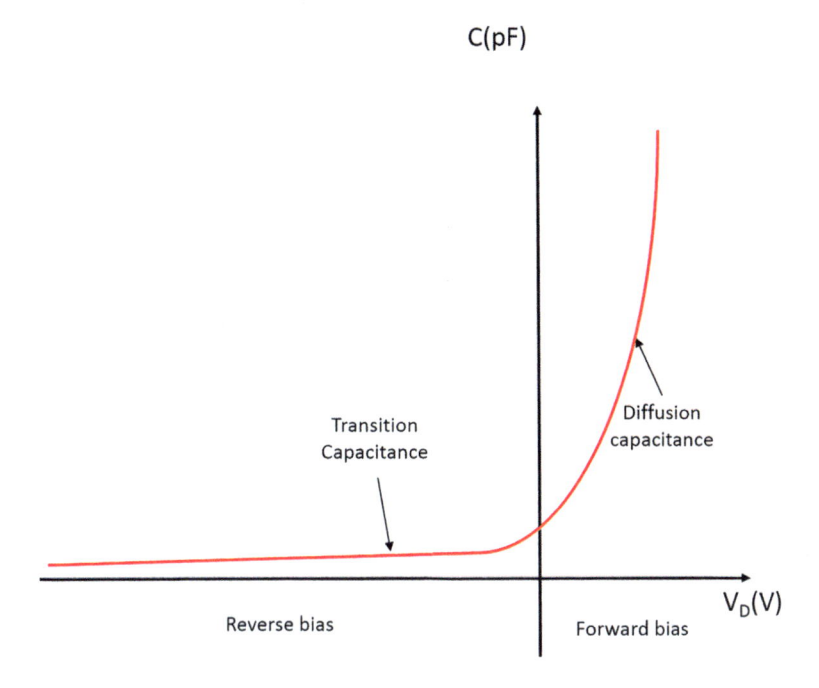

Fig. 15.4 C(V) curve characteristic of a $p-n$ junction

$$C_D = \frac{d\left(\tau I_s e^{V_D/V_{th}}\right)}{dV} = \frac{\tau I_D}{V_{th}} \qquad (15.5)$$

A generic waveform of $C(V)$ junction is represented in Fig. 15.4.

15.2.4 Load Estimation

The goal of this section is to give guidelines to estimate, the more accurately as possible, the load seen by a BBGEN. We are considering a digital body-biased area, i.e., a block made of standard cells. A characteristic of standard cells' layout is their regularity: height of N_{well} and P_{well} are constant on each row (or column depending on cells orientation). This leads to the particular arrangement described in Fig. 15.5. We name N_{col} as the N_{well} row number (also valid for the block's peripheral rows).

From this schematic, we can extract the area (A), the perimeter (P), and the multiplication factor n of the $N_{well}-P_{well}$ diode:

$$\begin{aligned} A_{nwell-pwell} &= h(Y_0 + X_{pw}) \\ P_{nwell-pwell} &= 2(h + Y_0) + 2(h + X_{pw}) \\ n_{nwell-pwell} &= 2(N_{col} - 1) \end{aligned} \qquad (15.6)$$

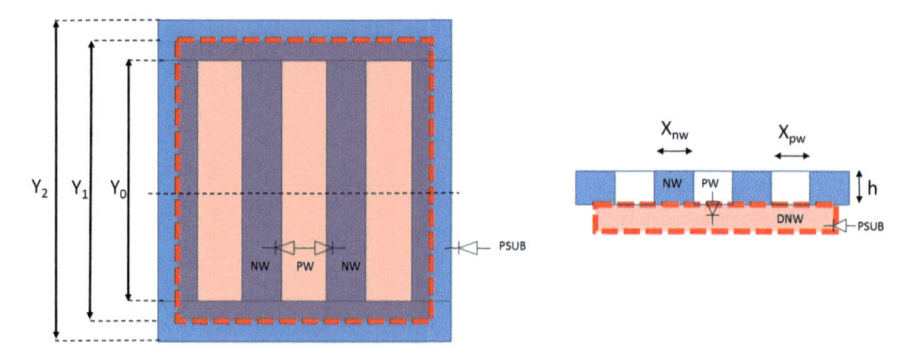

Fig. 15.5 Wells arrangement in digital side made with standard cells

The same method can be used for the diode formed by the P_{well}–Deep N_{well} junction:

$$\begin{aligned} A_{pwell-dnwell} &= X_{pw} Y_0 \\ P_{pwell-dnwell} &= 2(X_{pw} + Y_0) \\ n_{pwell-dnwell} &= N_{col} - 1 \end{aligned} \quad (15.7)$$

And the same for the Deep N_{well}–P_{sub} diode:

$$\begin{aligned} A_{dnwell-psub} &= Y_1(N_{col} X_{nw} + (N_{col} - 1) X_{pw}) \\ P_{dnwell-psub} &= 2(Y_1 + N_{col} X_{nw} + (N_{col} - 1) X_{pw}) \\ P_{dnwell-psub} &= 1 \end{aligned} \quad (15.8)$$

The last one is the peripheral diode N_{well}–P_{sub}:

$$\begin{aligned} A_{nwell-psub} &= h(Y_2 + N_{col} X_{nw} + (N_{col} - 1) X_{pw}) \\ P_{nwell-psub} &= 2(Y_2 + h) + 2(h + N_{col} X_{nw} + (N_{col} - 1) X_{pw}) \\ P_{nwell-psub} &= 2 \end{aligned} \quad (15.9)$$

All these calculations will be useful for the designer: they will allow to make a spice model including sized diodes. The model will help to validate performances of future BBGEN designs.

15.3 Specifications and Constraints for a Body-Bias Voltage Generator

In this section, we will address the main specifications to be known to design a coherent and reliable BBGEN circuit. Specifications can be placed in three main categories: timing, voltage range, and environmental constraints.

15.3.1 Timing

In timing specifications we find settling time and start-up time. Settling time could be critical when adaptive body-bias (ABB) technique is used to fit transistors V_{th} to real time needs. Indeed, in synchronous logic, it would be preferable to stop clocks during BB voltage transitions, and thus the shorter the transition is, the better it is from an operating system point of view, hence for the user. The maximum settling time (in μs/V) has to be defined as a function of the targeted application and will determine the architecture of our BBGEN. At last, the reader should keep in mind that settling is defined twice for each outputs: for rising voltage and for falling voltage of each BBGEN outputs (leading to four values). However, this can be simplified by choosing a symmetrical behaviour. Another point, when coming to timing, is start-up time: in applications that need to enable and disable BBGEN several times to save leakage, start-up time could be crucial. As an example let us consider a connected object that has to transfer data to a network: a wake up occurs periodically and it may be desirable to have the faster wake up possible.

So we saw that timing consideration will directly impact BBGEN output stage: we know that BBGEN's load is mainly capacitive, and consequently timing will be a matter of current capability (at least for settling).

15.3.2 Voltage Range

We are now entering voltage considerations. This specification item is double: a range must be defined for *vbbn* and *vbbp* outputs. These ranges will depend on the type of biasing we would like to apply (forward or reverse) and the effective amplitude we want to achieve. In a digital world, the top simulation is done at Verilog level. The way body-bias affects standard cells' characteristics is encoded in Verilog models (based on electrical characterization). There is no need to generate BB voltages out of the standard cells characterized range. In the same way, the designer should consider available supplies. He should think of the cost of generating a voltage higher than the supply voltage: it is easy to do but costs more area and consumption. In the common case where BBGEN has one positive output and one negative output, these outputs have not necessary the same range (in absolute values).

15.3.3 Environmental Constraints

The third but not the least important category is environmental constraints. In this category we can arrange all specifications imposed by a System on Chip environment:

- supply access resistances
- distance/resistance between BBGEN and body-biased area
- interface for controlling BBGEN
- available area and form factor
- testability possibility (pad availability, etc.)

To illustrate this, we can take [2]. In this paper a dual ARM CPU is body-biased using an embedded BBGEN. As the continue Adaptive Voltage Scaling (AVS) technique is used, a fast BBGEN is required. This implies big current spikes at the outputs of the BBGEN (and thus on supply), because this is equivalent to a fast variation on voltages of wells capacitances (CPUs WELLs). To achieve this, BBGEN has a dedicated supply ball and routing, and output metal lines are sized accordingly.

15.4 Body-Bias Voltage Generator Design

In this section the reader will find some information on embedded Body-Bias GENerator architectures, typical sides to use and more. A BBGEN is typically a programmable voltage generator. So we need a voltage reference (as bandgap circuit), a digital to analogue converter, a negative voltage generation, and output stages to drive the loads.

15.4.1 Voltage Reference

When coming to this topic, two schools can be considered: the BBGEN is a generic one that is part of standard library set, or the BBGEN is something that is integrated to specific power management strategy dedicated to one project. In the first case, it could be better to generate a voltage reference dedicated to (and inside) the BBGEN: in a "digital on top environment", properly propagating a central reference voltage could be a hard and risky task. In the second case, we consider that BBGEN is part

15 Body-Bias Voltage Generation

of a power management unit (PMU) that embeds voltage reference and deals with all analogue references and biasing of the circuit.

In both cases, a common voltage circuit is based on bandgap principle [3]. A reference voltage should always be independent of temperature and supply voltage. The way to achieve temperature independence is to combine a proportional to absolute temperature (PTAT) current (or voltage) to a complementary to absolute temperature (CTAT) current (or voltage), so that PTAT and CTAT variations are cancelled. PTAT and CTAT current (or voltage) can be obtained using properties of $p-n$ junctions. Generally we use bipolar transistors and specific properties to build circuits that produce CTAT and PTAT behaviours. A PTAT voltage can be obtained by comparing V_{BE} voltage of two bipolar transistors biased with an equal current and n factored area, or with n factored current and identical areas. Collector current I_C of a NPN bipolar transistor is

$$I_C = I_s e^{\frac{V_{BE}}{V_T}} \tag{15.10}$$

where $V_T = kT/q$ with k the Boltzmann's constant, T the temperature, and q the electronic charge; I_s is also a function of temperature.

Considering schematic of Fig. 15.6 we calculate δV_{BE} ($I_C = I$):

$$\begin{aligned} \Delta V_{BE} &= V_{BE1} - V_{BE2} \\ &= V_T \cdot \log_e \frac{I_C}{I_S} - V_T \cdot \log_e \frac{I_C}{nI_S} \\ &= V_T \left(\log_e I_C - \log_e I_s - \log_e I_C + \log_e n + \log_e I_s \right) \\ &= V_T \cdot \log_e n \\ &= \frac{kT}{q} \log_e n \end{aligned} \tag{15.11}$$

Fig. 15.6 Simple circuit to obtain PTAT Voltage

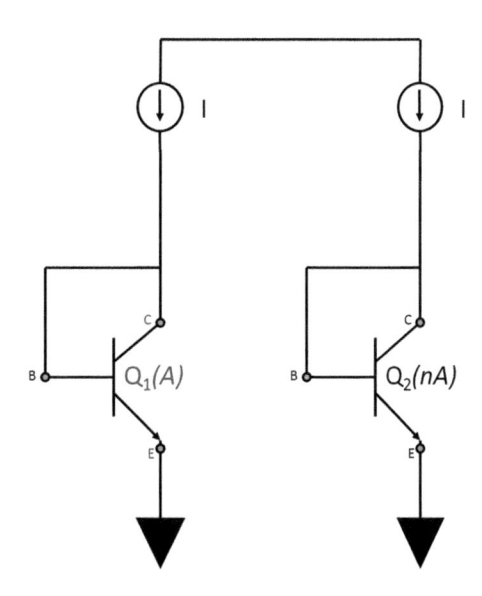

So we get something proportional to T by a factor $\log_e n \cdot k/q$. Next step is to build circuitry that gives the CTAT component. We consider again V_{BE} and from (15.10) we have:

$$V_{BE} = V_T \cdot \log_e \frac{I_C}{I_S} \tag{15.12}$$

It can be demonstrated that I_S is writable as:

$$I_S = I_0 \cdot \exp^{-V_{G0}} V_T \tag{15.13}$$

where I_0 is a process/geometry/temperature-dependent current and V_{G0} is the bandgap voltage (about 1.2 V). A generic expression for I_0 gives:

$$I_0 = A_{emitter} B T^r \tag{15.14}$$

where $A_{emitter}$ is the emitter area, B is a process-dependent constant, and r is a process-dependent quantity. In common modern CMOS and high-speed bipolar processes r is in the range [4; 6]. We can now rewrite (15.16) using (15.13) and (15.14):

$$V_{BE} = V_{G0} - V_T \log_e \left(\frac{A_{emitter} B T^r}{I_C} \right) \tag{15.15}$$

To establish CTAT behaviour of V_{BE}, let us calculate its derivative versus temperature:

$$\begin{aligned}
\frac{dV_{BE}}{dT} &= -V_T \frac{d}{dT} \left(\log_e \frac{A_{emitter} B T^r}{I_C} \right) \\
&= -V_T \frac{d}{dT} \left(\frac{A_{emitter} B T^r}{I_C} \right) \frac{I_C}{A_{emitter} B T^r} \\
&= -V_T \cdot r \frac{A_{emitter} B T^{r-1}}{I_C} \cdot \frac{I_C}{A_{emitter} B T^r} \\
&= -V_T \cdot r \cdot T^{-1} \\
&= -r \frac{kT}{qT} \\
&= -\frac{k \cdot r}{q}
\end{aligned} \tag{15.16}$$

Thus V_{BE} derivative versus temperature is a negative constant that is a proof of CTAT behaviour.

Figure 15.7 shows how to combine PTAT and CTAT voltage. From this schematic we can extract:

$$\begin{aligned}
V_{ref} &= I_1 V_1 + V_{BE1} \\
V_{ref} &= I_2 R_2 + I_3 R_3 + V_{BE2} \Rightarrow I_2 = \frac{V_{ref} - V_{BE2}}{R_2 + R_3}
\end{aligned} \tag{15.17}$$

15 Body-Bias Voltage Generation

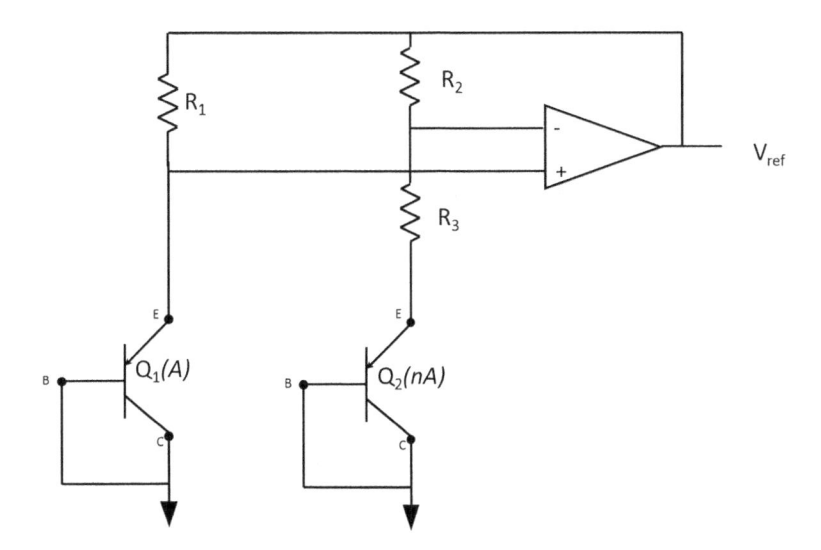

Fig. 15.7 Bandgap circuit that combines CTAT and PTAT with error amplifier

We want that both bipolar transistor have same current collector $I_C = I_1 = I_2 = I_{ref}$. Thus rearranging (15.17):

$$V_{ref} = \frac{R_1}{R_2+R_3}\left(V_{ref} - V_{BE2}\right) + V_{BE1}$$
$$\Rightarrow V_{ref} = \frac{\frac{-R_1}{R_2+R_3}}{1-\frac{R_1}{R_2+R_3}} V_{BE2} + V_{BE1} \quad (15.18)$$

if we note $\alpha = -R_1/(R_2 + R_3)$ (15.18) becomes:

$$V_{ref} = \frac{-\alpha}{1-\alpha} V_{BE2} + V_{BE1}$$
$$= \left(\frac{1}{1+\alpha} - 1\right) V_{BE2} V_{BE1} \quad (15.19)$$
$$= \Delta V_{BE} + \frac{1}{1+\alpha} + V_{BE2}$$

We finally obtained a reference voltage that is independent of temperature and supply voltage. This topology is easy to describe from a theory point of view. But from a process point of view, a more friendly topology exists and is described in Fig. 15.8. It is using "free" devices (that do not require additional process steps): PNP bipolar transistors. Nevertheless NPN transistor, when available (this is the case in FDSOI technology), is an opportunity to suppress error amplifier to gain power/area efficiency. The schematic of Fig. 15.9 illustrates this: transistors $Q1$ and $Q2$ provide the PTAT via $\Delta V_{be} = V_{be1} - V_{be2}$; whereas V_{be3} of $Q3$ is the CTAT term. PTAT and CTAT are combined through R_1 and R_2. It can be easily demonstrated that:

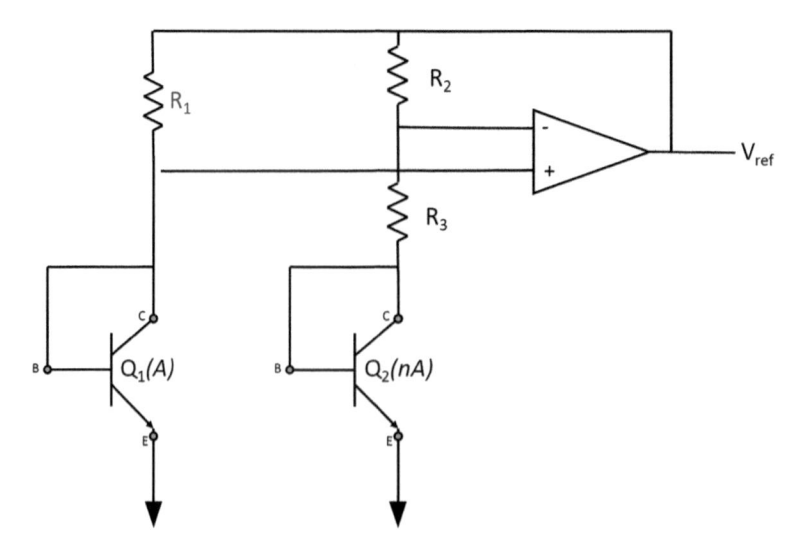

Fig. 15.8 Bandgap circuit that combines CTAT and PTAT with error amplifier

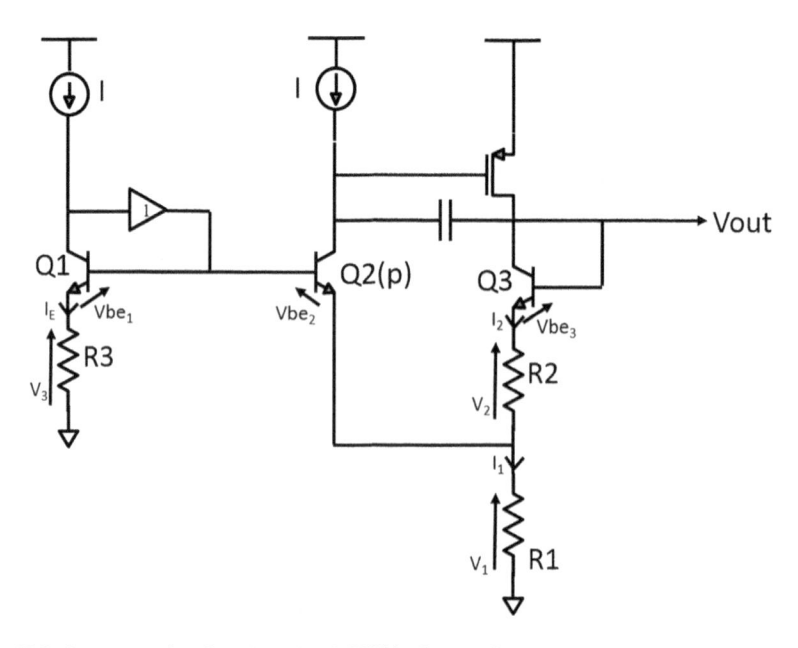

Fig. 15.9 Low power bandgap based on NPN bipolar transistors

15 Body-Bias Voltage Generation

$$V_{out} = \Delta V_{be} = \left(1 + \frac{R_2}{R_1}\right) + V_{be3} \quad (15.20)$$

only if $R_3 = R_1 // R_2$. Main advantage of this circuit is the small number of used devices and its consumption: it can be as low as 100–200 nA, leading to a total consumption (start-up and biasing circuits included) less that 1 µA.

To finish with this part, let us speak about the use of bipolar transistor in FDSOI technology. The designer will find in the design kit two bipolar devices: NPN type and PNP type, both vertical. They are implemented in hybrid area (without box isolation), meaning that they are similar to bulk vertical bipolar transistors. The only constraint will be the guard band to respect between hybrid and SOI areas.

15.4.2 Programmable Voltage Generation

Whatever the choice for reference voltage (external or internal to BBGEN), it will be used to generate programmable voltages for both paths ($vdds$ and $gnds$). To realize this function, a common circuit is used: a digital to analogue converter (DAC). Such a circuit is composed of a set of switches controlled by a binary decoder and a network of passive components: resistor or capacitor. There are several types of DAC, for a n bit DAC, we can consider:

- **String DACs**: built with a string of resistors (serial mount), this type of DAC is simple and fast for resolution lower than 9–10 bits. This architecture is inherently monotonic, directly controllable by digital circuitry, but requires 2^n resistors and 2^n switches for n bits. This may have a non-negligible area cost. At last, this kind of topology could have a high settling time for high n. A particular string DAC is the thermometer DAC [4] represented in Fig. 15.10. In this architecture

$$V_{out} = x \cdot \frac{V_{ref}}{2^n} \quad (15.21)$$

where x is the decoded input code and $V_{ref}/2^n$ the voltage across one resistor.

- **Binary-weighted DACs** use only one switch per bit and a resistors network. On one hand, we have voltage-mode binary-weighted DACs (as shown in Fig. 15.11): they are not inherently monotonic and are quite hard to manufacture for high resolution. Furthermore output impedance is not constant and change with input code. But this issue can be solved using a buffer, and output voltage is obtained using additive operational amplifier properties:

$$\begin{aligned} V_{out} &= -R_{out} I_{out} \\ &= -R_{out} (I_0 + I_1 + I_2 + I_3) \\ &= -R_{out} \left(B_0 \frac{V_{ref}}{R} + B_1 \frac{V_{ref}}{2R} + B_2 \frac{V_{ref}}{4R} + B_3 \frac{V_{ref}}{8R}\right) \end{aligned} \quad (15.22)$$

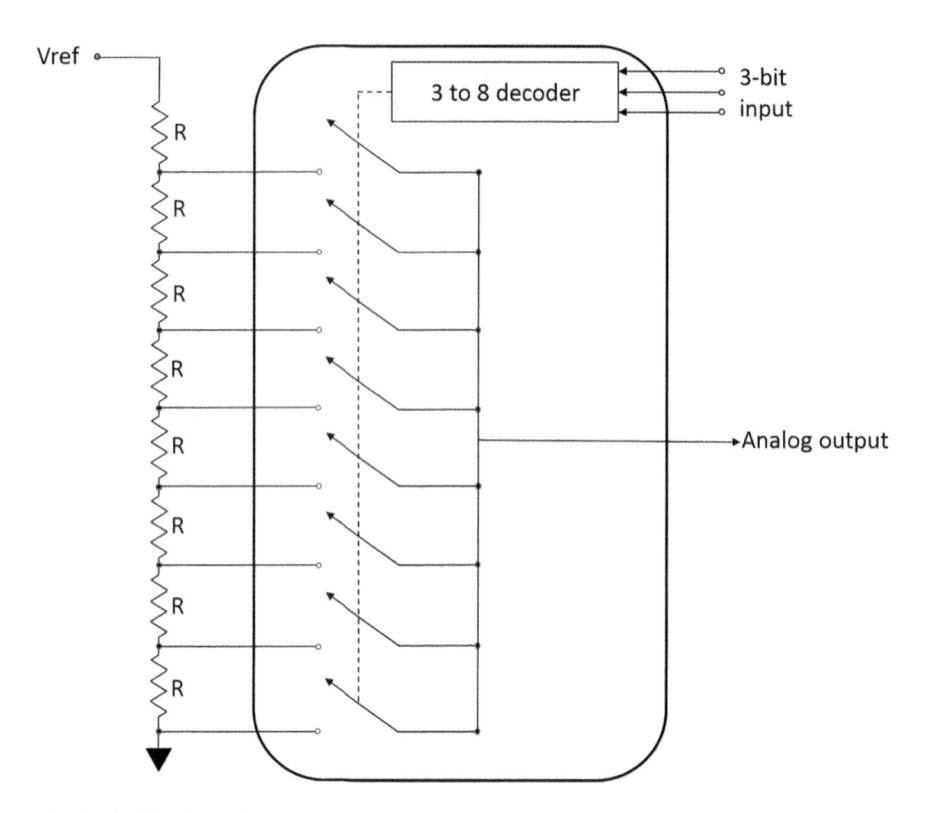

Fig. 15.10 Simplest voltage-output thermometer DAC: the Kelvin divider ("String DAC")

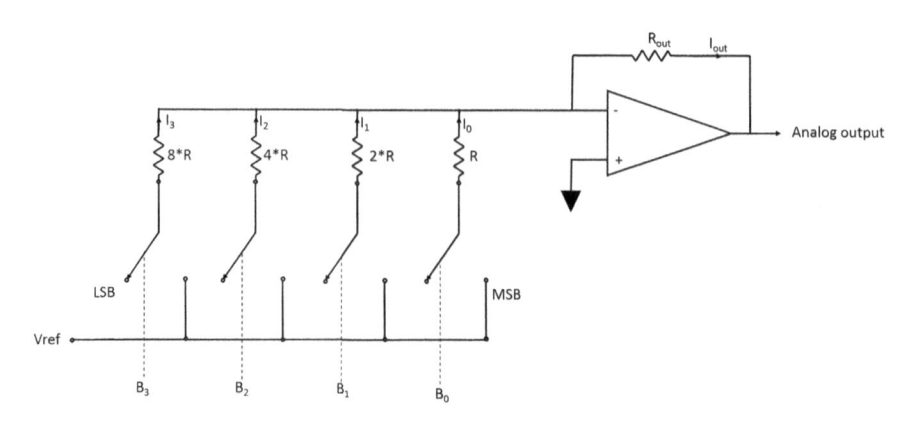

Fig. 15.11 Voltage-mode binary-weighted resistor DAC

15 Body-Bias Voltage Generation

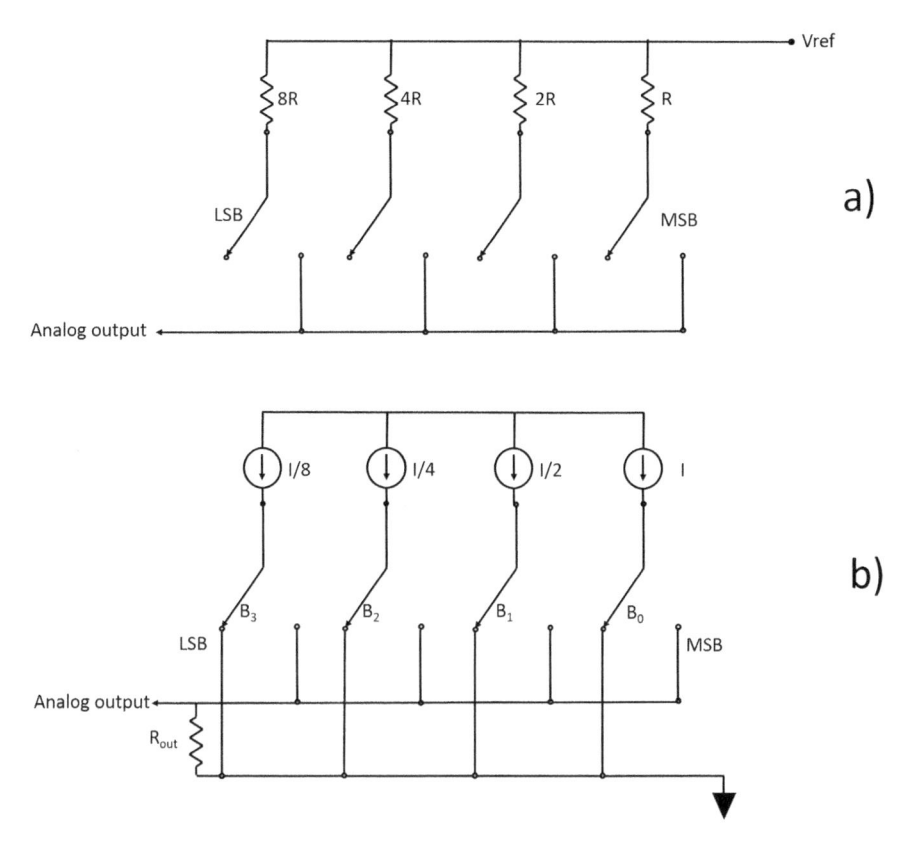

Fig. 15.12 Current-mode binary-weighted resistor DAC. (**a**) R as current source, (**b**) matched current sources

If we choose $R_{out} = R$, then Eq. (15.22) becomes

$$V_{out} = V_{ref}\left(B_0 + \frac{B_1}{2} + \frac{B_2}{4} + \frac{B_3}{8}\right) \qquad (15.23)$$

On the other hand we found current-mode binary-weighted DACs. It consists in n weighted current sources which may simply be resistors and a voltage reference in the ratio 1; 2; 4; 8; ...; 2^n (Fig. 15.12). In the case of current sources, output voltage can simply be written like:

$$\begin{aligned} V_{out} &= R_{out} I_{out} \\ &= R_{out}\left(B_0 \cdot I + \frac{B_1 \cdot I}{2} + \frac{B_2 \cdot I}{4} + \frac{B_3 \cdot I}{8}\right). \end{aligned} \qquad (15.24)$$

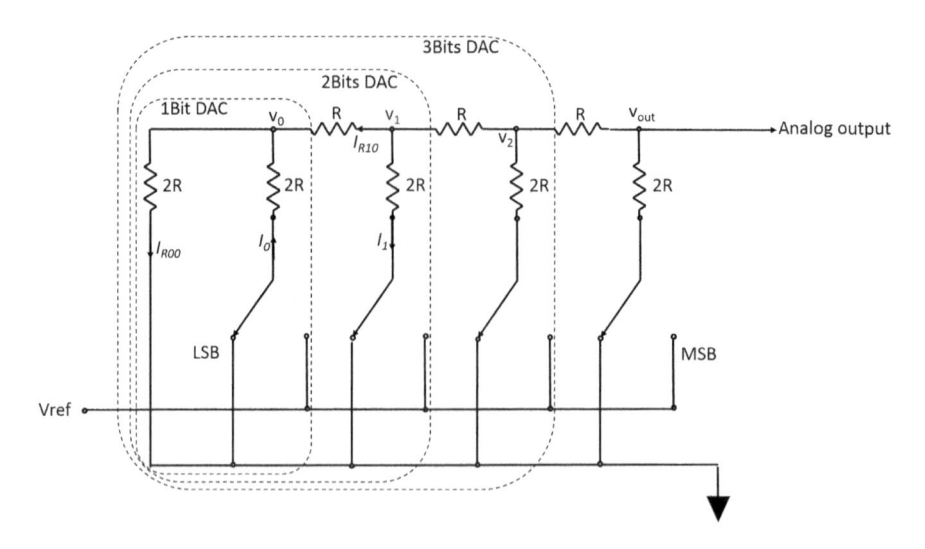

Fig. 15.13 Voltage-mode binary-weighted resistor DAC

This solution is easy to implement with current mirrors as sources, but precision and monotonicity can be impacted by bad matching and current leakages in switches and current sources transistors.

- **R2R DACs** use 2^n resistors that can have only two values R and $2R$ and n switches. Both voltage and current mode exist. In voltage-mode, the output is a voltage and the impedance is independent of code[5] (Fig. 15.13). To calculate output voltage, let us first calculate output for $n = 1$ DAC. Output node V_0 is easily calculated:

$$B_0 = 0 \Rightarrow V_0 = V_{ref}$$
$$B_0 = 1 \Rightarrow V_0 = V_{ref} \frac{2R}{2R+2R} = \frac{V_{ref}}{2} \quad (15.25)$$

The same reasoning can be applied to $n = 2$ DAC. We introduce B'_k that is equal to the product of logical term B_k by V_{ref}:

$$\begin{aligned} U_{R10} &= I_0 + I_{R00} \\ &= \frac{V_0 - B'_0}{2R} + \frac{U_{R0}}{2R} \\ &= \frac{2V_0 - B'_0}{2R} \end{aligned} \quad (15.26)$$

U_{R10} can also be expressed as:

$$U_{R10} = \frac{V_1 - V_0}{R} \quad (15.27)$$

Using (15.26) and (15.27) we obtain:

15 Body-Bias Voltage Generation

$$\frac{V_1 - V_0}{R} = \frac{2V_0 - B'_0}{2R}$$
$$\Leftrightarrow 2V_0 = V_1 + B0/2 \tag{15.28}$$

next we calculate current between B'_1 and V_1:

$$I_1 = \frac{B'_1 - V_1}{2R} \tag{15.29}$$

As I_1 equals I_{R10} we obtain (using (15.27) and (15.28)):

$$\frac{B'_1 - V_1}{2R} = \frac{V_1 - V_0}{R}$$
$$\Rightarrow V_1 = \frac{B0}{2} + \frac{B1}{4} \tag{15.30}$$

This last formula can be generalized to n bits DAC:

$$V_{n-1} = V_{ref} \sum_{k=0}^{k=n-1} \frac{B_k}{2^{n-k}} \tag{15.31}$$

To finish this short study, please note that maximum voltage of a such R2R network is given by $V_{max} = V_{ref}(2^n - 1)/2^n$.

In current-mode, output is a current and impedance varies with code: this is not the simplest solution to manage. A more popular topology for current mode is to use n matched current sources [6]. One can see in Fig. 15.14 that the DAC output is a voltage and that it has a fixed output impedance equal to R; two specifications that make it easy to use and implement. Output voltage calculation can be obtained with the same kind of equation than previously.

- **Switched capacitor DACs** are used where silicon area is a critical point. It uses n capacitors network with n switches. Reader must note that this topology needs a sample and hold circuit and additive switch to reset capacitors values periodically. A common example is given in Fig. 15.15. During phase ϕ_1 all capacitors are reset to 0 V. Then switches are controlled during phase ϕ_2. The equivalent Capacitor C_{eq} is given by:

$$C_{eq} = B_0 \cdot C + \frac{B1 \cdot C}{2} + \frac{B2 \cdot C}{4} + \frac{B3 \cdot C}{8} = C \cdot \sum_{k=0}^{n-1} \frac{B_k}{2^k} \tag{15.32}$$

Corresponding charge $Q_{eq} = C_{eq} \cdot V_{ref}$ is transferred in sampling capacitor C_s and output voltage is given by

$$V_{out} = \frac{C_{eq}}{C_s} V_{ref} = V_{ref} \cdot \frac{C}{C_s} \cdot \sum_{k=0}^{n-1} \frac{B_k}{2^k} \tag{15.33}$$

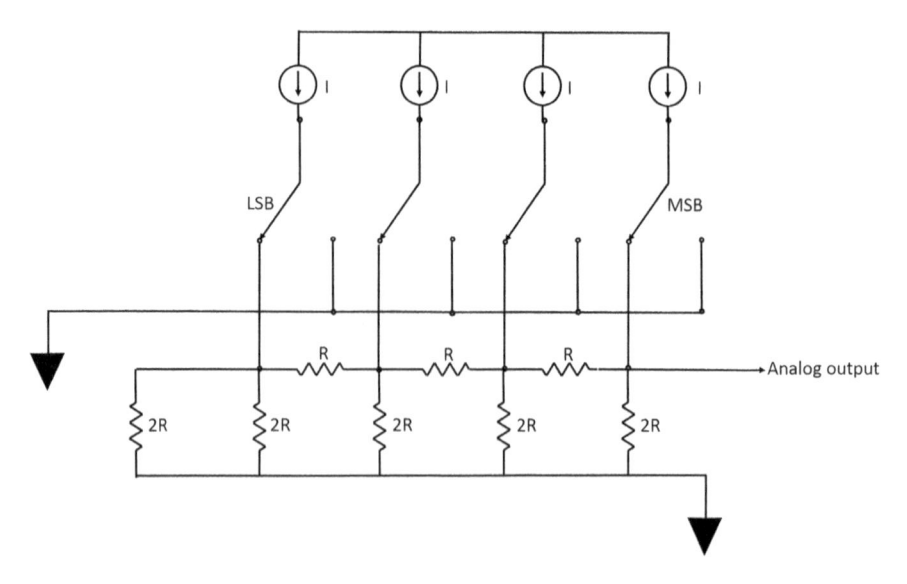

Fig. 15.14 Voltage-mode binary-weighted resistor DAC

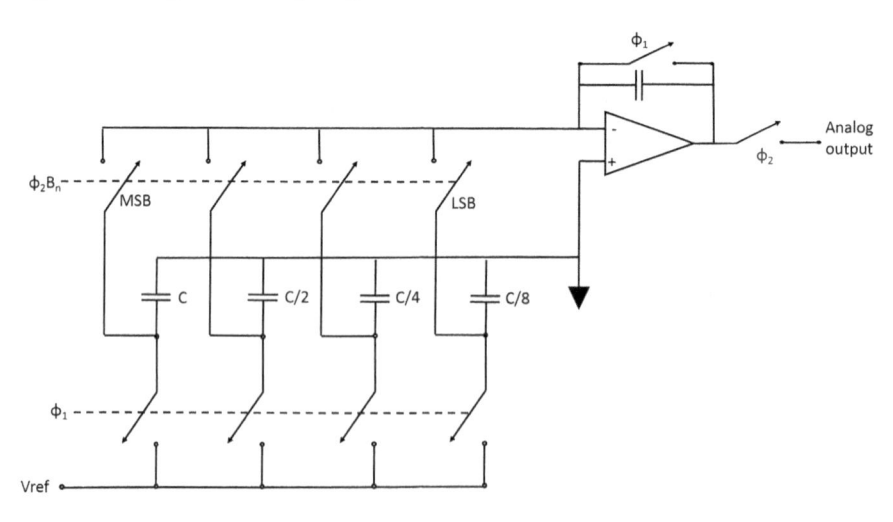

Fig. 15.15 Voltage-mode binary-weighted resistor DAC

We made a non-exhaustive list of possible DAC topology. In our application, we can suppose that DACs will be used in static state: for digital body-biasing purpose, generated voltage is not intended to have continuous change versus time. We have written previously that settling time at BBGEN outputs is an important specification point. But the DAC speed is generally not a concern with respect to output drivers slew rate. That is why DACs speed topic has not been considered. On the other hand, accuracy and ways to look after it must be touched on. Whatever the type of network you plan to use for your converter, you must pay attention to have the best matching as possible between your devices and choose their size to minimize process effects. This can be done by careful layout (using dummy devices, use large geometry, etc.). Monte Carlo simulation including global process variation (corners, applies to all devices) and local process variation (device parameter change applies to one device) will help designer to deal with process effects. DAC accuracy can also be affected by parasitics like switches impedance Z_{sw} (you should choose $R \gg Z_{sw}$) or charge injections due to switches. Indeed, charges C_{par} are stored in channel of transistors used as switches (you should check that charge stored in the network is large versus charge stored in C_{par}). Offset should also be treated carefully in amplifiers and buffers. Some offset cancellation techniques like auto-zeroing or chopping technique will be used to minimize offset effects. The accuracy of a DAC is defined by two parameters:

1. **Differential Non-Linearity** (DNL) defines difference voltage between two adjacent input code compared to ideal LSB (ILSB) value. In other words it compares each step to ILSB. It can be written as: $DNL_i = (V_{i+1} - V_i) / VILSB$ where $0 < i < 2^n - 2$. Ideal LSB can be calculated considering the full-scale range voltage V_{fsr}: $ILSB = V_{fsr}/2^n$
2. **Integral Non-Linearity** (INL) describes the deviation (in LSB or percent of V_{fsr}) of the transfer function (real DAC output) versus a straight line (ideal DAC response).

Some other specifications could be considered as crucial in other applications than BBGEN for digital. We can enumerate:

Noise: in resistors, we have thermal noise with a density $N_D^{thermal} = \sqrt{4kRT}$ in nV/$\sqrt{Hz}$, where k is Boltzmann's constant, R is resistance in Ohms, T is the temperature in Kelvin. Thermal noise is a white noise. Flicker noise, also called $1/f$ noise, occurs in transistor (and thus MOS capacitors) and some type of resistors. It is always associated with a DC current, in which some fluctuations are due to contamination in semiconductor materials. Flicker noise density is $N_D^{flicker} = KI\sqrt{\frac{1}{f}}$ in nA/$\sqrt{Hz}$, where K is a device constant, I is the DC current, f is the frequency. In resistors or capacitors network, each contribution can be cumulated (depending on topology) and even amplified by output buffer or sample and hold. Some noise cancellation technique exists, but will not be addressed here.

Resolution: defines the number of step available. For a BBGEN dedicated to digital domain, a step of 50 mV is sufficient, because for FDSOI, body effect is around 80 mV/V. That means that when you apply 1 V of BB, transistor V_{Th} will shift by 80 mV. Thus a 50 mV step of BB leads to 4 mV change for V_{Th}. If we choose $V_{fsr} = 1.3$ V, we need a resolution:

$$n = \log_e \left(V_{fsr}/50 \cdot 10{-3} \right) / \log_e(2) = 4.7 \Rightarrow n = 5\, bits \quad (15.34)$$

Speed: some application requires high speed: input code frequency change is high and analogue output must follow accordingly. For a BBGEN, a change every 1 μs seems to be a reasonable choice. All topologies described previously fit this need.

15.4.3 Output Stages

15.4.3.1 Analogue Output Buffer

In this part, we will suppose a generic BBGEN: that means that its output stage must be able to source and sink current (charge and discharge load). This specification will exclude some common topology used in power management like low drop regulators (LDOs), class A buffers, etc. A buffer is simply an amplifier used as a follower. Some specification of this buffer can be listed:

- supply voltage range
- output voltage range
- biasing current
- output load (current, capacitance)
- minimum phase margin
- DC gain over full input voltage range
- power consumption
- settling time

Many amplifier classes exist. AB class is particularly suited to our need. The main benefit of AB class amplifier is the ability to sink and source current, which is essential with a capacitive load. A good example of class-AB amplifier has been described by Hogervorst [7]. This is a two stage amplifier that has a rail-to-rail output: output range goes from 0 V to supply voltage vdd. Let us first have a look at push–pull output stage (Fig. 15.16). It consists of two common-source connected output transistors, MP_{out} and MN_{out}, which are directly driven by two in-phase signal currents, I_{in1} and I_{in2}. Transistors MP_{AB} and MN_{AB} form the floating class-AB control. The stacked transistors MP_{diode1}–MP_{diode2} and MN_{diode1}–MN_{diode2} are diode-connected and bias gates of the class-AB transistors MP_{AB} and MN_{AB}, respectively. The class-AB behavioural is performed keeping the voltage between gate of MP_{out} and MN_{out} constant. Transistor MP_{AB}–MP_{diode1}–MP_{diode2}–MP_{out} and MN_{AB}–MN_{diode1}–MN_{diode2}–MN_{out} create two trans-linear loops.

When the in-phase signal current sources, I_{in1} and I_{in2}, are pushed into the floating control branch, the current in MP_{AB} increases while the current in MN_{AB} decreases by the same amount. Consequently, the gate-voltages of both MP_{out} and MN_{out} increase. Thus the output stage pulls a current from the output node. This phenomena continues until the current through MP_{AB} is equal to I_{AB1}. At this point, the current of MP_{out} is kept at a minimum value, which can be set by W/L

15 Body-Bias Voltage Generation

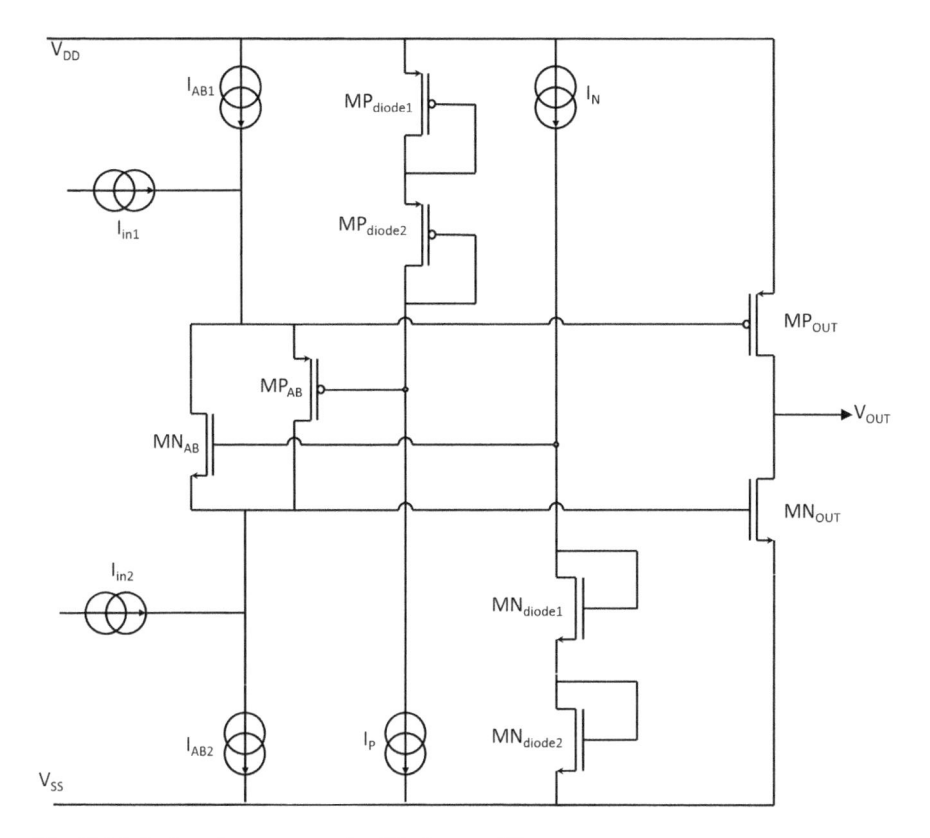

Fig. 15.16 Push–pull output stage for AB class amplifier

ratios of MP_{AB} and MN_{AB}. Note that the current in MN_{out} is still able to increase. A similar discussion can be held when input signals are pulled from the class-AB output stage.

A drawback of this topology is that the quiescent current in MP_{out} and MN_{out} depends on supply voltage variations. The supply voltage variations are directly applied, by the gate-source voltages of MP_{out} and MN_{out}, across the finite output impedances of MP_{AB} and MN_{AB}. The result is a power supply-dependent variation of the quiescent current.

The first stage is a dual differential pair: one made with NMOS transistors (MN_{dif1}, MN_{dif2}) and another pair made with PMOS transistor (MP_{dif1}, MP_{dif2}). The schematic of Fig. 15.17 details input stage and push–pull stage connection. NMOS pair is active $vdd/2$ to vdd whereas PMOS pair is active from $0\,\text{V}$ to $vdd/2$. This combination helps to reduce signal to noise ratio at extrema of dynamic range. A drawback of rail-to-rail input stage is that the variation of transconductance $g_{m,CM}$ over common-mode input range presents a bell curve. Such a variation could raise issues when coming to frequency compensation. A way to compensate $g_{m,CM}$ loss on extrema of the input voltage range is to increase current I_{Ptail} (I_{Ntail}) when

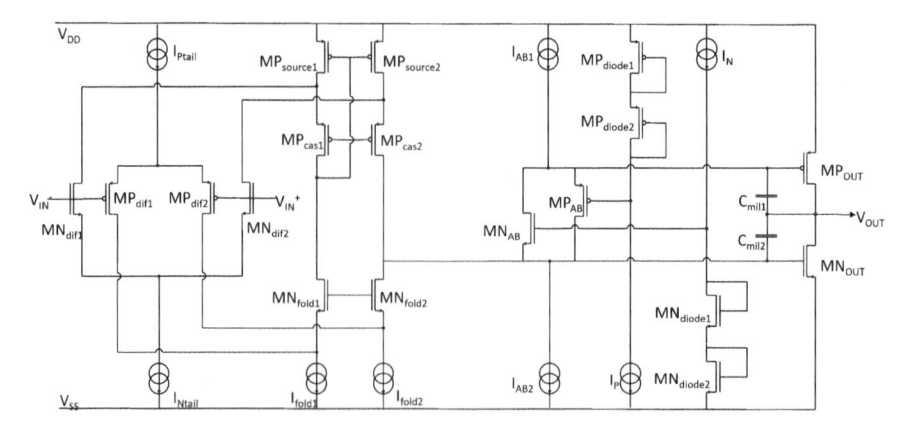

Fig. 15.17 Complete schematic of a class-AB amplifier, biasing circuit, and g_m compensation circuits is not represented

input voltage is low (high). The amount of current to add to tail sources can be determined by calculating $g_{m,CM}$:

$$\begin{aligned} g_{m,CM} &= \left.\frac{\partial i_{tail}}{\partial v_{in,CM}}\right|_{V_{in,CM}} \\ &= \left.\frac{\partial\left(\frac{W}{L}\frac{\mu C_{ox}}{2}(v_{in}-V_{TH})^2\right)}{\partial v_{in,CM}}\right|_{V_{in,CM}} \\ &= \sqrt{2K I_{tail}} \end{aligned} \tag{15.35}$$

with $K = \mu_N C_{ox} \frac{W}{L} = \mu_P C_{ox} \frac{W}{L}$, where μ is the mobility of the charge carriers, C_{ox} is the normalized oxide capacitance, and W and L are the width and the length of a transistor, respectively. The subscripts N and P refer to an N-channel or P-channel input transistor, respectively.

Thus we see that if we observe a reduction factor m of $g_{m,CM}$, we need to increase the tail current by a factor m^2. In other words, when V_{in} is near vss, a $(m^2 - 1)I_{Ptail}$ current source must be connected in parallel with I_{Ptail}. The same circuit must be added for I_{Ntail} side when V_{in} is near vdd.

The second part of input stage is built with transistors $MP_{source1}$, $MP_{source2}$, MP_{cas1}, MP_{cas2}, MN_{fold1}, MN_{fold2}, and two current sources I_{fold1} and I_{fold2} folded cascode stage and sum current coming from NMOS and PMOS pairs.

The current mirror, $MP_{source1}$, $MP_{source2}$, MP_{cas1}, MP_{cas2}, together with the folded MP_{cas2} cascodes, MN_{fold1}, MN_{fold2}, form a summing circuit. This summing circuit adds the signals coming from the complementary rail-to-rail input stage.

The connection between input and output stage is made by connecting output of first stage (between MN_{fold1}, MN_{fold2}, and MN_{fold2}) to floating class-AB control branch. The first stage injects a current (proportional to V_{in}) into output stage.

15 Body-Bias Voltage Generation

Lastly, Miller compensation is done placing two capacitors C_{mil1} and C_{mil1} between gates of MP_{out} and MN_{out} and output node.

The full schematic is presented in Fig. 15.17. A drawback of this design is the noise injected by I_{AB1} and I_{AB2} sources that is directly propagated to the output of the amplifier. But, in the case of a BBGEN, noise should not be an issue, because output voltage is applied to N_{Well} and P_{Well} that are very noisy because of digital cells activity placed in these wells.

This side can be used for positive output and negative output of BBGEN. On negative side class-AB amplifier has to be connected between ground and a negative rail supply V_{neg}; pin vdd of amplifier connects to ground and vss pin connects to v_{neg}. In this case designer should pay attention to wells biasing to avoid direct conduction in formed $p-n$ diodes. One can benefit of FDSOI flipped wells to avoid N_{well} negatively biased (using LVT PMOS, placed in P_{well}, instead of regular one, placed in N_{well}).

15.4.3.2 Digital Output Buffer

In the case of a digital output buffer, we consider the same output stage made of PMOS and NMOS transistor. The AB control is realized by a digital control side. PMOS and NMOS are on or off, they cannot be on at the same time. Regulation scheme of a such power stage will be treated in Sect. 15.4.3.4. The simple principle of this solution is shown in Fig. 15.18. When output voltage is higher than input voltage, MN_{OUT} is on and MP_{OUT} is off, whereas when output is lower than input, MN_{OUT} is off and MP_{OUT} is on. When output = input, both MN_{OUT} and MP_{OUT} are off.

15.4.3.3 Negative Voltage Generation

The common way to generate a negative voltage is to use a DC–DC converter with inverting property: it takes vdd as an input and provide $-vdd$ (ideally, without regulation: open loop) at output. Such converter uses switched capacitors topology as shown in Fig. 15.19.

In this circuit, four switches $S1$, $S2$, $S3$, and $S4$ are made with inverter: their output can be high or low. Switched capacitors are $Cfly_1$ and $Cfly_2$. They are named fly capacitors. Circuitry formed with $C1$, $C2$, $T1$, and $T2$ controls $S3$ and $S4$. $S3$, $S4$, $C1$, $C2$, $T1$, and $T2$ are in the negative voltage domain.

Operation can be described in two steps:

- **Step 1** When the input clock signal is low, $S1$ and $S3$ outputs are in high state: $Cfly_1$ is connected between vdd and ground and is charging at vdd. $S2$ and 4 output are in low state: $Cfly_2$ is connected between gnd and V_{neg} and is discharging into the output load. $T1$ gate (also connected to $S3$ input) is near V_{neg} and $T2$ gate (also connected to $S4$ input) is grounded. $T1$ is on and $T2$ is off.

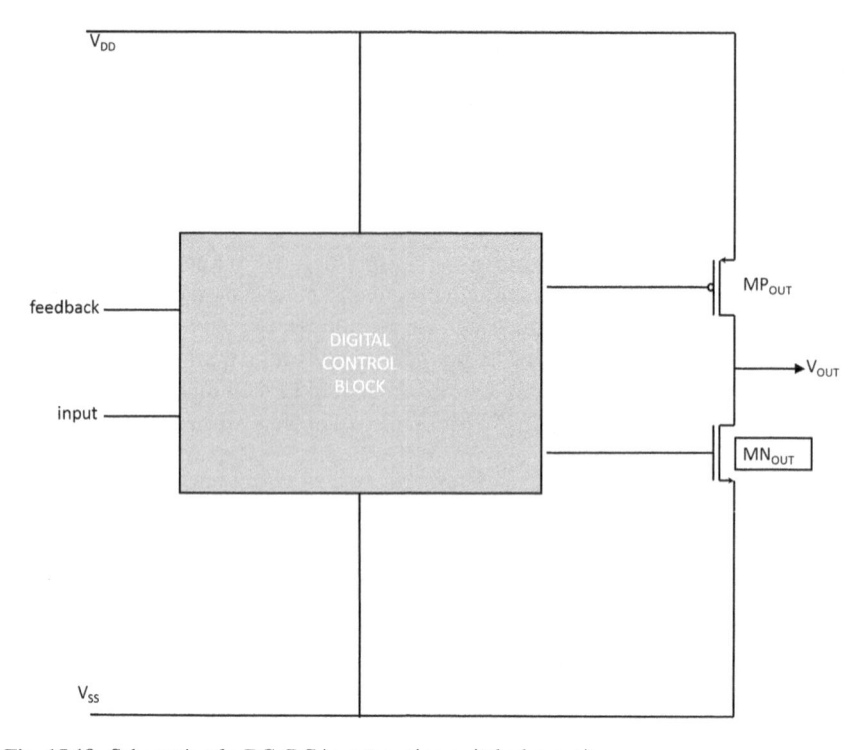

Fig. 15.18 Schematic of a DC–DC inverter using switched capacitors

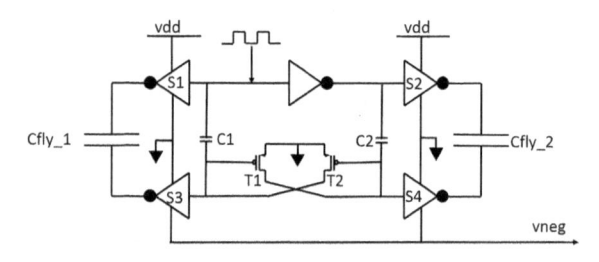

Fig. 15.19 Schematic of a DC–DC inverter using switched capacitors

- **Step 2** When the input clock signal is high, $S1$ and 3 outputs are in low state: $Cfly_1$ is connected between ground and V_{vneg} and is discharging into output load. $S2$ and 4 outputs are in high state: $Cfly_2$ is connected between vdd and ground and is charging to vdd. $T1$ gate is grounded and $T2$ gate is near V_{neg}. $T1$ is off and $T2$ is on.

Not represented in Fig. 15.19, a non-overlapping circuit is required on the clock signal to avoid that $Cfly_1$ and $Cfly_2$ are connected at the same time to output. This side must be inserted in place of an inverter between $S1$ and $S2$.

Thanks to charge conversion principle, connecting $Cfly$ (beforehand charged to vdd) between ground and output produces a negative voltage ideally equal to $-vdd$. Same principle applies to $C1$ and $C2$ allowing to control $S3$, $S4$, $C1$, $C2$, $T1$,

15 Body-Bias Voltage Generation

and $T2$ in the negative voltage domain. In practice, because devices are non-ideal and have leakage, some charge losses occur, and thus the output voltage is always superior to $-vdd$. $C1$ and $C2$ are on purpose separate components. If gates $T1$ ($T2$) were connected directly to $Cfly_1$ ($Cfly_2$), this would imply that gate voltages will be equal to output voltage (through $S3$ and $S4$) when in low state. The output voltage takes several clock cycles to go from 0 V to V_{neg} (it depends on output load). Adding $C1$ and $C2$ gives the maximum control on $T1$ and $T2$ in one clock cycle.

Switches sizing is an important step in designing this DC–DC converter. PMOS transistor of $S_{1,2,3,4}$ have to be big enough to fully charge $Cfly$ in half a clock cycle, whereas NMOS transistors size will impact output resistance of converter. Sizes cannot be increased so much, especially for $S3$ and $S4$: the bigger the switches are, the bigger must be $C1$ and $C2$. Indeed, in order to guaranty the most negative voltage on $S3$ and $S4$ input, we must have $C1 \gg (C_{S3} + C_{T1})$, where C_{S3} is the input capacitance of $S3$ inverter and C_{T1} the gate capacitance of $T1$.

A drawback of this topology is the conduction between NMOS and PMOS in $S_{1,2,3,4}$ during clock transitions. This phenomenon consumes charges in $Cfly$ during transfer and reduces the current efficiency of converter. One can maximize buffering to minimize clock transition times, but the issue is just minimized, not suppressed. A way to avoid it is to separate NMOS and PMOS gate signals: NMOS and PMOS are no more conducting at the same time. Limitations of this method are: a more complex logic control, at high frequency it might not be feasible. A big disadvantage of these circuits is the simulation time. A solution is to use a spice equivalent model. An example is given in Fig. 15.20.

In this schematic, R_{adj} modelizes the consumption of logic circuit (buffer, non-overlap circuit, etc.), and the output voltage is set by a DC voltage source $-V_{IN}(1 - par)$ with par in [0; 1] range: it represents charge losses due to leakage and parasitics, R_{OUT} is the output impedance of the charge pump:

$$R_{OUT} = \frac{1}{2C_{Fly}F_{CLK}} \quad (15.36)$$

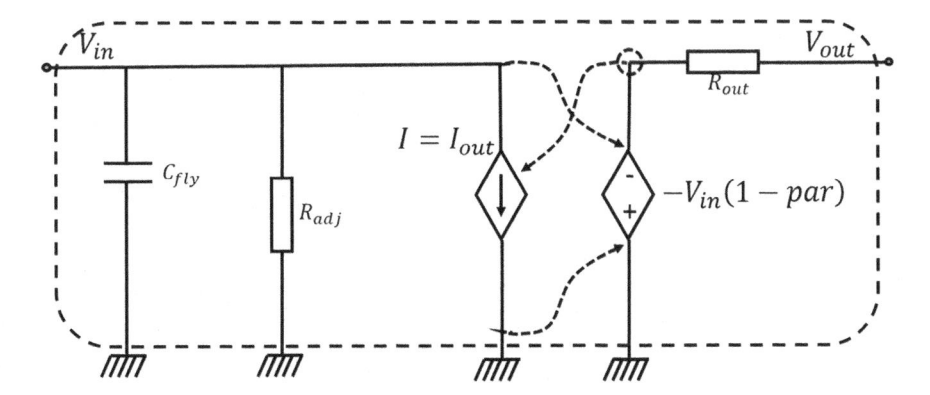

Fig. 15.20 Schematic of a spice model for a negative charge pump

with F_{CLK} the clock frequency. The factor 2 is coming from the fact that we have two fly capacitors charged in one clock cycle. This simple model is linear and reduces dramatically simulation time.

15.4.3.4 Loop Regulation for Clocked Power Stages

We name clocked power stage an output stage that requires a clock to operate. Regulation loop will act on this clock to control output voltage versus input reference voltage. Three cases are possible:

1. $V_{OUT} > V_{REF}$
2. $V_{OUT} < V_{REF}$
3. $V_{OUT} = V_{REF}$

Several types of loop regulation exist. The simplest is probably "go-no go" regulator. It consists in some comparators that compare the output value to the input voltage. One comparator can manage cases 1 and 2. But case 3 is not easy to support with comparators. A way to work around is to generate two references voltage $v_{REF+} = v_{REF} + \epsilon$ and $v_{REF-} = v_{REF} - \epsilon$. The three previous cases become:

1. $V_{OUT} > V_{REF} + \epsilon$
2. $V_{OUT} < V_{REF} - \epsilon$
3. $V_{REF} - \epsilon < V_{OUT} < V_{REF} + \epsilon$

Then with two comparators and by combining their output, it is easy to know in which case we are. This will help overall stability of the system and allow the clock to be stopped when we are in case 3. Indeed, we know that the BBGEN load is quite purely capacitive, and thus neglecting leakage, the output can stay at the same voltage for a while. Comparators should be fast enough to take decision before the output voltage goes to another case. It could be a good idea to anticipate this kind of issue by implementing trimming on ϵ. If comparators are too slow (or power stage settling time too large), the system will oscillate between region 1 and 2 and will never stop the clock. ϵ must be superior to mismatch between the comparators.

Another feedback loop control is the proportional integrated derivative (PID). The principle is to the calculate error between output and input voltages $e(t)$. Then we calculate the control signal $u(t)$:

$$u(t) = K_P e(t) + K_I \int_0^t e(\tau)d\tau + K_D \frac{de(t)}{dt} \qquad (15.37)$$

where K_P, K_I, and K_D are the PID constant. This constant has to be tuned to obtain a smooth and stable system. $u(t)$ can be used to control directly clock frequency or the duty cycle of the clock. The PID controller can be digitally implemented but requires some analogue to digital converters (ADC).

15 Body-Bias Voltage Generation

Lastly, we can address regulation with a low drop out (LDO) regulator. We saw previously that a DC–DC inverter output can be expressed as:

$$V_{OUT} = -\alpha V_{in} \qquad (15.38)$$

where α is between 0 and 1 and represents losses due to parasitics and leakage of used devices. So it is clear that controlling V_{IN} is a way to control V_{OUT}. This regulation technique has good precision but is limited in output range. On one end we have the minimum voltage supply of the charge pump, and on the end side we have the maximum voltage that LDO can provide, typically $vdd - V_{DS}^{pmos}$ where V_{DS}^{pmos} is the drain-source voltage of the pass transistor of the LDO circuit. Using the model of Fig. 15.20, the circuit is linearized and frequency stability analysis can be done considering schematic of Fig. 15.21.

This (simplified) small signal model allows to extract two dominant poles:

$$\begin{aligned} f_{p1} &= \frac{1}{2\pi(C_{load}//C_{fly})(R_{0-pass}//R_{load}+R_{out})} \\ f_{p2} &= \frac{1}{2\pi C_{par1}R_{0A1}} \end{aligned} \qquad (15.39)$$

The designer can choose to have a large C_{load} by adding an external capacitor. This will help for stabilization of the system by pushing second pole after the unity gain bandwidth (UGB) frequency. If this is not possible to add a such capacitor, a Miller compensation (RC type on pass transistor) can be added to generate a zero f_z around the UGB frequency, in order to give sufficient phase margin to the system. This is shown in Fig. 15.22.

15.4.4 BBGEN Architecture Examples

In this part, we will see two examples of how we could assemble blocks described in previous section. The goal here is to make the most generic generator, a BBGEN that could cover a wide range of applications. The first one (Fig. 15.23) uses two identical AB-class amplifiers for positive (N_{well} bias) and negative (P_{well} bias) outputs. An advantage of a rail-to-rail amplifier is design re-use: one covers [0; 1.3]

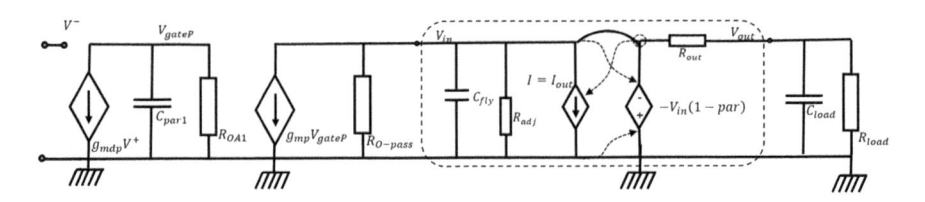

Fig. 15.21 Schematic of a small signal model for a negative charge pump with LDO regulation

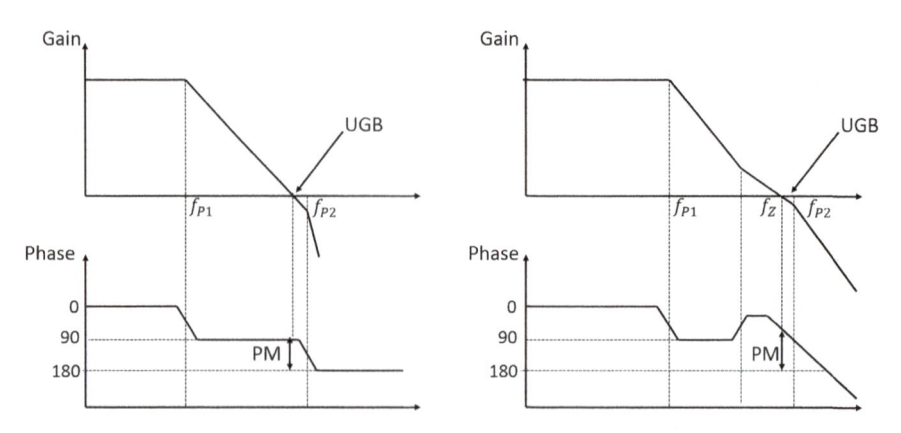

Fig. 15.22 Bode diagram of regulated charge pump system with large C_{load} (right), and with small C_{load} with Miller compensation (left)

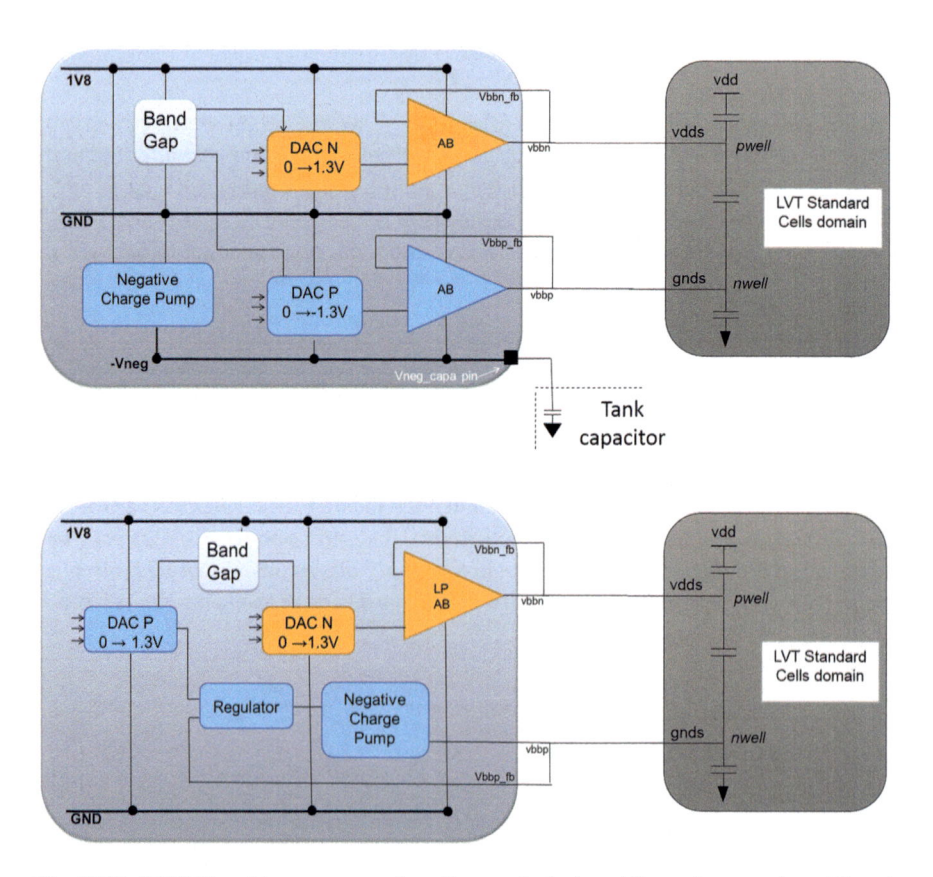

Fig. 15.23 BBGEN architecture examples. On top dual class-AB, on bottom class-AB and regulated charge pump

15 Body-Bias Voltage Generation

voltage range; and other covers $[-1.3; 0]$ voltage range which is equivalent to absolute range $[0.5, 1.8]$, with V_{neg} equal to -1.8 V. Two different DACs generate programmable reference voltages, and a bandgap makes the reference voltage. Programming is done through two buses of four or five bits (depending on the number of steps). Lastly a DC–DC inverter generates an internal negative rail voltage. This charge pump can be made of one charge pump in open loop, or two cascaded pump with a regulation to be sure to have exactly -1.8 V. To help the DC–DC to maintain a stable negative rail, an external tank capacitor is required when high current peaks are needed. We can now elaborate a second example where the AB-class amplifier on the negative side is replaced by a regulated negative charge pump.

In this second example there is no negative rail supply that means that all sides work in the positive voltage domain. Thus we need something to transpose the negative output voltage $vbbp$ into a positive voltage. This task can be performed simply using a resistor connected to $vbbp$ as shown in Fig. 15.24. Translated voltage V_{tr} is equal to $V_{neg} + RI$. As the V_{neg} voltage range is known, R and I can be chosen to force V_{tr} to be always positive. In practice I is produced by a current mirror, but as V_{tr} has the same voltage range than V_{neg}, drain to source of the two PMOS transistors could be very different leading to a bad current copy. To maintain $V1$ and $V2$ at the same voltage we add an error amplifier A_1 that controls transistor T_3. This loop fully compensates V_{neg} variation, maintaining $V1$ and $V2$ nearly identical.

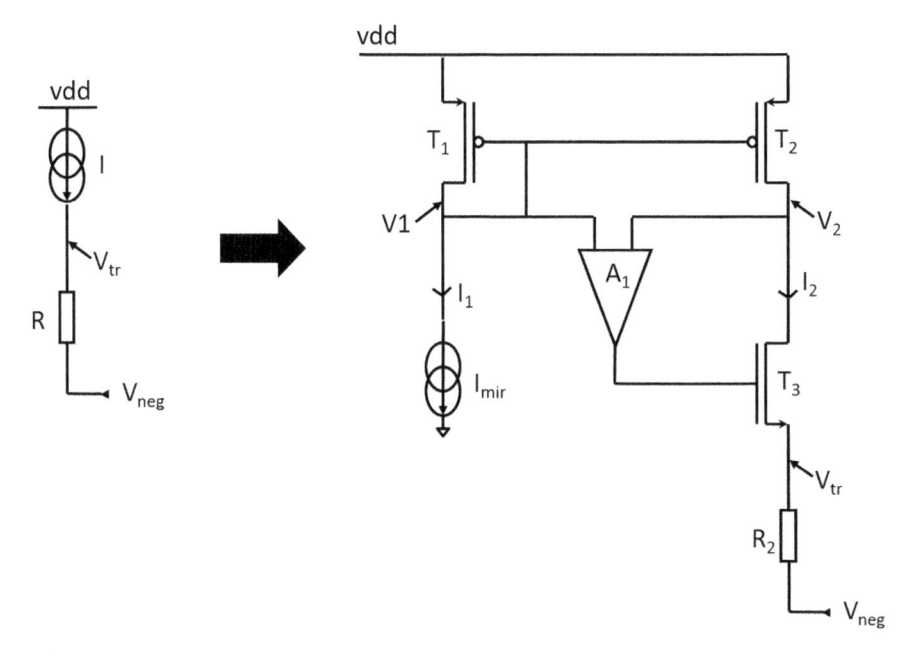

Fig. 15.24 Transposition of a negative voltage V_{neg} into a positive voltage V_{TR}; left: translation principle, right: physical implementation

15.5 Body-Bias Voltage Generator Implementation at System on Chip Level

15.5.1 Power Integrity Concern

In this part we will make some simple calculations to estimate the maximum current that BBGEN has to deal with. Let us first consider a settling time equal to 1 μs/V. We already detailed the load seen by BBGEN in Sect. 15.2. The order of magnitude of capacitors is $C = 1\,\text{nF/mm}^2$. Furthermore, we make the assumption that $vbbp$ and $vbbn$ have symmetrical variations so that $vbbp = -vbbn$. This allows to simplify the load model schematic as shown in Fig. 15.25.

In this particular case where $vpos$ travels $0\,\text{V} \rightarrow 1\,\text{V}$ and $vneg$ travels $0\,\text{V} \rightarrow -1\,\text{V}$, C_1 stores a charge $Q_1 = CV = C$, C_2 stores a charge $Q_2 = CV = 2C$, and C_3 stores a charge $Q_3 = CV = C$. Finally $vpos$ provides a total charge $Q = 3C$ and $vneg$ provides $Q = 3C$. That equivalence is explained and is true because $vpos$ and $vneg$ are symmetrical. Now we can consider the time constant τ of the formed $R_{out}C_{eq}$. Time to fully charge C_{eq} is roughly equal to 5τ. We are now able to calculate required BBGEN output impedance in order to be able to fit 1 μs/V settling time specification:

$$5\tau = 1\,\mu s \rightarrow R_{out} = \frac{1.10^{-6}}{5 \times 3.10^{-9}} = 66\,\Omega \quad (15.40)$$

With this output impedance value we evaluate the maximum current while C_{eq} is charging:

$$I_{max} = \frac{vpos}{R_{out}} \rightarrow I_{max} = \frac{1}{66} = 15\,\text{mA} \quad (15.41)$$

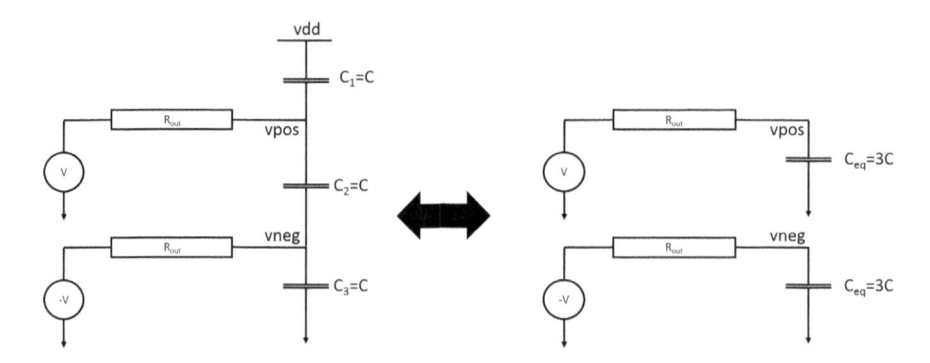

Fig. 15.25 Simplification of BBGEN load

This maximum current is for a 1 mm² load; this current must be proportionality sized to the actual load area.

For a large body-biased area, the current to provide can reach one hundred of mA: back-end designer should take some precautions to avoid damage to metallization and minimize voltage drop on $vpos$ and $vneg$ lines. Same thoroughness is mandatory for all BBGEN power-supply routing: vdd (which has to provide $2I_{max}$), external tank capacitor when present. Furthermore, such current spikes could disturb other analogue sides on the same power line.

15.5.2 Placement Concern

The main constraint on BBGEN placement will be power concern described previously. But routing between BBGEN and body-biased area has to be studied carefully: if the routing resistance is too large, settling time could increase out of specification range, and open loop response of output buffer could be significantly modified, leading to stability issues. Due to high current flowing in $vpos$ and $vneg$ lines, other signals nearby could be strongly disturbed because of parasitic capacitances between wires. It seems to be clear that placing BBGEN near the body-biased area is a good idea.

15.5.3 Testability

Testability for analogue IPs is not a standardized methodology in System on Chip world. For digital testability, standards exist, and the goal is to check that key cells in the design are functional. Test philosophy for analogue sides is different: we want to measure electrical specification like voltage reference, DAC output voltage, etc. This requires a pin and a pad to access from outside the chip to these electrical values. Designers can consider an analogue multiplexer to select which value to monitor. We can even go further and introduce debugging feature: ability to replace an internal voltage by an external one could help to debug in case of issue, like shown in Fig. 15.26.

15.6 Conclusion

In this chapter, we have covered several topics related to the design of a BBGEN. We focused on BBGEN that has to bias a digital area like a CPU or any island containing aligned standard cells. First, we have detailed and studied what could be a BBGEN load from an electrical point of view. Then we highlighted major constraints to take

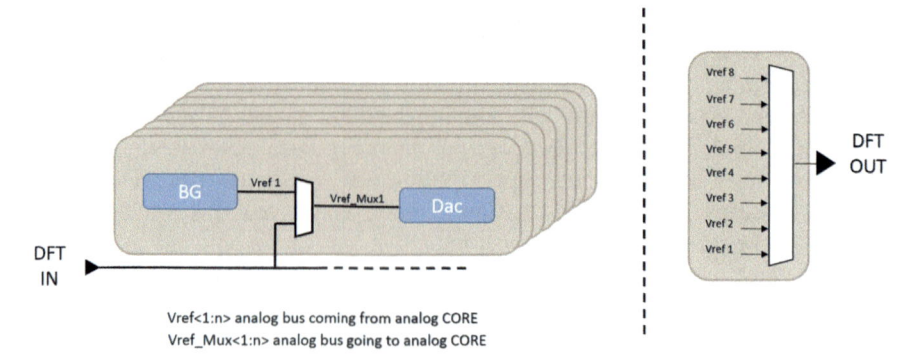

Fig. 15.26 Example of design for test and debugging strategy with eight paths

into account to write the design objective of our BBGEN. Once all these aspects are known, a non-exhaustive list of useful sides was made, as well as some examples on how to assemble them to build our generator. The last topic of this chapter was about how to implement such a circuit in a digital world: power integrity, placement, and design for test have been discussed.

References

1. W. Shockley, The theory of p-n junctions in semiconductors and p-n junction transistors. Bell Syst. Tech. J. **28**(3), 435–489 (1949). Equation 3.13 on page 454
2. D. Jacquet, F. Hasbani, P. Flatresse, R. Wilson, F. Arnaud, G. Cesana, T.D. Gilio, C. Lecocq, T. Roy, A. Chhabra, C. Grover, O. Minez, J. Uginet, G. Durieu, C. Adobati, D. Casalotto, F. Nyer, P. Menut, A. Cathelin, I. Vongsavady, P. Magarshack, A 3 GHz dual core processor ARM Cortex (tm)-A9 in 28nm UTBB FD-SOI CMOS with ultra-wide voltage range and energy efficiency optimization. IEEE J. Solid State Circuits **49**(4), 812–826 (2014)
3. A.P. Brokaw, A simple three-terminal IC bandgap reference. IEEE J. Solid State Circuits **9**(6), 388–393 (1974)
4. P.I. Wold, Signal-receiving system. U.S. Patent 1,514,753, filed November 1920, issued November 1924
5. B.D. Smith, Coding by feedback methods. Proc. IRE **41**, 1053–1058 (1953)
6. B.M. Gordon, R.P. Talambiras, Signal conversion apparatus. U.S. Patent 3,108,266, filed July 1955, issued October 1963
7. R. Hogervorst, J.P. Tero, R.G.H. Eschauzier, J.H. Huijsing, A compact power-efficient 3 V CMOS rail-to-rail input/output operational amplifier for VLSI cell libraries. IEEE J. Solid State Circuits **29**, 1505–1513 (1994)

Chapter 16
Digital Design Implementation Flow and Verification Methodology

Sébastien Marchal, Damien Riquet, and Sylvain Clerc

Digital Flow Steps and Vocabulary

The reader not familiar with digital flow and implementation steps and wording is invited to refer to Appendix B.

16.1 Specification and Engineering Test Body-Bias Prerequisites

This section will cover test aspects. It may be a surprise that this subject comes as an introduction to digital design flow and methodology but it is justified by the fact it governs static timing analysis strategy.

The test activity related to body-bias is split into two parts. The first one is the engineering test where the dies behavior is characterized in terms of voltage temperature and process range together with response to bias; this fundamental initial step defines a volume in the (process, voltage, temperature, age, body-bias)[1] space which should be included inside your design specifications.

The second part of testing is the production test where each die is tested and, according to the body-bias response characterized in the first step, a body-bias compensation is defined via on-chip programmable hardware. After the body-bias compensation is defined, each die has to be tested again to verify that it applies

---
[1]Further abbreviated (P, V, T, a, BB) space of static timing analysis.

S. Marchal (✉) · D. Riquet · S. Clerc
STMicroelectronics, Crolles, France
e-mail: sebastien.marchal@st.com

© Springer Nature Switzerland AG 2020
S. Clerc et al. (eds.), *The Fourth Terminal*, Integrated Circuits and Systems,
https://doi.org/10.1007/978-3-030-39496-7_16

the expected body-bias at selected test read points. Body-bias adds to usual circuit production test the following requisites:

- the electrical wafer sort (EWS) and in-package test time will be minimized as it is an important die cost aspect.
- test order is also driven by test time, test related to body-bias comes in order so that compensation test program is not executed on dies which would anyway fail.
- test read points are defined where gating specification parameters are checked, which include expected body-bias the die should apply.

An illustration of test steps is given in Fig. 16.1, in two compensation cases, process only (Fig. 16.1a) and process and temperature Fig. 16.1b, in both cases, the bias response is given by engineering characterization.

Then the production test follows the below steps, in the case of process compensation:

- The minimum bias needed to match the frequency specification is fused from an extrapolation of frequency-bias response at room temperature (middle point of Fig. 16.1a).
- The die is then tested at cold temperature to check that the frequency specification is met and checked at hot temperature to verify that frequency specification is met together with leakage ceiling.

In the case of process and temperature compensation:

- The frequency of device at room temperature is read.
- From this point, the minimum interpolation body-bias needed to match the frequency specification is extrapolated from temperature compensation law (middle point of Fig. 16.1b).
- The body-bias thus defined is fused into the hardware.

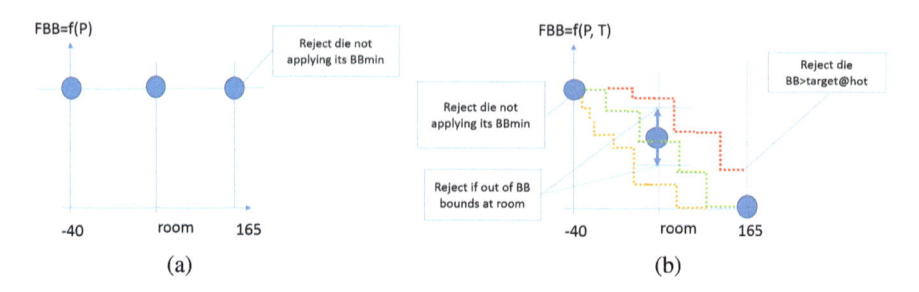

Fig. 16.1 Electrical wafer sort test read points in the cases of process only and process and temperature compensation. (**a**) Process compensation case. (**b**) Process and temperature case

- Then, the die is tested at cold, room, and hot temperatures to check that bias law is correctly applied. All dies failing to apply enough bias at cold,[2] too high bias at hot, or bias out of tolerance range at room are rejected.
- This rejection sorting is added to frequency and leakage gating of previous case.

Depending on the nature of the chip, the budget of test time and software availability at test, the search for body-bias points at room temperature can be done on functional, or at speed ATPG patterns, it can also be based on on-chip monitors response to gain test time or to work around patterns unavailability.[3]

Each product test will vary, from the two compensation examples detailed above. We would like the reader to be convinced that test conditions and static timing analysis verification strategy are tightly linked. The static timing analysis coverage will be exposed in the following sections (in Sect. 16.3).

16.2 Design Startup Caution

It is key for the architect to determine how the bias circuitry applies the body-bias, and then define validation scenarios and power intent accordingly. For example, if a cold startup needs body-bias to be applied prior to full-speed execution, then a low speed boot verification scenario needs to be put in place together with a frequency hopping hardware changing seamlessly the clock frequency when body-bias is ready.

Alternatively you may want to have your body-bias regulation module (aka: *the biaser*) not biased and let it control the reset of your main biased domain, when bias is ready. You will then need a separated non-biased power domain and in order your CAD places your biaser in that island.

This leads the following architecture recommendation[4]:

> Do not bias the biaser; and consider using dedicated RTL module and UPF power domain to let your CAD check that you do not.
> Alternatively, cover the startup with appropriate non-biased corners.

---

[2]Example of failure: the BBGEN charge pump may fail to deliver the amount of negative voltage needed at cold for a given die.

[3]The monitoring aspect is covered in Chaps. 14 and 12.

[4]One of us walked in this issue: we designed a CPU which was controlling its own bias on top of regular duty. The CPU needed about 3k cycles to boot and calculate it's bias. In some cold startup case the CPU was dying before it could determine the bias it needed.

16.3 Design Verification Principles and Corners Definition

Some design verification rules will be given, then we will apply these rules to different cases of compensation.

16.3.1 Design Verification Principles

Adding body-bias inside the design space adds one dimension to the circuit's verification corners definition which usually take into account process (P), voltage (V), temperature (T), and aging (a), this will be further referenced by PVTa. The design space of a circuit is defined as the space of points inside the PVTa space, the sign-off methodology defines extreme points of this space as the sign-off corners (SOF corners in short) where the static timing analysis will be executed.

Depending on whether the architect intends to either use body-bias for compensation, static or dynamic, or performance boosting, the body-bias addition will either restrict, translate, or extend design space.

As an illustration consider applying forward body-bias at slow-cold, it will lower the I_{on} difference between slow-cold and slow-hot and restrict design space volume. On the other hand if your design is planned to activate forward body-bias on demand depending on user performance requirement, you may have two frequency specifications associated with two body-bias voltages applied, in that case your design space has been extended, you will have to cover more cases in your static timing analysis scenarios.

Whichever the case depending on your application's specifications, let us state the sign-off rule:

> A circuit should be operated within its CAD static timing analysis operating range.

A consequence of this principle is that as long as your corners are checked, the design can move inside its operating space while remaining covered from timing analysis perspective, it is illustrated in Fig. 16.2.

16.3.2 Design Corners Definition

Definition of corners is provided by Foundries. The usage of process compensation has inevitable implications at design phase.

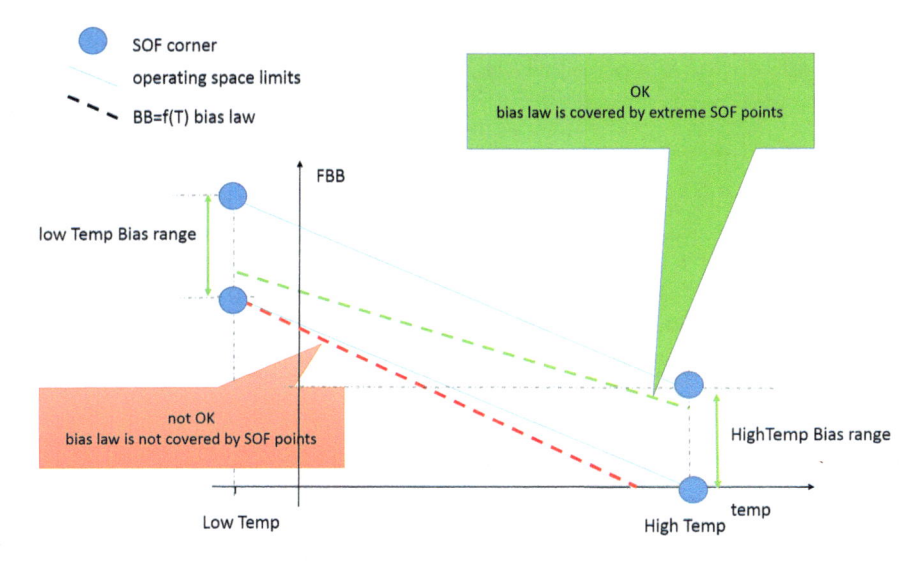

16 Digital Design Implementation Flow and Verification Methodology

Fig. 16.2 Static timing analysis design space in the case of temperature compensation, the blue lines delimit the allowed operating area, two couples of bias are defined at low and high temp, the dotted lines represent FBB as a function of T, the red version is wrong as it is outside the space delimited by verification corners, the green one is correctly covered

In summary to what follows: the presence of biasing is based on the very same principles as standard supply cases ones. It is a matter of identifying the updated slowest and fastest conditions, extending the foundry base corner definition. Foundry provides recommendation on the set of verification corners to use for a single supply design, without compensation. This recommendation boils down to providing the "slow" condition and the "fast" condition. Enough corners are provided to ensure the "slowest" condition is captured regardless of the timing paths topology. The same goes for the fast condition where enough corners need to be provided to ensure that the fastest case is found. An example is shown in Table 16.1 for 28 nm process.

To capture the slowest condition, the "Slow" and Min Vdd are associated together. This is obvious as any departure from one or the other necessarily produces faster timing. The argument does not apply to temperature and parasitic when you factor in the parasitic resistance dependency to temperature that is in competition with the transistor performance dependency to temperature. Also, cross talk does not make for an obvious winner among the four parasitics candidates. As such,

Table 16.1 28 nm Sign-off corners set

Intent	Transistor corner	Voltage	Temperature	Parasitics
Slowest	SS (Slow N, Slow P)	0.90 V	−40 °C and 125 °C	cworst, cbest, rcworst, rcbest
Fastest	FF (Fast N, Fast P)	1.05 V	−40 °C and 125 °C	cworst, cbest, rcworst, rcbest

the above table includes just enough T and RC combinations to ensure that the worst case scenario will necessarily be caught. The fastest case follows the same reasoning.

16.3.2.1 Which Corner for Setup Timing Check, for Hold Timing Check?

The previous table is particular in the sense that it does not tell where hold times or setup times need to be investigated. The answer is often "setup on slowest conditions," "hold on fastest conditions." This is not correct in all generality. Therefore, it should not be relied upon for verification and strangely it does not matter much. Commercial timing analysis tools perform both setup and hold analysis on any given corner. Therefore, regardless of any design reasoning done, the tool will prove you right or wrong. This being said, there are few observations that are useful to make.

Setup timing paths: are launched on one clock event and captured on the next clock event. The worst corner behavior of the path is therefore dictated by the data path. This means that the critical corner for setup is one of the slowest cases. It is still possible to have ill-behaved single cycle timing paths, but they are very unusual. If they do exist, they will behave like hold time in disguise.

Hold timing paths: hold time is a race between a clock and a data, with the requirement that the data must arrive after the clock. Determining which of the slowest of fastest corner will be the worst case is impossible. In most cases, the fastest condition is the worst one. But not always, it is possible for the worst corner of some paths to be in the slow condition. Therefore, for verification purposes, both the slowest and fastest condition must be looked at.

16.3.2.2 Design Verification Corners Definition Step by Step Method

The recommended steps to identify the corner for a particular compensation case are the following:

1. Identify the target slow condition. This case is always known beforehand as it is the condition for which the overall process compensation scheme is built on. In all generality, whatever the process compensation scheme, this corner is the transistor slow, Min Vdd max forward bias. Temperature needs to be looked at carefully if temperature compensation is used, but the base element does not change.
2. A chart is drawn that describes all possible extreme conditions the design needs to be operating on.
3. Adjust the chart taking into account what the chip engineering capabilities are. For instance, biasing compensation cannot have perfect accuracy and some margin is needed to enable the bias algorithm to find an applicable solution. Since circuit response to bias voltage is not steep the case of intermediate-to-extreme

corners cannot occur. For instance, there are enough bias steps to assume quasi-linear behavior. This is not true with other kind of compensation schemes like in voltage scaling.
4. By comparing their relative speed, corners are removed from this set to converge down to the set of corners for the functional mode
5. Then verify how the chip will boot and how the bias values will be set. This requires the bias algorithm to guess a bias that is always valid. If this case does not exist, some extra corners are needed to enable this boot sequence to succeed without bias.
6. Other modes, like scan, etcetera are defined to finish covering the design envelope.

16.3.2.3 Verification Corners Versus Implementation Corners

The requirements behind choosing verification corners and implementation corners are different. Commercial verification tools only have to calculate timings to provide an answer. On the opposite, implementation tools need to optimize the design while calculating timings. The consequence is that verification tools are very fast compared to implementation tools. Therefore, the requirement to choose verification corners is to ensure there are no escapes, while choosing implementation corners requires a balance between accuracy and speed of execution, this will be detailed in Sect. 16.4.1 below.

16.4 Physical Implementation

This section describes the specific addition to standard flow and details the implementation corners selection and the specific optimization steps to leverage forward body-bias effect.

16.4.1 Implementation Timing Corner Selection

Here we will describe how to select implementation corners. These corners are a subset of the exhaustive list of timing verification corners. We try to limit the number of timing corners used in place and route tools to the only ones that cover the majority of timing violations. They are commonly named "dominant" corners. Limiting the number of corners allows to keep the runtime and memory footprint of the EDA tools under control.

In a traditional digital implementation approach, the timing corner to be used during physical implementation is usually simple (Fig. 16.3):

Fig. 16.3 Illustration of usual dominant corners (circled) for P&R implementation

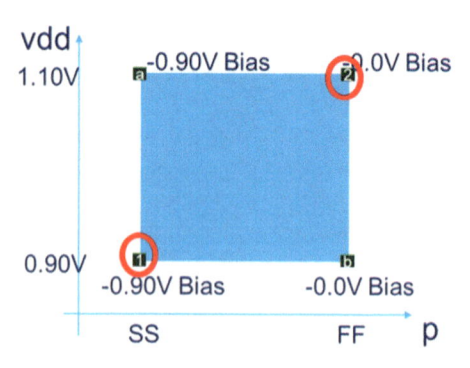

- For max frequency, we use min voltage, max RC corner, min or max temperature depending on whether the operating voltage is over or not the voltage of temperature inversion.
- For hold timing, max voltage, min RC corner, and min or max temperature corner should be used.

Introducing body-bias and in particular temperature compensation process makes this choice more complex and depends on the supply position with respect to V_{Tinv}.

The timing corners have to be combined with timing modes (Functional, test,...) resulting in an increase of the total number of scenarios the timing engine has to solve. Utilization of body-bias requires the introduction of a new variable in the design space as the substrate voltages must be taken into account.

It seems very easy as the speed of logic increases when forward body-bias is applied. This is true for standard process compensation techniques where the compensation deals with one single variable (process from slow to fast); in this case, body-bias compensation is just an extension of the traditional approach:

- Setup dominant corner is: min voltage, max wire capacitance corner, one extreme temperature corner, and minimal bias applied at worst corner.

With temperature and process compensation combined, the selection is becoming more complex, for worst timing corner. Body-bias is now compensating the timing change through temperature evolution. It means that a cell timing is now a function of temperature and bias, delay(Temp,Bias). Unfortunately, these two variables are not independent: body-bias performance boost depends on the temperature.

In addition, the resistance of wire is increasing with temperature, for technologies smaller than 40 nm, the wire delay is a significant contributor to the full timing of the logic path. This makes the behavior of the cell more complex. In the context of temperature inversion, temperature increase means less biasing, as the cell delay naturally decreases with temperature. But, at the same time, the wire delay increases due to positive temperature coefficient of the wire resistance. Those two effects operate in competition and identifying a dominant corner is difficult. Depending on the logic involved, some path will have there slowest condition for (MaxT,MinBias), while others will be limited by (MinT,MaxBias).

qFO4 (165C 0.3V FBB ref) or (-40C 0.9VFBB ref) vs P&R STA

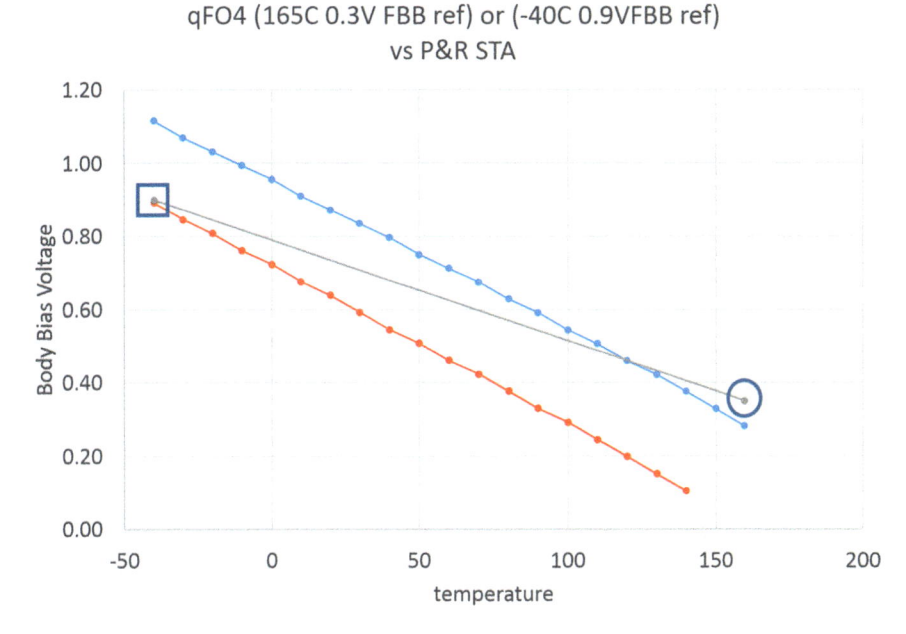

Fig. 16.4 Illustration of P&R corner selection. The two parallel lines represent the iso-delay characteristics anticipated from various ring oscillators which drove the definition of the STA corners. The plain line represents the trajectory of a path from a CPU that has in that case too much wiring contribution compared to the reference configurations. If only the square is used for implementation, the path will end up slower than the signoff corners and will generate violations unseen by implementation

For this case it is needed to include in the implementation corner the 2 temperature conditions: (slow process, worst wire capacitance, min vdd, max bias, min temperature) and (slow process, worst wire capacitance, min vdd, min bias, max temperature).

Not including these corners during physical implementation and introducing them only at final static timing analysis will make designer discover performance degradation only at late stages. There is a high risk that the tools are unable to fix the violations and to close timing due to too many critical paths not matching the expected performance.[5]

As an illustration, Fig. 16.4 illustrates the effect of using only slowest initial corner as the dominant corner for place and route, some un-optimized paths pop up during sign-off at slow process hot temperature min voltage zero bias just because the CAD tool could not optimize. This can be easily avoided by adding one extra corner in implementation.

---

[5] This is a convergence problem, see Appendix B, Sect. B.1.8.

16.4.2 Specific Addition to Standard Flow

The usage of body-bias requires a specific power grid to supply the PWELL and NWELL. As there is virtually no current to be flooding through, apart when bias voltage changes, the size of this specific power grid can be at minimum width. The only constraint is to ensure proper spacing to neighboring metal shapes as the voltage difference will be equal to $-Vbias + Vmax$. This voltage can lead to specific reliability issues (dielectric breakdown). This power grid needs to be connected to PWELL and NWELL thru specific well taps cells plus stacked vias connected to power pin of the strap cell up to power grid. In addition, the biased area must be drawn in a deep NWELL zone, this deep NWELL zone must be closed by a rounding Nwell to ensure perfect well isolation, and LVS clean layout.

16.4.3 Low-Power Flow

Handling body-bias in a design falls into, now of common practice, voltage handling inside a low-power digital flow.

Once the architect has carefully specified the body-bias supplies names, ports, and connection inside the circuits UPF, see Appendices B and C, the computer aided design tooling will handle the body-bias implementation without more exceptions than the one cited above: NWELL islands and specific spacing rules.

Appendix A
FD-SOI Process Flow

Figures A.1, A.2, A.3, and A.4 will give the reader an overview of the process steps for substrate and device manufacturing.

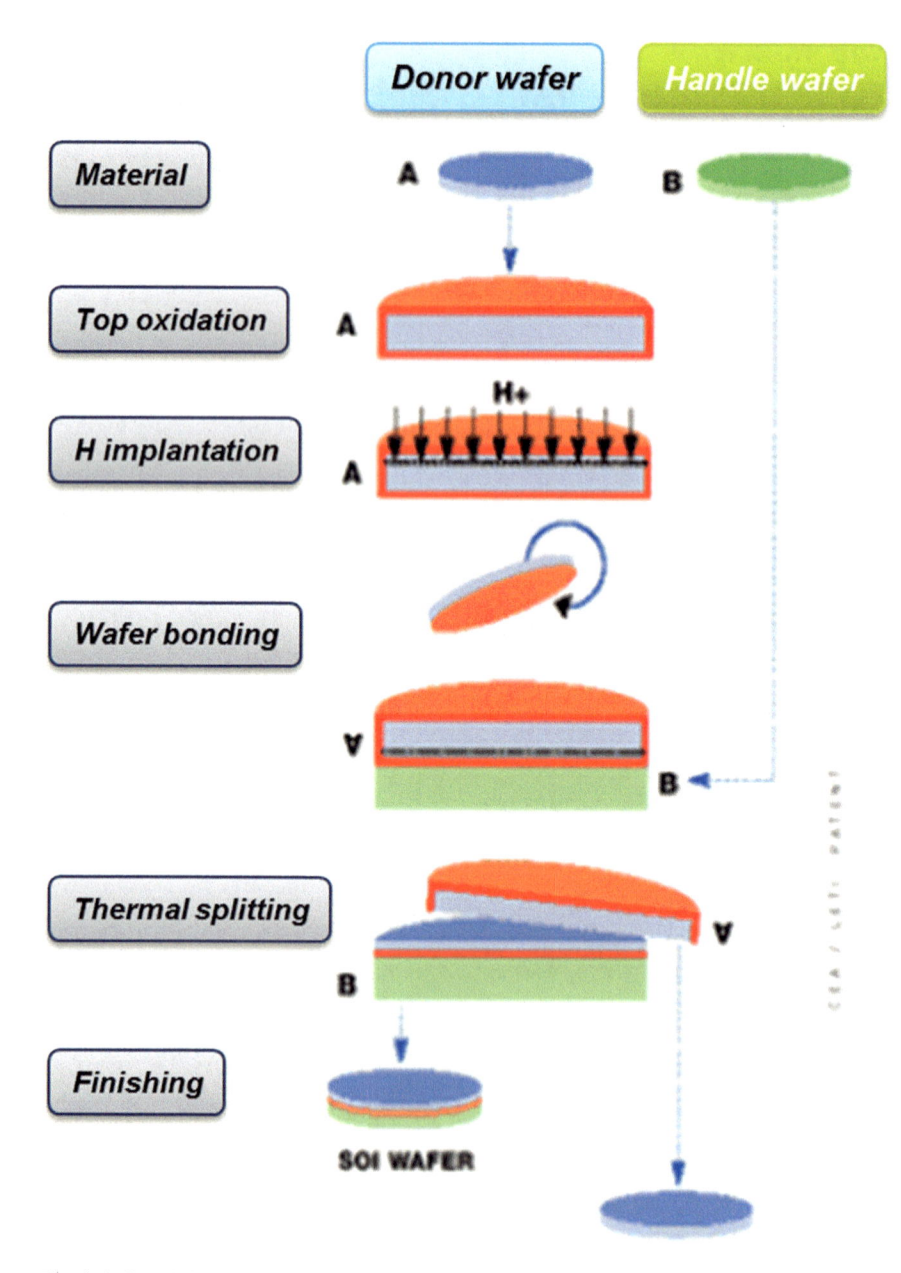

Fig. A.1 Step 1: FD-SOI substrate building

A FD-SOI Process Flow 397

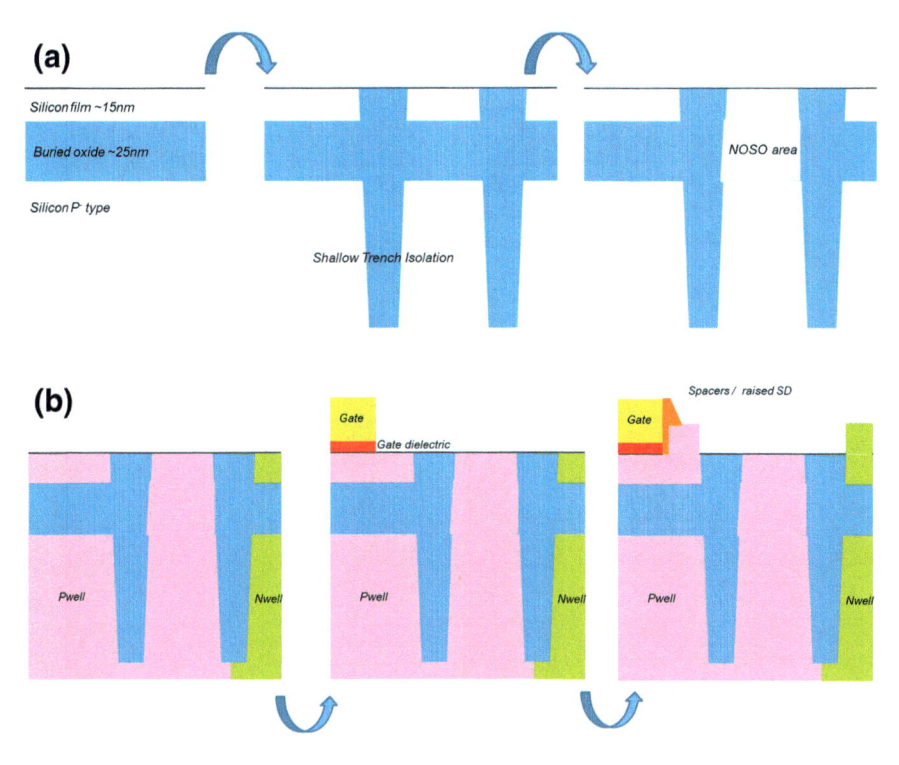

Fig. A.2 Step 2: Front-end of Line flow Oxyde steps (**a**), Gate and Well steps (**b**)

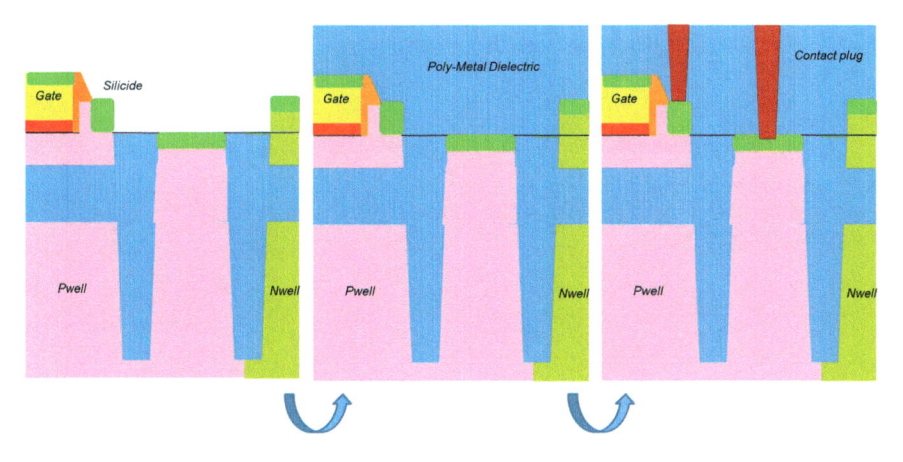

Fig. A.3 Step 3: Middle-end of Line flow

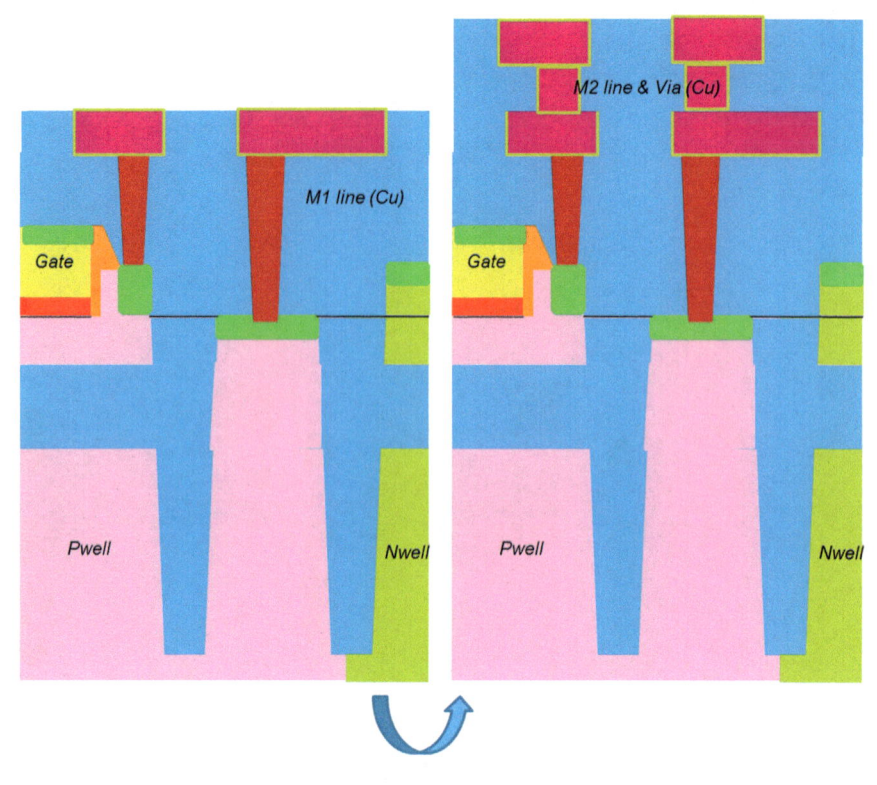

Fig. A.4 Step 4: BEOL-end of Line flow

Appendix B
Digital Implementation Flow and Terminology

The content of this appendix is mostly not Body-Bias specific, we will give the reader some usual[1] definitions and digital design procedures. However, the attention of the reader will be raised every time Body-Bias requires special action or attention which is detailed in Chap. 16. In this appendix, some definitions are referring to each other, we kindly ask the reader to make multi-pass reads to ensure no tautological deadlock reading conditions is met.

As circuits integrate more and more gates, concurrent multi-actor engineering is put in place and some incremental steps are executed to close the circuit design. These steps are interdependent and are repeated in a sequence called the 'digital flow', it will be detailed in Sect. B.3. Positive edge triggered Flip-Flop is abbreviated by FF, they are used in standard synchronous designs (see Sect. B.1.9) and alternatively named register or reg in tools reports. Sequential elements are design elements which sample their inputs, either the data of a register or some input at the digital interface of a hierarchical block, the sampling can be triggered by a clock edge or a clock level in case of level sensitive element. As will be further exposed, all digital design is based on satisfying constraints related to sequential elements sampling timing, the sampling signal is denoted as a clock.[2] The digital designers benchmark the results of their implementation in terms of Quality Of Result (QoR) which denotes how well the design constraints have been met (see Sect. B.2) and what is the amount of resources consumed to reach the final result.

The following notations will be used throughout this chapter:

- T_{cycle}, clock cycle period
- T_{CP2Q}, clock-to-output delay for a Flip-Flop

---

[1] Digital design experts this appendix is not for you!

[2] Strictly speaking, any signal connected to a sampling input of a sequential element is considered as a clock by CAD tools, it is part of the digital designer's duty to verify that no unintended assignment happens here, or said differently that some logic signals do not become a clock by accident.

- T_{comb}, time of combinational logic between two sequential elements, or between input and sequential elements, or between sequential element and output
- T_{launch} time when the data starts to propagate to the next element (clock edge of the startpoint)
- $T_{capture}$ time when the data is sampled at the next on sequential element at the end of the path (clock edge of the endpoint)
- T_{setup}, setup time, see Sect. B.2.1
- T_{hold}, hold time, see Sect. B.2.2
- T_{r_i}, clock capture path of FF index i
- T_{r_j}, clock launch path of FF index j

B.1 Digital Design Wording and Body-Bias

B.1.1 Quality of Results

Abbreviated QoR, this denotes the amount of resources consumed to achieve the specifications and compliance to Sign-Off (SOF) rules. Your design is a good candidate to apply Body-Bias if its QoR improves with it.[3]

B.1.2 Corner, Mode and Scenario

A mode is a state of your circuit where some inputs are stable and define a subset of functionality. An example is `scan shift` mode where the scan shift signal of your circuit is activated.

A `corner` is a set of operating conditions of your circuit, like process, voltage, temperature, ageing. In some cases it can encompass input slopes, output loads.

A `Scenario` is the cartesian product of Corner and Mode. The Sign-Off and Implementation are all based on Scenarios.

B.1.3 Temperature Inversion, V_{Tinv} and ZTC

CMOS transistors I,V characteristics depend on the competing effect of temperature and voltage on both V_T and mobility. These competing effects result in having a voltage point where they nullify themselves, this point is either called voltage of temperature inversion(V_{Tinv}) or zero temperature coefficient. In terms of digital design space this means that slow STA corners temperature depends on voltage:

---

[3] Refer to Sect. 14.1 to arbitrate with AVS.

- for low voltage systems, the slowest corners are at cold temperature
- for systems supplied at voltage above V_{Tinv}, the slowest corners are at high temperature
- for systems with wide operating range, all voltage-temperature cases need to be covered in STA constellation

B.1.4 Implementation and Sign-Off

The design implementation is transforming your design specification (most of the time coded in RTL) to manufacturing mask patterns. These masks and intermediate design descriptions are verified before being sent to manufacturing by various Sign-Off tasks. The initial steps of implementation are called Front-End, the last ones Back-End, which differ from software or process worlds Front-End and Back-End definitions, it will be detailed in Sect. B.3 below.

B.1.5 Functional Simulation and Static Timing Analysis

Static timing analysis and functional verification are two design verification tasks which complement themselves. Static timing analysis (STA) is a way to ensure coverage and productivity by covering all possible timing arcs of your design. Because it is exhaustive you are not exposed to a functional validation which would not cover all cases of usage and dataset, STA reads in the same constraints as the implementation tools do, this is a way to ensure consistency across all design steps. However because it covers all timings arcs it may report violations which do not correspond to real usage cases, these need to be filtered out by constraints exceptions.

Functional simulations are pattern based, here, the circuit behaviour is checked against real condition of usage in the correct sequence, it is realistic and application dependent. Because the usage sequence needs to be fulfilled it takes more runtime than STA. Functional simulations are used at the beginning of design to check that the design achieves the service it is intended to and at the end to debug design constraints with an emphasis on STA exceptions

In the Body-Bias context, if you have followed the recommendation of not 'biasing the biaser' from Chap. 16, there should be no timing arc detected during STA between your biaser and the biased islands. Further, appropriate design start-up sequence should be verified in simulations to ensure Body-Bias is present before the biased islands boots-up.

B.1.6 Clock Tree and Clock Skew

The clock tree is the high fanout net which propagates the clock to each sequential elements of your design. It is ideal, i.e. single stage during Front-End, and synthesized with real gates during Back-End's clock tree synthesis. The clock skew is the de-balancing between two clock tree endpoints. The skew of interest is the one between the launching sequential element and the capture sequencing element of the same path. The longer the skew, the more the logic path is susceptible to hold time violation. However in some cases time can be borrowed from neighbouring pipe-line stages to leave more computation time to a setup constrained path, this is called the `useful skew`.

B.1.7 High Fanout Nets

These are nets with a lot of fanout like `reset signals` scan shift control signals and of course the clock. High fanout nets are handled with special care during implementation either because their timing is critical or their integrity is critical.[4]

B.1.8 Convergence and Correlation

The Place and Route `convergence` is denoting the fact that timing in Place and Route is able to generate layout without Sign-Off violations (static timing analysis and physical). It is instrumental that appropriate corners are defined for implementation so that no unexpected violation appears during SOF, see Sect. 16.4 for that matter.

B.1.9 Sequential Design, the Digital Sheep Shepherd Problem

The digital design activity, since Alan Turing's contribution is all about handling a flow of binary information coming in the circuit, being transformed and then output to the external world. Design sequencing enables to discriminate the limits of incoming and outcoming digital information flow. Let us take the image of a shepherd[5] who needs to carry several groups of sheep across a corral. Our shepherd further needs to sort his sheep in several different boxes so that sheep groups

---

[4]No designer wants a cross talk induced glitch on an asynchronous reset, for example.

[5]May the persnickety reader forgive us, we are going to use sheep image to illustrate design constraints instead of equations.

composition is preserved. To match integrated circuit situation, we add an extra constraint that the sheep do not cross the corral at the same speed. That is the situation of a digital designer who needs to sort out binary information while preserving group composition.[6]

If our shepherd does not want sheep groups to mix, he needs corral gates to control when sheep start and end their journey. Closed gates will stop fast sheep from traveling further than expected, selective gate to gate opening delay will ensure which group of sheep travels and that everyone has arrived on time.

> Corral gate is the Shepherd's equivalent of sequential elements

To be sure that each group travels across the corral without mixing with the late sheep from the previous group, our shepherd will ensure that the next sheep group does not start until the corral is clear from the previous group's sheep, that's the hold constraint.

> Fast sheep of next sheep group mixing with slow sheep of previous group is the Shepherd's equivalent of hold constraint

To be sure that each sheep within a group has had enough time to travel, he will leave enough time between the launch corral entry gate opening and the capture corral gate closing, this is the setup constraint.

> Coral open gate to close gate is the Shepherd's equivalent to set up constraint

B.2 Digital Design Rules and Constraints Definition

For the two following sections on setup and hold constraint we will refer to Fig. B.1 below which displays a register-to-register timing path (STA nickname is reg2reg).

---

[6]The group composition preservation is the digital circuit sequence of operation or said differently the architectural state. Done with the persnickety reader.

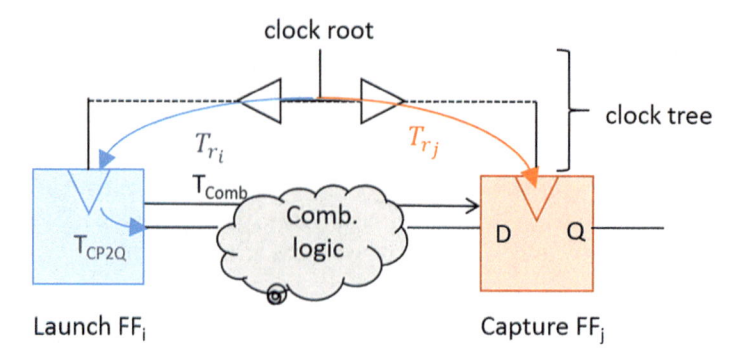

Fig. B.1 Register-to-register timing arc diagram, the root clock is divided in two diverging branches which have different delay (the skew), from [1]

B.2.1 Setup Constraint

In case of setup constraint, the design should meet its specs in frequency, said differently from Launch Flip-Flop to the Capture FF there should be enough time in any case so that the data at the input of the Capture FF is the correct one.

The timing condition to be met is that the sampling time of the capture FF ($T_{capture}$) occurs after the launch FF output has traveled across its logical cone, reached the input of capture FF early enough to provision the setup constraint ($T_{capture}$), with reference to Fig. B.2a:

$$T_{launch} \leq T_{capture}$$
$$T_{launch} = T_{r_i} + T_{CP2Q} + T_{comb} + T_{setup}$$
$$T_{capture} = T_{r_j} + T_{cycle}$$
$$T_{cycle} \geq T_{CP2Q} + \max(T_{comb}) + T_{setup} + T_{r_i} - T_{r_j} \quad (B.1)$$

The combinational time T_{comb} to consider in that case is the longest which can occur inside the logical cone (i.e.: max). In case the condition is not met, the data of the N+1 cycle is lost. Equation (B.1) is written as a constraint on T_{cycle} because it translates the microscopic FF wise constraint into macroscopic specification of circuit clock frequency. This setup constraint is said to be soft, in case the design is in violation, the user by lowering the clock frequency can recover functionality.[7]

---

[7] Provided that it is legal in the application, i.e.: your design does not depend on the other's clock.

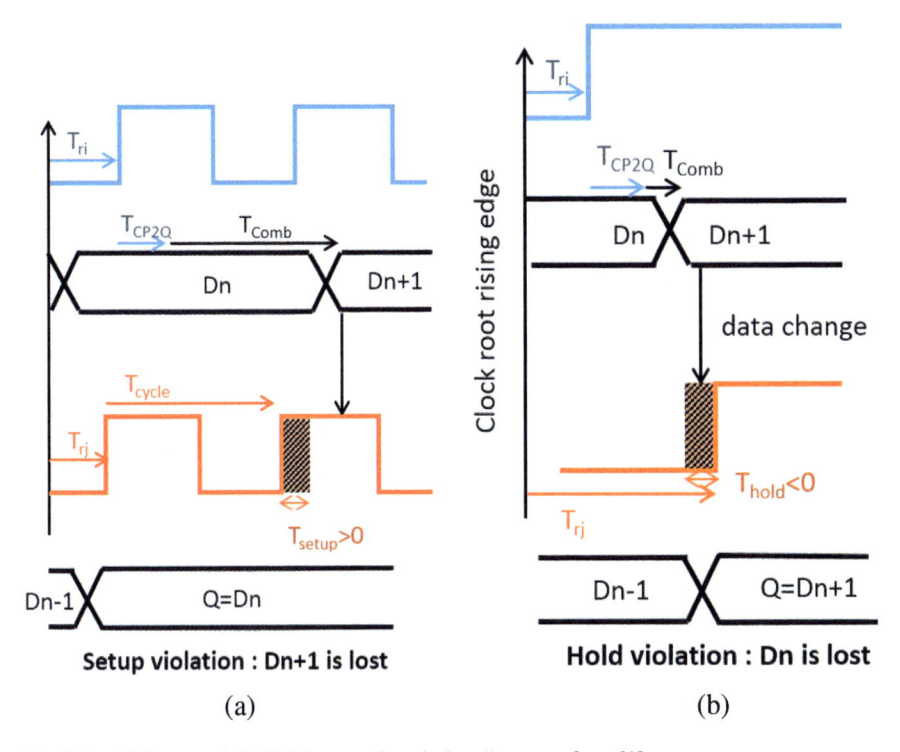

B Digital Implementation Flow and Terminology

Fig. B.2 (**a**) Setup and (**b**) Hold constraints timing diagrams, from [1]

Body-Bias can accelerate gates, it can solve setup violations directly, see Sects. 16.3, 16.4 and 3.2

B.2.2 Hold Constraint

This constraint denotes the sequencing coherence, as previously explained in the Shepherd image, we do not want our information groups to be mixed. The next cycle data from the launch FF should not reach the capture FF before the capture clock sampling window is closed which happens T_{hold} <u>after</u> the capture FF clock sampling (here rising) edge. In case this constraint is violated, the information stored in the capture FF is the next cycle one, i.e. instead of holding data of cycle N, it holds the data of cycle N+1. With reference to Fig. B.2b, the timing equations are:

$$T_{capture} \leq T_{launch}$$

$$T_{capture} = T_{r_j} + T_{hold}$$
$$T_{launch} = T_{r_i} + T_{CP2Q} + \min(T_{comb})$$
$$T_{r_j} - T_{r_i} \leq T_{CP2Q} + \min(T_{comb}) - T_{hold} \tag{B.2}$$

The combinational delay to consider in that case is the shortest which can occur in the logical cone (i.e. min). The constraint equation is rewritten with the clock delay difference between T_{r_j} and T_{r_i} which denotes the clock skew between the capture and launch FF.[8]

> Body-Bias accelerates device speed, it does not help or can even worsen the situation for hold violation in fast scenarios, however, it can indirectly help to fix hold violation constraints by lowering hold constraint variability in slow scenarios, see Sect. 3.2 and Table 3.2.

B.2.3 Clock Minimum Pulse Width Constraint

Min pulse width constraint reflects the time needed for charges to travel across the gates forming the latches of sequential elements, as an example for a C2MOS master latch clock transparent high, the minimum pulse width high reflects the time needed for the three state driver to toggle the internal node of the latch. Translated to a C2MOS master slave Flip-Flop, this will be transposed to a clock minimum pulse width low for the master latch and minimum pulse width high for the slave latch. A too fast clock or a clock tree edge erosion can lead to minimum pulse width violations. Body-Bias can help to fix this type of violation as it accelerates gate speed, but specific care must be taken in SOF corners definition for this purpose while checking the clock signal duty cycle is not degraded.

B.2.4 Maximum Fanout Load Capacitance, Signal Transition Rules

Maximum drive, fanout and capacitive load are static constraints attached to a given logic cell pin. It is a safeguard against charge sharing which induces cross talk weakness and slow transition and in the end bad Silicon-CAD correlation. Forward

---

[8] We will encounter this skew all across this appendix, together with Hold these two guys give digital designers nightmares.

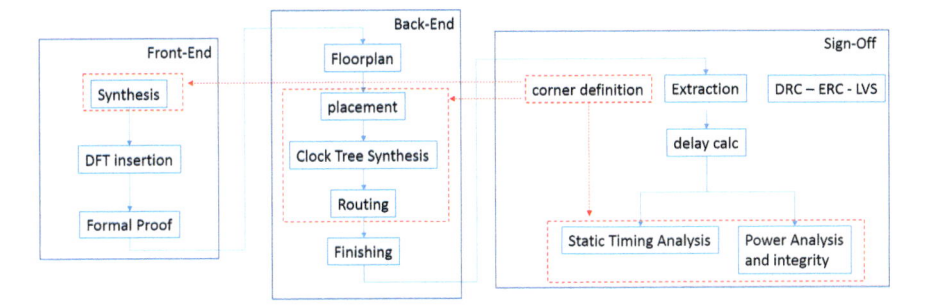

Fig. B.3 Digital implementation flow steps

Body-Bias cannot help for maximum capacitance or fanout rules but it can for maximum transition as it accelerates transitions.

B.3 Flow Steps

This part will describe the Digital Implementation flow steps (Fig. B.3) and points out the aspects which are impacted by Body-Bias, for which the reader will be referred to the corresponding sections of Chap. 16.

B.3.1 Front-End

This section deals with Register Transfer Language design entry fracturing into gate netlist and its validation.

B.3.1.1 Synthesis

It is the translation of your RTL intention into `gates` and macro mapping connected by nets.[9] Synthesis needs constraint to operate which is the dual macroscopic view of constraints detailed in this appendix. The hold constraint is not considered during this flow step because the clock tree is ideal. The minimal necessary set of constraints to start here is timing related and specifies clock frequency, input and output delays, input transitions output loads. On top of this, power domain description can be fed in (see power intent Appendix C below). Synthesis is executed at the slowest corner where clock frequency is difficult to reach.

---

[9] Hence the netlist term.

> Body-Bias can help during Synthesis as it accelerates slow device, see Sects. 3.2 and 16.3

B.3.1.2 DFT Insertion

Actual VLSI systems require specific embedded diagnosis hardware to filter out defective parts (caused by design or manufacturing) which would be too long to identify otherwise. It is embodied by scan chains for standard cell logic, BIST for memories and specific hard macros (like high speed interfaces). In scan chains no functional logic is present between sequential elements, it is highly susceptible to hold violations.

> Body-Bias can degrade your hold violations in fast corners QoR because of device acceleration, be sure it is correctly covered by your SOF scenarios, see Sect. 16.3.

B.3.1.3 Formal Proof

This Digital Flow step checks that two netlists or an RTL and a netlist are equivalent, this step is executed after the first Synthesis to compare the netlist result with initial RTL, and after Digital Flow macro steps: transition from Front-End to Back-End and Back-End to Sign-Off. Beyond detecting CAD tool errors[10] this step helps to debug constraints or ill-coded RTL. Body-Bias has no specific requirement here, as long as you follow the 'don't bias the biaser' rule.

B.3.2 Back-End

Following the previous Front-End macro step, here is the implementation Place and Route steps, gathered in the Back-End macro step. Computation workload is huge here, to minimize turnaround time, the cells views used include only the layers needed to make the routing and a subset of all verification corners will be used. This choice of what specialist call the `dominant corners` is key, it is detailed in Sect. 16.4.1.

---

[10] Highly uncommon, we have not seen that in 20 years on production software CAD tools.

B Digital Implementation Flow and Terminology

B.3.2.1 Floorplan

It is the first step of the Back-End implementation, it consists in placing macro-blocks, plan your power distribution network (taking the form of huge stripe of metal, called 'the power grid'), specify your IO ring, pre-route analog nets. The power is planned and pre-routed using IEEE1801 UPF specifications, see Appendix C.

> Care must be taken in case of externally applied Body-Bias or if you want to monitor the biasing nets voltage through analog IOs, negative voltage can turn on ESD protection diodes.

At this floorplan stage, you need to specify your Body-Bias islands which take the form of DNWELL areas surrounded by NWELL stripes, these islands need their WELL taps to be pre-routed to your Body-Bias supply nets. DNWLL and NWELL layout needs to be exported out of floorplan to be imported back at the last finishing stage but meanwhile, usual floorplan LVS and DRC must be run at the earliest stage of your project, implicit substrate connection can generate short circuits if not handled carefully.

> The floorplan task is where most Back-End Body-Bias implementation specific work needs to be done, see Sect. 16.4 chapter for details.

> It is highly recommended to run at the earliest stage of your project floorplan DRC and floorplan LVS to check that your Body-Bias specifications are met.

B.3.2.2 Placement

At this step the netlist gates are placed and a first estimation of delay induced by wire is done.

B.3.2.3 Clock Tree Synthesis

This is where your circuit's clock tree is generated, it is called synthesis because the tool tries to balance clock tree end points delay following your input constraints. These are all the sequential elements clock pins of your design.

B.3.2.4 Routing

During this step physical layout of interconnections is drawn by the routing tool, at this step some special net shape spacing may apply between your signal nets and Body-Bias supplies nets.

> Be sure the router enforce the specific spacing between circuit's signal nets and Body-Bias supplies.

B.3.2.5 Finishing

During all previous steps, to lower database size only routing layers were present in cells and macro views. During this step the all-layers physical views are used, DNWELL and NWELL shapes embodying your Body-Bias islands are red in. The final layout is also filled with density correction patterns. The Place and Route tool has modified your netlist from step to step, by inserting the clock tree cells and by changing cells to optimize timing, the final netlist is exported to be used for final verification.

B.3.3 Sign-Off

The Sign-Off macro steps gather the final verification steps, it is usually done with a different set of tools and less stringent tolerance tuning so as to close the verification.

B.3.3.1 Extraction

During this step the design interconnect parasitics induced by layout is computed and written to files.

B.3.3.2 Static Timing Analysis

The static timing analysis reads the design netlist in, the constraints and the parasitic delays and computes in an exhaustive manner all timing arcs of the design and checks that for each of them, the design constraints are met. STA uses no stimuli pattern as opposed to functional simulation.

B.3.3.3 Power and Voltage Drop Analysis

The power analysis step checks that the design matches its power specs with two aspects of importance: design integrity with respect to power grid sizing and power specification bounds. Voltage drop analysis is intended to check that your power grid network does not induce voltage drops caused by in-rush current of your design.

B.3.3.4 Design Rule Check and Layout Versus Schematic

These are the layout verification, the `Design Rules Check` (DRC) verifies that your circuit's layout fulfils the foundry manufacturing layout rules. The Body-Bias specific rules are WELL spacing, WELL taps spacing and signal nets spacing to WELL tap supplies (nets whose name contains 'gnds' and 'vdds' in Appendix C below).

The `Layout Versus Schematic` verifies that your circuit netlist, which corresponds to the analog designer's schematic, matches your layout. In digital there is not golden schematic reference as it is the case in Analog design because the tools modify the original netlist that is the reason why some formal proof and functional verification of the final netlist is required. In Body-Bias context, special care must be taken with WELL tap signals (the Body-Bias supplies) and implicit connection induced by NWELLs and substrate, this is illustrated in Fig. B.4 where a stripe of NWELL is missing, leading to a short circuit between the substrate and PWELL of the Body-Bias island.

Fig. B.4 Implicit PWELL short circuit between native substrate (right of the figure) and ill isolated PWELL Body-Bias island (left of the figure), the vertical orange NWELL shape does not close the DNWELL area (light orange)

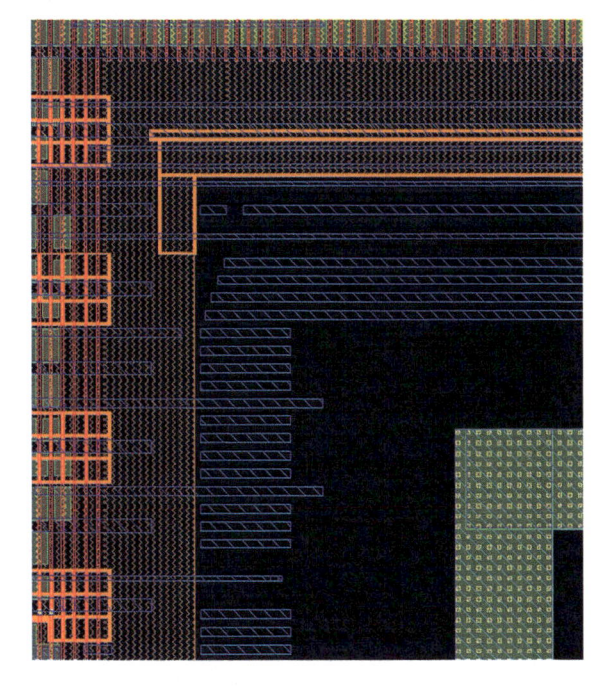

Appendix C
IEEE 1801 UPF Example

This appendix includes an example of IEEE 1801 UPF [2] with `vdd_CPU` being the supply of the biased domain of interest. The UPF standard specifies six functions for supply sets: *power, ground, pwell, nwell, deeppwell* and *deepnwell* with 'each function representing a potential supply net connection to a corresponding portion of a transistor'. In the case of Flipped-Well[1] in FD-SOI, `NMOS transistor body net` is connected to *nwell* and `PMOS transistor body net` is connected to *pwell*, see pwell/nwell statements below for `ss_pd_CPU`.

```
upf_version 2.0

set_design_attributes -elements {.} -attribute
                                    enable_bias true

set_scope /

#/ SUPPLY PORTS
create_supply_port gnds_BBGEN_core
create_supply_port vdds_BBGEN_core
create_supply_port vdd_1v8_core
create_supply_port vdd_comp
create_supply_port vdd_CPU
create_supply_port vdd_5V_BBGEN
create_supply_port vdd_1V8_BBGEN_core
create_supply_port vdd_top
create_supply_port GND
```

---

[1] For the digital designers who jumped directly to this section, kindly refer to Chap. 2 detailing Flipped-WELL.

```
#/ SUPPLY NETS
create_supply_net gnds_BBGEN_core
create_supply_net vdds_BBGEN_core
create_supply_net vdd_1v8_core
create_supply_net vdd_comp
create_supply_net vdd_CPU
create_supply_net vdd_5V_BBGEN
create_supply_net vdd_1V8_BBGEN_core
create_supply_net vdd_top
create_supply_net GND


#/ CONNECT SUPPLY PORTS TO NETS
connect_supply_net gnds_BBGEN_core -ports
                                  gnds_BBGEN_core
connect_supply_net vdds_BBGEN_core -ports
                                  vdds_BBGEN_core
connect_supply_net vdd_1v8_core -ports vdd_1v8_core
connect_supply_net vdd_comp -ports vdd_comp
connect_supply_net vdd_CPU -ports vdd_CPU
connect_supply_net vdd_5V_BBGEN -ports vdd_5V_BBGEN
connect_supply_net vdd_1V8_BBGEN_core -ports
                                  vdd_1V8_BBGEN_core
connect_supply_net vdd_top -ports vdd_top
connect_supply_net GND -ports GND


#/ SUPPLY SETS
create_supply_set ss_pd_top
-function {power vdd_top}
-function {ground GND}
-function {nwell GND}
-function {pwell GND}

create_supply_set ss_pd_comp
-function {power vdd_comp}
-function {ground GND}
-function {nwell GND}
-function {pwell GND}

create_supply_set ss_pd_CPU
-function {power vdd_CPU}
-function {ground GND}
-function {
```

```
vdds_BBGEN_core}
-function {

gnds_BBGEN_core}


#/ POWER DOMAIN including compensation engine,
         voltage and temperature sensor and BBGEN
create_power_domain pd_top-supply {primary ss_pd_top}
                                         -include_scope
create_power_domain pd_comp
-elements {
[...]
TC_LOGIC_inst/BBGEN_inst
TC_LOGIC_inst/VTSENS_inst
TC_LOGIC_inst/PMB_inst/u_P28_SENS_PMCONTROL_master
                                              _controller
TC_LOGIC_inst/PMB_inst}
-supply {primary ss_pd_comp}

create_power_domain pd_CPU -supply {primary ss_pd_CPU}
-elements {
TC_LOGIC_inst/CPU_inst
TC_LOGIC_inst/PROBE_TOP_inst
TC_LOGIC_inst/PMB_inst/... ### probes and monitors
                                              instances
[...]}


#/ POWER STATES
add_port_state
-state {top_vdd_ON_ss 0.90}
-state {top_vdd_ON_ff 1.15}
vdd_top

add_port_state
-state {comp_vdd_ON_ss 0.90}
-state {comp_vdd_ON_ff 1.15}
vdd_comp

add_port_state
-state {CPU_vdd_ON_ss 0.90}
-state {CPU_vdd_ON_ff 1.15}
vdd_CPU
```

```
add_port_state
-state {gnd_ON 0.00}
GND

#/ PST CREATION
create_pst PD_default_pst -supplies   {vdd_top
                 vdd_comp     vdd_CPU         GND      }
add_pst_state PD_default_pst_ss -pst PD_default_pst
-state { top_vdd_ON_ss    comp_vdd_ON_ss
                              CPU_vdd_ON_ss   gnd_ON }
add_pst_state PD_default_pst_ff -pst PD_default_pst
-state { top_vdd_ON_ff    comp_vdd_ON_ff
                              CPU_vdd_ON_ff   gnd_ON }
connect_supply_net vdd_CPU -ports "TC_LOGIC_inst/
                            PROBE_TOP_inst/DVDD"
connect_supply_net GND -ports "TC_LOGIC_inst/
                            PROBE_TOP_inst/DGND"
connect_supply_net vdd_CPU -ports "TC_LOGIC_inst/
                                  CPU_inst/vdd_CPU"
connect_supply_net vdds_BBGEN_core -ports
              "TC_LOGIC_inst/CPU_inst/vdds_ctrl"
connect_supply_net GND -ports "TC_LOGIC_inst/
                                  CPU_inst/gnd_CPU"
connect_supply_net gnds_BBGEN_core -ports
              "TC_LOGIC_inst/CPU_inst/gnds_ctrl"
connect_supply_net vdd_comp -ports "TC_LOGIC_inst/
                                BBGEN_inst/vdd1v0"
connect_supply_net vdd_1V8_BBGEN_core -ports
                 "TC_LOGIC_inst/BBGEN_inst/vdd1v8"
connect_supply_net vdd_5V_BBGEN -ports
                 "TC_LOGIC_inst/BBGEN_inst/vdd5v0"
connect_supply_net GND -ports "BBGEN_ctrl/gnd"
connect_supply_net gnds_BBGEN_core -ports
                 "TC_LOGIC_inst/BBGEN_inst/vbb_pos"
connect_supply_net vdds_BBGEN_core -ports
                 "TC_LOGIC_inst/BBGEN_inst/vbb_neg"
connect_supply_net vdd_comp -ports "TC_LOGIC_inst/
                                VTSENS_inst/DVDD"
connect_supply_net vdd_1v8_core -ports "TC_LOGIC_
                              inst/VTSENS_inst/AVDD"
connect_supply_net GND -ports "TC_LOGIC_inst/
                                VTSENS_inst/DGND"
connect_supply_net GND -ports "TC_LOGIC_inst/
                                VTSENS_inst/AGND"
```

```
## not BB specific trailer
[...]
```

References

1. M. Cochet, *Energy Efficiency Optimization in 28 nm FD-SOI: Circuit Design for Adaptive Clocking and Power-Temperature Aware Digital SoCs*. Doctoral Dissertation, 2016
2. IEEE Computer Society, *IEEE Standard for Design and Verification of Low-Power, Energy Aware Electronic Systems* (Revision of IEEE Std 1801-2013). https://doi.org/10.1109/IEEESTD.2016.7445797. https://ieeexplore.ieee.org/servlet/opac?punumber=7445795. Accessed 5 Nov 2018

Index

A
ABB, *see* Adaptive body-bias (ABB)
AB class amplifier, 372
Active devices
  dimensioning, 176–178
  layout optimization strategy, 178–181
  LVT NMOS transistors, 176
Adaptive body-bias (ABB), 27, 28, 31, 59, 71, 79, 280, 338, 343, 345, 346, 359
Adaptive clock generator, 284
Adaptive clocking systems, 267, 289
Adaptive voltage scaling (AVS), 4, 5, 9, 50–55, 59, 71, 74, 79, 81, 82, 267, 274–276, 280, 295–298, 331, 337, 338, 343, 345, 346, 360, 400
Age compensation, 114
Ageing compensation, 341–342
Ageing-induced VTH drift, 341
Amplification stage design
  BEOL interconnection, 142
  BEOL layout optimization, 142, 143
  drain and source, 142
  drain-to-source parasitic capacitances, 142
  fingers, 140
  fT/fmax performances, 144–147
  impact of dual gate, 142, 143
  maximum oscillation frequency, 144
  post-layout simulations, 144
  simulations, 140
  source topology, 140
  ST Microelectronics, 140
  total power consumption, 141
  transistor, 140, 141
  transistor layout, 142, 144
AM–PM measurements
  class A and B operation, 206
  digital pre-distortion, 207
  intermediary modes, 206
  large-signal S21 measurement, 205
  normalized S21 phase, 206
  wireless systems, 205
Analog, 87, 89, 91
  bandgap, 246
  design aspects, 3–4
Analog/RF transistor's behavior, 2
Analogue output buffer
  AB class amplifier, 372, 374
  digital output buffer, 375
  in-phase signal current sources, 372
  input stage, 374
  LDOs, 372
  loop regulation, clocked power stages, 378–379
  negative output, 375
  negative voltage generation, 375–378
  NMOS transistors, 373
  positive output, 375
  push–pull output stage, AB class amplifier, 373
  specification, 372
  spice model, 377
Assist techniques, 107, 110
Asymmetric bias, 343
Asynchronous pipelines, 119, 120
Automotive market, 351
AVS, *see* Adaptive voltage scaling (AVS)

B

Body-biased-controlled oscillator (BBCO), 225, 228, 229, 234, 237, 240
Back-end-of line (BEOL), 5, 178, 180, 187–189
Balanced topology implementation
  amplitude, 184
  cascaded twisted elementary cells, 185
  coupling coefficient, 183
  hybrid coupler, 182
  λ/4 transmission, 182
  quadrature hybrid coupler, 182
  unitary inductance, 184
  unitary twisted cell design, 184
Baluns, 181, 182, 198
Band diagram, 13
BBGEN, *see* Body-bias generator (BBGEN/BBG)
Better-than-worst-case, 334, 335, 346, 348
Bias controller
  ASIC flow based hardware centric, 327
  factors, 327
  fixed point, 330
  RTL, 329
  software centric, 327
Biaser, 387, 401, 408
Bias law, 73, 329, 330, 387
Bias temperature instability (BTI), 50, 343
Binary phase-shift keying (BPSK), 224–228, 231, 239
Binary-weighted DACs, 365
Bipolar junction transistor (BJT), 246
Bipolar transistor
  CTAT and PTAT behaviours, 361
  FDSOI technology, 365
  NPN, 361
  PNP, 363
Bitcell stability, 102, 104, 107, 113
BitLine margin (BLM), 102
BitLines (BL), 98
Bluetooth Low-Energy (BLE), 223
Body-Bias
  benefits, 59
  design limits
    design leakage ceiling, 73–77
    qFO4 LVT flippedWELL devices, 80
    thermal bound, 77–79
    timing ceiling, 79
  device acceleration, 82
  device fabrication and environmental conditions, 82
  digital compensation toolbox
    ageing compensation, 66
    asymmetric body-bias, 67
  dynamic power gain, 70
  engineering and deployment costs, 71
  leakage reduction, 68–70
  open loop bias law, 72–73
  process compensation, 64, 65
  temperature compensation, 62
  voltage compensation, 62–64
  yield/minimum operational voltage gain, 70–71
  digital performance boost, 80–81
  effects, 104–106
  logic performance benchmark method, 60–62
  optimization of design corners, 59
  trade-off, 82
  ultra-low voltage designs, 81–82
Body-bias compensated oscillator principle
  bias-compensated SED-THS
    body-bias, 249
    oscillator performance, 251
    supply-bias merged oscillator, 249–251
  uncompensated SED-THS
    basic principle, 248
    simulated Performance, 248–249
Body-bias generator (BBGEN/BBG)
  ABB, 280
  AB-class amplifiers, 379
  AVS, 280
  analogue output buffer (*see* Analogue output buffer)
  architecture examples, 379, 380
  block diagram, 280, 281
  controller, 280, 281
  DC–DC inverter, 381
  driver units, 280, 281
  environmental constraints, 360
  external BB solution, 351, 352
  implementation
    placement concern, 383
    power integrity concern, 382–383
    testability, 383, 384
  load model (*see* Load model)
  optimal body-bias, 280
  positive voltage domain, 381
  power consumption, 351
  programmable voltages, 365 (*see also* Programmable voltage generation)
  SS-SC controller
    comparators, 279
    topologies, 279
  timing, 359
  translated voltage, 381
  UTBB, 278
  voltage range, 359–360

Index

voltage reference (*see* Voltage reference)
Body-biasing, 87
    FD-SOI technology, 333
    post-silicon tuning technique, 333
    See also Voltage scaling
Body-biasing voltage
    fixed, 89–91
    variable, 87–89
Body-bias *vs.* voltage scaling
    ageing, 341–342
    closed-loop *vs.* open-loop, 346–349
    combination, 345–346
    multiple domain adjustment, 344–345
    process compensation, 336
    selective N *vs.* P adjustment, 343
    speed boosting, 343
    temperature compensation, 340–341
    variation compensation, 334–335
    voltage compensation, 336–339
Boot sequence, 391
Bulk technology, 3
Buried oxide (BOX), 2, 86
Burst pulse position modulation (BPPM), 224–227, 239

C
Calibration, 312–313
Canary-FF monitor, 318
Canary register, 311, 317, 319
Capacitances, 355–357
Capacitive neutralization (C_{neutro}), 187, 189, 190, 203
Carrier's energy, 54
Charge pump model, 377, 379–381
Chip yield, 94, 95, 101
Circuit designer, 2
Circuit implementation, temporary sensors
    bias generations, 251
    digital processing, 252–254
    noise analysis methodology, 254
    probe detailed implementation, 251–252
Circuits measurements
    measurement results, 159–161
    measurement setup, 155–159
    phase noise optimization, 165–166
    standalone transistor measurements, 152–155
    variability study
        distributed oscillators, 165
        distribution histograms, 162, 164
        DO architectures, 161
        oscillation frequency and output power, 161, 164

power consumption, 164
    slight layout modifications, 162
    Vgate and Vbody-biasing operation, 161, 162
Clocked power stage, 378
Clock generator
    DLL-based adaptive clocking, 282–284
    Free-running adaptive clocking, 284–285
    SS-SC converters, 282
Clock output (CLKOUT), 283, 284
Closed-loop
    control action, 347
    implementation, 348
    open-loop, 347
    product leakage, 348
    target frequency, 348
    worst-case timing corner, 348
Closed-loop timing compensation system
    architecture, 321
    body-bias, 305, 306
    building blocks, closed-loop control system, 305, 306
    compensation unit, 322–323
    control (*see* Control functions)
    delay monitoring, 306
    feedback loop, 305
    measurement, 324
    monitored circuit's propagation delay, 305
    product FMAX, 306
    sensor, 305
    speed monitors (*see* Speed monitors)
Closed-loop *vs.* open-loop, 346–349
CMOS circuits, 344
CMOS integration, 1
Common-gate (CG), 187
Common load topology, 353–354
Common-source (CS), 187
Compensation unit, 322–323, 328
Conduction film, 86
Control and status register (CSR), 269
Control functions
    proportional control system, 320
    proportional-derivative control, 320
    proportional-integral control, 320
    successive approximation, 319–320
Control unit, 280, 281
Core clock, 265
Core voltage, 264, 267
Corner selection, 391–393
CPR, *see* Critical path replica (CPR)
Critical path replica (CPR), 311–312, 314, 319
Current-matching loop
    BBN voltage, 232
    DAC refresh, 232

Current-matching loop (*cont.*)
  NMOS transistors, 232
  OTA output, 232
  PMOS transistors, 232
  thick-oxide 1.8V I/O transistors, 233
Current-mode binary-weighted DACs, 367
Currents, 354–355
Current-starving technique, 122–124
Custom canary register, 311
Custom SRAM, 265
Cutoff frequency, 144, 155, 156, 158

D
DAC, *see* Digital to analogue converter (DAC)
DC-DC converters
  implementation, AVS, 274
  interleaved approach, 275
  simultaneous switching designs, 276
  SS-SC, 276–278
  switched-capacitor designs, 276
  switched-capacitor voltage regulators, 274–276
DC-to-RF efficiency, 160, 161, 163–166
Deep NWell (DNW), 255
Delay element
  application, 119
  architecture
    body-biasing stage, 128, 129
    delay lines, 130, 131
    lead cells, 130
    on-chip control registers, 130
    topology and the control flexibility, 128
  capacitive shunting, 121
  cascading inverters, 120–121
  charge/discharge current, 126
  complexity, 119
  current-starving technique, 122–124
  delay lines, 131–133
  delay manipulation, 119
  design, 127–128
  qualitative comparison, 126
  rising and falling digital delay, 119, 120
  semi-static approach, 122
  signal processing function, 119
  thyristor-based delay element, 124–125
Delay locked loops (DLL), 119
Delay monitoring, 306
Design community, 2, 23
Design for test (DFT), 384
Design rules manual (DRM), 180, 188, 196, 199
Design space, 59, 60, 102, 103, 109, 111, 113, 388, 389, 392, 400

Design under test (DUT), 155, 307, 309, 312, 313, 315, 316, 319
Device parameters, 53
Device reliability
  categories, 50
  electric field, 50
  gate oxide breakdown, 50–52
  HCI, 53–55
  intrinsic degradation modes, 50
  NBTI, 52–53
  thinner oxide, 50
  transistor, 50
Differential non-linearity (DNL), 236, 371
Digital design aspects, 5
Digital design flow methodology (DDFM), 6
Digital design implementation
  body-bias regulation module, 387
  corners
    definition, 388, 389
    sign-off corners set, 389
    step method, 390–391
    timing paths, 390
    verification *vs.* implementation, 391
  design verification principles, 289, 388
  physical implementation
    low-power flow, 394
    specific addition to standard flow, 394
    timing corner selection, 391–393
  specification and engineering test, 385–387
Digital design rules and constraints
  clock minimum pulse, 406
  flow steps
    Back-End, 407–410
    Front-End, 407–408
    Sign-Off, 410–411
  hold constraint, 405–406
  register-to-register timing path, 403
  setup constraint, 404–405
  signal transition rules, 406
Digital differential CMOS, 247
Digital implementation flow
  clock skew, 402
  clock tree, 402
  convergence and correlation, 402
  corner, mode and scenario, 400
  design sequencing, 402–403
  digital designers benchmark, 399
  digital flow, 399
  functional simulation, 401
  high fanout nets, 402
  implementation and sign-off, 401
  QoR, 399, 400
  sequential elements, 399
  STA, 401

Index

temperature inversion, VTinv and ZTC, 400–401
Digitally controlled oscillators (DCO), 119
Digital output buffer, 375
Digital processing, 252–253
Digital to analogue converter (DAC)
  accuracy, 371
    accuracy, 371
    binary-weighted DACs, 365, 366, 368, 370
    current-mode binary-weighted resistor, 367
    DNL, 371
    INL, 371
    R2R, 368, 369
    simplest voltage-output thermometer, 366
    static state, 370
    string, 365
    switched capacitor, 369
Digital TX architecture
  back biasing, 225
  BPPM and BPSK modulation, 226
  burst position, 227
  duty-cycled nature, 225
  FBB generators, 225
  frequency synthesis, 225
  LO frequency, 227
  NMOS/PMOS current-matching, 225
  power consumption, 227
Dimensioning
  integration, 178
  intrinsic gate resistance, 178
  Metal1-Pcell, 177
  non-quasi-static effects, 176, 177
  sub-division advantage, 176
  transistor, 176
DLL-based adaptive clocking
  block diagram, 282
  CLKOUT, 283, 284
  clock generator, 284
  clock output signal, 283
  controller, 283
  signal paths, 284
  VCDL, 282
Drain-induced leakage (DIBL), 5
Driver units, 280, 281
DUT, *see* Design under test (DUT)
Duty-cycled frequency synthesis
  BBCO, 228
  BPSK, 227
  FBB, 229
  FLL, 229

NOR and NAND functions, 228
ROs, 228

E
Electrical models, 99
Electrical parameters, 101
Electrical value sensing, 86
Electrical wafer sort (EWS), 386
Electromagnetic extraction, 149
Electromagnetic (EM) simulation, 148
Electromigration, 198
Electrostatic discharges (ESD), 196
EM RLC extraction tools, 172
Endpoint monitors, 310–311, 317
Energy efficiency processor, 289
Energy efficient design, 5
Energy-efficient SRAMs, 274
Environmental constraints, 360
ESD protection
  circuit operation, 196
  MOS transistors, 197
  RF signal leakage blocking/ transformers, 198
  robust integration, 198
Extreme value theory, 103

F
Failure rate, 94–96, 102–107, 112
Fan-out-of-4 (FO4), 60
FD-SOI process flow, 395–398
FD-SOI technology
  analog performance enhancement
    bulk, 44
    cut-off frequency extraction, 37, 39
    drain current low frequency noise, 42
    drain current spectral density, 43
    electrons mobility, 37
    function of power supply, 36
    gain degradation, 40
    gate stack structure, 43
    mobility improvement, 36
    MOS junctions design, 39
    non-uniformity, 40
    pocket implantations, 40
    RON value *vs.* switch input voltage, 38, 39
    short-channel effect, 40
    source–drain architectures, 39, 40
    switching energy efficiency, 36
    threshold voltage mismatch (AVT), 41, 42

FD-SOI technology (cont.)
  transconductance improvement, 37, 38
  transistor parameters, 35
  Tsi, 38
AVS, 9
back gate biasing effect, 13
body-bias, 21–23
  capability, 20
  impact, 18
bulk-planar transistor, 9
bulk transistor, 12
capacitive coupling, 14
charge density, 15, 16
circuit performance, 12
CMOS technologies, 9
coupling capacitance system, 15, 16
device structure, 10
digital performance enhancement
  delay, 31
  dynamic power and propagation time co-optimization, 32, 33
  Energy.Delay product, 34, 35
  MOS transistor threshold voltage modulation, 34
  power analysis, 30
  power optimization, 28
  ring oscillator delay ratio vs. original point, 31, 32
  ring oscillator performance evolution vs. gate back, 31, 32
  static power, 31
  switching energy efficiency, 35
  witching energy efficiency plot, 35
doping level, 20
drawbacks, 9
drive current modulation vs. body-bias value, 21, 22
electron mobility, 17, 18
electrostatic, 17
  behavior, 15
  control, 12
FBB, 28–35
flip-well architecture, 19
front and back gate biasing condition, 16, 17
inversion layer, 17
junction design differences, 10, 11
leakage current of transistor, 22
macroscopic transistor scheme, 10, 11
MOS device, 15
Nwell and Pwell voltage, 20
physical structure, 10
polysilicon interface, 18

pseudo-3D structure, 12, 13
Raven-3 (see Raven-3)
Raven-4 (see Raven-4)
ring oscillator (RO), 23
scalability, 11
Si/Box interface, 15, 17
sub-threshold slope change vs. body-bias, 16, 17
superior electrostatic control, 10
surface potential equilibrium, 13, 14
thin silicon film, 12
threshold voltage evolution vs. transistor architecture, 18, 19
transistor parameters, 10, 21–23
transistor variability, 10, 23–28
two-dimensional electrode, 10
type of applications, 9
vertical MOS capacitance evolution, 16, 17
Feedback loop, 305, 378
Fin-FET, 1
5G mobile network
  FD-MIMO, 169
  FD-SOI CMOS technology, 169
  frequency spectrum, 169
  ITU WRC, 169
  mmW PA (see mmW power amplifier)
  WRC-19, 169
Fixed point, 330
Flicker noise, 41
Flip-well architecture, 19
Forward back-bias (FBB), 15, 28–35, 85, 225
  BBCO, 237
  BBP saturation, 238
  current-matching loop, 236
  FinFET CMOS, 239
  LO carrier frequency, 237
  NMOS/PMOS, 236
  SleepTalker SoC, 239
  TX power consumption, 239
  UWB channels, 238
Free-running adaptive clocking, 284–285, 292
Frequency-locked loop (FLL), 229
Fully depleted silicon on insulator (FD-SOI), 1

G
Gate induced drain leakage (GIDL), 3, 353
Gaussian distributions, 102
Giga operations per seconds scale computing (GOPS), 5
Grounded coplanar lines (CPW-G), 148
Ground path optimization, 199–200

Index

H
Hardware redundancy, 96
Hold timing, 390
Hold violation, 406, 408
Hot carrier injection (HCI), 50
Hwacha vector accelerator, 267, 271–272

I
IEEE 1801 UPF, 413–417
Impedance matching network implementation
  inter-stage and input matching, 194–196
  output matching network, 191–194
Importance sampling (IS) method, 102
Impulse-radio ultra-wideband (IR-UWB), 224, 225, 239, 240
Integral Non-Linearity (INL), 371
Integrated body-bias generator, 267, 269, 281, 298
Integrated circuit (IC) design, 98
Integrated voltage regulator, 268
International Communications Union (ITU), 170
International Technology Roadmap for Semiconductors (ITRS), 216
Internet-of-Things (IoT)
  functions, 223
  smart sensor, 224
  wireless communications, 223
Inter-stage/input matching
  differential impedance values, 195
  S1 and S2, 195
  single-to-differential conversion, 196
  S-parameters, 196
Inter-stage matching network, 172

L
Large-signal measurements
  body-biasing, 205, 208–210
  operating conditions, 203
  operation mode, 204
  PAE, 204
  power consumption, 205
  power gain graph, 203
Layout optimization strategy
  BEOL, 178, 180
  DRM and density rules, 180
  frequency bands, 178
  gate access, 178
  parasitic resistances, 180
  poly-silicon level, 180
  staggered structure, 180
LDO, *see* Low drop out (LDO)

Least significant bit (LSB), 245
Linear feedback shift register (LFSR), 226
Load estimation, 357–358
Load model
  capacitances, 355–357
  common load topology, 353–354
  currents, 354–355
  load estimation, 357–358
Local oscillator (LO), 225
Look-up table (LUT), 253, 257, 327
Loop feedback signal, 86, 87
Loop regulation, clocked power stages, 378–379
Low drop out (LDO), 244, 248, 249, 251, 252, 254, 255, 257, 338, 345, 379
Lower specification level (LSL), 100
Low frequency (LF), 99
Low frequency noise (LFN), 41
Low-VT (LVT) transistor, 85

M
Macroscopic parameters, 21
Mass-scale production, 101, 109
Maximum delay field, 131
Maximum oscillation frequency, 144
Millimeter-wave designs, 87
Millimeter-wave distributed oscillators
  amplification stages, 136
  applications, 135
  attractive properties, 135
  CMOS, 135
  efficient signal source, 135
  operation frequencies
    architecture, 137
    Barkhausen amplitude and phase criteria, 137
    drain phase relationship, 138
    gate/drain phase relationship, 138
    gate RLC network, 138
    oscillation frequency, 137, 139
    phase-shift correction, 138, 139
    standard CMOS transistor, 138
    tank inductance and capacitance, 138
    transmission lines, 137
  properties, 136
  SiGe, 135
  silicon-based, 135
  simplified n-stage, 136
Minimum coarse sensitivity, 131
Minimum fine sensitivity, 131
Minimum operating voltage, 70, 71, 101, 274
Mismatch, 95, 97, 104

mmW PA measurement
   active devices (*see* Active devices)
   AM–PM measurements, 205–207
   configurability, 172–175
   design flow, 171–172
   FD-SOI technology, 201
   large-signal measurements, 203–205
   measurements over frequency range, 207–208
   optimal operating point, 201–202
   power amplifier (*see* Power amplifier behavior)
   Small-signal analysis, 202–203
Mobile market, 351
Model to hardware correlation (MHC), 99–101
Monitor calibration
   canary-FF monitoring, 312
   multi-RO monitoring approach, 313
   per-die tuning, 313
   timing margin, 313–315
   types, 312
Monitor evaluation, speed monitors
   accuracy, 315
   acquisition's time, 315
   CPR, 315
   endpoint monitors, 317–319
   intrusiveness, 315
   monitor's activation, 315
   optimal monitoring, 315
   overhead, 315
   re-usability, 315
   RO, 316
   single *vs.* multi-valued monitor data, 316
   TRC, 316
Monte Carlo (MC) method, 94, 232, 233, 336, 337, 339, 340, 371
Multi-cell type path composition, 310
Multiple domain adjustment, 344–345

N
Negative bias temperature instability (NBTI), 44, 47, 49, 50, 52–54, 66, 113, 341–343
Negative BitLine (NBL) assist, 107
Negative charge pump, 377, 379
Negative voltage generation, 375–378
Noise analysis methodology, 254
Non-linear stability, 172
Non-overlapping circuit, 376
NWells (NW), 255

O
On-wafer variability statistical study
   high-gain/ high-linearity modes, 212
   PAE_{max} average value, 213
   RF path routing, 214
   standard deviation, 213
   variability risks, 214
Open loop compensation
   body-biasing, 347
   control action, 346
   design synthesis, 330–331
   full ASIC Flow, 329–330
   mixed ASIC flow, 328
   software, 328
Operating performance points (OPPs), 318, 345
Optimal body-bias, 280
Optimal operating point, 201–202
Output matching network
   amplification chain, 191
   balun bandwidth, 194
   body-biasing, 192
   differential-to-single conversion, 193
   intrinsic parameters, 192
   load-pull simulation, 191
   operating classes, 193
   optimal impedance, 192
Output power
   amplification chain, 186
   AM–PM value, 207
   balanced topology, 181
   body-bias voltage change, 160
   distribution histograms, 162
   high-linearity mode, 204
   lack of power amplifiers, 215
   low performances, 214
   oscillation, 161
   performance trade-off, 216
   power consumption, 205, 239
   PVT variations, 232
   small-signal input power level, 203
   source topology, 140
   stage layout, 188
   total transistor width, 140

P
Pass gate (PG), 44
PA topology
   advantages, 181
   amplification paths, 181

Index

PA topology (*cont.*)
  implementation (*see* Balanced topology implementation)
  robustness and reliability, 182
  supplementary power, 181
Performance boost, 80–81, 343, 388, 392
Phase locked loops (PLL), 119
Physical mechanism, 53
Place and route, 351, 391, 393, 402
Planar dual-gate device, 3
PMU, *see* Power management unit (PMU)
Post-silicon tuning technique, 333
Power-added efficiency (PAE), 181, 187, 192, 193, 203, 207
Power amplifier behavior, temperature variation
  large-signal measurements, 208–210
  on-wafer variability statistical study, 212–214
  small-signal analysis, 210–212
Power amplifier configurability
  FD-SOI body, 173
  geometrical parameters, 174
  I_D and V_{GS} reference, 175
  LVT NMOS transistor, 173, 174
  MOS device output resistance, 174
  multi-stage, 175
  pseudo-Doherty implementation, 173
  SoC control implementation, 172
  V_T variation, 173
Power amplifier topology
  balanced topology implementation, 182–186
  topology, 181–182
Power integrity, 382–383
Power management unit (PMU), 268, 270, 272, 293, 295, 297, 298, 352, 361
Power-performance-area (PPA), 6
Power stages design
  S1 power amplification stage, 190–192
  S2 power amplification stage, 186–190
Power supply, 23, 24
Power supply rejection ratio (PSRR), 249, 252
Process capability index, 100
Process compensation, 47, 64, 65, 246–247, 336, 352, 386, 388, 390, 392
Process voltage temperature (PVT), 87, 88, 104, 107, 111, 123, 231, 232, 252, 255, 316, 331, 348
Process voltage temperature and aging (PVTA), 87, 88, 305–307, 314, 319, 347, 348, 388
Programmable voltage generation

DAC (*see* Digital to analogue converter (DAC))
  noise, 371
  resolution, 371
  speed, 372
Proportional control system, 320
Proportional-derivative control, 320
Proportional-integral control, 320
Proportional integrated derivative (PID), 378
Proportional to absolute temperature (PTAT), 246
Pull down (PD), 44
Pulse generators (PG), 283
Pulse repetition frequency (PRF), 226, 239
Pulse-shaping digital PA
  architecture, 231
  BPSK-encoded data, 231
  efficiency, 231
  HSR, 232
  Monte-Carlo post-layout simulations, 232
  PSD, 229
  pulse shape temporal resolution, 230
  shaping functionality, 230
  spectral impact, 229
PVTAB, 314, 319

Q
Quality Of Result (QoR), 399, 400

R
Random device variability, 104
Raven-3
  annotated die micrograph, 267
  annotated floorplan, 265, 266
  bisynchronous FIFO, 265, 266
  block diagram, 264, 265
  converter topologies, 286
  core clock, 265
  core voltage, 264
  energy efficiency, 289
  features, 264
  generated clock, 265
  integrated system, 289
  IP blocks, 265
  multivoltage and multiclock design flow, 265
  oscilloscope traces, 286, 287
  replica path, 265
  Rocket Chip, 264
  simultaneous-switching converters, 286
  system conversion efficiency measurements, 288
  TRC, 288

Raven-4
  adaptive clocking systems, 267, 289
  annotated die micrograph, 270
  annotated floorplan, 269
  BBG, 300
  block diagram, 267, 268
  clocking systems, 268
  core voltage, 267, 289, 290, 293
  CSR, 269
  FBB, 269
  free-running adaptive clock generator, delay banks, 292
  hopping frequency, 294
  integrated body-bias generator, 298
  integrated voltage regulation, 267, 268
  measured system conversion efficiencies, 290
  n-well voltage, 298
  PMU, 268, 293–295, 297, 298
  p-well, 298, 299
  SoC, 269, 270
  SS-SC converter, 290
  SS-SC toggle clock, 293
  voltage-dependent frequency behavior, 269
  voltage dithering, 295, 296
Read ability (RA) failure, 99
Read stability (RS) failure, 98–99
Reference voltage, 360, 361, 363, 365, 378, 381
Register replication, 311
Regular-VT (RVT) transistor, 85
Regulation system, 318
Reserve body-bias (RBB), 15
Ring oscillator (RO), 10, 21, 228, 247, 307–309, 316
RISC-V processors
  free and open architecture, 270
  Hwacha vector accelerator, 271–272
  Raven-3, 270
  Raven-4, 270
  Rocket Chip, 271
  simplified pipeline, 271
  Z-scale, 272–273
Robust integration and reliability
  electromigration, 199
  ESD protection, 196–199
  ground return path optimization, 199–200
  safe operating area, 199
Rocket Chip, 264, 267, 271
R2R DACs, 368, 369

S
Safe design space, 95, 102, 103, 109, 111, 113
Safe operating area, 199

Schmitt Trigger, 123
Self-test capabilities, 96
Semi-static approach, 122
Sensor layout, 255
Setup timing, 390
Setup violation, 405
Short-circuit current, 122, 126, 127, 341
Sign-off (SOF), 327, 388, 400, 402, 406, 408
Simultaneous-switching switched-capacitor DC-DC (SS-SC), 276–279, 282, 286–290, 293–297
Single-cell type path composition, 310
Single-ended digital temperature sensor (SED-THS), 247
Single-port high-density (SPHD), 98
Single-Pwell architecture, 47
Single-p-well six-transistor (6T), 98
SleepTalker testchip, 235
SleepTalker TX SoC, 224
Small-signal analysis, 202–203, 210–212
SoC design in FD-SOI
  digital systems, 263
  Raven-3, 263–267
  Raven-4, 267–270
SoC integration, 234, 235
SoCs, see Systems-on-chip (SoCs)
Speed boosting, 343
Speed monitors, 307
  characteristics, 307
  CPR, 311–312
  endpoint monitors, 310–311
  monitor calibration, 312–315
  monitor evaluation (see Monitor evaluation, speed monitors)
  RO, 307
  TRCs, 309–310, 323
S1 power amplification stage
  C_{neutro}, 190
  common-source configuration, 190
  differential topology, 190
  EM simulations, 191
  interconnections parasitics, 190
S2 power amplification stage
  AluCap routing, 187
  cascode capacitance C_{cas}, 187
  C_{neutro}, 187, 189
  common-gate stage design, 188
  configurability, 186
  CS and CG, 187
  differential common-source, 188
  differential topology, 186
  IA metal layer, 188
  identical BEOL, 187
  inter-stage, 188

Index

operating class, 187
SRAM bit-cell
　bit-cell architecture evolution, 44, 45
　body-biasing structure, 47, 48
　carrier mobility and electrostatic control, 46
　characterized *vs.* bulk architecture, 46
　digital performance, 46
　NMOS transistors, 44
　operations limitation, 44, 45
　PU transistor, 47
　single-Pwell architecture, 47
　6T, 48
　Vmin, 45
　write assist, 49
SS-SC converter, 290
STA, *see* Static timing analysis (STA)
State of the art mmW PA comparison
　balanced topology, 220
　CMOS technology, 214, 221
　common-source topology, 215
　common-source transistors, 215
　Doherty configuration, 216
　FD-MIMO, 219
　FinFET technology node, 216
　high-linearity mode, 220
　ITRS FOM, 216, 218, 221
　linear performances, 214
　multi-gate topology, 214
　power consumption, 220
　robustness, 216
　SoC implementation, 215
　transformer-based Doherty topology, 215
Static noise margin (SNM), 47–49, 102, 113
Static random-access memory (SRAM)
　　functionality, 264, 268
　body-bias effects, 104–106
　circuit, 98–99
　expand-and-shrink trend, 96
　HF body-bias effects, 112–113
　HF effects, 106–108
　LF Gaussian model, 102–103
　measure/calculate, 97
　Nmax paradigm, 109–111
　silicon product
　　behavior alteration, 93
　　chip yield isolines, 95
　　electrical characteristics, 95
　　FD-SOI technology, 93
　　intellectual property (IP), 94
　　operating conditions, 94
　　safe design space, 95
　　semiconductor market, 94
　　small real-world devices, 94

　　yield influences, 93–94
　spice model, 99–101
　Vmin paradigm, 101–102
Static timing analysis (STA), 309, 312, 318, 319, 331, 393, 400, 401, 403, 410
Statistical models, 100
String DACs, 365
Sub-micron technology nodes, 171
Successive approximation control function, 319–320
Supply-bias merged oscillator, 249–251
Surface potential, 10, 12–14
Switched capacitor DACs, 369
Switched-capacitor designs, 276
Switched-capacitor voltage regulator (SCVR), 234, 274
Switches sizing, 377

T
TDC, *see* Time to digital converter (TDC)
Temperature coefficient of resistivity (TCR), 246
Temperature compensation, 28, 62, 72, 76, 81, 106, 324, 328, 340–341, 386, 389, 390, 392
Temperature inversion, 60, 340, 341, 343, 345, 392, 400–401
Temperature sensor
　circuit implementation, 251–254
　design/process compensation
　　analog bandgap, 246
　　digital differential CMOS, 247
　　resistor-based, 246
　　SED-THS, 247
　　thermal diffusivity, 246–247
　digital SoCs temperature monitoring, 244
　integrated requirements
　　accuracy and repeatability, 245
　　area, 245
　　calibration and references, 245
　　power consumption, 245
　　resolution, 245
　　voltage range, 245
　manufactured chip
　　sensor layout, 255
　　validation of calibration, 256–258
　principle (*see* Body-bias compensated oscillator principle)
　state-of-the-art, 258, 259
Testability, 383, 384
Thermal bound, 77–79
Thermal diffusivity (TD), 246–247
Threshold voltage behavior *vs.* body-bias, 21

Threshold voltage (VT) variation, 86
Thyristor-based delay element, 124–125
Time dependent dioxide breakdown (TDDB), 50
Time to digital converter (TDC), 120, 310, 312, 313, 316, 318
Time to failure (TTF), 50
Timing, 385, 387–393
Timing margin, 313–315
Timing specifications, 359
Total ionizing dose (TID), 344
Transceiver's signal path, 85
Transformer-based Doherty topology, 215
Transistor body, 3
Transistor channel, 1
Transistors, 85
Transistor variability, 23–28
Transmission line topology
    characteristic impedance, 148
    drain and gate lines, 151
    electromagnetic extraction, 149
    extracted RLCG model, 151
    linear approximation, 152
    measurement spectrum analyzer, 147
    microstrip, 147
    Momentum EM tool, 148
    oscillation frequency, 151–152
    RLCG model, 148
    state-of-the-art comparison, 166–167
Transmitter (TX)
    architectures (*see* Digital TX architectures)
    functions, 225
    ULP radios, 224
TRIG_CALIB signal, 227
Tunable replica circuit (TRC)
    clock generator, 282
        building blocks, 309
        classification, 309
        monitor evaluation, 316
        multi-cell type path composition, 310
        single-cell type path composition, 310
        speed monitors, 323
    output clock signal, 283
Tuning/trimming elements, 85

U

Ultra-low-power (ULP), 224, 225, 236
Ultra low voltage (ULV), 4, 19, 26, 81–82, 105, 107, 225, 321, 327, 328
Ultra-thin body and BOX (UTBB), 2
Ultra-thin body and buried oxide fully depleted silicon on insulator (UTBB FD-SOI), 126, 267

Unity gain bandwidth (UGB), 379
UPF, 382, 394, 409, 413–417
Upper specification level (USL), 100

V

Validation of calibration
    body-bias compensation, 256
    delay-temperature dependency, 256
    measured accuracy, 256–257
    oscillator's frequency, 256
    SoC integration, 257–258
Variation compensation, 334–335
Vector accelerator, 272
Vector execution unit (VXU), 272
Vector memory unit (VMU), 272
Voltage and temperature sensor (VTS), 329
Voltage compensation, 62–64, 336–340
    AVS, 337
    block diagram, 338
    body-biasing, 339
    impacts, 339
    voltage fluctuations, 336
Voltage-controlled delay line (VCDL), 282
Voltage dithering, 295
Voltage fluctuations, 336
Voltage range, 359–360
Voltage reference
    Bandgap circuit
        PTAT and CTAT voltage, 362, 363
    BBGEN, 360
    bipolar transistor, FDSOI technology, 365
    Boltzmann's constant, 361
    CTAT behaviour of *VBE,* 362
    CTAT component, 362
    low power bandgap, NPN bipolar transistors, 364
    PMU, 361
    process-dependent constant, 362
    PTAT voltage, 361
    temperature independence, 361, 363
Voltage scaling
    advantages, 343
    and body-bias, 345–346

W

Weibull distributions, 51
WELL devices, 80, 82, 409, 411
WordLine (WL), 98
World Radiocommunication Conference (WRC-19), 170
Worst-case bitcell, 96
Write ability (WA) failure, 98–99

Write margin (WM), 102
Write stability (WS), 99

Y
Yieldograms, 109

Z
Zero back bias (ZBB), 234
Zero temperature coefficient (ZTC), 244, 248, 249, 400
Z-scale, 272–273

Printed by Printforce, the Netherlands